METHODS IN MOLECULAR BIOLOGY

Series Editor
John M. Walker
School of Life and Medical Sciences
University of Hertfordshire
Hatfield, Hertfordshire, AL10 9AB, UK

For further volumes:
http://www.springer.com/series/7651

Functional Proteomics

Methods and Protocols

Edited by

Xing Wang

Array Bridge Inc., St. Louis, MO, USA

Matthew Kuruc

Biotech Support Group LLC, Monmouth Junction, NJ, USA

Humana Press

Editors
Xing Wang
Array Bridge Inc.
St. Louis, MO, USA

Matthew Kuruc
Biotech Support Group LLC
Monmouth Junction, NJ, USA

ISSN 1064-3745 ISSN 1940-6029 (electronic)
Methods in Molecular Biology
ISBN 978-1-4939-8813-6 ISBN 978-1-4939-8814-3 (eBook)
https://doi.org/10.1007/978-1-4939-8814-3

Library of Congress Control Number: 2018957271

This Humana Press imprint is published by the registered company Springer Science+Business Media, LLC, part of Springer Nature.
The registered company address is: 233 Spring Street, New York, NY 10013, U.S.A.

Preface

Over the past two decades, tremendous progress has been made in the field of proteomics, the purpose of which is to find systemic differences in protein populations. Once established, measurable protein markers can then help to define biological mechanism, disease, identify therapeutic targets, and offer better precision for personalized medical interventions.

Proteomics, like other "omics" analyses, is data driven and can generate unbiased protein profiles for a variety of end points that can contrast, for example, treated vs. untreated cell models or healthy vs. diseased tissue; it has provided us with more in-depth understanding of many biological systems and diseases. This progress in the field of proteomics parallels with the advancement in many analytical technologies, especially in mass spectrometry, which has been evolved from a less sensitive and qualitative tool to high sensitivity and quantitation system for protein analysis and characterization. Currently, system biology and proteomics in particular are advancing biology at two opposite but equally important polarities; one is the holistic understanding of a biological system, be it an organism, organ, tissue, or the human circulation system, and the other one is for single-cell analysis whereby biological heterogeneity can be minimized and a more discrete picture of biological processes can be modeled within a more homogeneous context. Having the tools and methods necessary to address these important studies will promote the significant impact that is foreseen in precision medicine and other biological fields.

In the most common view of proteomics, data is generally acquired after proteolytic processing of the parent proteomes. The derived peptides are then analyzed on instruments coupling Nano-Liquid Chromatography to Mass Spectrometry (LC-MS). Such instruments generate mass spectra of peptides and the MS2 spectra through further fragmentation, which can be compared to theoretical amino acid spectra definable through public gene repositories. Peptide sequence matches are thus computationally derived, and from that data, protein identifications are inferred. From such analyses, peptide markers can be used as surrogates for the gene products from which they are derived. Through differential expression analysis of these peptide markers, proteomics can thus help identify those gene products that define a phenotype. However the functions of the proteome, the driving force for almost all biological actions, are not adequately annotated through the current infrastructure of methods surrounding LC-MS sequence annotation. This book is intended to fill in this knowledge and technology gap with a specific collection of technologies that have been developed for the study of protein function at a proteome scale. In organizing the content of this book, the following points were taken into consideration: (1) It should bridge the understanding of biology from protein functions to other aspects of protein analysis, especially in post-translational modification, as most cellular proteins use this mechanism to carry out their unique role in cellular regulation. (2) The book should also act as a bridge to other levels of system biology research including genomics and metabolomics, so that the readers will gain a relatively complete picture of how one might study the biological system of their interest. (3) Technologies are categorized toward different aspects of protein functional analysis, so that readers can understand what is available to them in functional proteomics research. (4) Finally, the selection of technologies also takes into consideration the impact on current and future research in a variety of disease areas.

It is hoped that by using these novel technologies, new frontiers in biological research will be created, important drug targets can be identified, and clinically validated biomarkers and diagnostic tests can be developed. The aim of the editors of this book is to provide the most precise description of our technological capabilities in functional proteomics research and give our readers the tools they will need to create the new functional domains of our knowledge in the understanding of various biological systems.

St. Louis, MO, USA *Xing Wang*
Monmouth Junction, NJ, USA *Matthew Kuruc*

Contents

Contributors

JUHANI AAKKO • *Turku Centre for Biotechnology, University of Turku and Åbo Akademi University, Turku, Finland*

SHIVA AHMADI • *Institute for Biochemistry and Molecular Biology, University of Bonn, Bonn, Germany*

MAGNUS Ø. ARNTZEN • *Faculty of Chemistry, Biotechnology and Food Science, Norwegian University of Life Sciences (NMBU), Ås, Norway*

SANDRINE AROS-CALT • *Service de Pharmacologie et d'Immunoanalyse, Laboratoire d'Etude du Métabolisme des Médicaments, CEA, INRA, Université Paris Saclay, MetaboHUB, Gif-sur-Yvette, France; bioMérieux S.A., Marcy l'Etoile, France*

JOHN R. BARR • *Division of Laboratory Sciences, Centers for Disease Control and Prevention, Atlanta, GA, USA*

MATTEO BATTARRA • *Diabetes Research Institute, University of Miami, Miami, FL, USA*

MIROSLAV BERKA • *Faculty of AgriSciences, Department of Molecular Biology and Radiobiology, CEITEC—Central European Institute of Technology, Phytophthora Research Centre, Mendel University in Brno, Brno, Czech Republic*

KEREN BYRNE • *CSIRO Agriculture and Food, St Lucia, QLD, Australia*

ZHENG CAO • *Department of Laboratory Medicine, Beijing Obstetrics and Gynecology Hospital, Capital Medical University, Beijing, China*

JAVIER CARABIAS-SÁNCHEZ • *Proteomics Unit, Cancer Research Centre (IBMCC/CSIC/USAL/IBSAL), Salamanca, Spain*

FLORENCE A. CASTELLI • *Service de Pharmacologie et d'Immunoanalyse, Laboratoire d'Etude du Métabolisme des Médicaments, CEA, INRA, Université Paris Saclay, MetaboHUB, Gif-sur-Yvette, France*

MARTIN ČERNÝ • *Faculty of AgriSciences, Department of Molecular Biology and Radiobiology, CEITEC—Central European Institute of Technology, Phytophthora Research Centre, Mendel University in Brno, Brno, Czech Republic*

RU CHEN • *Department of Medicine, University of Washington, Seattle, WA, USA*

YUN CHEN • *School of Pharmacy, Nanjing Medical University, Nanjing, China*

MICHELLE L. COLGRAVE • *CSIRO Agriculture and Food, St Lucia, QLD, Australia*

LI-JUN DI • *Faculty of Health Sciences, Cancer Center, University of Macau, Macau, China*

HU DUAN • *State Key Laboratory of Proteomics, Beijing Proteome Research Center, National Center for Protein Sciences (PHOENIX Center, Beijing), Beijing Institute of Lifeomics, Beijing, China*

VINCENT G. H. EIJSINK • *Faculty of Chemistry, Biotechnology and Food Science, Norwegian University of Life Sciences (NMBU), Ås, Norway*

LAURA L. ELO • *Turku Centre for Biotechnology, University of Turku and Åbo Akademi University, Turku, Finland*

ANDREW EMILI • *Donnelly Centre for Cellular and Biomolecular Research, University of Toronto, Toronto, ON, Canada; Department of Biology, Boston University, Boston, MA, USA; Department of Biochemistry, Boston University, Boston, MA, USA; Center for Network System Biology, Boston University, Boston, MA, USA*

François Fenaille • *Service de Pharmacologie et d'Immunoanalyse, Laboratoire d'Etude du Métabolisme des Médicaments, CEA, INRA, Université Paris Saclay, MetaboHUB, Gif-sur-Yvette, France*

Jonatan Fernández-García • *Proteomics Unit, Cancer Research Centre (IBMCC/CSIC/USAL/IBSAL), Salamanca, Spain*

Manuel Fuentes • *Proteomics Unit, Cancer Research Centre (IBMCC/CSIC/USAL/IBSAL), Salamanca, Spain; Department of Medicine and Cytometry General Service-NUCLEUS, Cancer Research Centre (IBMCC/CSIC/USAL/IBSAL), Salamanca, Spain*

Rodrigo García-Valiente • *Proteomics Unit, Cancer Research Centre (IBMCC/CSIC/USAL/IBSAL), Salamanca, Spain*

Gaspard Gervasi • *bioMérieux S.A., Marcy l'Etoile, France*

Rafael Góngora • *Proteomics Unit, Cancer Research Centre (IBMCC/CSIC/USAL/IBSAL), Salamanca, Spain; Department of Medicine and Cytometry General Service-NUCLEUS, Cancer Research Centre (IBMCC/CSIC/USAL/IBSAL), Salamanca, Spain*

María Gonzalez-Gonzalez • *Proteomics Unit, Cancer Research Centre (IBMCC/CSIC/USAL/IBSAL), Salamanca, Spain; Department of Medicine and Cytometry General Service-NUCLEUS, Cancer Research Centre (IBMCC/CSIC/USAL/IBSAL), Salamanca, Spain*

Hana Habánová • *Faculty of AgriSciences, Department of Molecular Biology and Radiobiology, CEITEC—Central European Institute of Technology, Phytophthora Research Centre, Mendel University in Brno, Brno, Czech Republic*

Crispin A. Howitt • *CSIRO Agriculture and Food, Canberra, ACT, Australia*

Alba Hykollari • *Department für Chemie, Universität für Bodenkultur, Vienna, Austria*

Luca Inverardi • *Diabetes Research Institute, University of Miami, Miami, FL, USA*

Constance Jeffery • *Department of Biological Sciences, University of Illinois at Chicago, Chicago, IL, USA*

Christophe Junot • *Service de Pharmacologie et d'Immunoanalyse, Laboratoire d'Etude du Métabolisme des Médicaments, CEA, INRA, Université Paris Saclay, MetaboHUB, Gif-sur-Yvette, France*

Srikanth Kakumanu • *Focus Proteomics, Hudson, NH, USA*

T. S. Keshava Prasad • *Center for Systems Biology and Molecular Medicine, Yenepoya Research Centre, Yenepoya (Deemed to be University), Mangalore, India*

Zsuzsanna Kuklenyik • *Division of Laboratory Sciences, Centers for Disease Control and Prevention, Atlanta, GA, USA*

Matthew Kuruc • *Biotech Support Group LLC, Monmouth Junction, NJ, USA*

Lisa A. Lai • *Department of Medicine, University of Washington, Seattle, WA, USA*

Patricia Lamourette • *Service de Pharmacologie et d'Immunoanalyse, Laboratoire d'Etude du Métabolisme des Médicaments, CEA, INRA, Université Paris Saclay, MetaboHUB, Gif-sur-Yvette, France*

Alicia Landeira-Viñuela • *Proteomics Unit, Cancer Research Centre (IBMCC/CSIC/USAL/IBSAL), Salamanca, Spain; Department of Medicine and Cytometry General Service-NUCLEUS, Cancer Research Centre (IBMCC/CSIC/USAL/IBSAL), Salamanca, Spain*

Giacomo Lanzoni • *Diabetes Research Institute, University of Miami, Miami, FL, USA*

Sarah E. Leonard • *Chemical and Biomolecular Engineering, University of Illinois Champaign-Urbana School of Chemical Sciences, Champaign, IL, USA*

HAILI LI • *CSIRO Agriculture and Food, St Lucia, QLD, Australia; Institute of Animal Husbandry and Veterinary Science, Henan Academy of Agricultural Sciences, Zhengzhou, Henan, China*
PEIPEI LI • *Faculty of Health Sciences, Cancer Center, University of Macau, Macau, China*
LIANG LIU • *School of Pharmacy, Nanjing Medical University, Nanjing, China*
DANIEL MALZL • *Department für Chemie, Universität für Bodenkultur, Vienna, Austria*
YUAN MENG • *Faculty of Health Sciences, Cancer Center, University of Macau, Macau, China*
BRUNO H. MULLER • *bioMérieux S.A., Marcy l'Etoile, France*
SHENG PAN • *Institute of Molecular Medicine, University of Texas Health Science Center at Houston, Houston, TX, USA*
KATHARINA PASCHINGER • *Department für Chemie, Universität für Bodenkultur, Vienna, Austria*
RAY C. PERKINS • *New Liberty Proteomics Corporation, New Liberty, KY, USA*
SAMI PIETILä • *Turku Centre for Biotechnology, University of Turku and Åbo Akademi University, Turku, Finland*
SNEHA M. PINTO • *Center for Systems Biology and Molecular Medicine, Yenepoya Research Centre, Yenepoya (Deemed to be University), Mangalore, India*
PASCAL PONCET • *Allergy and Environment Team, Biochemistry Department, Armand Trousseau Children Hospital (AP-HP), Paris, France; Center for Innovation and Technological Research, Institute Pasteur, Paris, France*
REZA POURHAGHIGHI • *Donnelly Centre for Cellular and Biomolecular Research, University of Toronto, Toronto, ON, Canada*
J. ROBERT O'NEILL • *Cancer Research UK Edinburgh Centre, MRC Institute of Genetics and Molecular Medicine, The University of Edinburgh, Edinburgh, UK; Department of Clinical Surgery, Royal Infirmary of Edinburgh, Edinburgh, UK*
SWAPAN ROY • *Biotech Support Group LLC, Monmouth Junction, NJ, USA*
HÉLÈNE SÉNÉCHAL • *Allergy and Environment Team, Biochemistry Department, Armand Trousseau Children Hospital (AP-HP), Paris, France*
YOUCEF SHAHALI • *Razi Vaccine and Serum Research Institute, Agricultural Research, Education and Extension Organization (AREEO), Karaj, Iran*
GARY SMEJKAL • *Focus Proteomics, Hudson, NH, USA*
YASHWANTH SUBBANNAYYA • *Center for Systems Biology and Molecular Medicine, Yenepoya Research Centre, Yenepoya (Deemed to be University), Mangalore, India*
TOMI SUOMI • *Turku Centre for Biotechnology, University of Turku and Åbo Akademi University, Turku, Finland*
CHOO HOCK TAN • *Venom Research and Toxicology Laboratory, Department of Pharmacology, Faculty of Medicine, University of Malaya, Kuala Lumpur, Malaysia*
KAE YI TAN • *Protein and Interactomic Laboratory, Department of Molecular Medicine, Faculty of Medicine, University of Malaya, Kuala Lumpur, Malaysia*
NGET HONG TAN • *Protein and Interactomic Laboratory, Department of Molecular Medicine, Faculty of Medicine, University of Malaya, Kuala Lumpur, Malaysia*
ZACHARY TONG • *Department of Medicine, University of Washington, Seattle, WA, USA*
CHRISTOPHER A. TOTH • *Division of Laboratory Sciences, Centers for Disease Control and Prevention, Atlanta, GA, USA*
TINA R. TUVENG • *Faculty of Chemistry, Biotechnology and Food Science, Norwegian University of Life Sciences (NMBU), Ås, Norway*

INGRID M. VERHAMME • *Department of Pathology, Microbiology and Immunology, Vanderbilt University School of Medicine, Nashville, TN, USA*

DAVID WANG • *University of Iowa School of Medicine, Iowa City, IA, USA*

HONGYE WANG • *State Key Laboratory of Proteomics, Beijing Proteome Research Center, National Center for Protein Sciences (PHOENIX Center, Beijing), Beijing Institute of Lifeomics, Beijing, China*

LI WANG • *Faculty of Health Sciences, Cancer Center, University of Macau, Macau, China*

IAIN B. H. WILSON • *Department für Chemie, Universität für Bodenkultur, Vienna, Austria*

DOMINIC WINTER • *Institute for Biochemistry and Molecular Biology, University of Bonn, Bonn, Germany*

XIAOBO YU • *State Key Laboratory of Proteomics, Beijing Proteome Research Center, National Center for Protein Sciences (PHOENIX Center, Beijing), Beijing Institute of Lifeomics, Beijing, China*

LINA ZHANG • *Center for Translational Biomedical Research, University of North Carolina at Greensboro, Kannapolis, NC, USA*

QIBIN ZHANG • *Center for Translational Biomedical Research, University of North Carolina at Greensboro, Kannapolis, NC, USA; Department of Chemistry and Biochemistry, University of North Carolina at Greensboro, Greensboro, NC, USA*

Chapter 1

Making the Case for Functional Proteomics

Ray C. Perkins

Abstract

"Making the Case for Functional Proteomics" first differentiates the Functional Proteome from the products of genetic protein expression. Qualitatively, the prevalence of posttranslational modifications (PTMs) virtually insures that individual, functional proteins do not equate to their genetic expression counterparts. Quantitatively, considering the frequency of PTMs and a conservative estimate of the number of functional entities arising from protein interactions, the size of the Functional Proteome exceeds that of the human genome by at least two orders of magnitude. The human genome does not, cannot, map the Functional Proteome. Further, the collective genome of the human microbiome dwarfs the human genome. With these facts established, "Making the Case..." proceeds to examine Functional Proteomics (of which both "gene expression" and "epigenetics" are but parts of a larger whole) within the context of Systems Biology, concluding that functionally related networks comprise the dominant motif for biological activity. Creating just such a network focus is essential in not only expanding basic knowledge but also in applying that knowledge in the pragmatic efforts of drug and biomarker development. Outlines for development of drugs and biomarkers, as well as the realization of precision medicine, within a functional proteomics-based, network motif are provided. The chapter proceeds to asses both the knowledge base and the tools to fully embrace Functional Proteomics. Given the decades-long infatuation with the reductionism of genomics, it is not surprising that both the proteomics knowledge base and tools are assessed as poor to fair. However, even a minor shift in research funding and a renewed challenge to methods developers will rapidly improve the current situation. Adoption of the included "Roadmap" will realistically make the twenty-first century the century of a long-awaited revolution in biology.

Key words Protein, Gene, Genome, Proteome, Functional Proteome, Proteomics, Functional proteomics, Microbiome, Posttranslational modifications, Protein interactions, Epigenetics, Gene expression, Biological networks, Systems biology, Drug development, Biomarker development, Precision medicine

1 Introduction

The title of this chapter, "Making the Case for Functional Proteomics," can elicit a sense of the absurd. For example, it is difficult to name a single biological property or process that does not rely on the "function" of proteins. Basic properties such as size and shape of the organism, its interaction with the environment, and the existence of functionally specialized structures directly reflect the

Xing Wang and Matthew Kuruc (eds.), *Functional Proteomics: Methods and Protocols*, Methods in Molecular Biology, vol. 1871, https://doi.org/10.1007/978-1-4939-8814-3_1, © Springer Science+Business Media, LLC, part of Springer Nature 2019

properties of proteins. Large-scale processes such as metabolism, digestion, temperature maintenance, and organism reproduction all depend on the integrated functioning of proteins. At the small scale of transmembrane transport, pathogen detection, transcription, and translation, protein function is essential to the health and well-being of the organism. It is the latter of the noted small-scale activates—processes related to genetics—that creates the need for this chapter and this book: the reductionism and distraction of decades of DNA sequencing. In this context, "Making the Case for Functional Proteomics" is crucial.

This chapter interprets "Making the Case..." in two ways, one broad in scope and one more narrow though no less essential. The broad scope encompasses: (1) Specification of the "Functional Proteome," (2) The interrelationship of Functional Proteomics and Systems Biology, and (3) The potential utility of Functional Proteomics in areas of fundamental importance (e.g., Drug and Biomarker Discovery). Specification of the Functional Proteome includes the factual presentation and logic in determining the size and nature of the "Functional Proteome," especially as compared to output associated with the human genome. Examination of Functional Proteomics within the context of Systems Biology insures biological relevance, especially in the evaluation of the utility of Functional Proteomics for disease detection and treatment. The narrower interpretation of "Making the Case..." is an open-eyed evaluation of the existing knowledge that comprises Functional Proteomics. Accessibility of fundamental proteomic metrics, e.g., protein identity, localization, and activity, is assessed and metrics are indexed to pertinent methods. Comparison of fundamental metrics to available information sources and methods identifies areas of weakness and strength for both the methods per se and the status of existing coverage of the Functional Proteome.

Major headings and subheadings for "Making the Case..." are:

- The Functional Proteome in Relation to the Genome
- The Functional Proteome and Systems Biology
- Functional Proteomics Utility:
  - Biomarker and Diagnostic/Prognostic Discovery
  - Drug Discovery
  - Precision Medicine
- Functional Proteome Knowledge Base
  - Level of Knowledge/Ignorance
  - Proteomics Methods Appraisal
- Closing Comments

Summarizing key observations in this chapter it is noted that the Functional Proteome is orders of magnitude larger than the

Genome. Functional Proteomics is a natural context for Systems Biology regardless of the level of complexity. Biomarkers developed within a Functional Proteomics context are naturally multiplexed, selecting not for single proteins but for functional Networks and Pathways. Functional Proteomics naturally integrates with the forefront of drug discovery, phenotypic drug discovery. Unfortunately, given the enormous potential of Functional Proteomics as a generator of new knowledge and therapeutics, the existing knowledge base must be considered poor. Among other areas, this ignorance inhibits proteomic contributions to Precision Medicine. Fortunately the cure for ignorance is well known, comprising hard work, ingenuity, and commitment.

2 The Functional Proteome in Relation to the Genome

The gene COL1A1 on the minus strand of chromosome 17 starting at 50 million base pairs does not provide a primary basis for the structural integrity of the human body—the protein, collagen, does [1]. The gene PRSS1 on the plus strand of chromosome 7 starting at 143 million base pairs does not digest food—the protein, trypsin, does. The genes HBB, HBA1, and HBA2 located on chromosomes 11 and 16 do not serve as a blood circulation transport vehicle for oxygen in blood—the protein, hemoglobin, does. These examples can be even more instructive in establishing the non-relationship between genome and proteome. The translation products for both collagen and trypsin are not active. The translation product for collagen includes an N-terminus signal peptide and propeptides at both the N and C termini. All three peptides, totaling 406 amino acids, are excised from the translation product to produce the collagen alpha1 chain. Similarly, both an N-terminus signal peptide and N-terminus propeptide (required for activation) are excised from the translation product to produce Trypsin 1. Turning to the third protein in this example, as already noted, the subunits of hemoglobin are individually translated: three translation products combine to produce one functional protein. *For none of the proteins cited can the translational products be considered functional proteins.* The generality of this three-protein exercise is detailed throughout this chapter.

Defining the Functional Proteome has already begun and continues immediately below, maintaining comparisons to the genome. Mining publicly accessible resources, the number of expressed proteins, those arising from full-gene transcription and alternative expression, will be estimated. A second estimate of the number of distinct single-protein entities with human genetic origins is made, accounting for annotated frequencies of posttranslation modification. Included is a brief but telling estimation of the impact of human microbiota on the number of organism-wide proteins. Enumerating distinct protein entities is only the beginning. Included also is a summary of activities or events that lead to differential functionality, extending the scope and nature of the "Functional Proteome."

How many proteins are in the human body? This is the starting point and, surprisingly, finding a consensus number is tricky—estimates range from 250,000 to millions [2]. A straightforward exercise helps sort out relative orders of magnitude. Data for this exercise are those accessible at Uniprot [1], a public database with multilevel search criteria and flexible output formats. Uniprot distinguishes manually annotated ("Reviewed") data from data mined without review from literature searches or from other databases. Selecting for "Homo sapiens (Human) [9606]" as "Organism" at UniProt produces 160,566 hits, of which 20,239 are manually annotated or curated (search date: November, 2017). The data in the manually annotated base (aka "Swiss-Prot") include, among a host of other information, the number of posttranslational modification (PTM) products for individual proteins by kind and sequence location. Data was downloaded and collated for the four most frequently occurring single amino acid PTM's and the results are summarized in Table 1 ("Modified Residue" includes single amino acid modifications including phosphorylation, acetylation, hydroxylation, etc. "Chain" includes key proteolytic events such as "Signal," "Initiator," and "Propeptide." Proteolysis is extensively addressed in Chapter 20).

For each expressed protein, 6.1 distinct proteins exist (following other treatments, the existence of the protein as expressed is assumed, though as is later argued, may not represent an active form of the protein). With this number in hand along with the

Table 1
Posttranslational categories and incidence

PTM processing category	PTMs per protein
Crosslink	0.21
Modified residue	2.67
Glycosylation	1.01
Disulfide bond	0.88
Chain	0.33
Average PTMs per protein	**5.1**

This table summarizes the posttranslational modifications (PTMs) for an average protein, derived from data maintained by Uniprot. (Although glycosylation is technically a modified residue, it is common enough to be given its own category.) Assuming the unmodified protein and all of its modified forms are active, each protein-coding gene yields an average of six functionally distinct proteins number of genes and the number of alternative expression products per gene, the number of distinct proteins may be estimated. Pre-Human Genome Project (HGP) estimates of the number of human genes range into the millions, with most estimates between 40,000 and 140,000 [3]. As the HGP progressed the number progressively decreased and, even today, continues to decrease with recent reports placing the number of genes close to 19,000 [4]. The number of proteins expressed per gene is cited at 3.4 [5] (Uniprot data indicate an average of 2.5 proteins expressed per gene from alternative splicing alone, in reasonable agreement with published data). Assuming that alternatively expressed proteins are comparably subject to PTM processes, an estimate of the Functional Proteome size is:

$$(\text{Genome Size}) \times (\text{Number of Proteins per Gene})$$

$$\times \left(\text{Number of PTM's per Protein}\right)$$
$$= 329,460 \text{ proteins}$$

How then does the gene-expressed proteome relate to the Functional Proteome? The number of directly expressed proteins tallies to some 65,000 (19,000 genes × 3.4 proteins per gene). Each of the expressed proteins is subject, on average, to five post-translation modifications. Thus, given the high rate of posttranslational modification, it is likely that few or no proteins have zero modifications: the gene-expressed proteome comprises little or none of the Functional Proteome. Put more prosaically, the human genome does not map the human proteome either qualitatively or quantitatively. And it follows that...

$$\text{Protein}_{\text{Gene}} \neq \text{Protein}_{\text{Functional}}$$

The exercise thus far defines the number of individual proteins that comprise the human proteome, but this is just a starting point for defining the "functional" proteome. The exercise continues, then, with expanding the number of functional entities based on protein interactions with other biomolecules.

Protein interactions are at the heart of a high percentage of biological activity—including genetics. Proteins self-associate and hetero-associate, create complexes with multiple centers, associate with polynucleotides, and bind to membranes. In each case the interaction induces structural changes in all participating entities and, as function follows structure, the composite entity produced by any interaction must be classified as a distinct "functional" entity. The question, then, for purposes of the exercise, is in how many interactions is each protein a participant? As might be expected, estimates vary but a reasonable and conservative figure for binary protein interactions alone is five (a number also borne out by Uniprot data). Adding the fact of five interactions per

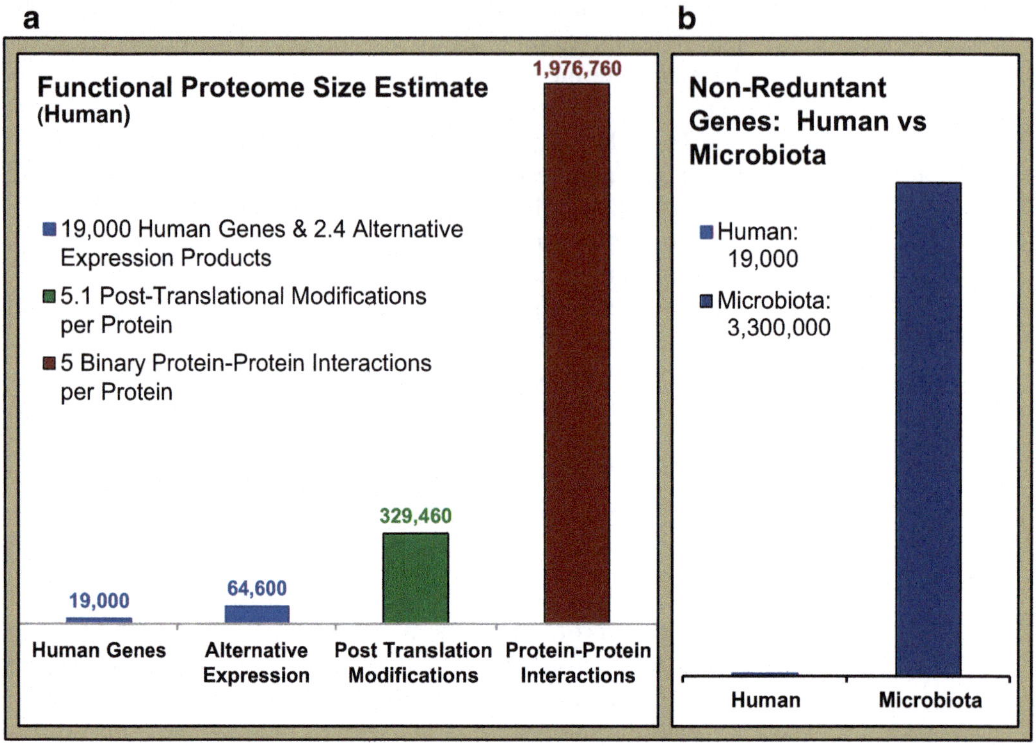

Fig. 1 (**a**) The challenge of proteomics lies in its fundamental, irreducible complexity. Taking into account alternative splicing and posttranslational modifications, a single human gene can produce an average of 12 unique proteins, all needing to be identified and catalogued. The true interest in proteins lies in their functions, however, and for that we must discern, quantify, catalogue, and compare all of their many interactions. With an average of five binary interactions for every protein (ignoring multi-center protein complexes and interactions with other types of molecules), we are confronted with a staggering sum of nearly two million to identify and study! Compare this to the measly 19,000 protein-coding genes in the human genome, and it is clear why reductionist approaches will never be up to the task. (**b**) The impact of the microbiome, based on its sheer size alone, must be significant. However, there is scarce data on the impact of non-infectious bacterial proteins on human proteins. (The effect of some infectious bacterial proteins are discussed in "Proteases: Pivot Points in Functional Proteomics")

protein which, when incorporated into the proteome size estimate of 308,142, results in an estimate for the Functional Proteome of (Fig. 1a):

$$\begin{aligned}&(\text{Number of Bound Entities} + \text{Free Protein})\\&\quad\times(\text{Number of Distinct Proteins})\\&\quad= 1{,}976{,}760\ \text{Functional Protein Entities}\end{aligned}$$

Interestingly, the exercise converges on the widely divergent estimates cited above that range from a quarter-million to millions. The difference is a consideration of individual entities vs. functional entities. Note that the relative cumulative size of the Functional Proteome is also a reflection of relative information content—a topic that will be revisited in the segment on Precision Medicine. Even now, the exercise is not yet complete given that the human organism comprises both genetic and proteomic contributions from more than one species—or more than one source depending on how human microbiota are considered. If now the microbiota resident in each human are to be included, at least one report notes that there are 3.3 million, nonredundant genes in the human microbiome (Fig. 1b). At a minimum the microbiota proteins that are not strictly intracellular functionally contribute to human biological activity. The collective "human" proteome could readily exceed a million distinct proteins and up to five million additional functional entities. Ignoring the microbiome altogether (in recognition of the paucity of proteomic data), the size of just the Functional Proteome is two orders of magnitude larger than the human genome. Even this is an understatement. In addition to interactions with other large molecules, proteins interact with, indeed are activated by, small molecules—small nucleotides, metabolites, peptides, lipids, and even water. Each of these interactions, when viewed from a protein function perspective, adds yet another member to the Functional Proteome. Even then the exercise is not technically complete as each conformation of a single protein is also a potential driver of activity (e.g., protein "folding" and "misfolding") and, thus, functionality. Distilling the manifestations of functionality that are biologically relevant is a significant challenge but one that must be accepted. The payoff is potentially enormous. When viewed as an information resource, the Functional Proteome is not only orders of magnitude larger than the genome, but the quality of the information reflects the instant to instant dynamic that reflects life.

Quantitatively and qualitatively, the vital activities that comprise a complex organism, that reflect health and disease, and that drive the interaction of that organism with its environment, cannot be defined or anticipated by their genome. The activity that is biology must focus on the actors: proteins, both free and bound.

3 The Functional Proteome and Systems Biology

This section of "Making the Case" is the context for the remainder of the chapter. The backdrop is, once again, decades of emphasis on the genome. The relatively recent expectation of genomics is expressed in the following attributed to Francis Collins (currently director of the US National Institutes of Health) in 2006 [6]:

> *"Essentially, we are now able to read our own instruction books. It is also a history book explaining how humans have evolved over time. It's a shop manual that describes with incredible precision how to build every cell in the human body. And it's a medical textbook containing insights that will help doctors predict and, eventually, cure disease."*

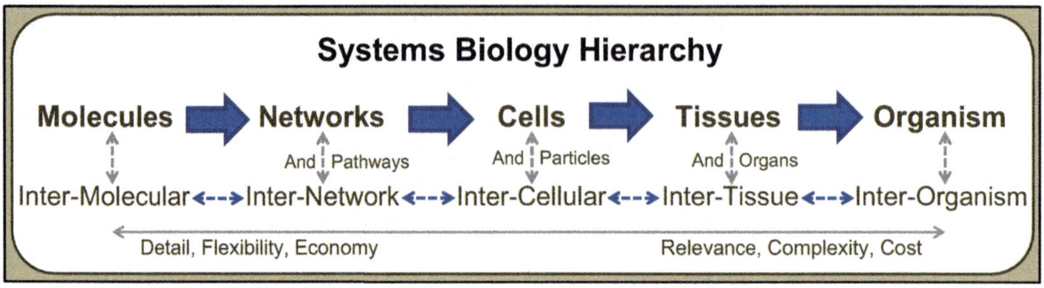

Fig. 2 The human body is so complex that we have traditionally approached it as a multilayered hierarchy of tissues, cells, and molecules. While this is a useful organizational tool, in reality these layers are not separate. They inform and influence each other. Therefore, our approach to studying them must not treat them as separate but as an integrated whole, Systems Biology. When investigating the detailed mechanisms of protein function, an awareness must be maintained of their cellular environment, and any deviations from biological conditions should be minimized and must be acknowledged. Testing on the organism level should be done with an eye toward how the interconnected pathways of molecular networks may cause unintended effects. More than anything else, a common vocabulary of materials identification and handling must be developed and maintained between the "disciplines" so that apples-to-apples comparisons can be made between studies

Now, little more than a decade later, the reality, a *"genome bubble"* [7] is much different [8].

> *"Having the complete gene set on the table, the knowledge of the genetic map and sequence is now considered by experts to be only a starting point for future research in biology and medicine,"*

In this aftermath of the "genome bubble," the context for continued research and translation of that research must change, and that change hinges upon the adoption of a more integrative and pragmatic paradigm, namely Systems Biology (Fig. 2) [9]. The nature and purpose of genes is accommodated as a part of a larger whole, as is "epigenetics," the functional activity that interfaces genes to the larger System. Systems Biology also encompasses earlier and ongoing schisms of, for example, the relative merit of in vitro versus in vivo research. Systems Biology not only acknowledges the value of multiple endeavors and test-material selection, it insists on their integration and synthesis: a tall order, to state the obvious.

Systems Biology then is a multilevel, zoom-in and zoom-out process, at one moment focused on single molecules and the next moment on the response of the organism to external stimuli. Systems Biology as an approach is the perfect context for multiple goals: increase in basic knowledge, understanding disease, development of new therapeutics and diagnostics, and the implementation of precision medicine. Systems Biology imposes a discipline and a context—"translatability," as will be further explored in the segment on Drug Discovery [10]. Molecular studies must be performed and analyzed in relation to their presumed context in the organism. Where molecular methods force a deviation from a physiological context, that deviation must be acknowledged and its potential qualification explicitly stated in terms understandable for the non-expert. Network, Cellular, Tissue, and Organism work must handle all materials, most especially proteins, in compliance with diverse molecular observations in order to avoid introducing artifacts. As with reports on Molecular activity, all area reports must state the qualifications of the results in approachable terms. Work performed at variable levels of complexity is both iterative and recursive—knowledge gained, e.g., on cells may well inform new work to be done on molecules. Without the knowledge gained, e.g., on Networks, work performed on Cells is ill-informed. There are no shortcuts. A series of related and ongoing efforts that include the author and colleagues, and multiple, international laboratories provides an example.

In 2015, New Liberty Proteomics (NLP) [11] was engaged to assess the ability of a library of test molecules to modulate the interaction of a peptide and a protein, the former associated with well-known disease pathology and the latter identified as a genetic risk factor for the same disease. Thus the Molecular assessment was, from the outset, connected to the Organism in the forms of disease manifestation and genetic predisposition to disease. Within the capabilities of the selected methodology (electron paramagnetic resonance spectroscopy and spin labeling) both the peptide-protein interaction and its modulation would, predictably, assume characteristic patterns. Upon completion of the screening, a portion of the data did, indeed, correspond to the expected pattern. However, that pattern was only one of four distinct apparent mechanisms of action exhibited by the library of molecules. Instead the library manifested four distinct apparent mechanisms of action: inhibition of peptide-protein binding, promotion of peptide-protein binding, and two of which altered the peptides conformation. Or so it seemed. NLP's work became a touchstone for additional biophysical work in other labs, which confirmed the two peptide-protein interaction modulations—some molecules inhibited the interaction and others promoted it (limitations of the second method employed missed the apparent changes in peptide conformation). These collective results drove support for further work in which additional, disease-connected proteins were incorporated into both biophysical laboratories. In the case of the NLP work, combinations of three or more proteins were examined, thus moving into the arena of Networks (NLP also pursued the peptide alone and confirmed earlier suspicions that some test molecules did indeed impact the peptide directly, though the precise interpretation was still elusive.). Work in both labs demonstrated once again diverse activity of the test library among multiple combinations of peptides and proteins, thus laying the foundation for other labs focused on Cells and Tissues continuing the progression toward Organism.

Work on both Cells and Tissues proceeded with a reduced set of molecules, identified from early biophysical results, and continued success was and is forthcoming. Thus the progression from Molecules (selected for direct connection to Organism) to Cells proceeds in a rational fashion despite the work being performed in labs with diverse locations and specialties. Incorporation of Organisms, mice in this instance, awaits further funding. Additional reference will, as appropriate, be made to this body of work in latter segments of this chapter.

As in the example, implementation of Systems Biology is intimately dependent on the selection of experimental observable(s). Optimistic views of the "instruction book" or "shop manual" potential of genomics have not proven viable in the face of ongoing data collection. Disease cannot be readily diagnosed nor are systems' level data forthcoming. Instead, given the multiplicity of environments, the variety of tasks performed, and the responsiveness to stimuli, one logical selection of molecules-as-observables is proteins. Categorically proteins are active drivers of digestion, metabolism, pathogen response, and muscle contraction. They also serve as primary structural elements, inter-tissue transport carriers, and storage devices. At a cellular level proteins control entry and egress of diverse molecules, recycle spent molecules and cells, and regulate genetic processes. At a molecular level, proteins engage in diverse interactions with small molecules, macromolecules, and membrane surfaces. They produce and are subject to multiple revisions, assume multiple conformations, and exist in multiple oligomeric states. Modifying protein activity is a primary goal and/or outcome of therapeutic treatment and assessing enzymatic activity is a dominant indicator for medical diagnostics. These latter examples—drug and diagnostic development—will be addressed in detail in subsequent segments of this chapter.

Implementation of Systems Biology in any context is challenging, even intimidating. The number of research specialties represented by the test-material-complexity hierarchy tallies into the hundreds. Publications within those specialties would correspondingly tally into the tens of thousands per year. Few if any single individuals exist with expertise in all areas, and cross-specialty communication is difficult and no funding practices accommodate such breadth and depth in academia or industry. Experiments on humans are, rightly, subject to regulation. Nonetheless Systems Biology establishes a logical, progressive, and self-correcting paradigm. Application and entry points for Functional Proteomics are works-in-progress as seen in the discussions below on Biomarker and Diagnostics Development, Drug Discovery and implementation of Precision Medicine.

4 Functional Proteomics Utility: Biomarker and Diagnostic/Prognostic Discovery

One working definition of "biomarker" is provided by the NIH [12], "*... a characteristic that is objectively measured and evaluated as an indicator of normal biological processes, pathogenic processes, or pharmacologic responses to a therapeutic intervention.*" Blood pressure, pulse rate, and body temperature are longstanding biomarkers, as is the analysis of body fluids such as urine [13]. Indeed, the laboratory analysis of urine in the twenty-first century echoes third century BC Hippocrates' noting the color—and taste—of urine. This also serves as an example of the close relationship between the object or property associated with a biomarker and available technology. Fever has long been noted as an indicator of disease. However, measurement of body temperature with precision and accuracy depends on access to a reliable thermometer—such as that of Galileo in 1592 or Fahrenheit in 1714. Given advances and changes in technology, available biomarkers of the twenty-first century range from system-level measurements, e.g., body temperature, to identification of pathogens to measurement of protein levels and gene sequences. The utility of validated biomarkers continues to expand. Biomarkers aid in diagnosis of disease, suggest therapy selection, and serve as monitors of treatment efficacy.

Diseases change the function of the organism, an observation that leads to Fig. 3a which provides a categorization of biomarkers by function. The function, e.g., of a diagnostic biomarker is disease differentiation given that the presence of a patient in a physician's office is prior evidence of disease per se. The function of a biomarker for selection of therapy assumes prior disease diagnosis for matching therapy to patient—the goal of precision medicine. Tracking disease progress is the function of a prognostic biomarker. Within this context there is no a priori reason to assume that a biomarker that differentiates disease also serves as a biomarker for therapy selection or for tracking disease progress. However, within the prevalent model of genomics, the biomarker equivalence for this apparent diversity is assumed: one gene to bind them. This is especially true for chronic diseases. In sharp contrast is the pragmatism of Functional Proteomics as the vehicle for biomarker development. Not only does the Functional Proteome encompass the diverse activity that is biology, it is immediately responsive to disease—and treatment—modification of that biology. In like manner, integration of a patient's manifestation of disease with known mechanisms of therapeutic action argues for a Functional Proteomics context. Similar arguments favor Functional Proteomics for tracking disease progress. Predictive biomarkers, those that putatively anticipate future disease risk, are included in Fig. 3a for the sake of completeness, and are discussed in the chapter segment on Precision Medicine.

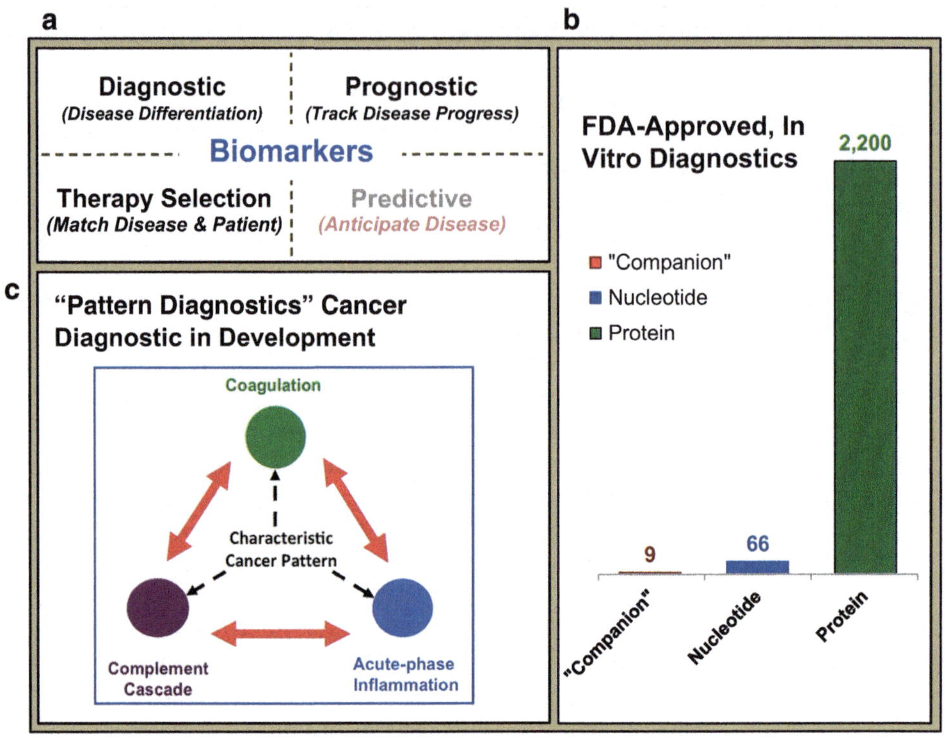

Fig. 3 (**a**) Biomarkers are, in essence, anything that we can measure that informs us about a person's health. Historically, we have used them to diagnose disease and to track disease progress. Recently, there has been an effort to use them as a means of determining which patients would be best served by which treatments, which has met with some success. There has also been an effort to use them to predict which diseases a patient is likely to be afflicted by, which has met with less success and more controversy. (**b**) In vitro diagnostic tests are regulated by the FDA, and tests of protein quantities and/or activities outnumber tests for gene variants 33 to 1. "Companion" diagnostics (of which there are only 9) are tests that select treatment-suitable patients already diagnosed with a specific disease condition, the hopeful beginnings of the modern Precision Medicine movement. (**c**) A recent study of cancer patients [13] detected a pattern of elevated serum levels of multiple proteins associated with Hemostasis, Inflammation, and the Complement System. For more information on the connections between these systems, *see* "Proteases: Pivot Points in Functional Proteomics"

This chapter segment addresses biomarker discovery and development within the paradigm of Systems Biology and the context of Functional Proteomics. Key areas of focus include biomarker relationship to disease diagnosis, prognosis, and selection of patient-appropriate therapy (aka precision medicine). Prediction or risk assessment is briefly addressed as a separate topic. Examination of existing diagnostic tests approved by the US Food and Drug Administration (FDA) is a useful starting point.

The battery of 59,707 in vitro diagnostic tests approved by the FDA [14] covers a wide range of "Test Systems": hormone and metabolite quantitation, detection and identification of pathogens, measurement of pH, etc. Proteins constitute 22% of the approved tests, dominated by enzyme activity with the balance being single protein quantitation (Fig. 3b). Many redundancies appear in the approved list as multiple companies provide diagnostic tools/services for a single assay. Overall, the number of distinct approved diagnostic assays totals some 10,000 of which some 2200 are protein assays. Separate listings exist for approved diagnostics that target nucleotides and the number of entries is much smaller: 99, of which a third are redundant for purposes of this accounting. Even shorter is the list of approved "Companion" diagnostics (tests that select treatment-suitable patients already diagnosed with a specific disease condition), which includes only nine entries, of which three are proteomic and six assess specific gene variations. Summarizing and accounting for redundancy in the listings, the FDA in vitro diagnostics lists include 2200 protein assays, 66 nucleotide assays, and 9 companion diagnostics. These numbers reflect both historical development of biomarkers and the changing face of medicine. They are also tests that are directly regulated. Other tests are certainly performed but are not subject to direct FDA regulation, tests that include cellular examination, e.g., for which expert execution and interpretation is required. For these tests, the FDA regulates only the laboratory's operation under the Clinical Laboratory Improvement Amendments (CLIA) guidelines. The tests performed, however, are not subject to rigorous validation or regulation.

Historically, in vitro diagnostics follow the progression from small molecule to protein to nucleotide. This reflects both the level of understanding over time and the available technology. Further, the object of analysis is almost exclusively singular—concentration of a single protein, a limitation of past technologies. The small number of assayed proteins (less than 1% of total human proteins) and nucleotides presages enormous opportunity.

A companion diagnostic is a test administered to select patients for clinical drug trials. If successful, the companion diagnostic is subsequently employed in patient treatment as well. The fact that only nine companion diagnostics have been approved severely limits progress in developing new therapeutics. All nine are either genetic tests or assays for a gene-associated protein.

The breadth of Functional Proteomics, armed with significant improvements in key technologies, far exceeds the scope of existing diagnostics—both in quality and quantity. Whereas the Functional Proteome numbers in the millions, only a few thousand proteins have been exploited to date. Furthermore, within those few thousand all focus on a single protein, whether via quantitation of concentration or assessment of enzyme activity. Within the paradigm of Systems Biology existing diagnostics address only a small fraction of the potential of Molecules alone. One opportunity for new development lies, therefore, in increasing the number of proteins examined while searching for more biologically relevant patterns. Technology now allows examination of large numbers of proteins and disease-related variation may be minimally examined at the level of both Molecular and Network complexity. This process is already showing signs of success as will be discussed in the segment on Diagnostics/Prognostics.

The Functional Proteome is conservatively estimated at two million functional entities comprising some 300,000 proteins (Fig. 1a)—reluctantly neglecting the microbiome, if with considerable trepidation. The Functional proteome is dynamic in multiple senses including posttranslational modifications and the interactions of proteins to other proteins, polynucleotides, small molecules, and cell membranes. It is also dynamic in that proteins are continually being "recycled" with lifetimes of individual proteins ranging from minutes to months. The Functional Proteome is multifaceted, plastic, and immediately responsive to internal changes and external stimuli. By comparison the genome is static ("epigenetic" and "gene expression" events are covered under the Functional Proteomics umbrella). Systems Biology is the organism milieu in which Functional Proteomics operates and responds. The combination is the context in which discovery and development of new Diagnostic biomarkers must occur.

As summarized above the majority of existing, protein-based in vitro diagnostics assess concentrations or activity of single proteins. Within a Systems Biology paradigm, existing in vitro diagnostics barely scratch the surface of the least organism-like stage: Molecules. No protein, indeed no biological molecule including genes, acts alone. Any deviation in the concentration, locale, or activity of one protein produces a ripple effect across multiple networks and, therefore, through the entire complex hierarchy to the organism. Any disease state therefore comprises alterations in numerous Molecules and Networks that in turn alter functionality at the Cellular, Tissue, and Organism levels. It follows that the minimum complexity for general discovery of Diagnostics is the Network level with a sound foundation laid at the Molecule level. This very approach—referred to by some as "pattern diagnostics"—actively engages many laboratories, academic and commercial, and one promising effort serves as an example [15].

The study is a classic "bottom-up" liquid chromatography-mass spectrometry (LC-MS/MS) approach followed by progressive focus on promising "hits" on human blood serum. Three factors distinguish the study from the outset:

1. No a priori expectations or modeling.
2. Removal of serum albumin.
3. Panel restricted to high-detectability proteins.

Willful naiveté is essential as prior expectations or model-based preconceptions consciously or unconsciously influence both design and analysis. The high concentrations of serum albumin (3.5–5.5 g/dL or 500–800 μM) obscure the detection of many serum proteins. Finally, though low-level proteins are of interest, the overriding goal in pursuit of a diagnostic-quality biomarker is repeatability and robustness: a basic set of proteins numbered in the hundreds versus thousands without the omnipresent confounding of serum albumin. The company, Biotech Support Group (BSG), LLC, elected to compare blood serum obtained from cancer patients versus normal controls, and details are presented elsewhere in this book. The outcome of their work, differentiation of cancer patients relative to controls, is of general interest for Functional Proteomics within a Systems Biology context.

The company has succeeded in demonstrating that the serum proteome from cancer patients (Fig. 3c) differs from age/sex-matched controls (biomarker deemed Stroma Liquid Biopsy). The analysis of their findings, however, goes far beyond creating a simple list of differentiating proteins. Proteins from three interconnected pathways or networks emerged. Further, differentiation is heavily dependent on posttranslational modifications and control mechanisms of those modifications. This provides a biomarker that is self-reinforcing with internal controls. Even more impressive is that the interconnectedness of the biomarker produces new knowledge—knowledge that potentially contributes on many fronts. A "bottom line": Functional Proteomics (in this instance concentration determination of multiple proteins) points not to single proteins but to functionally related groups of proteins—consistent with the behavior expected of biological systems. Thus even nominally Molecular studies can elucidate the behavior of more biologically complex entities—even the organism itself. And this only scratches the surface of Functional Proteomics' potential.

The complex hierarchy of Systems Biology presents numerous opportunities for application of Functional Proteomics in discovery and development of diagnostics and prognostics. Application at the Molecular level is an obvious example. For soluble protein sources (e.g., serum, plasma, cerebral spinal fluid, lymph, urine, extracellular fluid, cellular or tissue extracts) protein identification and concentration determination should be seen as just the starting point. As implemented by the Biotech Support Group, the first goal in discovery of a biomarker is development of an experimental protocol that produces robust and reproducible results (protein survey studies are not "cookbook" experiments). The standards for diagnostic discovery are much higher than for simple publication of results. How reproducible must results be? Once in use as a diagnostic, a "good" assay will be 90% "Sensitive" and 90% "Specific" implying false-negative and false-positive rates of 10%. The assay must be reproducible well inside the false reporting rates expected for the diagnostic itself. If, for example, the assay exhibits scatter of

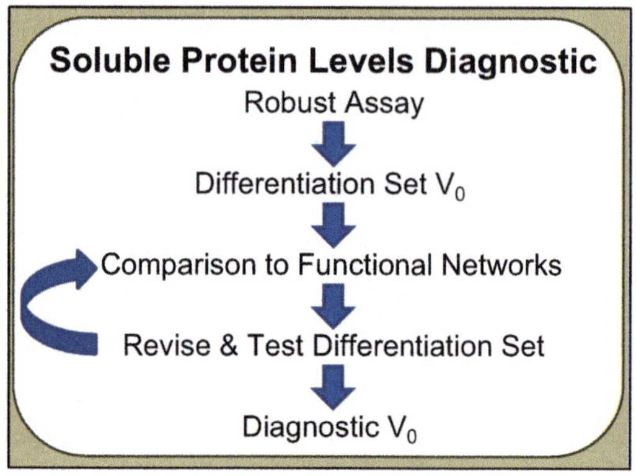

Fig. 4 Represented is a scheme to develop a reliable diagnostic based on protein concentrations or activity in bodily fluids. It is imperative that the selected assay be robust and reproducible from the outset, with metrics confined to fit those criteria. The assay should initially provide breadth so that a wide range of proteins are assessed ("shotgun"). Differences in test populations, should differences exist in reality, will suggest connections to biological networks. These "pattern-based" connections allow the assay test conditions to focus on the network(s) implicated in greater detail, i.e., selection of a progressively narrow set of test proteins. Lather, rinse, and repeat until populations differentiate with both high sensitivity and specificity: a candidate Diagnostic

a few percent of the mean for control samples, that scatter is a lower-bound for biomarker performance. Therefore, sound metrics are the first order of business. Assuming that such basic robustness is achieved and that test populations differentiate with sufficient precision, analysis of the differentiating proteins leads to consideration of the next level of complexity hierarchy: Networks (Fig. 4).

It is unlikely that an initial set of differentiating proteins—Differentiation Set V_0—is functionally unrelated. It follows that examination of the proteins in Differentiation Set V_0 will direct attention to one or more known Networks or pathways. Having inferred which Networks may be represented by Differentiation Set V_0, the potential exists for adding other proteins to a progressively focused group of proteins, a subproteome (*see* the Chapter 2, "Methods to Monitor the Functional Subproteomes of SERPIN Protease Inhibitors"). Iterating between subproteome selection and its ability to differentiate the targeted patient group converges on optimum subproteome selection and diagnostic quality simultaneously. The outcome for a successful exploration is Diagnostic V_0. Other options exist within this loop such as selection of various PTM products for a given protein, selection being based on known PTMs associated with network processes. The result of this process is designation of the zeroth version of a diagnostic candidate. While such an approach shows great promise, it only scratches the potential of Functional Proteomics as a driver of Diagnostics and Prognostics.

The example presented covers the most basic level of Functional Proteomics: determination of soluble protein identity and concentration coupled to known Networks or Pathways. Moving beyond this relatively simple strategy invokes shifts in Systems Biology complexity and/or experimental observables. Having discerned, e.g., that the concentrations of select, soluble proteins comprise an extended, network-focused diagnostic, a next logical refinement is assessment of protein interactions within those networks. From there, extending the strategy to include Cells is, at least hypothetically, a progression to increasingly physiological test materials. Observables at the cellular level expand to include protein-cell interactions and cell-associated protein distribution. However, each change in hierarchy complexity and experimental observable demands a clear-eyed assessment of methods currently available. These issues are the subject of the chapter segment on Proteomics Methods Appraisal below. Foreshadowing that analysis, methods at their current stage of development are a limiting step for full exploitation of Functional Proteomics—not just for diagnostic discovery but for drug discovery and precision medicine as well.

Closing this segment is a brief summary on the use of diagnostics to predict disease well in advance of clinical manifestation of that disease. This is certainly not a new concept and is, indeed, the underpinnings for widespread administration of cholesterol-lowering drugs to ameliorate or eliminate cardiovascular disease (the effectiveness of this ongoing clinical practice has been and is a subject for debate). Nonetheless the desire for early detection of disease persists. A couple of points must be made. The value of early detection only exists if disease-specific remedies or practices are immediately available, i.e., is the prediction actionable? Given that positive diagnostics and treatments for chronic diseases do not yet exist, any prescribed disease-avoidance remedy will, at best, be driven by the disease model du jour—as in the case of cholesterol-lowering drugs. These simple facts do not, however, temper administration of non-validated "risk assessment" tests, and this introduces the second point: *Administration of any diagnostic to a general population significantly raises the quality demands for any diagnostic* [16]. Consider, for example, a diagnostic with Sensitivity and Specificity values of 90. Within a demonstrably ill population these are acceptable numbers. Now imagine administration of the same test on a nominally well population. While missing 10% of an at-risk population may still be deemed acceptable, a 10% false-positive rate does considerable harm. The logic is inescapable. For any disease the relative incidence is a small fraction of the total population. Therefore in almost all cases a high false-positive value for any given test incorrectly identifies an at-risk population that far exceeds the true incidence of the disease itself. Current examples already exist for breast and prostate cancers [17] in which, for breast cancer, nine out of ten identified as at-risk will not have the disease. Unfortunately, in addition to the considerable emotional trauma induced by such errors, many women are unnecessarily subject to both invasive biopsies and exposure to hard radiation (ironically a procedure that increases risk of cancer). The bottom line is that the bar for Predictive Diagnostics must be set orders of magnitude higher than for Clinical Diagnostics. Unfortunately, in today's "wild west" practices, the reverse is true [16].

5 Functional Proteomics Utility: Drug Discovery

The productivity (number of new molecular entities per billion dollars invested) of the pharmaceutical industry has been in steep decline [18] for over half a century (Fig. 5a), leveling to less than 10% of its productivity in 1975 [19]. While this has had little

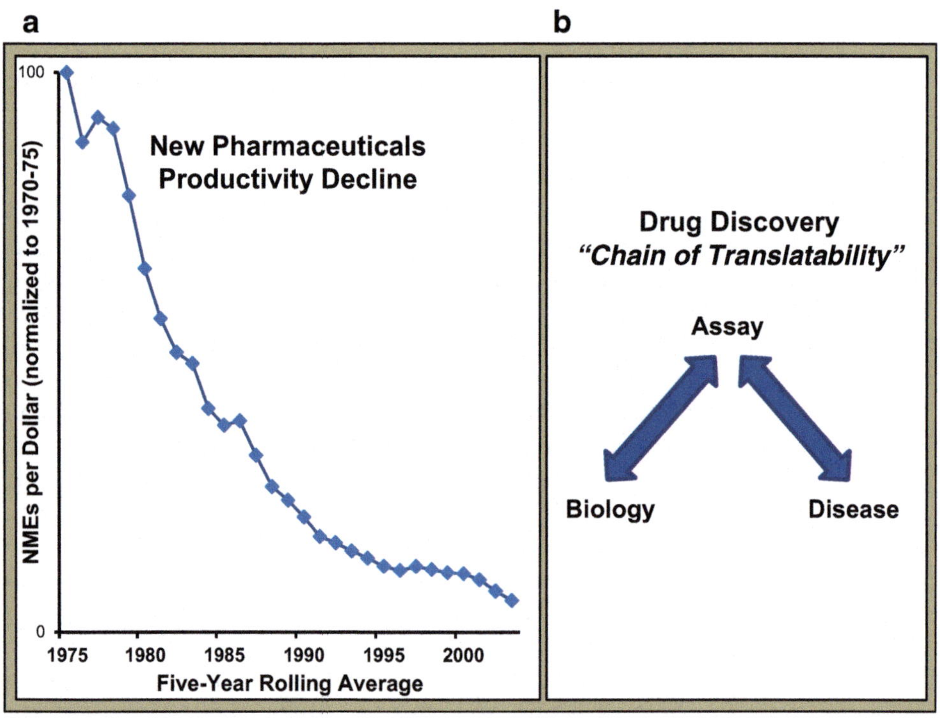

Fig. 5 (**a**) The efficiency of drug discovery research has plummeted to less than 10% of what it had been in 1975 (data adapted from reference 19). While many factors are at play, a significant shift in approach coincides with this period of decline: Target-based Drug Discovery (TDD). TDD is largely gene-centric, selecting "target" proteins based on gene mutations—a model that has failed. A resurgence of more phenotypic approaches is occurring, a resurgence highly intertwined with expanded characterization of the Functional Proteome. (**b**) A succinct summary of Phenotypic Drug Discovery approaches that emphasizes the fundamental importance of the Assay. The "Chain of Translatability" (reference 22) implies that the Assay must reflect and inform Biology and Disease pathology

influence on industry revenues the impact on healthcare has been predictably negative. No small part of this decline relates to a fundamental shift in the drug discovery paradigm from a phenotypic to a target-based approach. Historically, drug discovery followed a phenotypic approach in a protracted but largely successful effort to curb infectious disease. Would-be-drug molecules were tested for their ability to kill pathogen cells. Successful cell-killers were then tested in infected animals and molecules that were efficacious and safe in animals were subsequently tested in human patients. The success of those efforts, along with development and widespread use of vaccines, led to increased life span which, in turn, forced a shift in focus to diseases largely manifest with aging: cardiovascular disease, cancer, and neurodegenerative disease. However, translation of phenotypic drug discovery (PDD) methods proved elusive as cell and animal models for chronic disease did not exist. The stage was set for testing alternative discovery approaches and in less than a decade PDD declined to a small fraction of its former use. With equal doses of reductionist (and wishful) thinking and improvements in molecular evaluation technologies, the drug discovery paradigm shifted to a target-based drug discovery (TDD) approach. The thinking was (and is) that detrimental behavior of individual proteins lay at the heart of disease and that the detrimental behavior could be modified through "rational drug design." Some early success with ACE (angiotensin-converting enzyme) inhibitors hastened the transition from PDD to TDD. The more pervasive influence, however, in the switch from PDD to TDD was predicated on the possibility of full genome sequencing. According to the model, knowledge of the genome, specifically a single disease-related gene mutation, serves to identify the target protein whose activity could be modulated into a more acceptable state. The underpinnings of TDD rested (and rests) on the direct association of single gene mutations as origins for disease. Unfortunately the single gene mutation/single protein target model was generalized from experience with rare diseases, generally familial in origin and for which a Companion Diagnostic is obvious and effective. The reality for the majority of diseases, now decades following this errant generalization, is much different and much more complicated. A recently published meta-analysis is instructive [20].

Genomic data for 23 different cancers determined that 164,000 single-nucleotide polymorphisms (SNPs) exist—per cancer type. Attempts to functionally categorize the SNPs into diagnostically meaningful patterns were largely unsuccessful. Similarly, a host of apparent genetic mutations are catalogued for neurodegenerative and cardiovascular disease, none of which are definitive. The expectations for genomics as "*a shop manual that describes with incredible precision how to build every cell in the human body*" will not be met. By derivation, TDD has lost both its target source and primary rationale. Not surprisingly a resurgence of interest in returning to phenotypic drug discovery (PDD) has occurred, though TDD continues to dominate the pharma pipeline and much of funded research. Nevertheless, analysis of PDD outcomes in the clinic, despite low rates of PDD usage, is promising.

A 2011 survey [21] of drugs approved by the FDA between 1999 and 2008 indicated that phenotypic discovery methods were superior to target-based methods by 60% for "First-in-Class" drugs. The proportion was strongly reversed for "Follower" drugs. Noting the timeline, this is particularly impressive as the number of PDD-origin programs was vastly exceeded by those of TDD origin. Since the time of this report emphasis on developing and employing PDD has accelerated rapidly. Much, of course, remains to be done particularly for chronic diseases for which animal and cell models have proven elusive. However, it is appropriate to point out the essential compatibility of PDD with Functional Proteomics [22] within a Systems Biology context (underlining added for emphasis):

> *"Here, we propose the term chain of translatability to describe the presence of a shared mechanistic basis for the disease model, the assay readout and the biology of the disease in humans, as a framework for developing phenotypic screening assays with a greater likelihood of having strong predictive validity."*

The concept of "Chain of Translatability" (Fig. 5b) intimately connects PDD to Systems Biology and thus to Functional Proteomics. This is reinforced by realization that the analytical tools used in Functional Proteomics overlap with those used in PDD. The same is true for assay readouts. Thus, as was noted in 2004 [10], the core objectives of PDD and Functional Proteomics are, within a Systems Biology context, identical. Success in one is success in the other and the same is true for deficiencies and dead ends. To meet such a challenging and potentially beneficial goal, PDD efforts must extend into multiple arenas concurrently.

In keeping with historical success, most PDD efforts focus on activity at the Cellular level—note that Cellular is midway in the complexity hierarchy of Systems Biology. Historical precedent also drives a prominent use of stains/dyes in many, though not all, assays. One such example [23] is colorfully known as "... Cell Painting, which is a morphological profiling assay that multiplexes six fluorescent dyes, imaged in five channels, to reveal eight broadly relevant cellular components or organelles..." And "... automated image analysis software identifies individual cells and measures ~1,500 morphological features... ." "Cell Painting" is, therefore, a data-rich, "high-content" foundation for evaluating cellular modification by any number of biological or pharmacological modulators. Further, in keeping with theme of this book, the dyes (presumably) bind selected proteins and, thus, the readout reflects both the identity and locations of those proteins. The driving force of the assay is protein functionality. Still, the question naturally arises as to how such assays fit into the translatability triad of Assay-Disease-Biology. As of this writing a dominant "solution" for drug discovery is comparison of the "high-content" cellular patterns created by test library compounds to the patterns created by a reference library of molecules with "known" mechanisms of biological action. A "match" between a test molecule and reference molecule assumes a match in mechanism of action of the test molecule to the reference molecule. The connection, then, between the Assay and Biology (and Disease) is dependent on the reliability and suitability of the reference library, selection of the cell system, and the likely perturbations induced on cell activity by the dyes. Cell Painting and related approaches are powerful differentiators of members of a test library—but differentiation may or may not imply correct mechanism selection. Still, within the context of this book and chapter, it must be emphasized that the core responses of Cell Painting are driven by Functional Proteomics.

The cited approach is but one of hundreds of newly developing assay protocols in support of PDD. In concert are exciting developments in managing cells in increasingly biological structures. Developments also continue in isolation of an individual's stem cells and their conversion [24] to diverse organ-compatible cells. In this vein, enormous resources are being poured into so-called organs-on-a-chip [25]. These all augur well for relevant and translatable test beds for PDD. Even so, the predilection for starting in the middle of the Systems Biology complexity hierarchy—ignoring Molecules and Networks—is a mindset that must be reconsidered. A case in point arises out of the author's personal experience.

As cited above in the segment on Systems Biology, the author and colleagues engaged in successful library screenings to gauge modulation of the interaction between a pathology-related peptide and disease-risk-related protein [11]. Results demonstrated significant activity by members of the test library with 40% of the molecules expressing activity. However, in contrast to a simple binary interaction model, four distinct mechanisms of action were manifest, with some molecules apparently acting on the peptide alone. In the course of post-screening follow-up it was learned that the assay response for the peptide alone included contributions from not one but four distinct peptide entities—monomer plus three soluble oligomers. While assignment of these oligomers is still being made, it is known that they appear in variable concentrations as a function of total peptide concentration, concentrations selected to be within physiological ranges. Further, having adjusted data analysis procedures to visualize the response from the oligomers, the original protein partner was introduced. The data appear to clearly indicate two phenomena of both biological and pathological relevance: (1) Oligomeric peptides constitute the protein-interacting form. (2) Isoforms of the protein, one disease-related and one not, exhibit selective affinity for different peptide oligomers. Work is underway to reanalyze the library screening in accord with this new knowledge. Had the original screening work begun at the Cellular level, the Molecular activity would have been missed. More to the point of PDD practices, any impact of test molecules on the peptide screening readout would be misinterpreted and assignment of mechanisms of action compromised.

This example is telling within both PDD and Functional Proteomics contexts. Translatability to both biology and disease is established through choice of test materials, the peptide and protein. In a direct sense, those crucial connections define the assay. The assay itself directly tested for protein functionality, namely interaction of a peptide and protein. Once executed, the assay further expanded knowledge of the behavior of the peptide and connected to disease via differential and entity-selective binding to the protein. Finally as was mentioned above, the assay was immediately extended into assessments of modulation of protein networks and ex vivo application of the selfsame assay is in planning.

What, then, is the future of phenotypic drug discovery within the multiple contexts of Functional Proteomics, Systems Biology and Translatability? First, starting in the middle of the Systems Biology complexity hierarchy—Cells—may be a high-risk decision. Until and unless cell systems are developed that mimic chronic disease, the classic starting point for pathogenic PDD is problematic. By contrast, most diseases have identifiable relationship to one or more proteins so that a PDD development program can launch with the least complexity but still in a disease-relevant mode. As a bonus, molecular level screening is both faster and less expensive. Next, as with the example provided, extending a molecular study into network or pathway assays is, in principle if not in practice, straightforward. The combination of molecular and network findings will predictably inform both biology/disease and assay planning for cells. Extension, then, into original source material (patient-derived cells, fluids, extracts, etc.) is both abbreviated and as fully informed as can be made possible. A realistic bonus is the acquisition of knowledge that informs likely avenues for either a disease or companion diagnostic. Finally, the example cited points to yet another desirable outcome attainable through phenotypic approaches—selection of test molecules for their multiple or "polypharmacological" activities [26].

Within the complexity of any organism no molecule exhibits a single activity at all time points. The very functional attribute of any molecule that defines one activity, e.g., inhibition of a protein interaction, statistically matches to multiple, comparable sites on other proteins, cell, or nucleotides. Further, other attributes of the same molecule enable its ability to engage in additional activities. Thus multiple activities or functionalities are manifest by a single test-library molecule (or any molecule). Aspirin is perhaps the poster child for multiple, "mechanisms of action" combined with polypharmacological applications. Parallel multiple manifestations of disease also exist. At an organism level, disease is differentially manifest by combinations of symptoms. Aches and pains may or may not be accompanied by fever, fever may or may not be accompanied by intestinal distress, and all the above may or may not be accompanied by fatigue. At a molecular level, disease diagnosis derives from comparisons of multiple tests implying multiple manifestations of a single disease at a molecular level. Connecting the two realities—multiple activities of molecules and multiple manifestations of disease—combine in the concept of test molecule selection based on its polypharmacological activity. By its very conceptual strictures, target-based assessment is a non sequitur. Given its broader field of view, PDD more correctly and directly addresses multiple attributes simultaneously—and those multiple attributes are accessible through the Functional Proteome.

In closing this section on drug discovery, a watchword of sorts is forthcoming from the composite of this segment and the segment on Diagnostics: *A properly designed assay fills in gaps in the collective understanding of biology and disease, thereby supporting co-development of new therapeutics and diagnostics.* After all, both reflect properties of the same Organism.

6 Functional Proteomics Utility: Precision Medicine

Matching treatment to patient is the goal of Precision Medicine (PM) [27]. As many have observed, PM has long been in practice. Consider the cyclical relationship between disease diagnosis, therapy selection, and disease prognosis (Fig. 6b). A patient presents with apparent illness that may be attended by measurable attributes such as elevated temperature, blood pressure, or pulse rate. Upon questioning by the physician or nurse, other symptoms can often provide a working diagnosis. At this point multiple therapies may be available. Selecting the most appropriate is dependent not only on the diagnosis but also patient or family history with certain drugs and whether the patient is currently taking other medications. From this information a therapy, often a drug, is selected and the patient begins the recommended regimen. Disease progress is noted as the regimen proceeds, usually by the patient. If successful, i.e., symptoms are reduced or eliminated, the process ends except for any recovery from the therapy itself. If unsuccessful or if the patient reacts poorly to the selected therapy, an alternate therapy may be selected and the cycle continues. In the event that no therapy is successful the possibility of a more serious disease is considered, and expanded diagnostic procedures are brought into play. It should be noted that a full slate of Companion Diagnostics, a slate that includes the existing pharmacopeia, eliminates much of

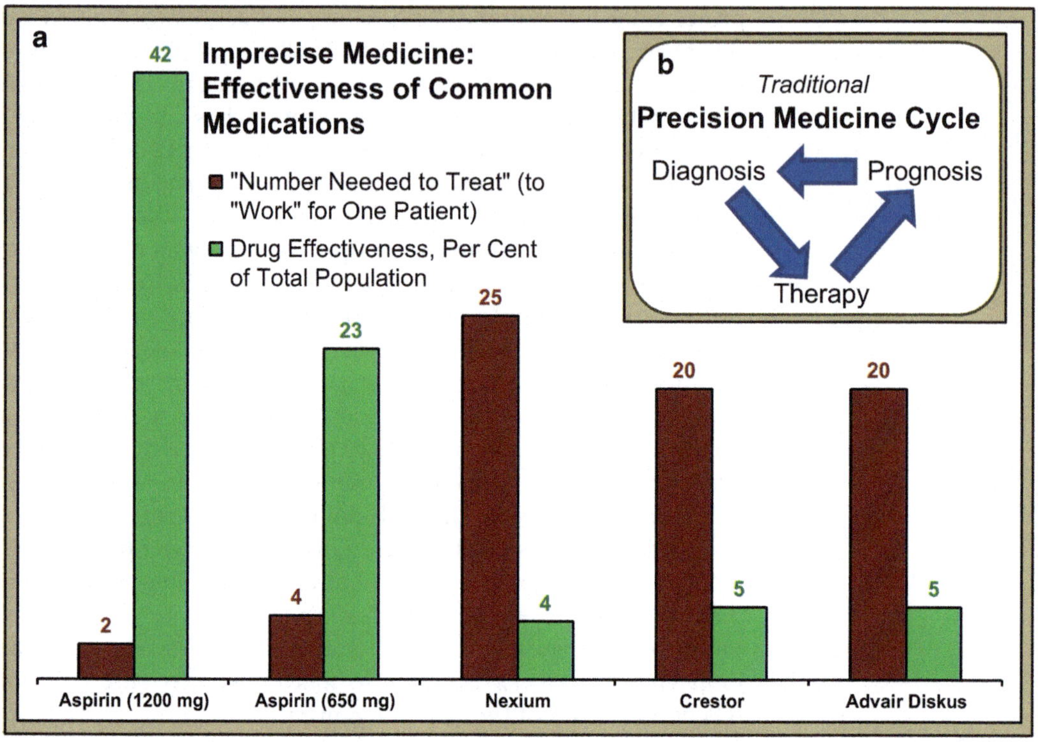

Fig. 6 (**a**) The drive for precision (meaning individualized) medicine is clear when looking at "Number Needed to Treat" figures for common medications. 1200 mg of Aspirin (equivalent to 2 extra-strength tablets or capsules) is only effective for 42% of the American population due to variations in personal biochemistry. For other popular medications a standard dose is effective for 5% or less. Knowing whether or not the patient is one of the lucky 5% is critically important. (**b**) The cycle of precision medicine has been practiced for millennia: Diagnose the patient. Determine what would be the most effective treatment and administer it. (Hope fervently that the patient suffers no serious ill effects.) Check the patient's progress and adjust treatment, or even diagnosis, according to the results. With the development of Companion Diagnostics there is the hope that the cycle can be made more efficient and minimize ill effects from inappropriate treatments

the treatment cycle. The benefits—reduced patient exposure to multiple drugs with multiple side effects, lower consumption of drugs, and the need for fewer visits to the clinic—could be considerable.

Precision Medicine as envisioned and heavily funded internationally in the twenty-first century is a more intensive and, at this point, genome-centered process. Before expanding on that process and "Making the Case" for Functional Proteomics, the relationship between the existing pharmacopeia and disease efficacy must be examined.

It is well known by medical professionals that not all drugs work for all people. This was acknowledged in the introduction to this section in the common practice of examination of family history and current drug usage. What is not so well known outside of the profession is the degree to which existing drugs don't work.

Consider available medications for pain relief, some available over-the-counter (OTC) and some via prescription (Fig. 6a). For OTC medications most people find relief from one but not another medication, or potential side effects of a medication affect some people but not others. The personally observed combination of pain relief and toleration of side effects define an individual's selection for pain relief. Studies have been made on the incidence of use of pain relievers, one such study being the being "The Oxford League Table of Analgesic Efficacy" [28]. The relevant category heading is "NNT" that stands for "number needed to treat." If, for example, the NNT is 2.4, then for every 2.4 people who take the medication, it only works for one person. This is the NNT for a most common pain medication, aspirin, at a 1200 milligram (mg) dosage (a common OTC tablet in the USA contains 325 mg). Another way to approach the NNT for a therapy is calculation of the fraction of the population effectively served by the medication. Aspirin, for example, at 1200 mg only works for around 42% of the population and only for 23% at 650 mg—two OTC tablets. These numbers seem high, but are actually consistent with a "good" drug. Examination of heavily prescribed [29] drugs such as Nexium®, Crestor®, and Advair Diskus® have NNTs of 25, 20, and 20, respectively, or only 4–5% of the diagnosed population. This presentation establishes a broader context for the twenty-first century discussion of PM, namely one in which existing medications are prescribed more rationally—or not at all.

Precision Medicine as touted by pharmaceutical companies, thousands of small companies and government funding agencies is intimately tied to highly publicized "successes" particularly in treatment of cancer. The presence of a particular gene mutation for patients diagnosed and previously treated for a particular cancer is selected for yet another treatment, one "targeted" at the protein expressed by that mutation. On average about a third of the selected population respond, with measureable declines in tumor size for an average of 3–8 months. At that time the cancer starts up again. Calculation of the NNT is bit tricky but, at best, is around three assuming that a 3–8 month relief for six-figure treatment costs is considered to have "worked." Detailed discussions on whether PM is proving successful within this context are available [16, 30, 31]. The purpose of the exercise here provides background for the true potential of PM if fully and properly implemented.

Full implementation of PM—matching therapy to patient for the whole of the pharmacopeia—requires knowledge of both the disease as manifest in the patient and the manifestation of therapeutic activity in the patient. A multitude of questions related to the functioning of the proteome arises, most not yet on any funding sources' view screen. How, e.g., does the Functional Proteome vary within a person over time? Or in relation to the wake/sleep cycle? Or in response to pathogens? Or as part of the development of chronic disease? Comparable questions can be posed for response to a medication. They can also be asked within the context of a single patient sitting in the consulting room awaiting diagnosis. The conclusion by all who consider the application of Functional Proteomics to PM is simple and telling: The necessary background knowledge does not exist. Why this is the case is abundantly apparent—a lack of focus in the face of hundreds of billions of dollars invested in sequencing the genome. For now, and in the foreseeable future, the overwhelming potential of Functional Proteomics in the rationalization of therapy selection is an untapped potential [9]. This thought provides a natural transition to a clear-eyed view of the Functional Proteomics knowledge base.

7 Functional Proteome Knowledge Base

Earlier chapter segments have established the need for refocusing attention on Functional Proteomics. The Functional Proteome is shown to be two orders of magnitude larger than the human genome. This is a conservative estimate as ubiquitous events that modulate protein activity and functionality, e.g., cofactor binding, are ignored, as is the significant but underexplored contribution of the microbiome to the Functional Proteome. Functional Proteomics encompasses all activity associated with genetics including transcription, translation, epigenetics, and the somewhat ill-defined "gene expression." Functional Proteomics is fertile ground for development of Biomarkers that lead to Diagnostics/Prognostics while simultaneously contributing new knowledge in both biology and disease. For comparable reasons, Functional Proteomics within a Systems Biology context is the natural paradigm for twenty-first century Drug Discovery: Phenotypic Drug Discovery. However, given the great potential of Functional Proteomics, completeness demands that two pragmatic questions be addressed:

1. How complete is the existing Functional Proteomics' knowledge base?
2. How effectively do existing methods contribute to that knowledge base?

Without the answers to these questions, efforts cannot be prioritized nor can a realistic strategy be developed. This chapter segment addresses both these key questions and ends with a prioritized roadmap to a highly desirable endpoint: a curated Functional Proteome and a full complement of robust tools to achieve that endpoint. Evaluating today's knowledge base relative to the desired endpoint is begun through comparison of efforts such as the "Human Proteome Project" (HPP) [32] to the data presented in Fig. 1a.

An ongoing goal of the HPP is creation of an organism-scale map of the human proteome, and the necessary inclusion of alternative expression, PTMs and protein interactions was recognized from the outset. However, data collected to date is gene-centric by conscious choice. Thus the current proteome knowledge base is largely defined by the smallest column in Fig. 1a representing some 19,000 distinct proteins. None would argue that such a knowledge base is complete or adequate but concerted efforts to expand that knowledge base are not being made. Therefore the databases upon which protein identification efforts rely restrict analysis to a highly limited, gene-centric misrepresentation of the proteome. Identification of proteins in a mixture that relies on gene-centric databases is almost certainly incorrect (as was argued above, gene-expressed proteins may not exist as active, functional entities). A last, ironic point must be made: the individual proteins tested to define the human proteome are not gene-expressed products at all, but rather are human-gene-sequence-compatible proteins subject to PTM processes active in the host cell from which the proteins were expressed. Despite the apparent negativity, this rendition of the proteomic knowledge base is the outcome expected from a gene-centric approach: an exchange of genome coverage at the expense of specificity.

The existing knowledge base, by design, relates the proteome to an admittedly incomplete reference set of data, namely the index provided by genes alone. However, that incompleteness, or lack of specificity, is progressively remedied through logical expansion of the reference set, an expansion that can take many forms. One of those forms is the ongoing, and never abandoned, protein-centric work that examines relatively simple protein systems but with a high degree of specificity. Thus the proteomic knowledge base expands in a reliable though piecemeal manner. Further, combining the two approaches—gene-centric/bottom-up/shotgun [33] and protein-centric/top-down/rifle [34]—can and is being used with great effectiveness. A case in point is the development of proteomic-based biomarkers discussed above [15]. Bottom-up approaches can provide, e.g., robust disease differentiation that lays the basis for subsequent highly specific selection of proteins that define that differentiation. This theme, the duality of coverage and specificity, is consistent for the whole of the Functional Proteomics knowledge base, encompassing not just the knowledge itself but also the methods by which that knowledge is derived.

An ideal Uniprot profile, e.g., for any given protein would include curated data on distribution of PTMs, structures for all possible isoforms and conformers, protein distribution at both tissue and cellular levels, concentrations for all possible locales, interactions with other molecules of any type within those locales, and . . . the list goes on. Such an idealized listing is unlikely nor is it truly required, but the breadth of proteomic attributes does

Table 2
Proteomic properties, methods, and proteome coverage

Proteomic property	Method examples	Methods' reliability	Coverage potential	Proteome coverage to date
Identity	Mass spectrometry	Good	Full	Limited
	Antibody	Fair	Limited	
Structure	X-ray	Good	Limited	Poor
	Nuclear magnetic resonance	Good	Limited	
Quantitation	Mass spectrometry	Fair-good	Full	Poor
	Antibody	Fair	Limited	
Localization	Immunofluorescence	Fair	Limited	Poor
	Fluorescent-protein tagging	Fair-good	Limited	
Activity	Yeast two-hybrid	Poor	Extremely limited	Poor
	Fusion	Poor		
	Surface affinity	Poor		

Our knowledge of proteomics is only as accurate and as deep as our data. Summarized here are the five basic protein properties. For each property is listed the most common methods used for measurement, graded in terms of the accuracy and precision of individual measurements (Methods' reliability), and their applicability to all proteins and protein interactions (Coverage potential). The final column indicates the degree to which any given property has been evaluated for the proteome (Proteome coverage to date). The lack of Functional Proteome Coverage To Date is the most serious impediment for advancing the frontiers in biomarker discovery and development of patient-specific therapies

establish the context for proteomic research and application of that research. Further, the list may be practically categorized in such a manner as to maintain generality while establishing a basic set of Functional Proteome attributes. It is against this basic set that the status of the Functional Proteomics knowledge base is appraised in parallel with the ability of existing methods to assess the key metrics of: Identity, Structure, Quantitation, Localization, and Activity (*see* Table 2).

7.1 Proteomics Methods Appraisal

Table 2 cross indexes key Functional Proteomics metrics to: Methods employed to assess the metric (Method Examples), the reliability of a given measurement by the Method (Methods' Reliability), the potential for the Method to address the whole of the Functional Proteome (Coverage Potential) for the Proteomic Property, and the extent to which all existing knowledge addresses the whole of the Functional Proteome for a given Proteomic Property. A couple of examples are helpful. The structural Method, X-ray, provides exquisitely detailed spatial information and is, therefore, a Reliable method. However, its ability to Cover the entire proteome is Limited by the inability to crystallize every protein in the proteome. By comparison, the Activity method for protein interactions, Yeast Two-Hybrid, is prone to a high incidence of false-positive and false-negative results (discussed in more detail below). Its Reliability is, therefore, poor and its ability to Cover the proteome is extremely limited.

The low assessments for existing Proteome Coverage derive from a combination of Methods limitations and lack of concerted effort. Summaries of Methods and their ability to contribute to fundamental protein Properties are provided immediately below.

7.2 Identity

The majority of "modern" efforts in proteomics has been and is focused on protein identification, especially for mixtures of proteins. Dominant in the effort is a variety of separation procedures ("Sep-Sci," gel electrophoresis, liquid chromatography, etc.) and analysis by mass spectrometry (MS). Indeed the use of MS is often (and wrongly) equated to proteomics and vice versa. Other identification methods include Edman sequencing, quantitative amino acid analysis, and antibody-based analysis (e.g., enzyme-linked immunosorbent assay or ELISA). Sep-Sci alone can be used for rough identification though molecular weight resolution is typically insufficient to differentiate closely spaced isoforms or the extent or kind of posttranslational modification. MS has somewhat similar issues for some approaches though not for others. So-called, "bottom-up" methods [33], typically applied to mixtures of proteins, rely heavily on comparison of post-digestion sequence patterns to databases of those sequences for individual proteins. Two issues must be, and are, acknowledged: (1) Algorithm-driven analyses can only assess identification in accord with members of the database (heavily weighted to proteins expressed by cDNA) and (2) "Sequence coverage" for any given identified protein can range from 30% to 99%, the range often reflecting relative concentrations of proteins in the mixture. In contrast to bottom-up approaches, several "top-down" MS approaches [34] are also available. As the name implies, whole proteins comprise the test sample. Top-down approaches can resolve closely spaced isoforms and can inform the degree and kind of PTM. As with bottom-up, however, if algorithmic comparison to existing databases is employed, those analyses are only trustworthy if the tested protein(s), including PTMs, is already in the reference database. Further, Top-down methods are typically applied to mixtures of fewer proteins. The bottom line for Sep-Sci/MS is that identification results are reasonable within the stated qualifiers. Confirmation, if needed, can be supplied by "orthogonal" methods such as Edman Sequencing. Then and only then can a non-qualified protein identification be reasonably certain.

In sharp contrast to instrument-based methods such as MS, antibody/antigen binding can be exploited to identify proteins [35]. Generally, antibodies are prepared and isolated for single-protein antigens. Subsequent binding to the antibody by a protein from a mixture of proteins is taken as evidence that the protein is the original protein/antigen. Inherently, the relationship between antibody and protein/antigen is a one-to-one—a single-protein assay. However, multiple antibodies can and have been incorporated into array panels with the net result that mixtures of proteins may be examined and antibody-selected proteins identified. In principle antibody-based assays and Sep-Sci/MS assays serve as orthogonal approaches for protein identification. In practice, both are seldom used in a single report and, in the few instances of rough comparability, the two approaches do not fully reinforce each other. Thus, it should not be assumed that Sep-Sci/MS and antibody-based identification are orthogonal methods. At the heart of all antibody-based work is the inherent fact that no antibody can be proven to be selective for a single antigen. Attempts to prove selectivity are, in the final analysis, attempts to prove a negative. On a practical note, antibody promiscuity should be assumed.

What's needed? The knowledge base for Identification is inconsistent, detailed for some proteins, nonexistent for others. Tissue distribution is a particular concern—one that must be addressed for even a rudimentary foundation for Systems Biology. A worldwide effort would be welcome. As for methods, two issues for MS methods stand out: (1) Expansion of the reference bases to more closely represent the actual proteome (inclusion of PTM's, e.g.) and (2) Resolution of the dependence on digestion methods for bottom-up approaches. For antibody work the overriding issue is selectivity and, as was noted above, proving absolute selectivity is not possible. However, demonstrating non-selectivity for key data is perhaps tenable. Reports from either major approach should not be considered definitive, but rather as direction indicators for more targeted methods.

7.3 Quantitation

Hand in hand with the need for protein identification is determination of protein concentration. The list of applicable methods for quantitation is comparable to those for identification, though execution differs. In the case of antibody-based work [35], addition of a reporter group to the antibody, e.g., as is done in preparation of ELISA assays, identification and quantitation occur in a single assay. For MS-based work [36], a variety of labeling approaches provide information on the relative concentrations of proteins in a mixture. These methods coupled to carefully prepared control mixtures can serve to provide some indication of absolute protein concentrations in the original source material. However, for optimum determination of absolute protein concentration coupled to dramatically improved detection of low-concentration proteins, application of so-called "Multiple Reaction Monitoring (MRM)" [37] is the approach of choice. It must be noted that MRM demands intensive method development and absolute concentrations are available only at the expense of "coverage."

What's needed? The knowledge base for Quantitation is inconsistent given its dependence on Identification as a prerequisite. A way forward is execution of MS labeling experiments which provide relative concentrations in concert with a select subproteome analyzed by MRM. Absolute concentrations from MRM provide, in principal, the appropriate adjustments to relative concentrations from the broader coverage of MS labeling results.

7.4 Structure

Structure and function are intimately related and only two methods dominate determination of protein structures: X-ray crystallography and "solutions" nuclear magnetic resonance spectroscopy (NMR) [38]. Of these two, X-ray structures listed with the Protein Data Bank (PDB) outnumber NMR structures listed by 10:1. Combined, the two methods account for 98.4% of all structures archived (cryo-Electron Microscopy, the object of a 2017 Nobel prize, accounts for only 1.4% of PDB structures) [39]. The quality of data forthcoming from these methods is outstanding—the issue is coverage of the proteome.

As its name implies, X-ray crystallography demands samples with significant, through-space uniformity. Coaxing a molecule into such a form is a combination of science, method, and art, and many proteins simply are not amenable to such forced uniformity. This fact limits the range of proteins that can be studied by X-ray. Coupled to this limitation is the inverse implication of coverage: proteins forced into an extended, uniform state almost certainly do not represent the multiplicity of states of proteins in solution. In the best of cases the crystallized structure may be deemed a dominant or active conformer, in the worst the structure may reflect no biologically active state. Solutions NMR largely avoids this potential pitfall but has limitations of its own.

Samples for solutions NMR range in concentration from 0.1 to 5 mM [40] and, for the overwhelming majority of protein work, contain proteins smaller than 50 kDa (slightly smaller than the average for human proteins (53 kDa) and slightly larger than the median (42 kDa)). For the most-practiced NMR methods as applied to proteins, half the human proteome is not accessible (NMR methods do exist to virtually eliminate the molecular weight limitation but are limited by magnet technology—and cost). For the 50% of human proteins accessible to solutions NMR, the information content reflects not only secondary structures but also partially inform variation in conformational states. Further, impacts on structure by solution variation and small molecule effectors may be assessed. Here, though, the second limitation of NMR, sensitivity/concentration, comes into play. With a preferred lower limit of 2 mM (1 mM = 10 mg/mL for a 10 kDa protein), NMR sample concentrations exceed the observed physiological concentrations of nearly all proteins. Further, a typical sample size of slightly over 0.5 mL, samples can be costly, a consideration that is amplified by the need to express proteins with local or large-scale isotopic substitution of one or more elements (C, H and N).

The "dark horse" for protein structure work may well be the peptide-protein interaction methods cited by the author in this chapter: electron paramagnetic resonance spectroscopy (EPR) coupled to spin labeling [41, 42]. Not only are interactions discernible, but intra-protein structure and individual amino acid mobility may be determined.

What's needed? The apparent solution to expanding the collective methods' capabilities is a wholly new method. Crystallization of proteins is a limiting factor for X-ray just as molecular weight is limiting factor for NMR. More widespread application of, e.g., cryo-Electron Microscopy may pick up some of the slack. As has been mentioned, EPR/spin labeling is an underexploited method, one that avoids the limitations of both X-ray and NMR. Improvement of proteome coverage can only be made with increased support.

7.5 Localization

Two widely divergent methodologies must be considered for assessing the status of protein localization: fluid-based and cell-based. Biological systems naturally afford access to extracellular fluid, examples being blood plasma, lymph fluid, cerebral spinal fluid, and interstitial fluid. For these cases methods such as those discussed under Identification and Quantitation apply. Fluid-based approaches can also be applied for selective isolation and lysis of subcellular organelles. Thus, within the qualifiers noted above, fluid-based localization studies are hypothetically amenable to examination. Cell-based localization of proteins demands entirely different approaches, exemplars being Immunofluorescence (IF) and fluorescent-protein tagging (FP) [43].

Distinct proteins are selected for either direct modification required for detection or as the antigen for protein-specific, fluorescently labeled antibodies. Images of cells with incorporated monitors reveal the localization of the labeled entities under a variety of conditions of interest. For either labeling approach certain cautionary flags are raised. Extracellular introduction of labeled proteins, whether FP or IF, cannot insure bio-relevant distribution, especially to intracellular organelles. Next, as with all labeling approaches the label itself may significantly alter behavior. In the specific case of IF, lack of antibody selectivity produces false-positive results. Given that they are by design active agents, antibodies may also shift equilibria in the direction of the captured antigen, thus perturbing the very system under observation. The impact of these concerns is quantified in a side-by-side study of 506 target proteins [43].

Comparative results for the two labeling methods were categorized as identical, similar, or dissimilar among different subcellular locales. Identical results ranged from 15 to 70% among those locales, with an average of approximately 40% (number estimated from visual inspection of graphic material). Identical plus similar results (similar implying overlap in localization of the two methods) yield ranges of 75–95%, and dissimilar results were observed for 10–25% of the tested proteins. While the sum of identical and similar results is encouraging, the relative number of both identical and dissimilar results is cautionary. Though not a conclusion of the study it's clear that the researchers favor labeled protein (FP) results over antibody localization (IF). For example, 37 (8%) of the antibodies gave no staining results even though RNA sequencing argued for the presence of those proteins. On the false-positive side, researchers suspected cross-reactivity of antibodies when target proteins existed in low concentrations.

What's needed? Localization of extracellular proteins is obtained via methods described above for Identification and Quantitation. Methods for localizing proteins in viable cells remain problematic, though anecdotal data favor a labeled protein approach over antibody detection of localization. More methods development is required prior to wholesale implementation of any given approach.

7.6 Activity

This segment primarily deals with the most prevalent protein activity, protein interactions—not simply binary protein-protein interactions but the whole of protein interactions. In accord with the analysis in this chapter, the dominant Functional Proteomics entities (83%) are bound proteins. Both Functional Proteomics and Systems Biology demand assessment of protein interactions in bio-relevant contexts. This is abundantly clear for translation into essential arenas such as Biomarker and Drug Discovery, and Precision Medicine. Effectiveness in these arenas comprises the ability to assess interactions among proteins (binary and groups), between proteins and polynucleotides and between proteins and membrane surfaces. Further, assessment must include the impact on those interactions of variations in solution conditions, and the introduction of effector molecules such as cofactors, substrates or new drug candidates. However, as two references attest, measurement even of protein-protein interactions is the weakest link in the methodology chain.

In 2002 a composite of existing data was analyzed [44]. Highlights from that study concluded that only 3% of reported interactions are supported by more than one method. That low "hit" ratio continues in a 2009 study [45] in which multiple methods agreed on only 8% of tested protein pairs—all of which were "known" interacting pairs. Further, from the 2009 study the best method test on known interacting pairs of proteins missed a full two-thirds of the interactions—a false-negative value of 66%. Such outcomes lead authors to conclude that, "... large datasets of protein-protein interactions vary enormously in their error rates and there is no simple way to compare different interaction data sets." Other recent reviews, noting the poor collective performance, opt for wholly comparative analyses of, e.g., homology, or in silico predictions of various sorts [46]. Indeed, many university courses take this approach, virtually eschewing the particulars of actual measurement as a confirmation of protein sequence comparisons. There is a circularity to such logic given that the basis for homology comparison is the poor performance of the methods from which the homology comparisons arise. Functional Proteomics cannot proceed under these conditions. A brief review of existing methods is, nonetheless, necessary.

The most employed method is yeast two-hybrid (Y2H) which features downstream activation of, e.g., a transcriptional event upon the interaction of two fusion proteins in the nucleus of a yeast cell. One portion of each fusion protein includes the sequence for one of two proteins to be tested. Since the downstream event is only triggered by interaction of the fusion proteins, a trigger of that event is presumed to represent interaction of the test proteins. In the 2009 report the best Y2H performance was 25%. Direct translation of Y2H methods to essential applications such as drug discovery is impossible. The majority of other existing protein interaction measurements include either capture/fusion of an interacting pair or analysis of interactions in which one test protein is bound to a surface. Among these are tandem affinity-tagged, coimmunoprecipitation and surface plasmon resonance. While all have application in limited circumstances, none afford the robustness or flexibility demanded by Functional Proteomics within a Systems Biology paradigm [46].

As was mentioned under Structure above one serious candidate, though little known, is the use of EPR and spin labeling. As demonstrated in the report on the author's ongoing work and elsewhere [41], the collective abilities of the method is amenable to detection of protein interactions [47] with any interacting partner and has been employed in test media of any complexity. Indeed at least one example exists that demonstrates its potential use in diagnosis of Duchenne Muscular Dystrophy. EPR's range of application, once translated, might be appreciable.

What's needed? The glaringly obvious answer is new methods. No one existing method stands forth as a standard by which other methods may be assessed. Nor do methods agree sufficiently to trust a variety of methods as checks and balances. It is time to focus on new approaches among them being EPR and spin labeling.

7.7 Prioritized Roadmap

Table 2 assesses the overall knowledge base of the Functional Proteome as poor, and that Methods in most cases have limited coverage potential. Any Roadmap must simultaneously focus on accelerated data collection in areas of strength while improving basic capabilities in areas of weakness. Areas of strength include the ability to Identify and Quantify proteins. This argues for aggressive support for both updating reference materials to include PTMs, etc., and widespread data collection. This work is foundational and receives high priority. Structural determinations may proceed in parallel along with the need for new methods development to address know limitations. Localization methods must, for now, be deemed developmental with Methods showing great promise but lacking in robustness and reproducibility. Proteomic Activity, as related to detection of protein interactions, is essentially moribund. Wholly new approaches, especially approaches that permit assessment in complex media, must be developed—now.

7.8 A Cautionary Lesson on the Use of Recombinant Proteins

As was briefly mentioned at the beginning of this chapter segment, recombinant proteins (the major source of laboratory proteins) are modified in accord with the posttranslational product of the host cell from which the protein is expressed. Bacterial cells will produce bacterial PTMs, e.g., and human cells will produce human PTMs. The degree and kind of PTMs are almost never specified or, as is likely, known by the supplier. Therefore, almost the whole of protein research that employs recombinant proteins cannot be fully specified. The content of those little vials opened thousands of times a day is simply not known. Researchers must insist on full protein specification, including PTMs (and protein content which is typically only 50% of the total mass delivered). *Finally, in-house protein sequencing must become commonplace.*

8 Closing Comments

Biology is the adaptation of chemical and physical processes and attributes in support of life. In this context, two classes of biological molecules warrant primary attention: polypeptides and polynucleotides—proteins and DNA/RNA. Two, simple statements define the relationship between the two, essential classes of biomolecules:

Without genes there are no proteins.
Without proteins there are no genetic events.

Addressing the first biological truism, it becomes increasingly clear that the word, "genome," is not singular. Considerable genetic variation exists not just among individuals but within a single individual. It's likely that a person's genome(s) varies over time as well. Addressing the second truism, few biological events occur without mediation by proteins including epigenetics, "gene expression," transcription, translation and DNA repair—which introduces a third truism:

Biology is change, all organisms change with each tick of the clock.

Proteins are agents of action, both driving change and responding to change. Clearly research must shift from the limited and passive perspective of genomics to the systemic and active perspective of Functional Proteomics. This chapter closes with a review of the demands of Functional Proteomics and how (or whether) those demands can (or cannot) be met. As prelude, the potential payoff for full characterization of the Functional Proteome is reviewed.

The literature is replete with reports of biomarkers potentially suitable for identifying druggable targets, diagnosing disease or selection of medication. Many are founded on the discovery of some previously unreported gene variation. It's a dark sort of consolation to realize that any given report has a 50/50 chance of being repeatable or that genomic variation is now known to be quite common. As a result no positive diagnostic exists for many diseases including Alzheimer's Disease, and truly predictive tests exist for no chronic disease. Both the focus and discipline must change. Functional Proteomics, even in this relatively unexplored stage, is already pointing the way forward in many areas. Even in the lowest level of Systems Biology, namely Molecular, diagnostic candidates are coming forward that not only differentiate disease but inform the nature of that disease [15]. In the forefront of drug discovery a renewed focus on phenotypic approaches include as a matter of course expansion of knowledge for biology and disease in addition to simply "screening" test library molecules [22]. Even further, assay demands include a coordinated search for reliable Companion Diagnostics to support the highly desired goals of Precision Medicine. If progress continues, the artificial boundaries between biological research and translation of that research will cease to exist. As they should. The question is, can that progress continue?

As with all scientific endeavors, progress relies as heavily on timing, resources, and will coupled to wisdom. The time for a focus on Functional Proteomics is now. This statement would not have been realistic even a few, short decades ago. Whereas access to proteins has traditionally been limited, the breakthrough of recombinant DNA affords today's researcher thousands of proteins available at the click of mouse. Standardization of the production of polyacrylamide gels didn't occur until 2004. The Nobel Prize that would launch the general application of NMR determination of protein structures would not be awarded until 1991. In 1980, only a few dozen protein structures had been determined by X-ray. Now, over 100,000 protein structures have been deposited in the PDB. As with NMR, MS has only recently evolved from a small molecule tool to one routinely used to identify multiple proteins simultaneously. Timing is apparently propitious, what about resources?

"Resource" in this context has at least two meanings. Staying first with the capability theme, the analytical resources required for Functional Proteomics is a mixed bag (*see* "Functional Proteome

Knowledge Base" above). The ability to identify and quantify proteins is reasonably good. Areas that require improvement are noted above with the most needed improvement being rationalization of reference bases and analysis algorithms. Both demands, while considerable, can be met with the other resource—funding. The payoff, a reference base of protein identification and quantification by PTM and by tissue distribution, provides benefits that can't yet be imagined. Of course, that's only scratching the surface of Functional Proteomics. Moving on to determination of protein structure the picture is less rosy. Dominant methods, while more than adequate in their niches, cannot readily be expanded outside those niches. Localization of proteins in viable cells may well spring from existing protein labeling methods, but more work is needed at the foundational level (adding the dimension of time is likely a requirement). The last proteomic property examined above, activity, is the most difficult to forecast. Existing protein interaction methods simply are inadequate to the task and, therefore, a significant proportion of the functional proteome—interacting entities—is currently out of reach. As has been mentioned earlier, new methods must be forthcoming for both Structural and Activity assessments, with one promising method in both arenas being EPR/spin labeling. Overall analytical resources are available to lay the foundation of Functional Proteomics, while development of latter stages require new methods development. Now for the second interpretation of resources—funding.

R&D spending by pharma and agencies such as the US National Institutes of Health (NIH) have grown tremendously over the last half-century. Although total spending has plateaued in the last decade, levels remain high: \$100 billion USD per year, or \$1 trillion over the decade (this does not include venture investment and sales within the gene sequencing market for which total spending is not readily available, although can be estimated at between \$500 and 1000 billion USD). This would seem to be adequate funding for a diversity of support but such is sadly not the case. Genomics, more specifically polynucleotide sequencing, has by any reckoning sucked the air out of the room. In the US, for example, would-be NIH applicants are advised that basic research in biochemistry has zero chance of being funded. Within such an environment, securing funds for a full-scale effort in Functional Proteomics would seem unlikely. Nonetheless there are indicators of optimism.

Two topics covered in this chapter are noteworthy. The early success of the approach described under "Biomarker and Diagnostic/Prognostic Discovery" above is but a harbinger of successes to come. The unbiased assembly of network-connected proteins is the future of diagnostics and prognostics. This optimism extends to the forefront of drug discovery for precisely the same reasons. The unbiased determination of test molecules' ability to modulate the phenotype is precisely in accord with the manifestation of disease. In both cases the concept of "Chain of Translatability" applies: discovery of both biomarkers and drug candidates must simultaneously add to our knowledge of biology and disease, and inform likely candidates for diagnostics of diverse application. This growing shift to a phenotypic basis for biomarker and drug discovery is accompanied by a multitude of new ways of handling cells and organelles in increasingly bio-relevant settings. A reevaluation of non-mainstream methods can open new vistas, as in the cited detection and characterization of heretofore unseen, disease-illuminating, functional entities. In step is continued improvement in isolation of stem cells and their conversion to multiple cell types—at the level of the individual. Even manufactured, patient-derived, organ-like structures are being developed at a rapid pace. All this progress, when coupled to a full arsenal of Functional Proteomics tools, augurs for the twenty-first century as the century of Biology.

The time has come and all the pieces are in place.

References

1. Pundir S, Martin M, O'Donovan C (2017) UniProt protein knowledgebase. Methods Mol Biol 1558:41–55. https://doi.org/10.1007/978-1-4939-6783-4_2
2. Savage N (2015) Proteomics: high-protein research. Nature 527:S6. https://doi.org/10.1038/527S6a
3. Pennisi E (2012) ENCODE project writes eulogy for junk DNA. Science 337 (6099):1159–1161. https://doi.org/10.1126/science.337.6099.1159
4. Ezkurdia I, Juan D, Rodriguez JM, Frankish A, Diekhans M, Harrow J, Vazquez J, Valencia A, Tress ML (2014) Multiple evidence strands suggest that there may be as few as 19 000 human protein-coding genes. Hum Mol Genet 23(22):5866–5878
5. Ponomarenko EA, Poverennaya EV, Ilgisonis EV, Pyatnitskiy MA, Kopylov AT, Zgoda VG, Lisitsa AV, Archakov AI (2016) The size of the human proteome: the width and depth. Int J Anal Chem 2016:7436849
6. Collins FS (2006) The language of god. Francis S. Collins on unveiling the human genome. Free Press, New York, p 1–3
7. Ball P (2010) Bursting the genomics bubble. Nature. https://www.nature.com/news/2010/100331/full/news.2010.145.html. https://doi.org/10.1038/news.2010.145
8. Gisler M (2010) The rise and fall of the human genome project. MIT Technology Review
9. Weston AD, Hood L (2004) Systems biology, proteomics, and the future of health care: toward predictive, preventative, and personalized medicine. J Proteome Res 3:179–196
10. Butcher EC, Berg EL, Kunkel EJ (2004) Systems biology in drug discovery. Nat Biotechnol 22:1253–1259. https://doi.org/10.1038/nbt1017
11. Perkins RC. Paul Kenis, Deborah Berthhold & Sarah-Ellen Leonard, University of Illinois, Urbana/Champaign; Jonathan Lee, recently of Eli Lily; and Ray Perkins, New Liberty Proteomics
12. Strimbu K, Tavel JA (2010) What are biomarkers? Curr Opin HIV AIDS 5(6):463–466
13. Berger D (1999) A brief history of medical diagnosis and the birth of the clinical laboratory. MLO Med Lab Obs 31(7). 28–30, 32, 34–40
14. FDA (2018.) In vitro diagnostics. https://www.fda.gov/MedicalDevices/ProductsandMedicalProcedures/InVitroDiagnostics/default.htm
15. Kuruc M (2017) Stroma liquid biopsy—biomarkers of the dysregulation of the serum proteome in cancer First presented at NJ cancer Retreat, May 25, 2017 New Brunswick, NJ USA. https://www.biotechsupportgroup.com/v/vspfiles/templates/257/pdf/NJ%20Cancer%20Retreat%20Stroma%20Liquid%20Biopsy%20Poster.pdf

16. Lowe D (2016) In the pipeline: precision oncology isn't quite there yet Science Translational Medicine weblog, Lowe D (2016). http://blogs.sciencemag.org/pipeline/archives/2016/09/12/precision-oncology-isnt-quite-there-yet
17. Gigerenzer G (2014) Risk savvy. Penguin Group, New York, NY
18. Booth B, Zemmel R (2004) Opinion: prospects for productivity. Nat Rev Drug Discov 3:451–456. https://doi.org/10.1038/nrd1384
19. Scannell JW, Blanckley A, Boldon H, Warrington B (2012) Diagnosing the decline in pharmaceutical R&D efficiency. Nat Rev Drug Discov 11. https://doi.org/10.1038/nrd3681
20. Lawrence MS, Stoianov P, Polak P, Kryukov GV, Cibulskis K, Sivachenko A, Carter SL et al (2013) Mutational heterogeneity in cancer and the search for new cancer-associated genes. Nature 499. https://doi.org/10.1038/nature12213
21. Swinney D (2013) Phenotypic vs. target-based drug discovery for first-in-class medicines. Clin Pharmacol Ther 93(4):299–301
22. Moffat JG, Vincent F, Lee JA, Eder J, Prunotto M (2017) Opportunities and challenges in phenotypic drug discovery: an industry perspective. Nat Rev Drug Discov 16:531–543. https://doi.org/10.1038/nrd.2017.111
23. Bray M-A, Singh S, Han H, Davis CT, Borgeson B, Hartland C, Kost-Alimova M, Gustafsdottir SM, Gibson CC, Carpenter AE (2016) Cell painting, a high-content image-based assay for morphological profiling using multiplexed fluorescent dyes. Nat Protoc 11:1757–1774. https://doi.org/10.1038/nprot.2016.105
24. Avior Y, Sagi I, Benvenisty N (2016) Pluripotent stem cells in disease modelling and drug discovery. Nat Rev Mol Cell Biol 17. https://doi.org/10.1038/nrm.2015.27
25. Esch EW, Bahinski A, Huh D (2015) Organs-on-chips at the frontiers of drug discovery. Nat Rev Drug Discov 14(4). https://doi.org/10.1038/nrd4539
26. Boran AD, Ivengar R (2010) Systems approaches to polypharmacology and drug discovery. Curr Opin Drug Discov Devel 13 (3):297–309
27. Ashley EA (2016) Towards precision medicine. Nat Rev Genet 17. https://doi.org/10.1038/nrg.2016.86
28. Bandolier (2007) The Oxford league table of analgesic efficacy. http://www.bandolier.org.uk/booth/painpag/Acutrev/Analgesics/lftab.html
29. Schork NJ (2015) Personalized medicine: time for one-person trials. Nature 520:609–611. https://doi.org/10.1038/520609a
30. Prasad V (2016) Perspective: the precision-oncology illusion. Nat Biotechnol 537(S63). https://doi.org/10.1038/537S63a
31. Brock A, Huang S (2017) Precision oncology: between vaguely right and precisely wrong. Cancer Res. https://doi.org/10.1158/0008-5472.CAN-17-0448
32. HUPO (2016) The human proteome project. https://hupo.org/human-proteome-project
33. Zhang Y, Fonslow BR, Shan B, Baek M-C, Yates JR (2013) Protein analysis by shotgun/bottom-up proteomics. Chem Rev 113 (4):2343–2394. https://doi.org/10.1021/cr3003533
34. Catherman AD, Skinner OS, Kelleher NL (2014) Top down proteomics: facts and perspectives. Biochem Biophys Res Commun 445 (4):683–693. https://doi.org/10.1016/j.bbrc.2014.02.041
35. Solier C, Langen H (2014) Antibody-based proteomics and biomarker research—current status and limitations. Proteomics 14 (6):774–783. https://doi.org/10.1002/pmic.201300334
36. Wasinger VC, Zeng M, Yau Y (2013) Current status and advances in quantitative proteomic mass spectrometry. Int J Proteomics 2013:180605
37. Wolf-Yadlin A, Hautaniemi S, Lauffenbuger DA, White FM (2007) Multiple reaction monitoring for robust quantitative proteomic analysis of cellular signaling networks. Proc Natl Acad Sci U S A 104(14):5860–5865. https://doi.org/10.1073/pnas.0608638104
38. Berman HM, Westbrook J, Feng Z, Gilliland G, Bhat TN, Weissig H, Shindyalov IN, Bourne PE (2000) The protein data bank. Nucleic Acids Res 28(1):235–242
39. Wang H, Wang J (2017) How cryo-electron microscopy and X-ray crystallography complement each other. Protein Sci 26(1):32–39. https://doi.org/10.1002/pro.3022
40. MSU 900 MHz NMR sample requirements. https://www2.chemistry.msu.edu/facilities/nmr/900mhz/MCSB_NMR_sample.html
41. Claxton DP, Kazmier K, Mishra S, Mchaourab HS (2015) Navigating membrane protein structure, dynamics, and energy landscapes using spin labeling and EPR spectroscopy. Methods Enzymol 564:349–387. https://doi.org/10.1016/bs.mie.2015.07.026

42. Yang Y, Ramelot TA, McCarrick RM, Ni S, Feldmann EA et al (2010) Combining NMR and EPR methods for Homodimer protein structure determination. J Am Chem Soc 132 (34). https://doi.org/10.1021/ja105080h
43. Stadler C, Rexhepaj E, Singan VR, Murphy RF, Pepperkok R, Uhlén M, Simpson JC, Lundberg E (2013) Immunofluorescence and fluorescent-protein tagging show high correlation for protein localization in mammalian cells. Nat Methods 10:315–323. https://doi.org/10.1038/nmeth.2377
44. von Mering C, Krause R, Snel B, Cornell M, Oliver SG, Fields S, Bork P (2002) Comparative assessment of large-scale data sets of protein–protein interactions. Nat Biotechnol 417:399–403
45. Braun P, Tasan M, Dreze M, Barrios-Rodiles-M, Lemmens I, Yu H, Sahalie JM, Murray RR, Roncari L, A-Sd S, Venkatesan K, Rual J-F, Cusick ME, Pawson T, Hill DE, Tavernier J, Wrana JL, Roth FP, Vidal M (2009) An experimentally derived confidence score for binary protein-protein interactions. Nat Methods 6 (1):91–97. https://doi.org/10.1038/nmeth.1281
46. Rao VS, Srinivas K, Sujini GN, Kumar GNS (2014) Protein-protein interaction detection: methods and analysis. Int J Proteomics 2014:12. https://doi.org/10.1155/2014/147648
47. Klare J (2013) Site-directed spin labeling EPR spectroscopy in protein research. Biol Chem 394(10):1281–1300. https://doi.org/10.1515/hsz-2013-0155J

Chapter 2

Methods to Monitor the Functional Subproteomes of SERPIN Protease Inhibitors

Swapan Roy and Matthew Kuruc

Abstract

Conformational variants of the unique family of protease inhibitors annotated as SERPINs are most often underrepresented in proteomic analyses. This limits understanding the complex regulation that this family of proteins presents to the networks within the protease web of interactions. Using bead-based separation provided by a family of proteomic enrichment products—notably **AlbuVoid™** and **AlbuSorb™**, we demonstrate their utility to satisfy investigations of serum SERPINs. We also suggest their use to develop functional profiles of the SERPIN proteoforms, and how those can establish relationships to disease phenotypes, gene mutations, and dysregulated mechanisms.

Key words SERPIN, SERPIN function, Functional proteomics, SERPIN mechanism, SERPIN biomarkers, SERPIN proteoforms

1 Introduction

The balance and regulation of proteolytic activity within serum is essential to blood based biomarker discovery and possibly to therapeutic intervention. Changes in blood components often reflect acute responses to thwart external stresses, such as coagulation when skin is severed, or inflammatory response during microbial infections. These fast-acting responses are controlled by proteolytic cascades, essentially modifying functionality by the controlled degradation of protein structures. While necessary for acute response, persistent activation of these proteolytic cascades can lead to chronic conditions. So, there is a balance and regulation of these proteolytic cascades which is necessary to keep aberrant proteolysis controlled.

This is done through systemic regulatory protein factors, called protease inhibitors or antiproteases. It is now quite apparent that the influence of inhibition can be just as important as zymogen activation in rapid switch cascades controlling subnetworks within the protease web [1]. One such example of this web's complexity, is

Xing Wang and Matthew Kuruc (eds.), *Functional Proteomics: Methods and Protocols*, Methods in Molecular Biology, vol. 1871, https://doi.org/10.1007/978-1-4939-8814-3_2, © Springer Science+Business Media, LLC, part of Springer Nature 2019

that one substrate (Neutrophil Elastase) for Alpha-1-Antitrypsin (the inhibitor), can activate the inactive zymogen proMMP-2, a metalloproteinase involved in tumor invasion and angiogenesis [2]. So it becomes necessary to consider that inhibitors are themselves being regulated under different and often complex means of regulation. Within this context, therein lies the special case of the SERPIN superfamily of protease inhibitors.

1.1 The SERPIN Superfamily of Suicidal Inhibitors

The SERPIN family of suicidal serine protease inhibitors plays an integral role in regulating a wide variety of biological activities, and represent 2–10% of circulating plasma proteins. SERPINs regulate coagulation, hormone transport, complement and inflammation, angiogenesis, and blood pressure along with many other pathways. Among the key regulators in blood serum, SERPINA1 (also known as α1-antitrypsin) protects lung tissue from neutrophil elastase, SERPINC1 (also known as antithrombin) controls coagulation proteases, SERPING1 (also known as plasma C1 inhibitor) regulates complement activation, and SERPINF2 (also known as α-2-antiplasmin) inhibits plasmin and regulates fibrinolysis [3, 4].

This unique family of protein inhibitors has been associated with progression or remission of cancer and so they may become valuable biomarkers for therapeutic or diagnostic use. Of clinical utility, prostate-specific antigen (PSA), also known as kallikrein-3, is commonly used as a biomarker for prostate cancer. However, the kallikrein protease family of proteins is of very low abundance in plasma, making observation and quantification difficult. Nevertheless, PSA is regulated by the SERPIN inhibitor family; in men with prostate cancer the ratio of free (unbound) PSA to total PSA is decreased, suggesting a greater role of inhibitory capacity in cancer.

By way of these examples, rather than focusing proteomic discovery efforts on low-abundance proteins like the tissue kallikreins, it may be advantageous to profile much higher abundance Tissue Kallikrein inhibitors like SERPINA5 (Protein C Inhibitor), SERPINA3 (Antichymotrypsin), and SERPINA4 (Kallistatin), to better understand underlying disease mechanisms and potentially generate new biomarkers. However, the role of SERPINs in these critical junctures is rarely straightforward as would be the case for more simplistic binary binding inhibition. For functional interpretation, reliance on strict abundance measurements, such as data that might be derived by ELISA or quantitative LC-MS, does not differentiate the subpopulations of the seemingly opposing outcomes of the SERPIN interaction with its target protease.

This is because SERPINs differ from all other families of protease inhibitors in having a complex mechanism of action that involves a drastic change in their shape, forming the basis of a suicidal substrate inhibition mechanism [3, 4]. The reactive center loop (RCL) extends out from the body of the protein and directs binding to the target protease. The protease cleaves the SERPIN at

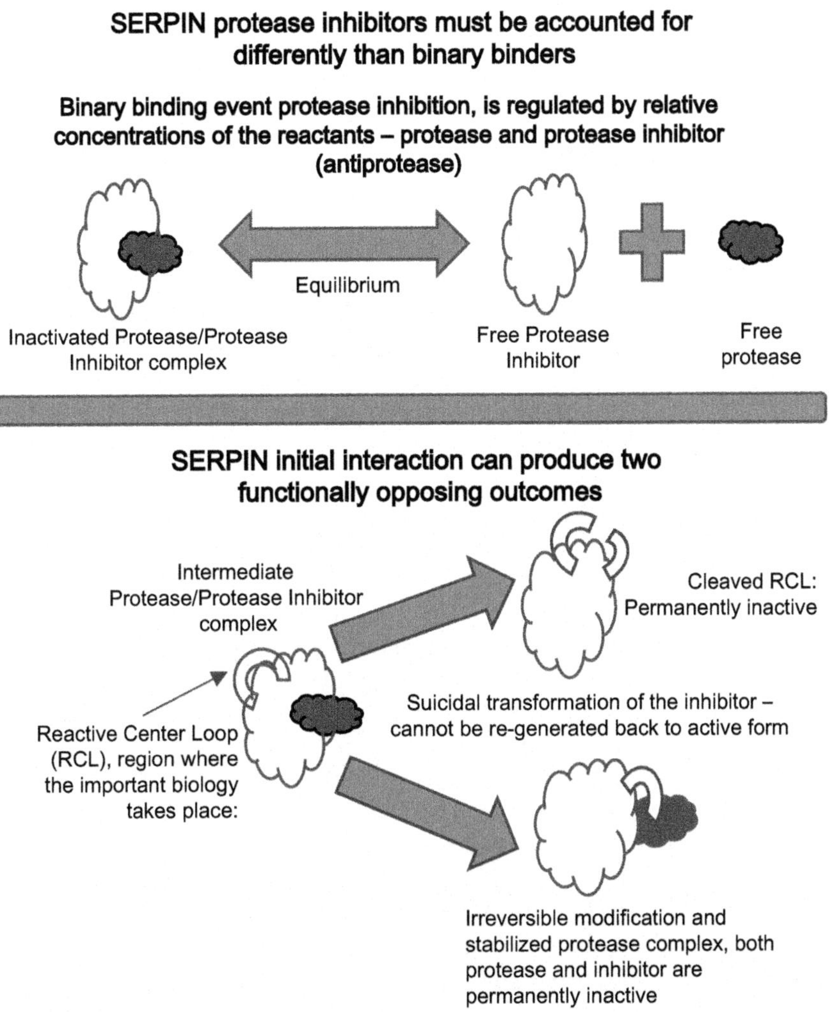

Fig. 1 SERPIN protease inhibitors must be accounted for differently than binary binders

the reactive bond site within the RCL, establishing a covalent linkage between the carboxyl group of the SERPIN reactive site and the serine hydroxyl of the protease [4]. The resulting inactive serpin-protease complex is highly stable, and the structural disorder induces its proteolytic inactivation. As a consequence, the protease is permanently inhibited and functionally inactivated. Nevertheless, the story does not end there for the inhibitor, as after the initial interaction with the substrate protease, one of two possible outcomes can occur, Fig. 1.

One possible outcome is driven by covalent modification permanently inactivating the inhibitory capacity as the SERPIN peptide reactive bond region is irreversibly bound to the protease, and thus cannot be reconstituted back to an active form. The second possible outcome is a permanently inactive variant of the SERPIN as the peptide RCL region is cleaved and can no longer bind target substrates [4]. As a result, even minor changes in the structure due to genetic variation and posttranslational modifications can modify the function of SERPINs and give rise to a variety of clinical presentations. Some 200 different mutations in serpins are known to result in disease [5]. In particular, mutations affecting antithrombin confer a predisposition to thrombosis, those affecting C1 Inhibitor confer a predisposition to angioedema, and those affecting antiplasmin confer a predisposition to hemorrhage. Interestingly, an alternative function is made possible by a mutation in which the methionine in the RCL region of Alpha-1-Antitrypsin is replaced by an arginine converting its function as an inhibitor of neutrophil elastase to a highly effective inhibitor of the coagulation proteases, the consequence of which is life-threatening hemorrhagic disease [6].

Mutations can affect function throughout the sequence. However, the most common loss of serpin function from mutation are those affecting the mobile hinges of the molecule within or near the RCL. These lead to spontaneous changes in conformation that allow either the insertion of the intact reactive loop into the main β sheet, resulting in the formation of an inactive "latent" form, or the insertion of the loop of one molecule into the β sheet of the next, resulting in the formation of polymers. Polymerization occurs in Alpha-1-Antitrypsin with the common Z variant mutation, leading to decreased secretion from the liver into the circulation, resulting in emphysema and cirrhosis [7]. Amino acid substitution in the RCL region is the likely event transforming the non-inhibitory serpins. Posttranslational modifications at the RCL region such as oxidation of methionine in Alpha-1-Antitrypsin have also been proposed as a source of dysfunction [8].

So understanding the underlying mechanisms, contributions from genetic wiring or environmental stresses, and their relationships with aberrant proteolysis is necessary to characterize disease. Functional proteomic analyses offer a new lens of observation to examine the resulting conformational variants that can be reported as potential biomarkers of disease phenotypes. As an example, one such inhibitor SERPINA1, known more commonly as Alpha-1-Antitrypsin (AAT), has several isoforms observed in plasma using 2-DE, and often serves as a model for conformational diseases [5, 9]. Circulating levels of AAT are between 1.2 and 2 mg/mL in healthy persons, but are known to increase during acute phases of inflammation and infection. Its function and activity is controlled by the many variants attributable to its conformational nexus of features; the term "proteoform" is often used to describe such conformational variability and we adopt that term here.

Other reports observe that the conformational properties of AAT have multiple effects on tumor cell viability and diverse roles in tumorigenesis, suggesting such isoforms may display a specific basis for diagnosis of cancer and neurodegenerative disorders [8, 10, 11]. Yet, most often in proteomics, all subpopulations of SERPINs are rolled into and counted as one homogeneous population. As a result, the regulation, balance and dynamism within these systems and its impact on the protease web of disease progression cannot be properly investigated, and indeed conclusions based on such measurements may be very misleading.

So methods that account for important distinctions among the many subpopulations generated by conformational variants within this superfamily of proteins are considered in this chapter. Specifically, a functional proteomic investigation of the two seemingly opposing outcomes of the initial inhibitor-protease interaction can be monitored:

1. The total amount of potentially inhibitory SERPIN activity as reported by an intact RCL region.
2. A transformed subpopulation of the inhibitor, as reported by a cleaved RCL region permanently inactivating its inhibitory potential.

1.2 New Methods to Functionally Profile SERPINs

By combining unique strategies of binding and voiding high-abundance proteins, we can observe different subpopulations with characteristic binding biases. We have previously reported for Alpha-1-Antitrypsin that the resultant cleaved-RCL proteoform and the uncleaved-RCL proteoform are very distinctive subpopulations, separated by **AlbuVoid™**, and reported at the peptide feature level by LC-MS [12]. In this chapter, we consider how Albumin Removal products—**AlbuVoid™** and **AlbuSorb™** (Biotech Support Group LLC, Monmouth Junction, NJ, USA), can help to functionally profile and unravel this complex biology of the SERPIN superfamily of proteins.

Through a proprietary polymer coating, 50 μm porous silica beads are crosslinked and passivated. This is the foundation of the **NuGel™** surface chemistry. Mixed-mode of binding interactions form the basis of general nonspecific protein adsorbents or beads with weak affinity or imperfect fit interactions. In this way, binding behavior is very different from classical high affinity binding which demands near perfect fits. Under protein saturation conditions, progressive displacement provides a separation bias towards or against select proteins. As a result, all derivative **NuGel™** products were empirically characterized to meet the needs of the application, for example, **AlbuVoid™** to selectively void (not bind) Albumin with special bias toward the vast majority of the remaining low-abundance serum proteome on the bead. Two **NuGel™** based products support Albumin Removal:

1. **AlbuSorb™** and **AlbuSorb™ PLUS** (also binds immunoglobulins) for selective binding of Albumin.

2. **AlbuVoid™** for negative selection or voidance of Albumin with consequent enrichment of the remaining serum sub-proteome on the bead.

So, while other proteomic methods might observe this:
Past observations:

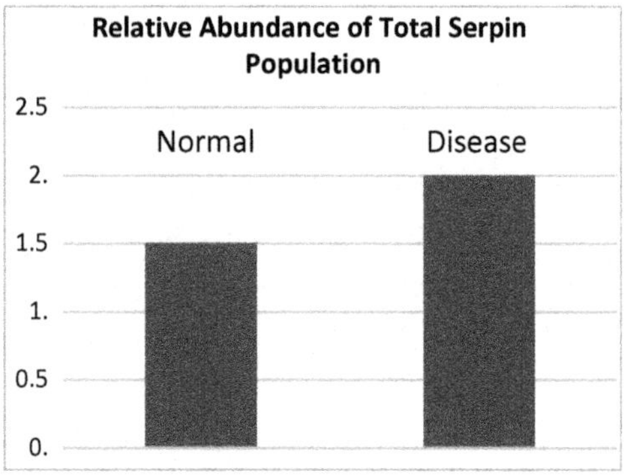

We describe methods to observe this:

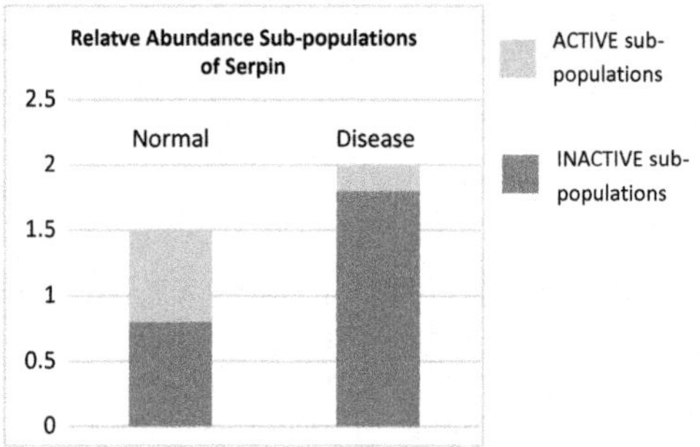

In this hypothetical case, the ratio of the ACTIVE subpopulation vs. the INACTIVE subpopulation is greatly altered in disease, whereas simple abundance measurements of the total population would not be very informative (*see* **Notes 1** and **2**). In the following Table 1, we report on the SERPINs observable by LC-MS and how they bias toward **AlbuVoid™** and **AlbuSorb™**, as measured by spectral counting.

We suspect that conformational changes associated with the cleavage of the reactive bond confer more or less binding affinity to the nonspecific interactions with our beads. Such cleavage stabilizes the SERPIN structures; **AlbuVoid™** binding especially biases toward unstructured proteins, and we have previously reported the SERPINA1 (Alpha-1-Antitrypsin) RCL-intact proteoform binding favorably over the RCL-cleaved proteoform [12]. Noteworthy is that several non-inhibitory SERPINs A6-8, all bind poorly to **AlbuVoid™**, supporting evidence for the role of conformational stability in binding biases.

Table 1
Serum SERPINs observable by LC-MS

Protein ID	Also known as (conc.)	Function	AlbuVoid™ bead bound S.Cts.	AlbuSorb™ flowthrough (unbound) S.Cts.	Reactive (RCL) bond site	Notable variants
SERPINA1	Alpha-1-Antitrypsin (AAT) (1–2 mg/mL)	Inflammation, elastase inhibition	59 (strong bias toward RCL-intact proteoform)	519	Met382-Ser383	Z variant {Glu366 → Lys366} deficiency syndrome, Pittsburgh variant {Met382 → Arg382} life-threatening bleeding
SERPING1	Plasma protease C1 inhibitor (0.25 mg/mL)	Regulates complement cascade, levels rise ~2-fold during inflammation	51	63	Ala465-Arg466 chymotrypsin, Arg466-Thr467	
SERPINA3	Antichymotrypsin (100–500 μg/mL)	Apoptosis, Alzheimers, inflammation	86	117	Leu383-Ser384	
SERPIND1	Heparin cofactor II (40–80 μg/mL)	Coagulation, thrombin inhibitor activated by heparin	124	28	Leu463-Ser464	
SERPINA8	Angiotensinogen (AGT) (40–60 μg/mL)	Angiotensin I precurser, blood pressure regulation, non-inhibitory	4	62	None	Disulfide bond is labile, near 40:60 ratio with the oxidized disulfide-bonded form
SERPINC1	Antithrombin, ATIII (0.12 mg/mL)	Inhibits thrombin, regulates coagulation, angiogenesis, heparin cofactor	58	79	Arg425-Ser426	Mutations/variants can lead to increased risk of thrombosis, alter functional heparin and thrombin binding domains

(continued)

Table 1 (continued)

Protein ID	Also known as (conc.)	Function	AlbuVoid™ bead bound S.Cts.	AlbuSorb™ flowthrough (unbound) S.Cts.	Reactive (RCL) bond site	Notable variants
SERPINF1	Pigment epithelium-derived factor, PEDF (20–175 μg/mL)	Neurotrophic factor, non-inhibitory	45	0	None	
SERPINA4	Kallistatin (20 μg/mL)	Kidney function, inflammation	45	0	Phe388-Ser389	Cleavage at the reactive site by tissue kallikreins
SERPINF2	α-2-antiplasmin (60–80 μg/mL)	Fibrinolysis, inhibitor of plasmin and trypsin	10	39	Arg403-Met404 plasmin, Met404-Ser405 chymotrypsin	Alanine insertion at the reactive site promotes serious bleeding disorders
SERPIN-A10	Z-dependent proteinase inhibitor (1–2 μg/mL)	Coagulation regulation	23	0	Tyr408-Ser409	Tyr408 → Ala408 loss of inhibition
SERPINA5	Protein C inhibitor (5 μg/mL)	Coagulation, inflammation	13	0	Arg373-Ser374	Variants near or at the reactive bond alter inhibition of thrombin activity
SERPINA6	Corticosteroid-binding globulin (60–80 μg/mL)	Hormone transport, non-inhibitory	0	26	None	
SERPINA7	Thyroxine-binding globulin (15 μg/mL)	Hormone transport, non-inhibitory	0	17	None	

2 Materials

Items required	Reagent
AlbuVoid™ beads	Manufacturer supplied
Binding buffer AVBB, *PH 6.0*	Manufacturer supplied
Wash buffer AVWB, *PH 7.0*	Manufacturer supplied
SpinX centrifuge tube filters	Manufacturer supplied
Trypsin, DTT, iodoacetamide	Not supplied

3 Methods

For this chapter, we shall consider only the workflow supporting **AlbuVoid™**, but LC-MS workflows supporting **AlbuSorb™** would be similar, taking into account which fractions would contain the majority of Albumin and which do not, Fig. 2.

The workflow follows the **AlbuVoid™ LC-MS On-Bead** sample prep method following the manufacturer's protocol. In brief, 50 μL serum is prepared by adding a binding buffer, then applied to the **AlbuVoid™** beads, and washed. All steps are performed within a microfuge spin-filter format. Albumin is most especially voided out, while the majority of the remaining serum proteome is retained on the bead. After the final wash, reduction, alkylation, and Trypsin digestion all take place on the bead.

For best results—the serum should be clear and free of colloidal material. We recommend first filtering through a 0.45 μm syringe-type filter before beginning the prep.

In bold are the **AlbuVoid™***LC – MSOn – Beadkitcomponents.*

1. Weigh out 25 mg of **AlbuVoid™** bead in a spin-tube (**0.45 μ SpinX centrifuge tube filter supplied**).
2. Add 125 μL of **Binding Buffer AVBB**. Vortex for 5 min at room temperature followed by centrifugation at 1500 × *g*. Discard the supernatant.
3. Repeat **step 2**.
4. Condition clarified serum by adding 100 μL of **AVBB** to 50–100 μL of the Serum. Using a syringe-type micro-filter, clarify the serum. Add sample to the **AlbuVoid™ beads** in **step 3**. Vortex for 10 min and then centrifuge for 5 min at 10,000 × *g*.
5. Discard the albumin filtrate.
6. To the beads, add 250 μL of **Wash Buffer AVWB**. Vortex for 5 min and centrifuge for 4 min at 10,000 × *g*. Discard the Wash.

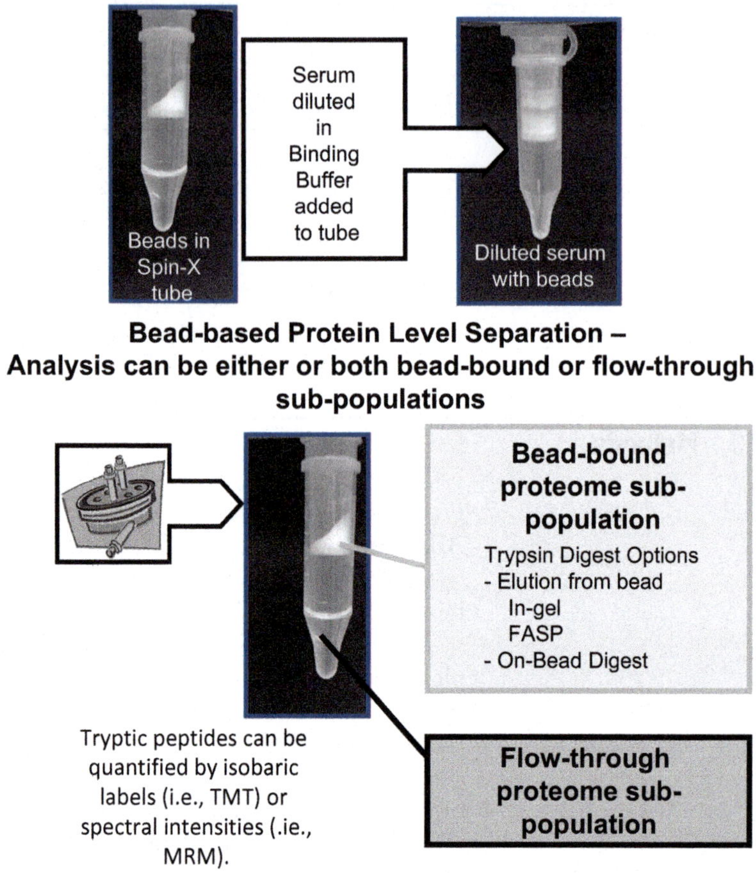

Fig. 2 Enrichment/depletion option for serum proteome separations

7. Repeat **step 6** two times.

   The AlbuVoid™ beads are now enriched with albumin-depleted low-abundance proteins. For LC-MS sample preparation, the on-bead digestion protocol is as follows. Option—the proteins can be eluted with 0.25 M Tris, 0.5 M NaCl, pH 10 (*see* **Note 3**).

8. After the final wash steps from **step 7** from the enrichment, add 10 μL 100 mM DTT + 90 μL **Wash Buffer AVWB**, vortex 10 min, incubate ½ h at 60 °C.

9. After cooling, add 20 μL 200 mM Iodoacetamide and 80 μL **Wash Buffer AVWB**, incubate in dark for 45 min at room temp.

10. Centrifuge at 10,000 × *g* (microfuge max setting) for 5 min, and discard supernatant.

11. Add 40 μL sequencing-grade trypsin (0.4 μg/μL, in 50 mM acetic acid) + 60 μL **Wash Buffer AVWB** to the beads. Digest

overnight (maximum) at 37 °C or other suitable time period (s).

12. Centrifuge at 10,000 × *g* (microfuge max setting) for 5 min and retain peptide filtrate.
13. To further extract remaining peptides, add 150 μL 10% formic acid, vortex 10 min, centrifuge at 10,000 × *g* (microfuge max setting) for 5 min, and add this volume to the first volume.
14. Total is about 250 μL. Prepare to desired final concentration. Store at −80 °C until LC-MS/MS.

Example of LC-MS reporting features.

After TMT labels (Proteome Sciences plc, Surrey, UK) labeling, the peptides are pooled and analyzed with a single LC-MS/MS 3 h gradient run using nanoRSLC system interfaced with a Thermo Scientific™ Q Exactive™ HF (Thermo Scientific) instrument, using data-dependent acquisition with resolution of 60,000, followed by MSMS scans (HCD 30% of collision energy) of 20 most intense ions, with a repeat count of two and dynamic exclusion duration of 60 s (Table 2).

The amino acid region of the RCL is 368–392, so the adjacent RCL tryptic peptide at Lys367, highlighted in gray, serves as a good comparison between the observable serum subpopulations, Fig. 3.

Bead Bound—The subpopulation of proteins that bind and are observed by the **AlbuVoid**™ *methods.*

Flow-through (unbound)—The subpopulation of proteins that flow-through the **AlbuVoid**™ *beads, unbound.*

Table 2
SERPINA1 (AAT) TMT ratio: pooled pancreatic cancer/pooled normal

		Sp Ct = peptide spectral counts		Bead bound		Flow-through (unbound)		Serum untreated	
SERPINA1 (AAT) Peptide region	**Start**	**Amino acid sequence**	**End**	**TMT ratio**	**Sp Ct**	**TMT ratio**	**Sp Ct**	**TMT ratio**	**Sp Ct**
Adjacent RCL Tryptic	360	AVLTIDEK	367	0.35	9	1.78	21	1.53	14
RCL cleaved	368	GTEAAGAMFLEAIPM	382			1.05	7	1.16	23
RCL intact	368	GTEAAGAMFLEAIPM SIPPEVK	389	0.77	5	1.75	1	1.34	50
RCL cleaved	383	SIPPEVK	389			1.45	27	1.44	18
Total all peptide features				0.54	132	1.57	372	1.44	460

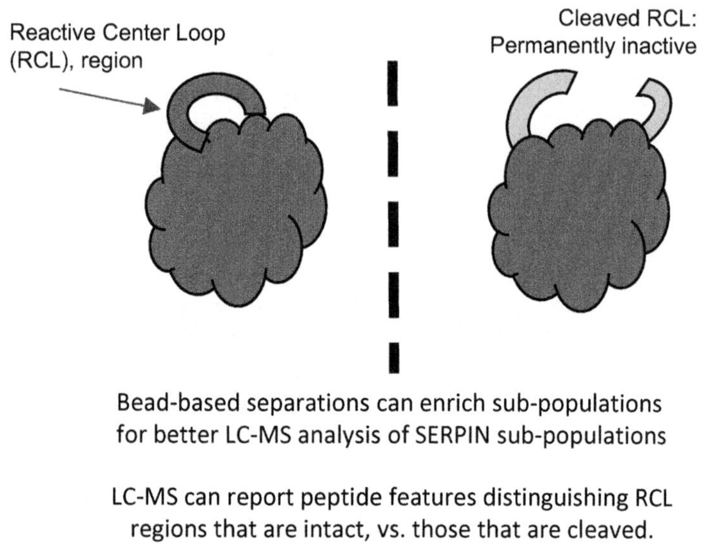

Fig. 3 SERPIN LC-MS reporting features

Untreated—The total population of proteins that are observable in serum without any sample enrichment, that is without the use of **AlbuVoid™**.

Highlighted in dark gray is the RCL intact peptide. Highlighted in light gray are the two RCL peptides that are cleaved at Met382, during suicidal substrate interaction; note that these peptides were not observed in the Bead-Bound fraction. These data suggest that the overall SERPINA (AAT) population is dominated by the subpopulation *up*-regulated and collected in the Flow-through fraction of **AlbuVoid™**, and this same subpopulation dominates the analysis when untreated sera is investigated. Such would be the case in acute AAT up-regulation commonly observed with malignancies and inflammation. However, using our methods we distinguish a subpopulation enriched by the bead and reporting with the bound fraction, as being severely *down*-regulated with cancer! While this observation may have potential biological significance, no conclusion about the particular cancer-specific proteoform uncovered can be made at this time (*see* **Note 4**). Nevertheless, from a biomarker perspective, this serves an additional multiplier benefit; that is the ratio of the two subpopulations report Adjacent RCL Tryptic peptide region as unbound/bound (1.78/0.35) = 5. As isobaric label ratios in discovery methods can sometimes compress the reporting difference, this ratio may become much greater once more targeted quantitative methods are developed, a prospect for future tests.

4 Notes

- Bead-based proteomic enrichment methods as described can support the functional and structural proteomic analyses necessary to characterize these conformational subpopulations so that they may become useful biomarkers for disease. It should nonetheless be recognized that the RCL reporting methods described here only work for RCL regions where the cleavage site is *non-Tryptic*, and these must be entered as special peptides into LC-MS computational workflows. "neXtProt: a knowledge platform for human proteins," provides a useful web-based resource for annotating RCL cleavage sites [13]. To distinguish RCL regions where the cleavage site is Tryptic, it becomes necessary to differentiate those sites that are cleaved in vivo by those cleaved ex vivo. Several methods have been developed for this purpose and generally fall under methods called N-terminomics [14]. This is an area for future investigations.
- Classical high-abundance proteins like the SERPIN superfamily (i.e., Alpha-1-Antitrypsin) are often overlooked as potential biomarkers of disease. Yet discoveries certainly can rest in the data-rich features of the diverse variety of conformational and proteoform variants associated with many of the classical serum proteins. When considering these mid- to high-abundance proteins, disease differentiation can be obtained through the discreet quantification of the multiple subpopulations available to measure. The methods described in this chapter can begin to unravel and sort these variant subpopulations so that LC-MS peptide reporting features, and potentially other functional reporting features (i.e., substrate turnover), can distinguish these proteoforms with more functional details. It is our intention that these methods will lead to characteristic disease profiles, which can then be compared and evaluated for eventual biomarker utility.
- Many trypsin digestion protocols have been developed to improve the reproducibility and, in some cases, reduce the digestion time necessary for LC-MS analysis [15]. While we have shown methods that adapt **AlbuVoid**™ for on-bead digestion, the bead-based enrichments described here are nonetheless compatible, after elution from the beads, with other common digestion methods, such as filter-aided (FASP) solution methods and post-electrophoresis, in-gel methods.
- By using the peptide reporting features of the RCL peptide regions within SERPIN inhibitors, both "potentially active" and "permanently inactive" proteoforms are now distinguishable, adding a new level of proteomic characterization to the underlying mechanisms of disease. As one example, hereditary

dysfunction of SERPINA1 (Alpha-1-Antitrypsin) has been previously determined as a risk factor for cancer [16]. As many proteins within the SERPIN family proteins are of moderate- to high-abundance quantities in serum (10 + µg range/mL), depleted functionality would impose severe dysregulation to a normal and healthy individual. Several of the key regulators in the Coagulation Pathway such as SERPINA10 (Z-dependent Proteinase Inhibitor) and SERPINA5 (Protein C Inhibitor) have notable genomic variants that alter their inhibitory function [13]. These might therefore be risk factors for disease. So using the methods described in this chapter, hereditary genomic factors that associate with SERPIN function can be further investigated.

References

1. Fortelny N, Cox JH, Kappelhoff R, Starr AE et al (2014) Network analyses reveal pervasive functional regulation between proteases in the human protease web. PLoS Biol 12(5): e1001869
2. Shamamian P, Schwartz JD, Pocock BJ et al (2001) Activation of progelatinase A (MMP-2) by neutrophil elastase, cathepsin G, and proteinase-3: a role for inflammatory cells in tumor invasion and angiogenesis. J Cell Physiol 189(2):197–206
3. Law RH, Zhang Q, McGowan S et al (2006) An overview of the serpin superfamily. Genome Biol 7(5):216
4. Khan MS, Singh P, Azhar A et al (2011) Serpin inhibition mechanism: a delicate balance between native metastable state and polymerization. J Amino Acids. https://doi.org/10.4061/2011/606797
5. Carrell RW, Lomas DA (2002) Alpha1-antitrypsin deficiency—a model for conformational diseases. N Engl J Med 346(1):45–53
6. Owen MC, Brennan SO, Lewis JH et al (1983) Mutation of antitrypsin to antithrombin: α1-antitrypsin Pittsburgh (358 Met→Arg), a fatal bleeding disorder. N Engl J Med 309 (12):694–698
7. Sifers RN (1992) Z and the insoluble answer. Nature 357(6379):541
8. Janciauskiene S (2001) Conformational properties of serine proteinase inhibitors (serpins) confer multiple pathophysiological roles. Biochim Biophys Acta 1535(3):221–235
9. Mateos-Cáceres PJ, García-Méndez A, Farré AL et al (2004) Proteomic analysis of plasma from patients during an acute coronary syndrome. J Am Coll Cardiol 44(8):1578–1583
10. Wang Y, Kuramitsu Y, Yoshino S et al (2011) Screening for serological biomarkers of pancreatic cancer by two-dimensional electrophoresis and liquid chromatography-tandem mass spectrometry. Oncol Rep 26(1):287–292
11. Zelvyte I, Sjögren HO, Janciauskiene S (2002) Effects of native and cleaved forms of α1-antitrypsin on ME 1477 tumor cell functional activity. Cancer Detect Prev 26(4):256–265
12. Zheng H, Zhao C, Roy S et al (2016) The commonality of the cancer serum proteome phenotype as analyzed by LC-MS/MS, and its application to monitor dysregulated wellness. Poster presented at the AACR annual meeting 2016 conference, New Orleans, LA, USA, April 17–20 2016
13. Lane L, Argoud-Puy G, Britan A et al (2011) neXtProt: a knowledge platform for human proteins. Nucleic Acids Res 40(D1):D76–D83
14. Lai ZW, Petrera A, Schilling O (2015) Protein amino-terminal modifications and proteomic approaches for N-terminal profiling. Curr Opin Chem Biol 24:71–79
15. Zheng H, Zhao C, Qian M et al (2015) AlbuVoid™ coupled to on-bead digestion-tackling the challenges of serum proteomics. J Proteom Bioinformatics 8(9):225
16. Sun Z, Yang P (2004) Role of imbalance between neutrophil elastase and α1-antitrypsin in cancer development and progression. Lancet Oncol 5(3):182–190

Chapter 3

Two-Dimensional 16-BAC/SDS Polyacrylamide Gel Electrophoresis of Mitochondrial Membrane Proteins

Gary Smejkal and Srikanth Kakumanu

Abstract

The substitution of the reverse polarity benzyldimethyl-*n*-hexadecylammonium chloride (16-BAC) polyacrylamide gel electrophoresis (PAGE) for isoelectric focusing (IEF) in the first dimension of electrophoresis improves the solubility of extremely hydrophobic proteins and their recovery compared to conventional 2D IEF/SDS PAGE. The acidic environment of 16-BAC PAGE has also been shown to better preserve the labile methylation of basic proteins such as the histones. Several improvements of the 2D 16-BAC/SDS PAGE method are collectively described here with particular emphasis on the separation of mitochondrial membrane proteins of low molecular mass. Lowering the 16-BAC concentration 50-fold in the gel and buffers decreases the formation of mixed 16-BAC/SDS micelles, which otherwise interferes with the separation of very low molecular mass proteins in second dimension SDS PAGE, and consequently improved the resolution of mitochondrial membrane proteins in the 10–30 kDa range.

Key words Benzyldimethyl-*n*-hexadecylammonium chloride, Cationic detergents, Membrane proteins, Mitochondria, Polyacrylamide gel electrophoresis, Proteins, Sodium dodecylsulfate, Two-dimensional gel electrophoresis, Transmembrane domains

Abbreviations

16-BAC	Benzyldimethyl-*n*-hexadecylammonium chloride
CMC	Critical micelle concentration
DTT	Dithiothreitol
HED	Hydroxyethyl disulfide
IEF	Isoelectric focusing
KDS	Potassium dodecylsulfate
PAGE	Polyacrylamide gel electrophoresis
PMSF	Phenylmethylsulfonyl fluoride
SDS	Sodium dodecylsulfate
TCEP	Tris (2-carboxyethyl) phosphine
TMDs	Transmembrane domains

Xing Wang and Matthew Kuruc (eds.), *Functional Proteomics: Methods and Protocols*, Methods in Molecular Biology, vol. 1871, https://doi.org/10.1007/978-1-4939-8814-3_3,

1 Introduction

Two-dimensional gel electrophoresis, as we've come to know it, combines IEF with orthogonal sodium dodecylsulfate (SDS) PAGE. The resolution of IEF is best exemplified in several landmark publications by Klose et al. [1, 2] which reported the separation over 10,300 proteins from mouse tissue using very large format IEF/SDS PAGE. The resolution of IEF is further increased using immobilized pH gradients (IPGs) which are capable of separating protein charge isoforms differing by only 0.001 pI units, a resolution an order of magnitude higher than carrier ampholyte generated pH gradients [3].

However, the strict requirement of nonionic or zwitterionic detergents limits the compatibility of many detergents with IEF. Many extremely hydrophobic proteins are insoluble in the detergents typically used for IEF, or if they are initially solubilized, they may precipitate near their isoelectric point and are excluded from the second dimension analysis. Klein et al. [4] showed the separation of peripheral membrane proteins from *Halobacterium salinarum* by conventional IEF/SDS PAGE, but failed to separate integral membrane proteins having multiple transmembrane domains (TMDs) that were irreversibly precipitated and trapped in the first dimension IPG strip. Kalinowski et al. [5] reported that CHAPS, a zwitterionic detergent commonly used in IEF, solubilized only 52% of proteins from *Corynebacterium glutamicum* membranes, compared to SDS which solubilized 90% of the membrane proteins. The use of SDS to initially solubilize proteins from *Chlorobium tepidum* membrane fractions, when followed by acetone precipitation to remove SDS, enabled IEF and more than doubled the number of proteins identified by IEF/SDS PAGE, compared to fractions solubilized with Triton X-100 [6].

The use of SDS in both dimensions of orthogonal PAGE (2D SDS/SDS PAGE) exploits the anomalous migration of extremely hydrophobic proteins at different gel concentrations [7]. While soluble proteins typically bind 1.4 times their mass of SDS, some membrane proteins can bind as much as 4.5 times their mass [8], in which case, the protein itself constitutes less than 20% of the total mass of the nascent SDS-protein complex [9]. Hence, the migration of membrane proteins inconsistent with their true molecular mass can be explained, at least in part, by detergent-induced shifts in both molecular mass and charge density. Recently, Meisrimler and Luthje [10] used SDS/SDS PAGE as second and third dimensions in a three-dimensional electrophoresis scheme that identified over 10% more membrane proteins with multiple TMDs from plant microsomes than IEF/SDS PAGE. Several other variations of SDS/SDS PAGE have been described [11, 12].

MacFarlane [13, 14] first described 16-BAC PAGE to preserve base labile protein methylation of platelet and promyelocyte proteins during electrophoresis, and soon afterward published the first methods for 2D 16-BAC/SDS PAGE [15] and preparative 2D 16-BAC/SDS PAGE [16]. Hartinger et al. [17] later applied 2D 16-BAC/SDS PAGE for the analysis of proteins in purified membranes from synaptic and clathrin-coated vesicles. Using 2D 16-BAC/SDS PAGE, Zahedi et al. [18] identified 42 mitochondrial membrane proteins not isolated by 2D IEF/SDS PAGE proteins, including the cytochrome-c oxidase subunit I which contains 12 TMDs. They later introduced 16-BAC PAGE in tube gels to eliminate the need to excise gel lanes [19].

These seminal papers spawned over three decades of publications on 16-BAC PAGE and 2D 16-BAC/SDS PAGE with very few improvements of the initial descriptive. Hence, some early mistakes have been repeated for decades. For example, nearly every paper ever published on 2D 16-BAC/SDS PAGE follows the earlier convention of staining the first dimension gels with Coomassie to guide lane excision prior to second dimension SDS PAGE. This necessarily precipitates the proteins in the first dimension gel and assumes that they will be completely resolubilized during the brief SDS equilibration that precedes second dimension. Other proteins that are soluble under such acidic conditions are not fixed in the gel and are washed out, unfortunately risking "throwing away the baby with the wash water." Despite this, the overnight staining of first dimension gels [18], and even their storage in staining solution [17], before second dimension PAGE have been reported. Using radiolabeled proteins, Hartinger et al. [17] estimated a 10% loss of the total protein when stained gels were transferred to second dimension. To the contrary, we have experienced the nearly complete loss of low molecular mass mitochondrial membrane proteins (in the 10–40 kDa range) from Coomassie-stained 16-BAC gels, compared to control gels that were immediately transferred to second dimension SDS PAGE without prior staining.

The separation of proteins by 2D 16-BAC PAGE is based on the differential binding of the 16-BAC and SDS detergents to proteins. Unlike SDS, the binding of 16-BAC to proteins is not well characterized (Fig. 1). From Ferguson plot analysis of 16-BAC protein derivatives, radically different slopes for high and low molecular mass proteins are obtained [20]. The lack of a common Y intercept in these plots implies that, unlike SDS, a constant net charge density is not obtained for all proteins. This is the prerequisite for separations based purely on molecular size with no influence by charge, such as in SDS systems. The differential binding of 16-BAC to some proteins is exemplified in the case of ovalbumin, which consistently exhibits mobility similar to other proteins half its size in 16-BAC PAGE [13, 20].

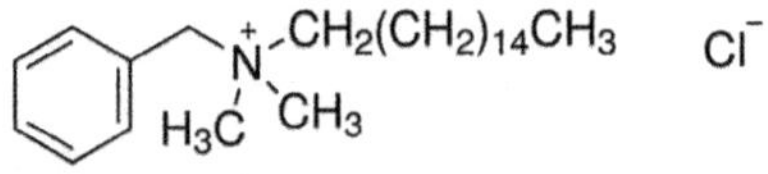

Fig. 1 Chemical structure of benzyldimethyl-*n*-hexadecylammonium chloride. Molecular weight 396.1

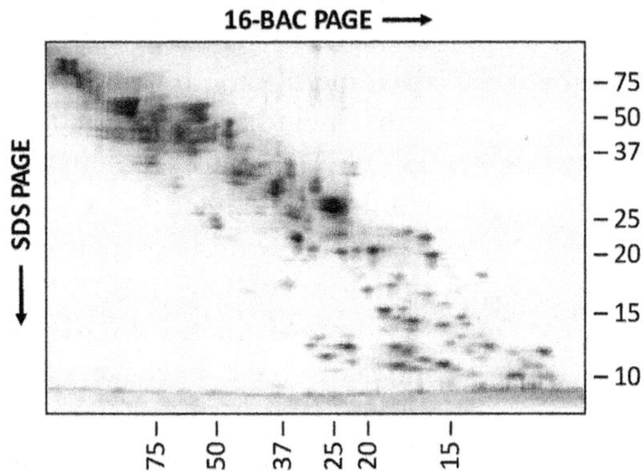

Fig. 2 2D 16-BAC/SDS PAGE of mitochondrial membrane proteins from porcine myocardium. First dimension 16-BAC PAGE was 12% polyacrylamide concentration. Second dimension SDS PAGE was 15% polyacrylamide concentration. Gel concentrations were selected to maximize separation of proteins in the 10–30 kDa range. Gels were stained overnight in KUMASI

In another important publication, Kramer [20] determined the critical micelle concentration (CMC) of 16-BAC in the PAGE buffers and demonstrated that the detergent concentration could be lowered 50-fold in the gel and running buffers (A molar excess of 16-BAC was kept in sample buffer to ensure the complete saturation of proteins.). Working at concentrations near or below the CMC of 16-BAC significantly decreases the formation of mixed 16-BAC/SDS micelles, which otherwise interfere with the separation of very low molecular mass proteins in second dimension SDS PAGE (Fig. 2). Consequently, the separation of membrane proteins in the 10–30 kDa range is improved, even at high polyacrylamide gel concentrations where micelles are sieved. In theory, it should be possible to exclude 16-BAC entirely from the stacking and resolving gels, relying on the constant influx of free detergent from the anode buffer to maintain detergency.

2 Materials

2.1 Isolation of Mitochondrial Membrane Proteins

1. 100 mM phenylmethylsulfonyl fluoride (PMSF) in 100% isopropanol (*see* **Note 1**).
2. Mitochondria isolation buffer (MIB): 20 mM Tris–HCl pH 7.4, 250 mM sucrose, 10 mM potassium fluoride, 2 mM EGTA, 1 mM sodium vanadate, 1 mM PMSF.

3. Chloroform.
4. Methanol.

2.2 First Dimension 16-BAC PAGE

Buffers and concentrated buffer stocks can be prepared in advance without the addition of 16-BAC. All buffers are filter sterilized (e.g., Millipore Steriflip or similar) and can be stored up to 3 months at 4 °C. The 16-BAC detergent is added to working solutions immediately before use and used the same day.

1. Concentrated (4×) resolving gel buffer: 300 mM KH_2PO_4 pH 2.1 (*see* **Note 2**).
2. Concentrated (4×) stacking gel buffer: 500 mM KH_2PO_4 pH 4.1 (*see* **Note 3**).
3. 125 mM 16-BAC stock solution (*see* **Note 4**).
4. Concentrated (10×) electrode buffer: 1.5 M glycine, 500 mM ortho-phosphoric acid (*see* **Note 5**).
5. Working (1×) electrode buffer: 150 mM glycine, 50 mM ortho-phosphoric acid, 0.05 mM 16-BAC (*see* **Note 6**).
6. 9 M urea (*see* **Note 7**).
7. AG 501-X8 mixed bed ion exchange resin (*see* **Notes 7** and **9**).
8. Bond-Breaker™ 500 mM Tris (2-carboxyethyl) phosphine (TCEP) solution, neutral pH (Thermo-Fisher Scientific, 77720).
9. 50% glycerol.
10. Concentrated (100×) Pyronin Y tracking dye: 1 mg/mL Pyronin Y in water.
11. 1× sample buffer: 4 M urea, 50 mM 16-BAC, 10 mM TCEP, 10.5% glycerol, 0.005% Pyronin Y (*see* **Note 8**).
12. 29.2% acrylamide, 0.8% methylene bisacrylamide solution (*see* **Note 9**).
13. 80 mM ascorbic acid (*see* **Note 10**).
14. 5 mM ferrous sulfate (*see* **Note 11**).
15. 30% hydrogen peroxide.
16. 25% isopropanol.
17. Reflection™ Dual Vertical Electrophoresis System (Galileo Biosciences, 85-1614).
18. Precision Plus Protein™ Standards (Biorad, 1610374).
19. Stainless steel tissue slicing blade, 22 cm length.

2.3 Neutralization and SDS Equilibration

1. Neutralization Buffer: 375 mM Tris HCl pH 8.8, 3 M urea, 5% glycerol, 0.001% bromophenol blue (*see* **Note 12**).
2. Dithiothreitol (DTT).
3. SDS Equilibration Buffer: 375 mM Tris HCl pH 8.8, 3 M urea, 2% SDS, 5% glycerol, and 0.001% phenol red. Solid 50 mM DTT is added immediately before use (*see* **Note 13**).
4. Polypropylene reagent reservoirs, 60 mL.
5. Medium thickness filter paper, 100 × 20 mm.

2.4 Second Dimension SDS PAGE

1. Criterion Dodeca Cell Vertical Electrophoresis System (Biorad, 165-4130).
2. Criterion 15% Tris–HCl precast polyacrylamide gels (BioRad, 345-0019) or Criterion empty cassettes (*see* **Note 14**).
3. SDS PAGE electrode buffer: 25 mM Tris, 192 mM glycine, 0.1% SDS pH 8.3.
4. 30% ethanol, 10% acetic acid.
5. KUMASI stabilized colloidal Coomassie staining solution (Focus Proteomics, FPKS-001).
6. 1 M sodium azide.

3 Methods

3.1 Isolation of Mitochondrial Membrane Proteins

Mitochondria were isolated as described by Lee et al. [21]. Porcine heart was resected within minutes of euthanasia and placed immediately on wet ice. The tissue was dissected within 1 h of collection. All steps were performed at 4 °C.

1. Remove connective tissue and fat and course grind the myocardium in a food grinder.
2. Suspend the macerate in five volumes of MIB.
3. Blend the suspension in a food processor for three times 30 s.
4. Centrifuge the suspension at 650 RCF for 10 min and filter the supernatant through multiple layers of cheese cloth.
5. Resuspend the remaining pellet in additional MIB and repeat **steps 3** and **4**. Combine the two supernatants.
6. Centrifuge the pooled supernatants and centrifuge at 14,000 RCF for 20 min.
7. Homogenize the resulting pellet in a ground glass homogenizer with Teflon plunger.
8. Centrifuge at 400 RCF for 8 min to pellet cellular debris. Transfer the supernatant to a new tube and centrifuge for 14,000 RCF for 20 min.

9. Repeat **steps** 7 and **8** until the supernatant is clear and the pellet is beige in color.
10. Resuspend the mitochondrial pellet in a small volume of MIB and determine the protein concentration using the Lowry method. Adjust the sample volume with MIB so that protein concentration is 20 mg/mL. The mitochondria can be stored at −80 °C for later analysis.
11. Prior to isolation, dilute the proteins to 3 mg/mL in MIB. Add four volumes of methanol, one volume of chloroform, and three volumes of water, vortexing vigorously after each addition.
12. Centrifuge at 14,000 RCF for 2 min.
13. Remove the top aqueous layer and add 400 μL of methanol. Centrifuge at 14,000 RCF to pellet the precipitated proteins.
14. Aspirate the supernatant and allow the precipitated protein pellet to air-dry making sure to not overdry the pellet as it will be very difficult to solubilize for electrophoresis. Pellets can be stored at −80 °C for later analysis.

3.2 16-BAC PAGE

The Fenton reaction as modified by MacFarlane [13] is used to catalyze the polymerization reaction. Generally, 16-BAC gels are run on the same day they are cast. If necessary, the resolving gel can be cast on the first day, then overlaid with 1× resolving buffer and stored overnight at room temperature. The stacking gel should be cast within a few hours of running the gel on the second day. All steps are performed at room temperature.

1. Assemble two 16 × 14 cm glass plates and two 0.8 mm thick spacers using the alignment pins of the upper buffer chamber and clamp securely. The entire assembly is transferred to the casting stand and cammed into place for leakproof casting (Fig. 3).
2. Prepare the resolving gel as prescribed in Table 1. Combine all components except for the catalysts in a 50 mL screw cap centrifuge tube. Cap securely and mix by gentle inversion taking care not to introduce bubbles.
3. Add sequentially the absorbic acid, $FeSO_4$, and H_2O_2. Cap securely and mix by gentle inversion following each addition.
4. Immediately pour the polymerization mixture into the gel cassette to within 3 cm of the top of the notched glass plate. Carefully overlay the gel with 400–600 μL of 25% isopropanol (*see* **Note 15**).
5. Allow the gel to polymerize for 15–20 min (*see* **Note 16**).

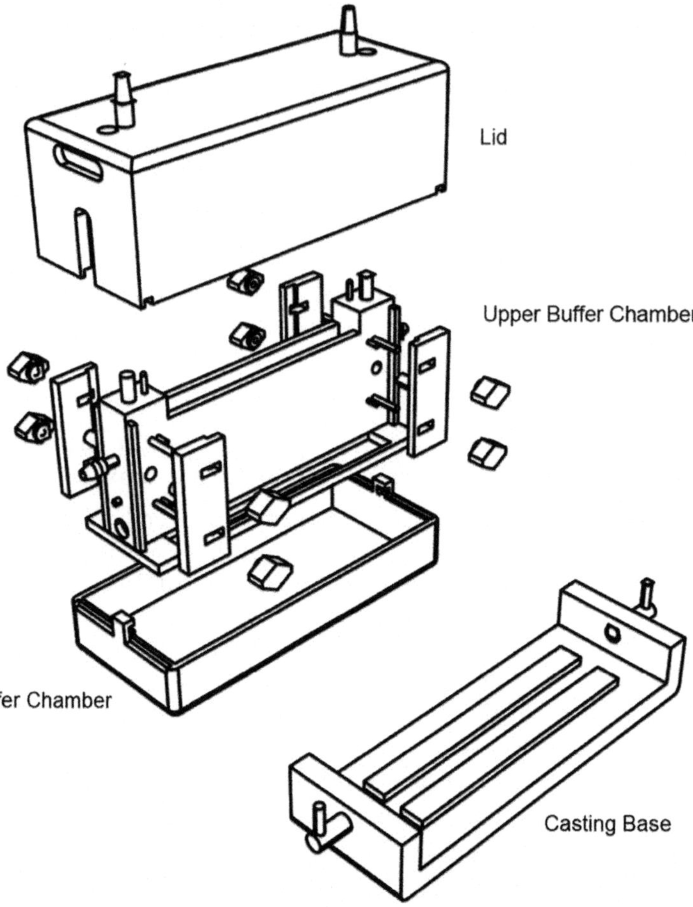

Fig. 3 Exploded view of the Galileo Biosciences large format electrophoresis system and leakproof gel casting system. Glass plates and spacers are clamped to the upper buffer chamber during casting and are not removed until electrophoresis is complete. This avoids the "flexing" of gels which can result in disadherence of the gel from the glass surface and the formation of bubbles between the gel and glass

6. Decant the isopropyl overlay and wash the gel surface twice with water. (If storing the gel overnight, overlay with 1X resolving buffer.)
7. Prepare the stacking gel as prescribed in Table 2. Combine all components except for the catalysts, cap securely, and mix by gentle inversions. Add sequentially the absorbic acid, $FeSO_4$, and H_2O_2. Cap securely and mix by gentle inversion following each addition.
8. Fill the remaining space in the gel cassette with stacking gel and insert the 0.8 mm thick comb taking care not to trap air bubbles. Allow to polymerize for at least 1 h.
9. Remove the comb and forcefully flush any unpolymerized solution from the wells with water using a transfer pipette.

Table 1
16-BAC PAGE resolving gel composition

Stock solution	Final concentration	Volume
29.2% acrylamide, 0.8% bisacrylamide	12%	20 mL
9 M urea	2.6 M	14.9 mL
4× resolving gel buffer pH 2.1	75 mM	12.5 mL
125 mM 16-BAC	0.05 mM	20 μL
Water	–	–
80 mM absorbic acid	4 mM	2.5 mL
5 mM ferrous sulfate	8 μM	80 μL
30% hydrogen peroxide	0.002%	3.3 μL
	Total volume	50 mL

Table 2
16-BAC PAGE stacking gel composition

Stock solution	Final concentration	Volume
29.2% acrylamide, 0.8% bisacrylamide	4%	1.35 mL
9 M urea	3 M	3.35 mL
4× stacking gel buffer pH 4.1	75 mM	2.5 mL
125 mM 16-BAC	0.05 mM	4 μL
Water	–	2.3 mL
80 mM absorbic acid	4 mM	0.5 mL
5 mM ferrous sulfate	8 μM	16 μL
30% hydrogen peroxide	0.002%	1 μL
	Total volume	10 mL

10. Fill the upper and lower buffer chambers with electrode buffer. Flush the water from the wells with electrode buffer using a transfer pipette.

3.3 Sample Preparation

16-BAC is dissolved in the sample buffer on the same day of the analysis. Protein samples are prepared immediately before electrophoresis. Hartinger et al. [17] observed protein degradation and a loss of resolution when samples were stored in the 16-BAC sample buffer.

1. Dissolve the mitochondrial pellet (100–200 μg total protein) in 25 μL of 16-BAC sample buffer. Incubate at 60 °C for 10–15 min or until completely dissolved (*see* **Note 17**).
2. Dilute the protein standards at least tenfold in 16-BAC sample buffer.
3. Centrifuge the samples and standards at 14,000 RCF for 5 min. Apply 20 μL of supernatant to each well, leaving a blank lane between samples when possible.
4. Connect to the power supply in reverse polarity. Commence electrophoresis at 50 mA constant current until the Pyronin Y dye has migrated 8–10 cm into the resolving gel.

3.4 Neutralization and SDS Equilibration

1. Immediately following electrophoresis, disassemble the upper buffer chamber and open the gel cassette like a book using one of the spacers to leverage. One plate should release while the gel remains adhered to the other plate.
2. Using the stacking gel and Pyronin Y dye front as reference, excise a slice from the center of each lane using the tissue slicing blade. For best results, the gel slice should not exceed 4 mm in width. Trim off the stacking gel and excess resolving gel below the Pyronin Y.
3. Transfer each gel slice to a clean polypropylene reagent reservoir and incubate for 2× 2 min in 5 mL of neutralization buffer.
4. Incubate each gel slice for 2× 10 min in 5 mL of SDS equilibration buffer.

3.5 Second Dimension SDS PAGE

1. Flush the surface of the second dimension gel with water to remove residual storage buffer. Fill the upper buffer chamber with SDS PAGE electrode buffer.
2. Soak a 100 × 20 mm filter paper in SDS PAGE electrode buffer.
3. With a spatula, position the equilibrated gel strip on top of the second dimensional gel. Place the saturated filter paper on top of the equilibrated gel strip and gently press down to keep the gel strip in intimate contact with the second dimension gel (Fig. 4).
4. Connect to the power supply in normal polarity. Commence electrophoresis at 40 mA constant current for 10 min. Turn the power off and remove the filter papers. Continue electrophoresis until the phenol red dye front has migrated to within a few millimeters of the bottom of the gel.

3.6 KUMASI Staining

1. Fix the gel for at least 1 h in 100 mL of 30% ethanol, 10% acetic acid (*see* **Note 18**).

Fig. 4 Second dimension SDS PAGE of excised lane from first dimension 16-BAC PAGE. SDS equilibrated gel strips were held in intimate contact with the second dimension gel with a buffer-saturated filter paper wick (arrow). This eliminates the need for an agarose gel overlay

2. Decant the fixative and stain the gel overnight in 100 mL of KUMASI stain (*see* **Note 19**).
3. Rinse the gel for 30 s in 30% ethanol, 10% acetic acid.
4. Incubate the gel at least 20 min in 100 mL of activated enhancer solution from the KUMASI staining kit. Contrast is improved with overnight incubation in the enhancer solution.
5. If gels are to be stored for more than 1 week, add 10 mM sodium azide (*see* **Note 20**).

4 Notes

1. The half-life of PMSF is 55 and 35 min at pH 7.5 and 8.0, respectively, in aqueous buffers [22]. Stock solutions prepared in 100% isopropanol are stable for months.
2. To make 4× resolving gel buffer, dissolve 8.2 g KH_2PO_4 in 150 mL water and adjust to pH 2.1 with 1 N HCl. Adjust volume to 200 mL. Filter sterilize and store at 4 °C for up to 3 months.
3. To make 4× stacking gel buffer, dissolve 13.6 g KH_2PO_4 in 150 mL water and adjust to pH 4.1 with 1 N HCl. Adjust volume to 200 mL. Filter sterilize and store at 4 °C for up to 3 months.
4. Prepare fresh 16-BAC solution daily. To make 125 mM stock solution, dissolve 50 mg 16-BAC (Millipore-Sigma, B-4136) in 10 mL water.

5. To make 10× electrode buffer, dissolve 56 g glycine in 450 mL water. Add 28.8 mL 85% ortho-phosphoric acid and adjust volume to 500 mL. Do not adjust pH. Filter sterilize and store at 4 °C for up to 3 months.
6. Use 100 mL of concentrated (10×) buffer per liter of working electrode buffer. Dissolve 20 mg solid 16-BAC per liter immediately before use.
7. Dissolve 24.3 g urea in 45 mL water in a 50 mL screw cap centrifuge tube. Add 0.5 g AG 501-X8 mixed bed ion exchange resin (Biorad, 143-7424) and incubate 30–60 min with gentle nutation. Filter sterilize, store at room temperature, and use within 3 days.
8. Prepare sample buffer fresh daily. Admix 4.5 mL 9 M urea and 3.5 mL 30% glycerol. Add 0.2 g 16-BAC and dissolve by nutation. Add 50 μL 500 mM TCEP and 50 μL of 100× Pyronin Y and adjust volume to 10 mL.
9. Acrylamide and methylene bisacrylamide are potent neurotoxins. To minimize hazards, use premixed 30% acrylamide-bisacrylamide solution (BioRad, 161-0159). Add 0.5 g AG 501-X8 mixed bed ion exchange resin to 45 mL 30% acrylamide-bisacrylamide solution in a 50 mL screw cap centrifuge tube and incubate 30–60 min with gentle nutation. Filter sterilize and store at 4 °C for up to 2 months.
10. Prepare 80 mM ascorbic acid fresh daily. Dissolve 140 mg absorbic acid in 10 mL water.
11. Prepare 5 mM $FeSO_4$ fresh daily. Dissolve 70 mg $FeSO_4$ in 50 mL water.
12. Bromophenol blue is a pH indicator dye that transitions from yellow to blue indicating when the gel strip is neutralized. Another more important role of the neutralization step is to remove residual KH_2PO_4 from the gel prior to SDS equilibration since K+ ions would otherwise drive the formation of insoluble potassium dodecylsulfate (KDS). The Krafft point of KDS is 36 °C.
13. DTT has a half-life of approximately 1.4 h at pH 8.5 at room temperature. Desiccated DTT solids should be dissolved in the SDS equilibration buffer and used within 1 h. The SDS equilibration buffer can be prepared in advance (without DTT), filtered sterilized, and stored at −20 °C for at least 2 months. DTT is ineffective as reducing agent below pH 7.0 where only about 1% of the thiol groups of DTT are in the reactive thiolate form [23].
14. Gels can be hand cast at any desired gel concentration into Criterion empty cassettes as described in detail by Smejkal and Bauer [9].

15. The gel can be "misted" with 25% isopropanol from a spray bottle.
16. Gel polymerization rates are affected by temperature and relative humidity. If the gel polymerizes in less than 10 min, discard and prepare a new gel using 10% less of each catalyst.
17. Avoid heating the sample higher than 60 °C. Asp-Pro linkages are susceptible to hydrolysis at acidic pH and elevated temperature.
18. For improved staining, fix overnight to completely remove SDS from the gel.
19. The fixative and KUMASI staining solutions can be reused at least two times.
20. Gels stored in enhancer solution with sodium azide added are stable for years. We have observed improved contrast with no significant loss of sensitivity in gels stored for 9 years.

References

1. Klose J, Kobalz U (1995) Two-dimensional electrophoresis of proteins: an updated protocol and implications for a functional analysis of the genome. Electrophoresis 16:1034–1059
2. Klose J, Nock C, Herrmann M, Stühler K, Marcus K, Blüggel M, Krause E, Schalkwyk LC, Rastan S, Brown SDM, Büssow K, Himmelbauer H, Lehrach H (2002) Genetic analysis of the mouse brain proteome. Nat Genet 30:385–393
3. Hamdan M, Righetti PG (2005) Proteomics today: Protein assessment and biomarkers using mass spectrometry, 2D electrophoresis, and microarray technology. Wiley & Sons, Hoboken, NJ, pp 219–265
4. Klein C, Garcia-Rizo C, Bisle B, Scheffer B, Zischka H, Pfeiffer F, Siedler F, Oesterhelt D (2005) The membrane proteome of Halobacterium salinarum. Proteomics 5:180–197
5. Kalinowski J, Wolters D, Poetsch A (2008) Proteomics of Corynebacterium glutamicum and other Corynebacteria. From Corynebacteria: genomics and molecular biology (Burkovski A, ed). Caister Academic Press, Norfolk, pp 56–77
6. Aivaliotis M, Corvey C, Tsirogianni I, Karas M, Tsiotis G (2004) Membrane proteome analysis of the green-sulfur bacterium Chlorobium tepidum. Electrophoresis 25:3468–3474
7. Moller AJB, Witzel K, Vertommen A, Barkholdt V, Svensson B, Carpentier S Mock HP, Finne C (2011) Plant membrane proteomics: challenges and possibilities. Sample preparation in biological mass spectrometry. Springer, Heidelberg, pp 411–434
8. Rath A, Glibowicka M, Nadeau VG, Chen G, Deber CM (2009) Detergent binding explains anomalous SDS-PAGE migration of membrane proteins. PNAS 106:1760–1765
9. Smejkal GB, Bauer DJ (2012) High speed isoelectric focusing of proteins enabling rapid two-dimensional gel electrophoresis. Gel electrophoresis: principles and basics. Intech, Rijeka, pp 157–170
10. Meisrimler CN, Lüthje S (2012) IPG-strips versus off-gel fractionation: advantages and limits of two-dimensional PAGE in separation of microsomal fractions of frequently used plant species and tissues. J Proteome 75:2550–2562
11. Rabilloud T (2010) Variations on a theme: changes to electrophoretic separations that can make a difference. J Proteome 73:1562–1572
12. Miller M, Ivano Eberini I, Gianazza E (2010) Other than IPG-DALT: 2-DE variants. Proteomics 10:586–610
13. Macfarlane DE (1983) Use of benzyldimethyl-n-hexadecylammonium chloride (16-BAC), a cationic detergent, in an acidic polyacrylamide gel electrophoresis system to detect base labile protein methylation in intact cells. Anal Biochem 132:231–235
14. Macfarlane DE (1984) Inhibitors of cyclic nucleotide phosphodiesterases inhibit protein carboxyl methylation in intact blood platelets. J Biol Chem 259:1357–1362

15. Macfarlane DE (1986) Phorbol diester-induced phosphorylation of nuclear matrix proteins in HL60 promyelocytes. Possible role in differentiation studied by cationic detergent gel electrophoresis systems. J Biol Chem 261:6947–6953
16. Macfarlane DE (1989) Two dimensional benzyldimethyl-n-hexadecylammonium chloride sodium dodecyl sulfate preparative polyacrylamide gel electrophoresis: a high capacity high resolution technique for the purification of proteins from complex mixtures. Anal Biochem 176:457–463
17. Hartinger J, Stenius K, Högemann D, Jahn R (1996) 16-BAC/SDS-PAGE: a two-dimensional gel electrophoresis system suitable for the separation of integral membrane proteins. Anal Biochem 240:126–133
18. Zahedi RP, Meisinger C, Sickmann A (2005) Two-dimensional benzyldimethyl-nhexadecylammonium chloride/SDS-PAGE for membrane proteomics. Proteomics 2005 (5):3581–3588
19. Zahedi RP, Moebius J, Sickmann A (2007) Two-dimensional BAC/SDS-PAGE for membrane proteins. In: Bertrand E, Faupel M (eds) Subcellular proteomics: from cell deconstruction to system reconstruction. Springer, Dordrecht, pp 13–20
20. Kramer ML (2006) A new multiphasic buffer system for benzyldimethyl-n-hexadecylammonium chloride polyacrylamide gel electrophoresis of proteins providing efficient stacking. Electrophoresis 27:347–356
21. Lee I, Salomon AR, Yu K, Samavati L, Pecina P, Pecinova A, Huttemann M (2009) Isolation of regulatory-competent, phosphorylated cytochrome c oxidase. Methods Enzymol 457:193–210
22. James GT (1978) Inactivation of the protease inhibitor phenylmethylsulfonyl fluoride in buffers. Anal Biochem 86:574–579
23. Singh R, Whitesides GM (1995) Reagents for raid reduction of disulfide bonds in proteins. Techniq Protein Chem VI:259–266

Chapter 4

Systematic Glycolytic Enzyme Activity Analysis from Human Serum with PEP Technology

David Wang

Abstract

A functional proteomics technology was used to systematically monitor metabolic enzyme activities from resolved serum proteins produced by a modified 2-D gel separation and subsequent Protein Elution Plate, a method collectively called PEP. Both qualitative and quantitative differences in the metabolic enzyme activity could be detected between cancer patient and control group, providing excellent biomarker candidates for cancer diagnosis and drug development. This technology has a wide range of applications; it can be used for rapid functional protein purification and characterization as well as drug target identification and validation. The ability for the PEP technology to efficiently separate and recover functional proteins makes it useful for the analysis of any proteins and its variants; this is especially advantageous for enzyme families with large number of enzymes such as protein kinases, phosphatases, proteases, and metabolic enzymes.

Key words Functional proteomics, 2-D gel electrophoresis, Protein purification, Biomarkers, Protein elution plate (PEP), Cancer diagnosis, Drug target identification

1 Introduction

In the last decade, many new technologies have been utilized for biomarker discovery with significant progress. Each of these technologies has focused on a different type of biological entity such as circulating tumor cells (CTC), extracellular vesicles, micro-RNAs and cancer-derived cell-free DNA or circulating tumor-derived DNA (ctDNA) [1–9]. However, several fundamental issues such as tumor heterogeneity, plasticity, and diversity of cancer stem cells (CSC) make biomarker discovery and development a challenging endeavor. The variation introduced during sample collection and storage and the lack of robust validation approach once biomarker leads are identified further complicate biomarker development [10–19]. As a result of these hurdles, there are currently no United States FDA-approved serum tests for early detection of the disease. Given the considerable public health importance of breast cancer, it

Xing Wang and Matthew Kuruc (eds.), *Functional Proteomics: Methods and Protocols*, Methods in Molecular Biology, vol. 1871, https://doi.org/10.1007/978-1-4939-8814-3_4,

is crucial to quickly identify new biomarkers with the potential to enhance early diagnosis and to predict patient prognosis, drug resistance development, and treatment choice. Blood-based biomarkers have great potential in cancer screening and their role could extend further from general population risk assessment to treatment response evaluation and recurrence monitoring [20–27]. The rich content of diverse cellular and molecular elements in blood, which provide information about the health status of an individual, makes it an ideal compartment to develop noninvasive diagnostics for cancer. However, despite a large literature collection related to biomarkers for common cancers, blood-based diagnostic tests that inform about the presence of cancer at an early stage and predict treatment response have been difficult to develop.

For the past decade, proteomics has been used for the discovery of potential biomarkers from human fluids including serum. So far, most efforts in proteomics has been focused on the identification and sequence annotation of the proteome by mass spectrometry analyses of peptides derived through proteolytic processing of the parent proteome. In such a manner, thousands of proteins have been identified from human serum (www.serumproteome.org). It is generally recognized that sequence annotation alone cannot capture this vital information, so new strategies are necessary. Two-dimensional (2-D) Gel Electrophoresis is a powerful technology to separate complex protein samples. In the first dimension called isoelectric focusing (IEF), the proteins are separated based on their isoelectric points (pI), proteins with as little as 0.02 unit pI differences could be separated, making it a high-resolution method. In the second dimension, the proteins are separated based on their molecular size. Because 2-D Gel Electrophoresis is using two orthogonal parameters (charge and size) for separation and displaying the proteins in a two-dimensional manner, it is one of the most powerful technologies in protein separation. In a large format gel, more than 10,000 proteins could be separated and detected with information on their relative abundance and post-translational modification acquired simultaneously. Because of these advantages, 2-D Gel Electrophoresis has been used widely in proteomics studies. However, in a typical 2-D Gel Electrophoresis, the proteins are denatured by the addition of reagents to disrupt disulfide bonds (DTT or β-mercaptoethanol), chemicals to prevent disulfide bond formation (iodoacetamide) and high concentration of SDS (typically 1%). To keep the proteins active in 2-D Gel Electrophoresis, a few modifications were made in the current PEP technology. First, no reducing reagent is used in the isoelectric focusing step, keeping the disulfide bonds in the proteins intact. Secondly, iodoacetamide is omitted from the process. Thirdly, much reduced SDS concentration is used in the SDS-PAGE (from 1% to 0.1%) or no SDS was used at all, again trying to maintain enzymatic activity. Recent studies indicated that many different enzyme families from a wide variety of organisms are active in the presence of SDS such as protein kinases, protein phosphatases, proteases, and oxido-reductases.

In addition of method modification, a high-resolution Protein Elution Plate (PEP) was designed. The small format PEP has 384-wells matching the current 384-well microplate dimension for ease of sample processing. For the large format PEP, the plate is composed of 4 384-well PEP, thus having 1536 wells. In both the large and small format PEP, a membrane with molecular cutoff of 6000 Dalton is attached that will allow the electric current and small charged molecules to pass through but collect proteins with molecular weight large than 6000 Dalton in the PEP wells. Furthermore, a special solution was developed for the PEP to reduce protein diffusion after the proteins are transferred from the gel to the PEP. After transferring the solutions from PEP to a deep-well master plate, the enzyme activity or protein function can be analyzed using part of the sample from the master plate and purified protein can be verified using SDS-PAGE in standard condition and identified using mass spectrometry. It is hypothesized that the levels and distributions of certain enzyme functions in serum could produce proteomic features and collective profiles which reflect physiological changes of an individual and can serve as possible biomarkers or diagnostic parameters [28–38]. In this chapter, we summarized the use of PEP technology for the systematic analysis of metabolic enzyme from human serum. We believe that the identification and validation of those functional proteins from human serum could lead to the development of biomarkers for cancer diagnosis. The PEP technology can also be used for the discovery of functional biomarkers for other diseases as well as drug target identification and drug safety evaluation (Fig. 1).

2 Materials

2.1 Chemicals

All the chemicals were purchased from MilliporeSigma (St. Louis, MO). Isoelectric focusing (IEF) unit that is capable of running IEF at different length is from Bio-Rad (PROTEAN IEF Cell, Hercules, CA). Spectrophotometer Plate Reader capable of reading 384-well plates with a wide wavelength selection and fluorescence reading is the SPECTRAMax Plus from Molecular Devices (Sunnydale, CA). Semi-Blot unit for protein transfer was Bio-Rad's Trans-Blot SD Semi-Dry Transfer Cell. AlbuVoid™ serum protein enrichment beads were from Biotech Support Group (Monmouth Junction, NJ). Protein Elution Plate (PEP) is a product of Array Bridge (St. Louis, MO).

1. *SDS-PAGE gels*: Customer can choose any format SDS-PAGE gel to run the sample. For 1-D gel, preferentially the loading

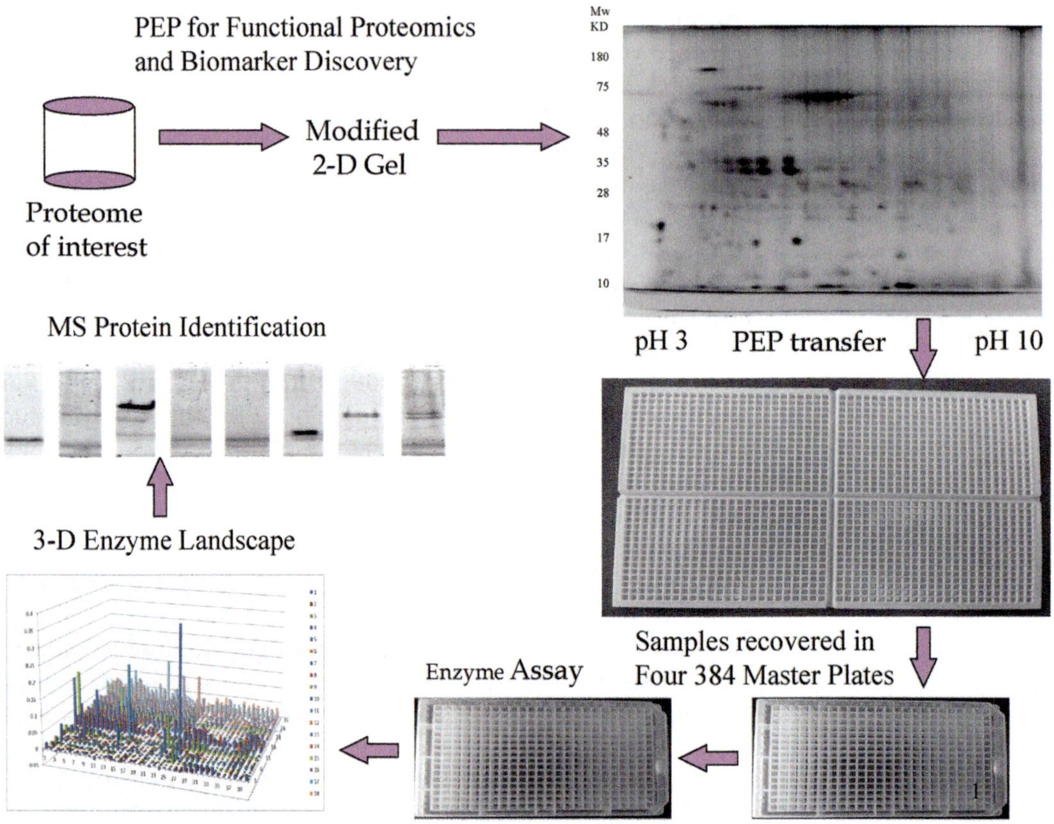

Fig. 1 Diagram of the PEP Technology (adapted from Wang DL et al., PLoS One, 2015, 10(3) with permission)

capacity of each well should be 15 μL or more. Gels from Bio-Rad (Criterion 10–20% 18 well Tris–HCl gel, catalog number: 345-0043) was used in our studies. For 2-D gel separation, Criterion 10–20% IPG + 1 well Tris–HCl gel (Bio-Rad, catalog number: 345-0107) or similar gels from Invitrogen, etc. can be used for the protein separation.

2. *Isoelectric focusing strips*: Immobilized pH gradient (IPG) strips to run IEF can be purchased from either Bio-Rad (Catalog Number: 163-2014 for 11 cm IPG strips and 163-2033 for 18 cm IPG strips) or GE Health Life Sciences (Catalog Number: 18101661 for 11 cm pH 3–10 Immobiline Dry Strips; 17123501 for 18 cm, pH 3–10 Nonlinear Immobiline Dry Strips).

3. *Electrolyte*: Electrolyte used for the IEF gel can be purchased from either Bio-Rad (Bio-Lyte buffer, pH 3–10, catalog number: 163-2094) or GE Health (Pharmalyte pH 3–10, catalog number: 17-0456-01).

4. *Protein staining components*: If protein staining is required, the following conditions can be used: gels after electrophoresis first fixed in fixing solution (10% acetic acid, 10% ethanol in Mili-Q water) for 1 h, then stained in SYPRO Orange (Invitrogen, catalog number: S6650) or other fluorescence dye overnight in Mili-Q water, dilute the fluorescence dye as recommended by the manufacturer.
5. *Single- and multi-channel micro-pipettes* with disposable tips to accurately dispense volumes 5–250 μL. Plastic tubes (i.e., 1.5–15 mL) for sample dilution. Reagent reservoirs for sample addition.

2.2 Glycolytic Enzyme Activity Assay

The glycolytic enzymes from human serum were detected by measuring the first enzyme in the glycolytic pathway, hexokinase. By using a beef liver extract to provide a low and basal level of glycolytic enzymes, any additional enzymes from the PEP-eluted serum sample can be detected by the increased hexokinase activity. Therefore the measurement of the glycolytic enzyme activity from PEP samples was calculated by the increased hexokinase activity from the basal level of beef liver extract instead of the total hexokinase activity.

Hexokinase activity can be monitored by a cascade reaction as follows:

Hexokinase

Substrates added {***D***-Glucose + ATP}
→ Products {***D***-Glucose 6-Phosphate + ADP}

G-6-PDH

D-Glucose 6-Phosphate + β-NADP
→ 6-Phospho-D-Gluconate + β-NADPH

In the final assay solution, glucose was at 216 mM; $MgCl_2$ at 7.8 mM, ATP at 0.74 mM, and NADP at 1.1 mM. 25 μL of this enzyme assay solution was mixed with 25 μL of sample from the Master Plate (described below) and the enzyme activity was monitored by the increased 340 nm absorbance from the reduction of NADP to NADPH. The readings at different time points such as 0, 1, and 2 h were recorded for both the normal serum and breast cancer patient serum sample. However, in lieu of purified G-6-PDH used for the hexokinase assay, 0.25 mg/mL beef liver protein was used as the source of glucose-6-phosphate dehydrogenase (G-6-PDH). The assay thus reports the additive contributions of the endogenous hexokinase activity present in the beef liver extract, and any exogenous activity from the presence of test sera protein in the PEP plate, which may influence the reduction of NADP (the reporting signal). In light of the ambiguities that may arise from

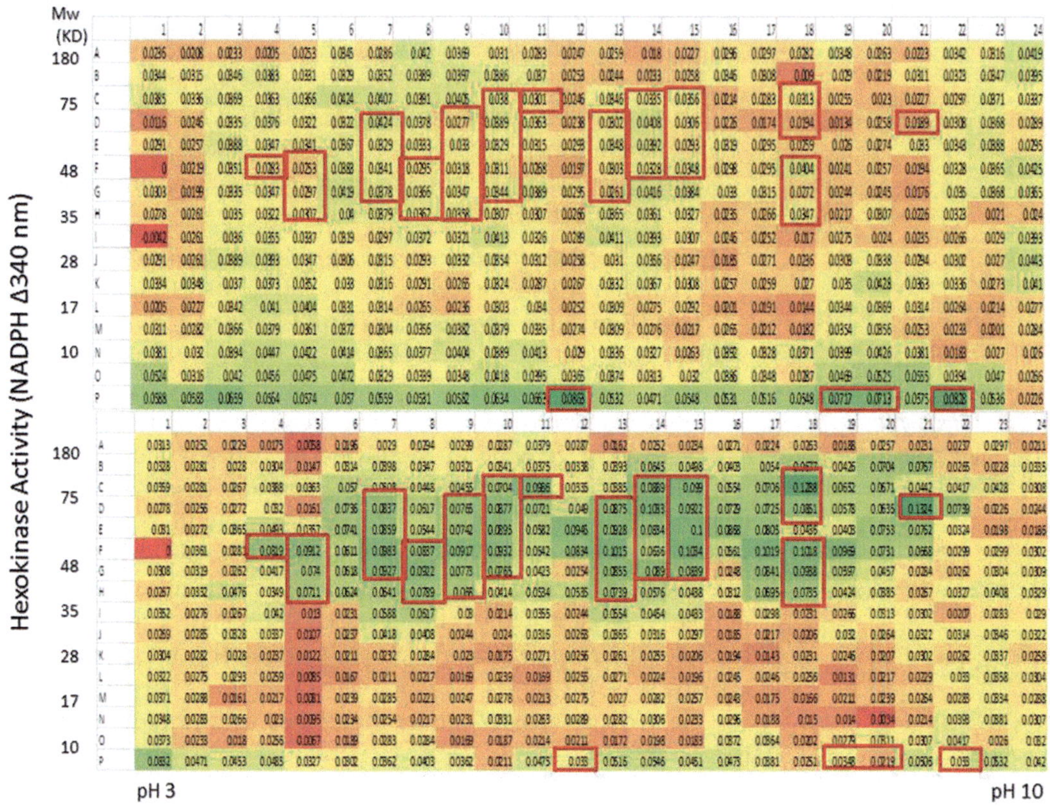

Fig. 2 Measurement of hexokinase activity from normal and breast cancer serum (adapted from Wang DL et al., BMC Biomarker Research, 2017, 5(11) with permission)

such a reporting system, the primary goal of this investigation was to generate sufficient signal intensities and activity features which could be monitored and compared between the two samples types within an "omics" context. Therefore, this broader spectrum assay was chosen that could potentially detect the activities of hexokinase and downstream glycolytic enzymes and other cross-regulating proteins from the test sera (Fig. 2).

2.3 Supplied Components from the PEP Universal Protein Purification Kit (Small Format PEP) from Array Bridge Inc. (Catalog Number: AB-000401)

1. *384-well PEP plate.* A PEP plate is provided. The plate was treated with a special solution to reduce the binding of the transferred protein and increase the recovery efficiency.
2. *384-well mater plate.* A deep-well plate is provided to contain samples recovered from the PEP plate.
3. *384-well enzyme assay plate.* A standard 384-well polypropylene plate is provided for enzyme assay to identify which wells contain the protein of interest.
4. *10× protein transfer buffer (50 mL).* Buffer used for running the modified SDS-PAGE or the second dimension of 2-D gel,

and also used to wet the filter papers for the transfer of proteins from the gel to the PEP plate.

5. *10× PBS (10 mL)*. Buffer used for the pretreatment of the master plate and fill in each well of the Master Plate with 50 μL of PBS.
6. *Standard SDS-PAGE sample buffer (0.5 mL)*. Solution used in sample treatment for the standard SDS-PAGE to check enzyme fraction purity.
7. *PEP plate protein recovery buffer (25 mL)*. Solution used in the PEP plate to recover proteins eluted from the gel and prevent protein diffusion.
8. *Plate sealer.* For sealing the Master plate and the enzyme assay plate during the purification process. Kit AB-00402 (two).
9. *Filter papers.* Used to form a sandwich in the protein transfer process.

2.4 Instruments

Gel electrophoresis unit includes power supply and gel unit.

Isoelectric focusing unit that is capable of running IEF at different length, an example of such unit is the Bio-Rad PROTEAN IEF Cell (Catalog Number: 165-4000).

Spectrophotometer Plate Reader capable of reading 384-well plates with a wide wavelength selection and fluorescence reading.

Semi-Blot unit for protein transfer such as Bio-Rad's Trans-Blot SD Semi-Dry Transfer Cell (Catalog Number: 170-3940).

3 Methods

3.1 Sample Treatment

High concentration of salt will interfere with the isoelectric focusing step. If the protein concentration is less than 5 mg/mL and the salt concentration is more than 100 mM, it is recommended to dialyze the samples in 5 mM phosphate buffer, pH 7.2 before use or using desalting column to reduce the salt concentration.

3.2 Running the IEF Gel First Followed by Running the Native or Modified SDS-PAGE

1. It is suggested to use the 11 cm IPG strip (Bio-Rad, catalog number: 163-2033) for the IEF. To rehydrate one IPG strip, 225 μL of solution is needed. It is suggested to use 200 μL of samples with up to 200 μg total protein, add urea to a final concentration of 8 M, add 2 μL of Ampholyte such as Bio-lyte (Bio-Rad, catalog number: 163-2094). If the protein sample can be lyophilized, then the lyophilized sample can be dissolved into a sample solution with 8 M urea and 0.5% Bio-lyte.
2. The solution is first added to a rehydration tray, the IPG strip is taken out from the storage, and the plastic cover is peeled off. The side with the dried gel surface is facing down to make

contact with the sample solution in the rehydration tray. Please make sure to let the whole IPG strip making full contact to the sample solution. Add enough mineral oil to cover the IPG strip to prevent evaporation and rehydrate the sample overnight at room temperature (sometimes, a 6 h rehydration is enough for the IEF, this is especially important if the enzyme of interest is not stable at room temperature).

3. After rehydration, the IPG strip is taken out from the rehydration tray and the attached mineral oil is carefully removed with a Kim wipe paper.
4. In the IEF tray, carefully wet two pieces of Electrode Wick (Bio-Rad, catalog. No. 165-4071) and put on the metal wire in one lane. Carefully lay down the IPG strip face down, and gently push the IPG strip so that it can make a close contact to the filter paper-covered metal wire. Add enough mineral oil to cover the IPG strip to prevent the evaporation.
5. Put the IEF tray cover to the try followed by close the IEF unit cover (Bio-Read Protean IEF Unit).
6. In the first step, set the voltage gradient from 0 to 8000 V for 4 h; in the second step, set at constant voltage at 8000 for 24 h. The gel will actually run overnight, but the minimum voltage-hours are 30,000 for a good 2-D separation.
7. After the IEF is completed, turn off the unit, and carefully take out the IPG strip and use Kim wiper paper to remove the mineral oil from the IPG strip. Put the IPG strip into a rehydration tray and incubate in Tris-Glycine transfer buffer supplied in this kit, incubate for 10 min to remove the urea, and allow the SDS to bind to the proteins (if the enzyme is sensitive to SDS, the incubation can be carried out in Tris-Glycine only, this will also produce acceptable protein resolution).
8. Take out a Bio-Rad Criterion gel and remove the plastic comb, use Milli-Q water to rinse the flat well. Put the gel into the running unit and fill both the lower and upper tank with Tris-Glycine-SDS buffer (if the enzyme is sensitive to SDS, only Tris-Glycine buffer will be used).
9. Carefully lay down the IPG strip in the IPG well with the acidic side always on the left side when facing the gel. Load 5 μL of unstained protein standard in the protein standard well (the well next to the acidic end of the IPG strip).
10. Run first at 80 voltages for 15 min followed by 120 voltages until the dye front from the protein standard is about 0.5 cm from the bottom of the gel (it is important to run 80 voltages for 15 min to allow as much of the proteins in the IEF gel getting into the second dimensional gel).

3.3 Protein Transfer After the 2-D Gel

1. While the SDS-PAGE is still running, put the PEP plate in a tray and add 50 μL of the protein recovery solution to each well of the plate with a multiple channel pipette, there will be some overflow of the solution during this step, it is fine. If an eight channel pipette is used, the solution could be dispensed every other row. For example, in the first round, add solutions to row A, C, E, and so on; in the second round, add solution in the row B, D, F, and so on. Cover the tray to minimize evaporation.
2. When the dye front is about 0.5 cm from the bottom of the gel, stop the running and carefully take out the gel from the gel cassette and rinse with Milli-Q water followed by adding 200 mL of the transfer buffer (supplied with the kit) in the tray. Wet four pieces of the transfer filter paper (supplied with the kit) completely in a different tray and lay down two pieces on the metal plate of the Semi-Dry Trans-Blot (Bio-Rad or similar Semi-Dry Trans-Blot from other manufacturers).
3. Lay the PEP plate on top of the filter papers followed by carefully lay the gel on top of the PEP plate and make sure the upper left corner of the gel align with the upper left corner of the PEP plate.
4. Lay another two pieces of transfer filter paper on top of the gel to form a sandwich (from the bottom it should be filter papers, PEP plate, gel and filter papers again).
5. Cover the sandwich assembly with the other metal plate of the Semi-Dry Trans-Blot, and transfer the proteins with constant voltage at 20 voltages for 60 min. It is shown that under this condition, the proteins in the gel will be efficiently transferred into the PEP plate, longer protein transfer is not recommended.
6. While the gel is transferring, condition the 384-well deep-well plate by adding 100 μL PBS in each well (if protein kinase or protein phosphatase assays are performed, a phosphate-free buffer such as Tris–HCl should be used to minimize interference from the buffer). This treatment will improve the protein recovery in later steps for enzyme activity analysis and mass spectrometry protein identification. After 30 min treatment, completely empty the solution from each well and refill the well with 50 μL PBS (for protein kinase assay or any other assay where phosphate is interfering, Tris–HCl buffer or other buffer of choice could be used).
7. When the protein transfer is completed, turn off the power, take off the Semi-Dry Trans-Blot cover and release the top metal plate. Wait for 10 s before lifting the top metal plate (this is important to let some air in so that the solutions in the PEP plate will not be sucked out to cause proteins in one well

over flow to adjacent wells). After removing the metal plate, carefully lift the two pieces of filter paper followed by remove the gel (sometimes the filter papers and gel will stick together, in this case lift both parts together). When removing the gel, it is important to remove it from left to right, it should be point out that the specific composition of the PEP transfer buffer will reduce the protein diffusion. Carefully take the PEP plate without the two transfer paper on the bottom of the PEP plate and put in a tray.

8. Use multiple channel pipettes to transfer the recovered protein solution from the PEP plate to the deep-well Master Plate in the corresponding columns. If using an eight channel pipette, set the transfer volume at 45 μL to make sure most of the solution in the well is transferred. The transfer will start at column 1 from the left side of the PEP plate, and the wells with odd numbers (row A, C, E, and so on) is first transferred followed by transfer of the wells from the even number wells in the first column (rows B, D, F, and so on). Repeat the process until all the samples from the PEP plate are transferred to the 384-well Master Plate.

3.4 Glycolytic Enzyme Analysis

1. After the transfer of samples from the PEP plate to the deep-well Master Plate, the Master Plate should be used immediately (preferred) for glycolytic enzyme analysis. Multiple enzymes can be analyzed from samples collected since the total volume in each well of the Master Plate is about 90 μL (50 μL buffer plus 40–45 μL sample transferred from the PEP plate). Refer to Subheading 2.2 for details of the glycolytic enzyme assay protocol. Use a spectrophotometer to measure the glycolytic enzyme activities at 340 nm. Before the assay readout, some wells of the enzyme assay plate may contain bubbles because of the SDS in the protein transfer buffer (one technique to avoid bubble is to set the dispensing volume smaller than the aspirating volume so that the pipette will not cause bubbles when dispensing). Use a transfer pipette tip to pinch the bubbles to get rid of them before the reading; this will reduce the interference from the bubble.
2. When reading the enzyme assay plate, use pipette to remove the solutions from the well P24 (lower right corner well in the 384-well plate) and use this well as blank for the reading. It is recommended to read at least 3 data points such as 0, 60, and 120 min and save the reading in separate files.

3.5 Data Transformation and Analysis

1. Export the data set from the three readings (0, 60, and 120 min) to an Excel file (if not already in this format).
2. In Microsoft Excel, subtract the 60 min readings from each well from the corresponding readings at 0 min to obtain the

data set for the 340 nm absorbance difference which reflect the glycolytic enzyme activities from the serum proteome. Use Excel Heat Map to display the enzyme activity in a 384-well table or use Insert function, and select the 3-D display to build the graph of this data set in 3-D display.

3. Subtract the 120 min readings from each well from the corresponding readings at 0 min to obtain the data set for the 340 nm absorbance difference which reflect the glycolytic enzyme activities from the serum proteome for the second data set.

3.6 Protein Purity Confirmation (Optional)

1. If the enzyme testing showed that some wells have the enzyme activity of interest, the next step is to test the purity of the protein in that well. Collect all the samples from the wells with enzyme activity in a siliconized microcentrifuge tube, dry down the solution, and resuspend into 20 μL of Milli-Q water. Take 10 μL and mixed with 10 μL of SDS sample buffer (this sample buffer is a 2× SDS-PAGE sample buffer with 20 mM DTT), incubate at 37 °C for 60 min.
2. Load on a SDS-PAGE gel and run the gel as in Subheading 2 of this protocol.
3. Fix the gel in a gel-fixing solution (10% each of ethanol and acetic acid in Milli-Q water) for at least 2 h.
4. Rinse with distilled water and stain the gel in Sypro Ruby or other fluorescence dye overnight.
5. The next day, remove the staining solution; wash the gel twice with distilled water followed by incubation in the distilled water for 5 min with moderate shaking.
6. Take the gel image with a CCD camera such as the Bio-Rad ChemiDoc.
7. Save the image in tiff file for later image processing. The gel image will tell whether the protein is pure or not.

3.7 Mass Spectrometry to Identify the Protein of Interest

1. If the gel staining in Subheading 3.6 shows that the fraction with enzyme activity is pure, the 10 μL Milli-Q water resuspended sample in Subheading 3.6, **step 1** can be submitted for mass spectrometry analysis (sometimes fraction with more than one protein bands can be submitted for MS analysis, and the identity of the protein can be assigned by bioinformatics effort based on protein homology, it is unlikely that more than one protein from the preparation share the same type of enzyme activity, for example, GAPDH).
2. Alternatively, if there is enough protein to be seen in Subheading 3.6, **step 7** with the fluorescence staining, the protein band can be excised and sent for MS analysis.

4 Notes

1. Allow diluted reagents and buffers to reach room temperature (18–25 °C) prior to starting the assay. Once the assay has been started, all steps should be completed in sequence and without interruption. Make sure that required reagents and buffers are ready when needed. Prior to adding to the plate, reagents should be mixed gently (not vortexed) by swirling.
2. Avoid contamination of reagents, pipette tips and wells. Use new disposable tips and reservoirs, do not return unused reagent to the stock bottles/vials, and do not mix caps of stock solutions.
3. For some enzymes, 8 M urea might be too strong for its enzymatic activity; in this case, 3 M urea and 2% CHAPS will be used in the IEF gel. If the presence of SDS also inactivates the enzyme, the second dimension separation can use non-denaturing gels. The gel resolution needs to be tested for the modified conditions before PEP elution and enzyme assay.
4. Sometimes the IEF does not have to run as high as 8000 Voltage as was suggested by the manufacturer, it has been found that setting the highest voltage at 5000 can also achieve good protein separation.

Acknowledgments

I would like to thank Array Bridge Inc. for the supply of the PEP Universal Protein Purification kits and the opportunity to carry out this research in its laboratory. I would also like to thank Dr. Liang Li for providing the breast cancer and normal people sera in this research.

References

1. Dos Anjos Pultz B et al (2014) Far beyond the usual biomarkers in breast cancer: a review. J Cancer 5(7):13
2. Li J et al (2002) Proteomics and bioinformatics approaches for identification of serum biomarkers to detect breast cancer. Clin Chem 48(8):9
3. Chan MK, Cooper JD, Bahn S (2015) Commercialisation of biomarker tests for mental illnesses: advances and obstacles. Trends Biotechnol 33(12):12
4. Chung L et al (2014) Novel serum protein biomarker panel revealed by mass spectrometry and its prognostic value in breast cancer. Breast Cancer Res 16:R63
5. Henderson MC et al (2016) Integration of serum protein biomarker and tumor associated autoantibody expression data increases the ability of a blood-based proteomic assay to identify breast cancer. PLoS One:11(8)
6. Ingvarsson J et al (2007) Design of recombinant antibody microarrays for serum protein profiling: targeting of complement proteins. J Proteome Res 6:10
7. Lee JS, Magbanua MJM, Park JW (2016) Circulating tumor cells in breast cancer: applications in personalized medicine. Breast Cancer Res Treat 160:411–424

8. Mehan MR et al (2014) Validation of a blood protein signature for non-small cell lung cancer. BMC Clin Proteomics 11(32):12
9. Ross JS et al (2003) Breast cancer biomarkers and molecular medicine. Expert Rev Mol Diagn 3(5):13
10. Ross JS et al (2004) Breast cancer biomarkers and molecular medicine: part II. Expert Rev Mol Diagn 4(2):20
11. Surinova S et al (2015) Prediction of colorectal cancer diagnosis based on circulating plasma proteins. EMBO Mol Med 7(9):13
12. Yezhelyev MV et al (2007) In situ molecular profiling of breast cancer biomarkers with multicolor quantum dots. Adv Mater 19:6
13. Kirmiz C et al (2007) A serum glycomics approach to breast cancer biomarkers. Mol Cell Proteomics 6:13
14. Harsha HC et al (2009) A compendium of potential biomarkers of pancreatic cancer. PLoS Med 6(6):6
15. Kaskas NM et al (2014) Serum biomarkers in head and neck squamous cell cancer. JAMA 140(1):7
16. Wang C-H et al (2015) Current trends and recent advances in diagnosis, therapy and prevention of hepatocellular carcinoma. Asian Pac J Cancer Prev 16(9):10
17. Alexander H et al (2004) Proteomic analysis to identify breast cancer biomarkers in nipple aspirate fluid. Clin Cancer Res 10:11
18. Ma S et al (2016) Multiplexed serum biomarkers for the detection of lung cancer. EBio Med 11:9
19. Evens MJ, Cravatt BF (2006) Mechanism-based profiling of enzyme families. Chem Rev 106:23
20. Wang DL et al (2015) Identification of multiple metabolic enzymes from mice cochleae tissue using a novel functional proteomics technology. PLoS One 10:e0121826
21. Wang DL et al (2017) Identification of potential serum biomarkers for breast cancer using a functional proteomics technology. Biomark Res 5:11
22. Sun Z et al (2016) Identification of functional metabolic biomarkers from lung cancer patient serum using PEP technology. Biomark Res 4:11
23. Sun Z, Yang P (2004) Role of imbalance between neutrophil elastase and a1-antitrypsin in cancer development and progression. Lancet Oncol 5:9
24. Wang X et al (2015) Bead based proteome enrichment enhances features of the protein elution plate (PEP) for functional proteomic profiling. Proteomes 3:13
25. Amorim M et al (2016) Decoding the usefulness of non-coding RNAs as breast cancer markers. J Transl Med 14:15
26. Mabert K et al (2014) Cancer biomarker discovery: current status and future perspectives. Int J Radiat Biol 90(8):18
27. Surinova S et al (2015) Non-invasive prognostic protein biomarker signatures associated with colorectal cancer. EMBO Mol Med 7:13
28. Orla T et al (2011) Metabolic signatures of malignant progression in prostate epithelial cells. Int J Biochem Cell Biol 43:8
29. Teicher BA, Marston WL, Helman LJ (2013) Targeting cancer metabolism. Clin Cancer Res 18(20):9
30. Araujo EP, Carvalheira JB, Velloso LA (2006) Disruption of metabolic pathways—perspectives for the treatment of cancer. Curr Cancer Drug Targets 6:77–87
31. Bryksin AV, Laktionov PP (2008) Role of glyceraldehyde-3-phosphate dehydrogenase in vesicular transport from golgi apparatus to endoplasmic reticulum. Biochemistry 73:7
32. Cairns RA, Harris IS, Mak TW (2011) Regulation of cancer cell metabolism. Nat Rev Cancer 11:11
33. Chaneton B, Gottlieb E (2012) Rocking cell metabolism: revised functions of the key glycolytic regulator PKM2 in cancer. Trends Biochem Sci 37(8):7
34. Chang C-H et al (2015) Metabolic competition in the tumor microenvironment is a driver of cancer progression. Cell 162:13
35. Chiaradonna FR et al (2012) From cancer metabolism to new biomarker and drug targets. Biotechnol Adv 30:30–51
36. Favaro E et al (2012) Glucose utilization via glycogen phosphorylase sustains proliferation and prevents premature senescence in cancer cells. Cell Metab 16:14
37. Ledford H (2014) Metabolic quirks yield tumour hope. Nature 508:2
38. Anderson NL, Anderson NG (2002) The human plasma proteome: history, character, and diagnostic prospects. Mol Cell Proteomics 1:23

Chapter 5

A Protein Decomplexation Strategy in Snake Venom Proteomics

Choo Hock Tan, Kae Yi Tan, and Nget Hong Tan

Abstract

Snake venoms are complex mixtures of proteins and peptides that play vital roles in the survival of venomous snakes. As with their diverse pharmacological activities, snake venoms can be highly variable, hence the importance of understanding the compositional details of different snake venoms. However, profiling venom protein mixtures is challenging, in particular when dealing with the diversity of protein subtypes and their abundances. Here we described an optimized strategy combining a protein decomplexation method with in-solution trypsin digestion and mass spectrometry of snake venom proteins. The approach involves the integrated use of C_{18} reverse-phase high-performance liquid chromatography (RP-HPLC), sodium dodecyl sulfate-polyacrylamide gel electrophoresis (SDS-PAGE), and nano-electrospray ionization tandem mass spectrometry (nano-ESI-LC-MS/MS).

Key words Snake venom, Protein decomplexation, Venom separation, Reverse-phase high-performance liquid chromatography, Tandem mass spectrometry, Venomics

1 Introduction

The advent of proteomics has greatly facilitated the investigation of venom protein composition in a high-throughput and comprehensive manner. For a decade, the term "venomics" has been used with increasing popularity to denote venom-related "-omics" studies including snake venom proteomics [1, 2]. Prior to the venomic era, bioassay-guided protein purification was the main platform available to identify and characterize proteins in a snake venom, but this method was akin to finding pieces of puzzle at a time and complete protein profiling of the snake venom was hardly possible. The application of proteomics and bioinformatics has now enabled the study of the global profiling of venom proteins in great details, even for components that exist in a very low amount [3, 4]. This revolutionizing breakthrough by venomics has propelled the growth of knowledge tremendously on the various aspects of snake venom research including venom evolution, envenomation

Xing Wang and Matthew Kuruc (eds.), *Functional Proteomics: Methods and Protocols*, Methods in Molecular Biology, vol. 1871, https://doi.org/10.1007/978-1-4939-8814-3_5,

pathophysiology, antivenom production, and in toxin-based drug discovery [5–7].

However, snake venoms are complex mixtures of proteins and peptides which are inherently variable [8, 9]. The depth of proteomic findings for a venom also tends to vary with different experimental protocols, equipment or techniques used, thus posing a challenge when one intends to collate and compare global venomic data. To obtain as much useful proteomic information from a venom, it is therefore important that the protocol provides good resolution of the proteins [10]. This can be achieved through a protein separation method prior to the mass spectrometric analysis, as widely adapted in many venomic studies [11–13]. Protein separation is typically achieved by either a gel-based method such as SDS-PAGE or two-dimensional gel electrophoresis (separation by protein differences in isoelectric point and molecular mass), or liquid chromatography using various columns (separation by protein differences in ionic charges, hydrophobicity or molecular mass) [14–18]. Often, the chromatographic separation method, in particular the use of C_{18} reverse-phase column, is preferred over the gel-based method for better protein resolution and advantage in the estimation of protein abundance based on peak areas (the area under the curve) [13, 19]. Venom proteins bind to the reverse-phase column (stationary phase) through hydrophobic interaction, and in general the more hydrophobic proteins will bind stronger to the C_{18} beads in the column. The mobile phase is composed of an aqueous blend of water with a miscible, polar organic solvent, e.g., acetonitrile, delivered under high pressure. The flow of the mobile phase elutes the venom proteins following a stepwise increase of acetonitrile concentration over an extended time. Proteins are collected into different fractions as they are eluted, and the proteins can be visualized on SDS-PAGE. The protein fractions are then subjected to liquid chromatography tandem mass spectrometry and data mining for protein identification and proteome construction. Thus far, we have reported a number of quantitative snake venom proteomics based on this approach and have found that the results provide good functional correlation and insights into the complexity of snake venoms and toxins.

2 Materials

2.1 Snake Venom Samples

Freeze-dry snake venom samples and store at −20 °C until use.

2.1.1 Reverse-Phase High-Performance Liquid Chromatography (RP-HPLC)

Stationary phase: Reverse-phase HPLC column LiChroCART® 250-4 LiChrospher® WP 300 (Merck, USA), or any equivalent column.

Mobile phase: **Equilibrium buffer** (Eluent A): 0.1% trifluoroacetic acid (TFA) in HPLC grade water. Add 1 mL TFA into 999 mL of HPLC grade water. **Elution buffer** (Eluent B): 0.1% trifluoroacetic acid (TFA) in HPLC grade acetonitrile (ACN). Add 1 mL TFA into 999 mL of HPLC grade ACN.

2.1.2 Sodium Dodecyl Sulfate-Polyacrylamide Gel Electrophoresis (SDS-PAGE)

30% (weight/volume) Acrylamide/Bis-acrylamide (29.2%:0.8%) solution: Weigh 29.2 g of Acrylamide monomer and 0.8 g of Bis-acrylamide (cross-linker) and transfer to a 100 mL Scott bottle containing 50 mL of ddH_2O. Add a magnetic stirring bar (20 × 6 mm) and allow the mixture to mix for 10 minutes. Make up the solution to 100 mL with ddH_2O. Store at 4 °C.

Sodium dodecyl sulfate (SDS)-containing resolving gel buffer: 1.5 M Tris–HCl, pH 8.8. Weigh 181.7 g Tris–HCl and 4.0 g of SDS. Transfer to a 1 L Scott bottle and add ddH_2O to a volume of 900 mL. Mix and adjust the pH with HCl. Make up the solution to 1 L with ddH_2O. Store at 4 °C (*see* **Note 1**).

SDS-containing stacking gel buffer: 0.5 M Tris–HCl, pH 6.8. Weigh 60.6 g Tris–HCl and 4.0 g of SDS. Prepare a 1 L solution as described in the previous step. Store at 4 °C (*see* **Note 1**).

10% (w/v) ammonium persulfate (APS): Weigh 30 mg of APS and transfer to a 1.5 mL centrifuge tube. Add 300 μL ddH_2O into the tube and dissolve it completely (Freshly prepared).

N,N,N,N′-Tetramethyl-ethylenediamine (TEMED): Store at 4 °C.

Electrophoresis buffer: 0.025 M Tris–HCl, pH 8.3, 0.192 M glycine, 0.1% SDS.

Sample loading buffer (1×): 62 mM Tris–HCl (pH 6.8), 2.3% (w/v) SDS, 5% (w/v) beta-mercaptoethanol, 0.005% (w/v) bromophenol blue, 10% (w/v) glycerol.

Gel staining and fixing solution: 0.2% (w/v) Commassie blue R-250, 40% (v/v) methanol, 10% (v/v) acetic acid in ddH_2O.

Gel destaining solution: 5% (v/v) methanol, 7% (v/v) acetic acid in ddH_2O.

2.1.3 Protein Digestion (In-Solution Tryptic Digestion)

Trypsin stock (0.1 μg/μL): Add 200 μL of ddH_2O to 20 μg lyophilized trypsin with 1 mM HCl.

Digestion buffer: 50 mM ammonium bicarbonate.

Reducing buffer: 100 mM dithiothreitol (DTT).

Alkylation buffer: 100 mM iodoacetamide (IAA).

2.1.4 Peptides Extraction and Desalting

Materials:

Millipore ZipTip® C_{18} Pipette Tips were supplied by Merck (USA).

Solution:

Wetting solution: 50% acetonitrile (ACN).

Equilibrium/wash solution: 0.1% formic acid (FA).

Elution solution: 0.1% FA in 50% ACN.

3 Methods

3.1 Protein Separation

3.1.1 Reverse-Phase High-Performance Liquid Chromatography (RP-HPLC)—Shimadzu LC-20AD HPLC System (Japan)

1. *System equilibration:* Attach a C_{18} column (LiChroCART® 250-4 LiChrospher® WP 300) to the HPLC system. Equilibrate the C_{18} column with Eluent B for 30 min, followed by Eluent A for 30 min.
2. *Sample preparation:* Weigh 2 mg of lyophilized venom and transfer into a 1.5 mL centrifuge tube. Add 200 μL of 0.1% TFA into the tube and centrifuge at 10,000 × *g* for 12 min at 4 °C. Transfer the supernatant to a new centrifuge tube.
3. *Sample fractionation:* Inject 200 μL supernatant into the injection loop (at loading position). Elute the venom sample with a linear gradient of 5% Eluent B for 10 min, 5–15% Eluent B over 20 min, 15–45% Eluent B over 120 min, and 45–70% Eluent B over 20 min (*see* **Note 2**). Monitor the venom protein elution by UV absorbance at 215 nm. The fractionation is conducted at room temperature (20–24 °C).
4. *Fractions collection:* Collect all protein fractions (according to absorbance measurement) manually. Freeze-dry all fractions obtained and stored at −20 °C until use.

**Figure 1 shows a typical C_{18} RP-HPLC profile of a cobra venom under the above experimental conditions.*

3.2 Protein Visualization

3.2.1 15% Sodium Dodecyl Sulfate-Polyacrylamide Gel Electrophoresis (SDS-PAGE)

1. Mix 4.5 mL of SDS-containing resolving gel buffer, 3.0 mL of acrylamide mixture, and 1.5 mL of ddH_2O in a 15-mL centrifuge tube. Add 100 μL of 10% APS and 10-μL of TEMED right before gel casting. Cast the gel immediately within a 7.25 cm × 10 cm × 1.5 mm gel cassette (*see* **Note 3**). Layer the stacking gel with isopropanol (*see* **Note 4**).
2. Prepare stacking gel by mixing 1.25 mL of SDS-containing stacking gel buffer, 0.7 mL of acrylamide mixture, and 3.05 mL of ddH_2O in a 15-mL centrifuge tube. Add 80 μL of 10% APS and 8 μL of TEMED right before the gel casting. Insert a 10-well gel comb immediately without introducing air bubbles.
3. Reconstitute the lyophilized fractions collected from RP-HPLC (Subheading 3.1.1, **step 4**) with ddH_2O and determine the protein concentration of each fraction with

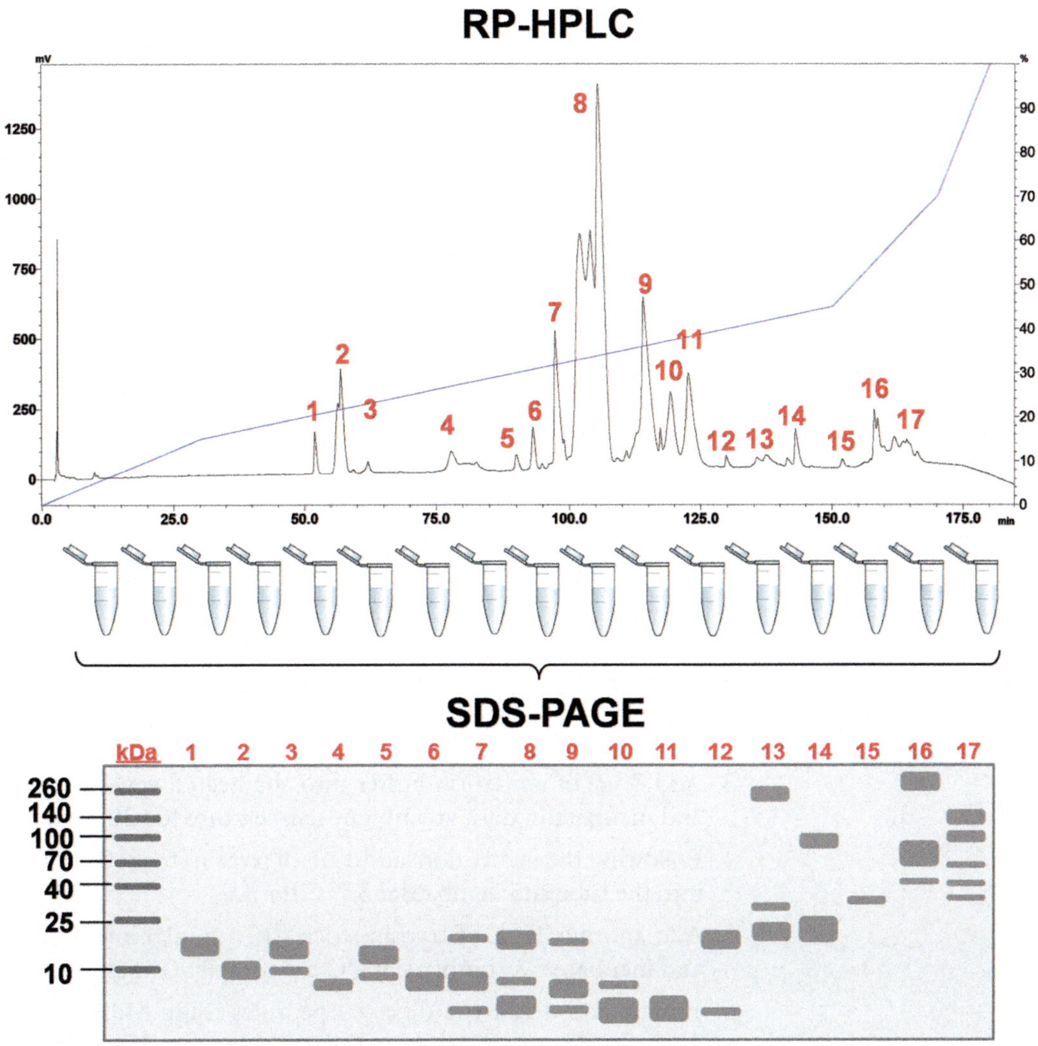

Fig. 1 Reverse-phase HPLC fractionation of snake venom using LiChrospher® WP 300 C_{18} column (upper panel) following the chromatographic condition: 5% B for 10 min, 5–15% B for 20 min, followed by 15–45% B for 120 min and 45–70% B for 20 min. The chromatographic fractions are collected manually at 215 nm absorbance and the lyophilized fractions are further electrophoresed on SDS-PAGE (lower panel, under reducing conditions). Protein marker is used for molecular weight calibration. The protein bands are visualized by Coomassie blue staining

NanoDrop Spectrophotometer (Thermo Scientific, USA). Add sample loading buffer to venom fractions (5–50 μg) in a one-to-one volume ratio (keep the total volume less than 20 μL). Place the mixture in boiling water for 10 min and cool the sample to ambient temperature. Centrifuge the sample at 6000 × *g* for 30 s to bring down the condensate.

4. Remove the gel comb prior to sample loading. Load protein marker on the left side of the gel and heated samples in the subsequent wells. Electrophorese the samples at 90 V until the front dye reaches the bottom of the gel.
5. Following electrophoresis, remove the gel cassette from the electrophoresis system. Pry the gel plates with the gel remover and rinse the gel with ddH_2O. Transfer the gel carefully to a container and stain with Coomassie Brilliant Blue R-250 staining solution for 15 min. Destain the gel with destaining solution until the gel background is clear and scan the gel with a gel scanner.

**A schematic drawing of the SDS-PAGE of the protein fractions was shown in Fig. 1.*

3.3 Protein Identification

3.3.1 Protein Digestion (In-Solution Tryptic Digestion) and Peptides Extraction

1. Venom fractions from RP-HPLC (Subheading 3.1.1, **step 4**) are reconstituted in ddH_2O. Aliquot approximately 5 μg venom proteins (estimated using Nano-Drop Spectrophotometer) in 10 μL from each reconstituted fraction, and add into a 1.5 mL centrifuge tube, respectively.
2. Mix 15 μL of digestion buffer and 1.5 μL of reducing buffer to the centrifuge tube and heat the mixture at 95 °C for 5 min. Cool the sample to ambient temperature.
3. Add 3 μL of alkylation buffer into the heated venom samples and incubate in dark at ambient temperature for 20 min.
4. Following the incubation, add 1 μL of trypsin stock (0.1 μg/μL) into the tube and incubate at 37 °C for 3 h.
5. Add another 1 μL of trypsin stock (0.1 μg/μL) into the tube and incubate overnight at 30 °C for complete digestion.
6. Extract and desalt the digested peptides using Millipore ZipTip® C_{18} Pipette Tips (Merck, USA). Aspirate 10 μL wetting solution using ZipTip for three times, followed by aspirating the equilibrium solution for three times. Next, aspirate and dispense the digested samples using equilibrated ZipTip for ten times to allow binding of peptides onto the C_{18} resins of ZipTip. Wash the peptide-bound ZipTip with washing solution (aspirating and dispensing for three times) to remove salt content.
7. Lastly, elute the peptides from the C_{18} resins of ZipTip by aspirating and dispensing for ten times in a new centrifuge tube containing 10 μL of elution solution. Lyophilize the extracted and desalted peptides and store at −20 °C. These tryptic peptides will be subjected to mass spectrometry analysis.

**The workflows are illustrated in Fig. 2.*

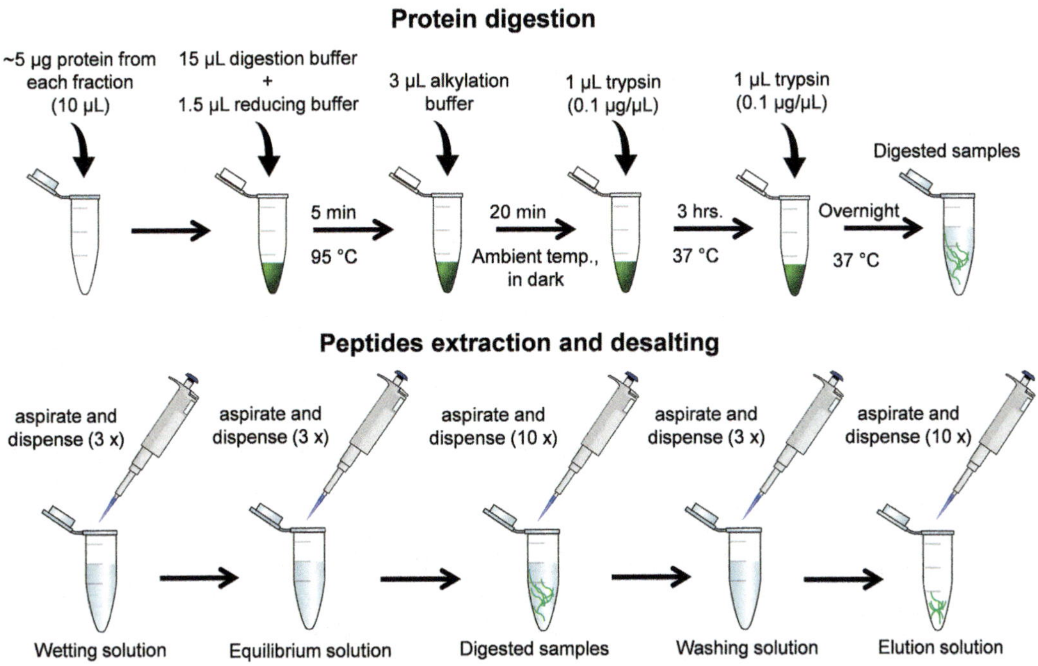

Fig. 2 Step-by-step workflows for protein digestion (upper panel) and peptides extraction and desalting (lower panel) protocol

3.3.2 Nano-Electrospray Ionization-Liquid Chromatography Tandem Mass Spectrometry (ESI-LC-MS/MS) and Data Mining

1. Perform the detection analysis using 1260 Infinity Nanoflow LC system (Agilent, Santa Clara, CA, USA) that is connected to Accurate-Mass Q-TOF 6550 series with a nano-electrospray ionization source.
2. Reconstitute the lyophilized peptide analytes in 7 µL of 0.1% formic acid in ddH_2O. Subject the peptide analytes to HPLC Large-Capacity Chip Column Zorbax 300-SB-C18 (160 nL enrichment column, a 75 µm × 150 mm analytical column with 5 µm particles) (Agilent, Santa Clara, CA, USA).
3. Adjust the injection volume to 1 µL per sample, using a flow rate of 0.4 µL/min, with a linear gradient of 5–70% of solvent B (0.1% formic acid in 100% acetonitrile).
4. Flow the drying gas at a rate of 11 L/min at temperature 290 °C. Set the fragmentor voltage at 175 V and the capillary voltage at 1800 V. Acquire the mass spectra using Mass Hunter acquisition software (Agilent, Santa Clara, CA, USA) in a MS/MS mode with an MS scan range of 200–3000 m/z and MS/MS scan range of 50–3200 m/z.
5. Extract the data with MH^+ mass range between 50 and 3200 Da and process with Agilent Spectrum Mill MS Proteomics Workbench software packages version B.04.00 against merged database incorporating both non-redundant NCBI

database of Serpentes (taxid: 8570) and in-house transcript database (*see* **Note 5**).

6. Specify the fixed modification to carbamidomethylation and oxidized methionine as a variable modification.
7. Validate the identified proteins or peptides with the following filters: protein score > 20, peptide score > 10, and scored peak intensity (SPI) > 70%.
8. Filter the identified proteins to achieve false discovery rate (FDR) < 1% for the peptide-spectrum matches.
9. Consider all results showing 2 or greater than 2 "distinct peptide" for identification.

3.4 Protein Quantitation

3.4.1 Relative Abundance and Venom Protein Quantitation

1. Estimate the relative abundance of protein in venom fractions by the peak area measurement using Shimadzu LCsolution Software Version 1.23 (Shimadzu, Japan).
2. Implement the relative abundance (%) obtained from peak area measurement (area under curve) to all collected fractions that show protein band(s) in SDS-PAGE.
3. Estimate the relative abundance (%) of each individual venom protein in a fraction based on its mean spectral intensity (MSI) relative to the total MSI of all proteins in the fraction identified through ESI-LC-MS/MS (Subheading 3.3.2).

$$\text{Relative abundance of a protein in an HPLC fraction (\%)} = \frac{\text{Mean spectral intensity of protein in a fraction}}{\text{Total spectral intensity of a fraction}} \times 100\%$$

4. Estimate the relative abundance of individual protein in a venom by multiplying the % area under curve with the relative abundance obtained based on mean spectral intensity (**step 3**).

$$\text{Relative abundance of a protein in a venom (\%)} = \text{\% AUC of a fraction} \times \text{Relative abundance of a protein in a fraction (\%)}$$

5. Accumulate the relative abundance of protein (%) according to the protein identity and family for the characterization of venom proteome.

4 Notes

1. SDS tends to precipitate in cold (below 15 °C). SDS-containing buffers may need to be warmed prior to use.
2. The linear gradient stated in above Subheading 3.1 is an optimized protocol to fractionate venoms of elapid snakes such as

cobras (*Naja* sp.). Readers are advised to adjust and optimize the elution protocol for venom samples from snakes other than cobra (*Naja* sp.).

3. The percentage of polyacrylamide gel prepared depends on the target protein(s) to be visualized. High percentage gel (15–18%) is suitable for separating the low molecular weight proteins (< 20 kDa), whereas low percentage gel provides a better separation for the higher molecular weight proteins. In general, the proteins for most snake venoms can be separated and visualized on a 15% gel.
4. 4.5 mL of resolving solution mixture is required in preparation of a resolving gel. We found that isopropanol works better than water in layering the resolving gel. Tilt the gel casting holder slightly (20° up and down) before leaving the gel to solidify for an even distribution of gel level.
5. The in-house transcript database is created using data obtained from venom-gland transcriptomic study. This use of a transcript database is optional in venomic studies, but can be incorporated with up-to-date non-redundant NCBI dataset of Serpentes (taxid: 8570) to provide a more complete database for mass spectra matching in protein identification.

References

1. Lomonte B, Fernández J, Sanz L, Angulo Y, Sasa M, Gutiérrez JM, Calvete JJ (2014) Venomous snakes of Costa Rica: biological and medical implications of their venom proteomic profiles analyzed through the strategy of snake venomics. J Proteome 105(Supplement C):323–339. https://doi.org/10.1016/j.jprot.2014.02.020
2. Calvete JJ, Sanz L, Angulo Y, Lomonte B, Gutiérrez JM (2009) Venoms, venomics, antivenomics. FEBS Lett 583(11):1736–1743. https://doi.org/10.1016/j.febslet.2009.03.029
3. Tan CH, Tan KY, Lim SE, Tan NH (2015) Venomics of the beaked sea snake, *Hydrophis schistosus*: a minimalist toxin arsenal and its cross-neutralization by heterologous antivenoms. J Proteome 126:121–130. https://doi.org/10.1016/j.jprot.2015.05.035
4. Tan KY, Tan NH, Tan CH (2018) Venom proteomics and antivenom neutralization for the Chinese eastern Russell's viper, *Daboia siamensis* from Guangxi and Taiwan. Sci Rep 8(1):8545. https://doi.org/10.1038/s41598-018-25955-y
5. Tan KY, Tan CH, Fung SY, Tan NH (2015) Venomics, lethality and neutralization of *Naja kaouthia* (monocled cobra) venoms from three different geographical regions of Southeast Asia. J Proteome 120:105–125. https://doi.org/10.1016/j.jprot.2015.02.012
6. Gutiérrez JM, Lomonte B, León G, Alape-Girón A, Flores-Díaz M, Sanz L, Angulo Y, Calvete JJ (2009) Snake venomics and antivenomics: proteomic tools in the design and control of antivenoms for the treatment of snakebite envenoming. J Proteome 72(2):165–182. https://doi.org/10.1016/j.jprot.2009.01.008
7. Vetter I, Davis JL, Rash LD, Anangi R, Mobli M, Alewood PF, Lewis RJ, King GF (2011) Venomics: a new paradigm for natural products-based drug discovery. Amino Acids 40(1):15–28. https://doi.org/10.1007/s00726-010-0516-4
8. Tan KY, Tan CH, Chanhome L, Tan NH (2017) Comparative venom gland transcriptomics of *Naja kaouthia* (monocled cobra) from Malaysia and Thailand: elucidating geographical venom variation and insights into sequence novelty. PeerJ 5:e3142. https://doi.org/10.7717/peerj.3142
9. Augusto-de-Oliveira C, Stuginski DR, Kitano ES, Andrade-Silva D, Liberato T, Fukushima I,

Serrano SM, Zelanis A (2016) Dynamic rearrangement in snake venom gland proteome: insights into *Bothrops jararaca* intraspecific venom variation. J Proteome Res 15 (10):3752–3762. https://doi.org/10.1021/acs.jproteome.6b00561

10. Calvete JJ (2014) Next-generation snake venomics: protein-locus resolution through venom proteome decomplexation. Expert Rev Proteomics 11(3):315–329. https://doi.org/10.1586/14789450.2014.900447
11. Tan CH, Wong KY, Tan KY, Tan NH (2017) Venom proteome of the yellow-lipped sea krait, *Laticauda colubrina* from Bali: insights into subvenomic diversity, venom antigenicity and cross-neutralization by antivenom. J Proteome 166:48–58. https://doi.org/10.1016/j.jprot.2017.07.002
12. Alape-Giron A, Sanz L, Escolano J, Flores-Diaz M, Madrigal M, Sasa M, Calvete JJ (2008) Snake venomics of the lancehead pitviper *Bothrops asper*: geographic, individual, and ontogenetic variations. J Proteome Res 7 (8):3556–3571. https://doi.org/10.1021/pr800332p
13. Wong KY, Tan CH, Tan KY, Naeem QH, Tan NH (2018) Elucidating the biogeographical variation of the venom of *Naja naja* (spectacled cobra) from Pakistan through a venom-decomplexing proteomic study. J Proteome 175:156–173. https://doi.org/10.1016/j.jprot.2017.12.012
14. Faisal T, Tan KY, Sim SM, Quraishi N, Tan NH, Tan CH (2018) Proteomics, functional characterization and antivenom neutralization of the venom of Pakistani Russell's viper (*Daboia russelii*) from the wild. J Proteome 183:1–13. https://doi.org/10.1016/j.jprot.2018.05.003
15. Petras D, Sanz L, Segura A, Herrera M, Villalta M, Solano D, Vargas M, Leon G, Warrell DA, Theakston RD, Harrison RA, Durfa N, Nasidi A, Gutierrez JM, Calvete JJ (2011) Snake venomics of African spitting cobras: toxin composition and assessment of congeneric cross-reactivity of the pan-African EchiTAb-Plus-ICP antivenom by antivenomics and neutralization approaches. J Proteome Res 10(3):1266–1280. https://doi.org/10.1021/pr101040f
16. Tan NH, Fung SY, Tan KY, Yap MKK, Gnanathasan CA, Tan CH (2015) Functional venomics of the Sri Lankan Russell's viper (*Daboia russelii*) and its toxinological correlations. J Proteome 128:403–423. https://doi.org/10.1016/j.jprot.2015.08.017
17. Tan CH, Fung SY, Yap MK, Leong PK, Liew JL, Tan NH (2016) Unveiling the elusive and exotic: Venomics of the Malayan blue coral snake (*Calliophis bivirgata flaviceps*). J Proteome 132:1–12. https://doi.org/10.1016/j.jprot.2015.11.014
18. Dutta S, Chanda A, Kalita B, Islam T, Patra A, Mukherjee AK (2017) Proteomic analysis to unravel the complex venom proteome of eastern India *Naja naja*: correlation of venom composition with its biochemical and pharmacological properties. J Proteome 156:29–39. https://doi.org/10.1016/j.jprot.2016.12.018
19. Tan CH, Tan KY, Yap MK, Tan NH (2017) Venomics of *Tropidolaemus wagleri*, the sexually dimorphic temple pit viper: unveiling a deeply conserved atypical toxin arsenal. Sci Rep 7:43237. https://doi.org/10.1038/srep43237

Chapter 6

Fractionation Techniques to Increase Plant Proteome Coverage: Combining Separation in Parallel at the Protein and the Peptide Level

Martin Černý, Miroslav Berka, and Hana Habánová

Abstract

Peptide spectral libraries enable targeted identification and quantitation of low-abundance proteins in a complex plant proteome. Here we describe parallel protein and peptide fractionation techniques to improve plant proteome coverage and facilitate construction of spectral libraries.

Key words Plant proteomics, Protein fractionation, Peptide fractionation, C18, SCX, PEG

1 Introduction

Proteins may form up to 20% of total cellular weight, and rough estimates predict that this corresponds to a range of two to four million proteins per cubic micron [1]. However, most of these proteins belong to only a few, highly abundant, protein families and the difference in concentration between a low-abundance protein and a highly abundant protein within a single cell can easily be five to six orders of magnitude [2]. The dynamic concentration range is further expanded in multicellular organisms. For instance, the average human body consists of ca 37 trillion cells which can be grouped into at least 200 different cell types. Proteome complexity is further increased by posttranslational modifications. This overall complexity represents a significant obstacle to proteome analyses and even the rapid development in mass spectrometry instruments that we have seen in recent years cannot address all these issues. For this reason, proteome fractionation is the best approach if a reasonable level of proteome coverage is to be achieved. However, fractionation requires a relatively large amount of starting material, which is not always readily available, and the methods are time-consuming and constitute a limitation for quantitative analyses. This problem can be circumvented by the targeted methods

Xing Wang and Matthew Kuruc (eds.), *Functional Proteomics: Methods and Protocols*, Methods in Molecular Biology, vol. 1871, https://doi.org/10.1007/978-1-4939-8814-3_6, © Springer Science+Business Media, LLC, part of Springer Nature 2019

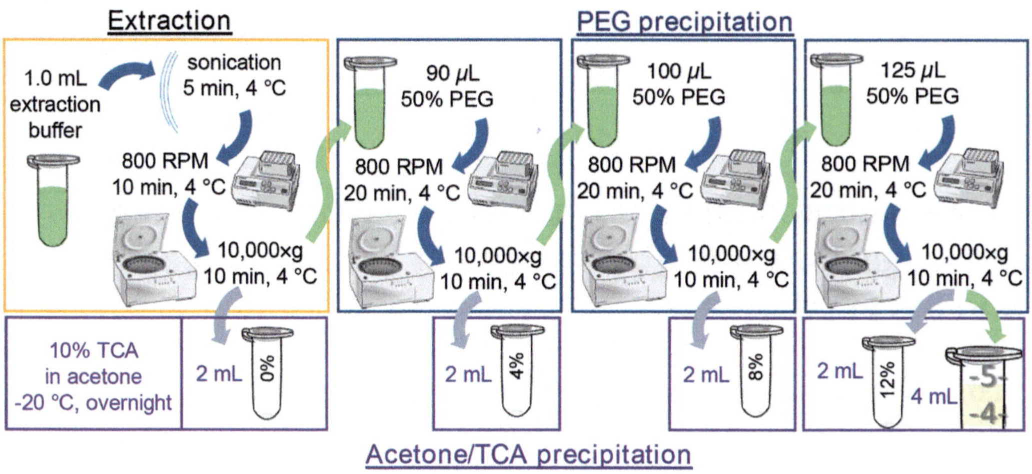

Fig. 1 Native extraction and PEG fractionation

selected/multiple reaction monitoring (SRM/MRM) and/or sequential window acquisition of all theoretical spectra (SWATH). Both these techniques improve detection limits but require the availability of a reference peptide spectral library [3]. Here, we present a protocol with which to obtain data for building such a library. This protocol employs protein precipitation and parallel fractionations at the protein (Figs. 1 and 2) and peptide levels (Fig. 3): it comprises inexpensive nondenaturing polyethylene glycol (PEG) fractionation [4, 5], protein precipitation by low pH and acetone followed by phenol re-extraction [6], mass-based separation on sodium dodecyl sulfate polyacrylamide gel electrophoresis (SDS-PAGE) and charge-based separation by isoelectric focusing [7], high pH C18 peptide fractionation [8], and strong cation exchange (SCX) peptide fractionation [9, 10]. Although this workflow is optimized for ca 1 g of starting material, the amount can be reduced and the protocol adapted for a smaller-scale experiment. In addition, it can be combined with fractionations at the tissue level, subcellular enrichment and techniques that improve the detection of lower abundance proteins by means of immunodepletion of abundant proteins or proteome equalization [11–13].

2 Materials

Always wear laboratory gloves for self-protection and to prevent sample contamination. Prepare solutions using ultrapure solvents, preferably of LC-MS grade quality.

2.1 Homogenization

1. Mixer Mill MM 400 (Retsch), stainless steel grinding jars and milling balls (*see* **Note 1**).

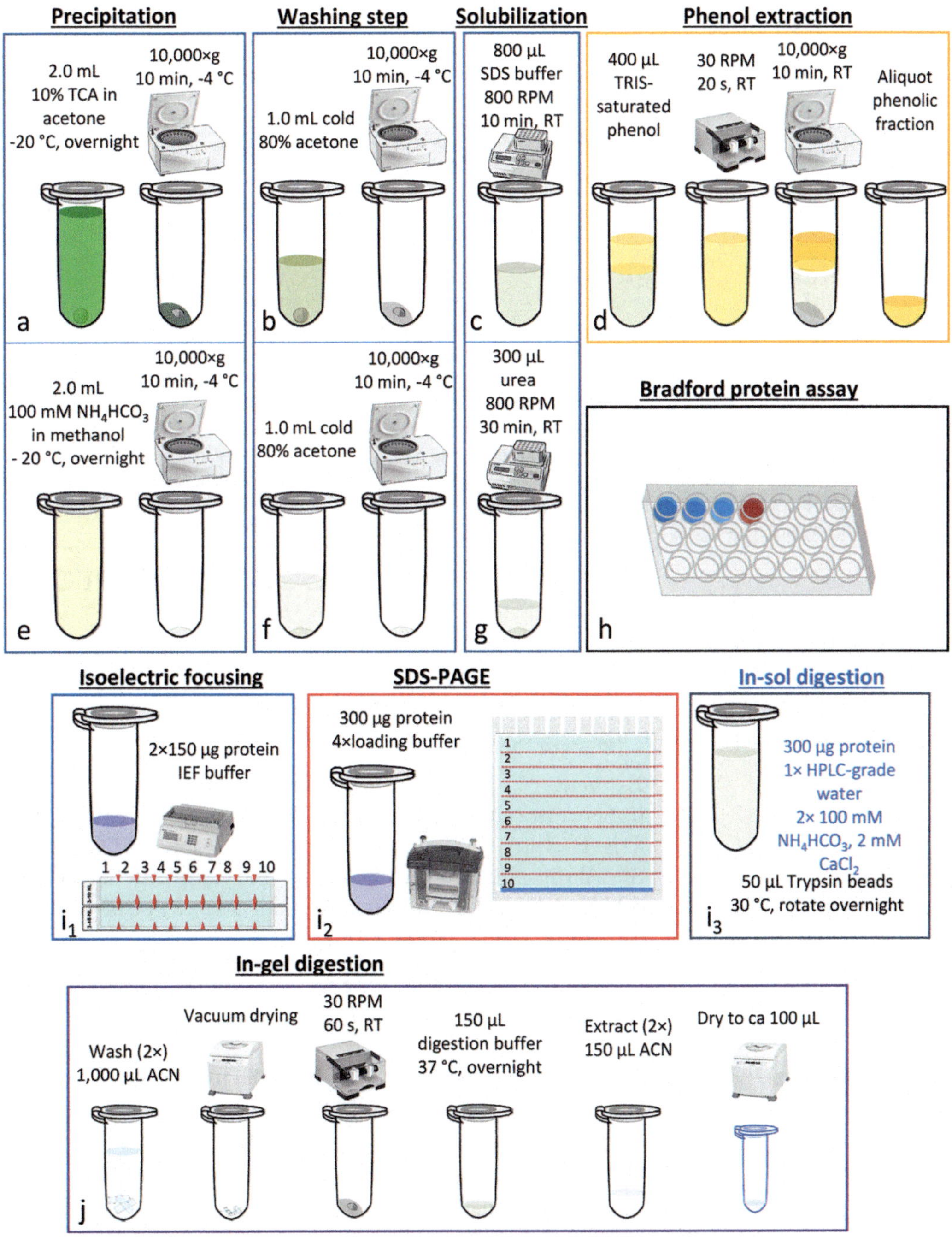

Fig. 2 Denaturing protein extraction and protein digestion. (**a–g**) Protein extraction and purification. (**h**) Determination of protein concentration, (**i_{1-2}**) protein separation, and (**i_3**, **j**) digestion

2. Liquid nitrogen.
3. 2.0 mL Eppendorf LoBind tubes or similarly coated low-protein-binding microcentrifuge tubes.

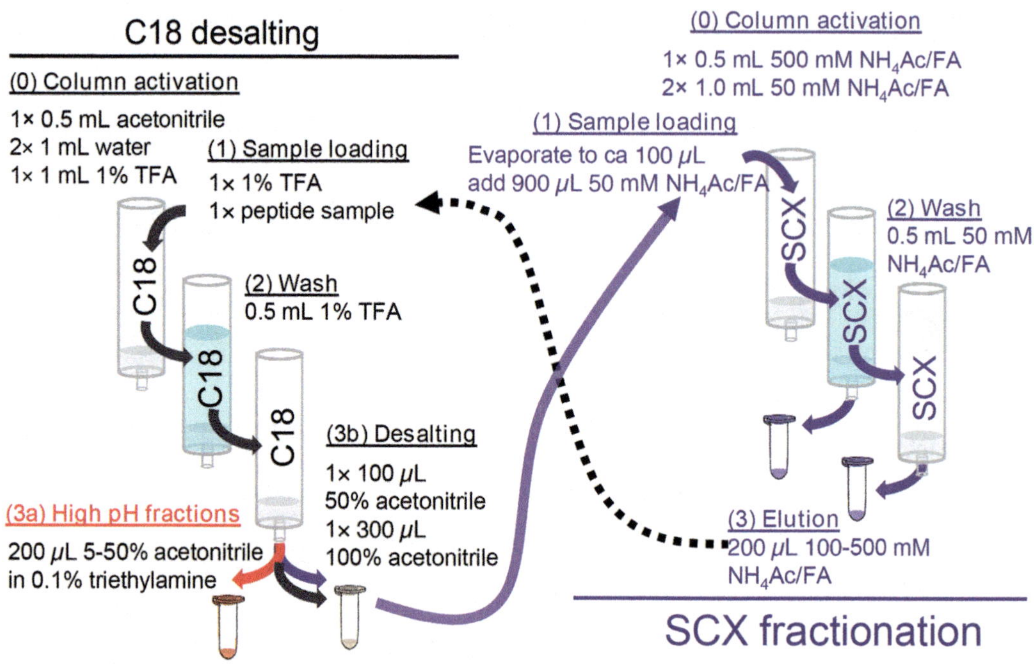

Fig. 3 Peptide desalting and fractionation.

2.2 Native Extraction and PEG Fractionation

1. Eppendorf Thermomixer R (*see* **Note 2**).
2. Ultrasonic bath.
3. Extraction buffer: 20 mM $MgCl_2$, 1% (v/v) β-mercaptoethanol, 1 mM EDTA, 2% (v/v) IGEPAL, 0.5 M Tris–HCl, pH 7.8. Prepare 50 mL and store at 4 °C (*see* **Note 3**). Aliquot into 1 mL portions and supplement each with 50 μL of Protease Inhibitor Cocktail (Merck) prior to use.
4. Polyethylene glycol 4000 (PEG).
5. 2.0 mL and 5.0 mL Eppendorf LoBind tubes or similarly coated low-protein-binding microcentrifuge tubes.

2.3 Denaturing Acetone/TCA/Phenol Extraction

1. 10% (w/v) Trichloroacetic acid (TCA) in acetone. Prepare 1 L and store at −20 °C (*see* **Note 4**).
2. 80% (v/v) acetone in water. Prepare 250 mL and store at −20 °C.
3. SDS buffer: 2% (w/v) SDS, 30% (w/v) sucrose, 5% (v/v) β-mercaptoethanol, 5 mM ethylenediaminetetraacetic acid (EDTA), 100 mM Tris–HCl, pH 8.0. Prepare 50 mL and keep at 4 °C; shelf life is more than a month.
4. TE-saturated phenol—phenol saturated with 10 mM Tris–HCl buffer pH 8.0 and 1 mM EDTA.
5. Thermomixer.

6. Retsch mill with an adapter for 2.0 mL tubes.
7. 100 mM ammonium acetate in methanol; prepare 500 mL and store at −20 °C.
8. 100 mM ammonium bicarbonate, 8 M urea in water; prepare 100 mL, keep at 4 °C (*see* **Note 5**).
9. IEF solubilization solvent: 7 M urea, 2 M thiourea in water, 2% (w/v) CHAPS, 90 mM dithiothreitol. Prepare 50 mL, aliquot into 5 mL tubes, and store at −20 °C (*see* **Notes 3** and **5**).
10. Bradford Reagent (Merck), bovine serum albumin standard, a 96-well microplate, and a microplate reader (*see* **Note 6**).

2.4 In-Solution Digestion

1. Ammonium bicarbonate buffer: 50 mM NH_4HCO_3, 2 mM $CaCl_2$, 8% (v/v) acetonitrile. Prepare 100 mL, keep at 4 °C.
2. Vertical rotator, incubator.
3. Immobilized Trypsin (Promega) (*see* **Notes** 7 and **8**).

2.5 Protein Separation and In-Gel Digestion

2.5.1 Isoelectric Focusing

1. IPG strips: 7 cm ReadyStrips with immobilized nonlinear pH gradient 3-10 (Bio-Rad).
2. Ampholytes pH 3-10 (Bio-Rad).
3. PROTEAN IEF Cell unit and focusing tray (Bio-Rad).
4. Paper wicks: electrode wicks suitable for isoelectric focusing.
5. Mineral oil.
6. Scalpel.
7. 1.5 mL Eppendorf LoBind tubes or similarly coated low-protein-binding microcentrifuge tubes.

2.5.2 SDS-PAGE

1. Mini-PROTEAN cell and power supply (Bio-Rad).
2. Precast Mini-PROTEAN TGX gel, 4–20%, 10 wells, 50 μL (*see* **Note 9**).
3. Running buffer: 25 mM Tris–HCl, 192 mM glycine, 0.1% SDS, pH 8.3. Prepare 1 L using 3 g of Tris (base), 14.4 g of glycine, and 1 g of SDS, do not adjust the pH; store at 4 °C.
4. 4 × Loading buffer: 10% (w/v) SDS, 20% glycerol, 10 mM dithiothreitol, 0.05% (w/v) bromophenol blue, 200 mM Tris–HCl, pH 6.8. Prepare 20 mL, aliquot into 1.5 mL tubes, and store at −20 °C.
5. Thermomixer.
6. Scalpel.
7. 1.5 mL Eppendorf LoBind tubes or similarly coated low-protein-binding microcentrifuge tubes.

2.5.3 In-Gel Digestion

1. Acetonitrile.
2. SpeedVac Evaporator (Thermo Scientific).
3. Retsch mill with an adapter for 2.0 mL tubes and milling balls.
4. Digestion buffer: Dissolve 20 μg of a sequencing-grade trypsin (e.g., Promega) in 3.0 mL of ammonium bicarbonate buffer (Subheading 2.4). Prepare on ice and use immediately for protein digestion. This amount is sufficient for 20 samples obtained from isoelectric focusing and SDS-PAGE separations.
5. Thin-walled 0.5 mL PCR tubes.

2.6 Peptide Desalting

1. 1% TFA: 1% (v/v) trifluoroacetic acid (TFA) in water. Prepare 200 mL, keep at room temperature (*see* **Note 4**).
2. 50% (v/v) acetonitrile in water. Prepare 15 mL, cover with aluminum foil or store in the dark at room temperature.
3. VersaPlate, collection plate and C18 (25 mg) tubes (Agilent), vacuum pump (*see* **Note 10**).
4. Thin-walled 0.5 mL PCR tubes.
5. SpeedVac Evaporator (Thermo Scientific).
6. Quantitative Colorimetric Peptide Assay kit (Thermo Scientific), a 96-well microplate and a microplate reader.

2.7 Peptide Fractionation

1. VersaPlate, collection plate, C18 (25 mg) and SCX (50 mg) tubes (Agilent), vacuum pump (*see* **Note 10**).
2. 1% TFA (prepared in Subheading 2.6).
3. Concentration series of acetonitrile in 0.1% (v/v) triethylamine: Prepare 2 mL of 0.1% (v/v) triethylamine in water and 2 mL of 0.1% (v/v) triethylamine in acetonitrile. Mix 50, 75, 100, 125, 150, 175, 200, and 250 μL of triethylamine in acetonitrile with the appropriate volume of triethylamine in water to obtain 1 mL of each stock solution (5–50%).
4. 0.5% formic acid: Dilute formic acid (FA) with water to produce 15 mL of 0.5% (v/v) FA, store at room temperature.
5. Concentration series of ammonium acetate in 0.5% FA: Prepare ca 1.5 mL of 500 mM ammonium acetate in 0.5% FA (dissolve 38.5 mg of ammonium acetate per 1 mL of 0.5% FA), and then dilute this stock by mixing 50, 100, 250, 300, and 400 μL with the necessary amounts of 0.5% FA to obtain 0.5 mL volumes of 50–400 mM stock solutions.

3 Methods

3.1 Homogenization

1. Homogenize ca 1 g of plant tissue using liquid nitrogen and an MM400 Retsch mill with a prechilled stainless steel grinding jar and milling balls (*see* **Note 1**).

2. Mill at 30 Hz for 60 s or until a fine powder is produced; keep the jars cold to prevent samples from melting. Wear protective gloves and use appropriate equipment for handling liquid nitrogen.
3. Aliquot ca 250 mg samples into 2.0 mL Eppendorf LoBind tubes and store the aliquots at −80 °C.

3.2 Native Extraction and PEG Fractionation

1. Take a homogenized aliquot, place it on ice and add 1.0 mL of extraction buffer.
2. Sonicate at 4 °C for 5 min and then incubate in the Thermomixer at 4 °C, 800 rpm for 10 min.
3. Centrifuge for 10 min (10,000 × *g*, 4 °C), transfer the supernatant to a new 2.0 mL tube, and keep it on ice. Mix the pellet with 10% (w/v) TCA in acetone and then follow **steps 2–9** of the denaturing extraction protocol (Subheading 3.3.).
4. Add 90 μL of 50% (w/v) PEG solution to give a final concentration of ca 4% (w/v), incubate in the Thermomixer (4 °C, 800 rpm, 20 min), then centrifuge for 10 min (10,000 × *g*, 4 °C), collect the supernatant, and process the pellet as in the previous step.
5. Repeat the supernatant precipitation in a stepwise PEG concentration gradient, adding 100 and 125 μL of 50% (w/v) PEG to give, respectively, a 8 and a 12% (w/v) mixture. Collect pellets, mix with 10% (w/v) TCA in acetone, and transfer the last supernatant into a 5.0 mL tube and precipitate it with 4.0 mL of 10% (w/v) TCA in acetone, then proceed to the second step of the denaturing extraction (Subheading 3.3).

3.3 Denaturing Acetone/TCA/Phenol Extraction

1. Resuspend an aliquot of frozen homogenized tissue with 10% (w/v) TCA in acetone (fill to 2.0 mL), adding a stainless steel milling ball to facilitate sample solubilization.
2. Precipitate the total protein overnight at −20 °C (*see* **Note 11**).
3. Centrifuge the sample for 10 min (10,000 × *g*, 4 °C) to bring down the precipitate.
4. Wash the pellet with 80% (v/v) acetone, centrifuge again at 10,000 × *g* for 10 min, and then resuspend in 0.8 mL of SDS buffer. Incubate in the Thermomixer at 800 rpm and room temperature for 10 min.
5. Remove the milling ball, add 400 μL of TE-saturated phenol, and shake in a Retsch mill for 20 s at 30 rpm (*see* **Note 1**).
6. Centrifuge the mixture for 10 min (10,000 × *g*, 20 °C) and aliquot the top (phenolic) layer into three 2.0 mL LoBind tubes.
7. Precipitate overnight in ice-cold 100 mM ammonium acetate in methanol (−20 °C).

8. Centrifuge samples for 10 min to collect protein pellets (10,000 × *g*, 4 °C), wash pellets with 1.0 mL 80% (v/v) acetone in water, remove all solvent, and dry on air for 5 min.
9. Dissolve protein pellets in the Thermomixer with 300 μL (i) 100 mM ammonium bicarbonate, 8 M urea for in-solution digestion, or (ii, iii) IEF solubilization solvent for isoelectric focusing and SDS-PAGE. Incubate at 30 °C, 800 rpm for 30 min (*see* **Note 5**) and estimate protein concentration by means of a Sigma-Aldrich Bradford assay in microplate format. The isolation method should yield at least 500 μg of protein per aliquot (*see* **Notes 12** and **13**).

3.4 In-Solution Digestion

1. Dilute 300 μg of protein with an equal volume of water and two volumes of ammonium bicarbonate buffer.
2. Add 50 μL of immobilized trypsin beads (Promega, *see* **Note 7**) and incubate overnight on a rotator at 30 rpm in an incubator at 30 °C (*see* **Note 8**).

3.5 Protein Separation and In-Gel Digestion

3.5.1 Isoelectric Focusing

1. Dilute 300 μg of protein to a final volume of 260 μL with IEF solubilization solvent (if needed), add 1.3 μL of ampholytes (pH 3–10), and load onto two 7 cm 3–10 NL IPG strips (Bio-Rad) in a rehydration tray.
2. Overlay with mineral oil and rehydrate overnight at room temperature.
3. Wet four paper wicks in water, transfer IPG strips into the focusing tray, and put wet paper wicks between gel and electrode to prevent their making direct contact. Overlay again with mineral oil.
4. Isoelectrically focus proteins at 20 °C in a PROTEAN IEF Cell unit (Bio-Rad) in six steps: 150 V (20 min), 300 V (20 min), 600 V (20 min), 1500 V (20 min), 3000 V (20 min), and 4000 V up to 12,000 Vh.
5. Put the IPG strips onto a clean filter paper with the gel facing up to dry off the mineral oil.
6. Align one IPG strip above the other, cut gels vertically into ten equal fractions, and collect them in 1.5 mL LoBind tubes.

3.5.2 SDS-PAGE

1. Assemble the Mini-PROTEAN cell employing a Mini-PROTEAN TGX precast gel (4–20%, 10 wells, 50 μL) and add 700 mL of running buffer (*see* **Note 9**).
2. Mix 300 μg of protein with 4× loading buffer (3:1), incubate at 95 °C for 10 min (Thermomixer), spin down (1000 × *g*), and load into wells (30 μg per well).
3. Connect the Mini-PROTEAN to its power supply and separate proteins with the following settings: 100 V (10 min) followed

by 150 V (30 min). The bromophenol blue line should be ca 1 cm above the end of the gel.

4. Disconnect the electrophoresis apparatus and carefully open the gel cassette. Remove empty parts of the gel with a clean scalpel blade, cut gel horizontally into ten equal pieces. Cut each fraction into smaller pieces and collect it into a 1.5 LoBind tube.

3.5.3 In-Gel Digestion

1. Wash gel pieces twice with 1.0 mL acetonitrile and dry samples down in a SpeedVac Evaporator.
2. Use a Retsch mill and stainless steel milling balls to obtain a fine powder (*see* **Note 14**). Place samples on ice, add 150 μL digestion buffer, incubate for 15 min, then transfer tubes to 37 °C and incubate overnight.
3. Extract peptides twice with 150 μL acetonitrile, collect extracts into 0.5 mL thin-walled PCR tubes, and dry down to ca 100 μL in a SpeedVac Evaporator (*see* **Note 15**).

3.6 Peptide Desalting

1. Mix a peptide sample 1:1 with 1% TFA, shake, and clarify the sample by centrifugation (10,000 × *g*, 5 min).
2. Wash a C18 SPE column with 0.5 mL 100% acetonitrile, 2 × 1 mL of water, 1 × 1 mL of 1% TFA.
3. Load the acidified sample solution onto the SPE column (*see* **Note 16**).
4. Wash the column twice with 0.5 mL 1% TFA.
5. Elute peptides in two steps: 100 μL 50% (v/v) acetonitrile in water, followed by 300 μL of acetonitrile; collect the flow-through into 0.5 mL thin-walled PCR tubes, dry samples down in a SpeedVac Evaporator to ca 40 μL (*see* **Note 15**), and estimate peptide concentration by Quantitative Colorimetric Peptide Assay (Thermo Scientific).

3.7 Peptide Fractionation

1. Mix the peptide sample from in-solution digestion 1:1 with 1% TFA, shake, and clarify the sample by centrifugation (10,000 × *g*, 5 min).
2. Wash two C18 SPE columns with 0.5 mL of 100% acetonitrile, 2 × 1.0 mL of water, 1× with 1.0 mL of 1% TFA.
3. Divide the acidified sample equally between two SPE columns, wash once with 1% TFA and then proceed to high pH or SCX fractionation.

3.7.1 High pH C18 Fractionation

1. Wash bound peptides with 1.0 mL of water (*see* **Note 16**).
2. Elute peptides with a stepwise gradient of acetonitrile in 0.1% triethylamine. Load, successively, 200 μL each of 5%, 7.5%,

10%, 12.5%, 15%, 17.5%, 20%, 25%, and 50% acetonitrile, and collect fractions into 0.5 mL thin-walled PCR tubes.

3. Dry down to ca 20–30 μL in a SpeedVac Evaporator (*see* **Note 15**) and estimate peptide concentration by Quantitative Colorimetric Peptide Assay (Thermo Scientific).

3.7.2 Peptide SCX Fractionation

1. Elute bound peptides with 100 μL of 50% (v/v) acetonitrile in water followed by 300 μL of acetonitrile, collect fractions into 0.5 mL thin-walled PCR tubes.
2. Dry down to ca 100 μL in a SpeedVac Evaporator (*see* **Note 15**) and dilute in 1.0 mL of 50 mM ammonium acetate in 0.5% FA.
3. Wash an SCX SPE column with, successively, 0.5 mL of 500 mM ammonium acetate in 0.5% FA and 2 × 1.0 mL of 50 mM ammonium acetate, then load the peptide sample and collect the flow-through into a 0.5 mL thin-walled PCR tube (the first fraction).
4. Elute peptides with a stepwise gradient of ammonium acetate in 0.5% FA. Load, successively, 200 μL each of 100, 250, 300, 400, and 500 mM ammonium acetate, and collect fractions into 1.5 mL tubes (*see* **Note 17**).
5. Dilute the collected eluates with 800 μL of 0.1% (v/v) FA in water; desalt peptides on a C18 SPE column as described in Subheading 3.6.

4 Notes

1. The Retsch mill employed in our protocol for homogenization and phenol extraction can be substituted with a standard mortar and pestle, and a vortex mixer, respectively.
2. We recommend working in a cold room or placing the Thermomixer in a refrigerator.
3. You may wish to consider replacing the foul-smelling β-mercaptoethanol with 20 mM dithiothreitol (DTT), but this solution must be prepared fresh or stored in aliquots at −20 °C (the half-life of DTT at pH 8.0 and room temperature is only hours [14]).
4. TCA and TFA are strong acids; wear appropriate protection.
5. The dissolution of urea is endothermic, so prepare urea directly in a closed 100 mL flask and use a magnetic stirrer at room temperature. Avoid heating; any temperature above 30 °C increases urea decomposition, produces cyanate, and results in protein carbamylation. Prepare this solution every 4 weeks or store aliquots at −20 °C.

6. Pipette 2 μL of each of the protein samples and corresponding blanks (ammonium bicarbonate and IEF buffers) and then rapidly overlay with 200 μL of the Bradford Reagent. This will provide sufficient mixing and the assay can be carried out immediately with reasonable precision and reproducibility.
7. The immobilized trypsin can be substituted with a standard sequencing-grade trypsin, but the immobilized form is less prone to self-cleavage than its counterpart and does not require the user to work on ice. However, take care to mix the slurry of beads properly in order to obtain a homogeneous mixture for pipetting.
8. Some samples (e.g., seed storage tissues) contain protein inhibitors of proteases that may interfere with in-solution digestion. If the peptide yield is lower than expected, increase the trypsin:protein ratio and consider predigestion with Lys-C or a modification of the digestion buffer (e.g., an increase in the acetonitrile concentration). The digestion step can be speeded up by ultrasound, microwave, and heating treatment; however, the buffer contains urea and the sample is likely to suffer from nonenzymatic modification by carbamylation.
9. Precast gels are convenient but the available gradient range may not be sufficient for your protein extract. You may consider using a prestained protein ladder and running a test PAGE in order to determine the optimal gel gradient, and/or readjusting the positions of the lines when slicing the gel. We do not recommend staining of the material for digestion, as the stain would then need to be removed and washing the gel could result in depletion of smaller proteins.
10. The vacuum manifold that we use for the VersaPlate solid phase extraction can be substituted with a pipette, e.g., an Eppendorf 1 mL pipette fits well on the SPE tubes.
11. Precipitation overnight is not mandatory but a shorter time will have a negative effect on the protein yield. However, a longer storage time will not decrease the yield, and we have seen excellent results with samples stored for more than 6 months at −20 °C. Note that some nonenzymatic posttranslational modifications may still occur and affect the quality of the protein sample.
12. Based on our experiments, the expected protein yield from 250 mg fresh weight is >1500 μg for plantlet, seedling and leaf tissue, at least 1000 μg for root tissue and >2500 for seed extracts (species such as *Arabidopsis thaliana*, *Solanum lycopersicum*, *Nicotiana tabacum*, *Hordeum vulgare*, *Pisum sativum*, and *Quercus robur* [6, 7, 12]).
13. Note that our routine protocol does not include cysteine alkylation and the resulting proteome library is thus missing most

cysteine-containing tryptic peptides. The thiol side chains of cysteine residues are highly susceptible to posttranslational modifications and we prefer to avoid these in a quantitative analysis. Cysteine alkylation can be included prior to the digestion step. For in-gel digestion, incubate homogenized gels in 0.5 mL of 100 mM DTT in IEF buffer for 30 min, 800 rpm, room temperature, then centrifuge, wash pellets with 0.5 mL acetonitrile, and resuspend the pellets in 0.5 mL of 100 mM iodoacetamide in IEF buffer. Incubate at 800 rpm for 30 min in the dark (iodoacetamide is light-sensitive). Centrifuge, wash with acetonitrile, and dry on the SpeedVac evaporator. Alkylation for in-solution digestion: Add DTT from a 250 mM aqueous stock solution to the protein dissolved in urea/ammonium bicarbonate to a final concentration of 10 mM DTT and incubate for 30 min at room temperature, 800 rpm. Add iodoacetamide (250 mM in water) to a final concentration of 30 mM; incubate in the dark for 30 min at 800 rpm. Quench iodoacetamide with a further addition of DTT, using the same volume as for iodoacetamide.

14. Do not add milling balls to only partially dried gel pieces, as they will stick and the milling will be ineffective. Milling of fully dried gel pieces is rapid and does not lead to excessive heat production. In our experience, it significantly improves peptide recovery rates and facilitates uniform distribution of trypsin in the digest.

15. Volatile acetonitrile will evaporate and its concentration will not interfere with C18 binding. Monitor the evaporation and once the liquid volume reaches ca 100 μL (in-gel digests) or ca 40 μL (desalting), remove the sample from the SpeedVac Evaporator. Try to avoid drying of the samples, as this limits peptide recovery due to peptide aggregation and peptide-surface interaction [15]. If you are unsuccessful and the samples are fully dried, reconstitute in 4% (v/v) acetonitrile in water by sonication and carefully wash the surface of the tube. Thin-walled PCR tubes are more suitable as these will improve sonication efficiency.

16. Smaller peptides and hydrophilic peptides (e.g., some phosphopeptides and glycopeptides) will be lost in this step. To improve sample recovery, you may employ, e.g., graphite columns [16]. However, these peptides will not be retained by C18 during the LC-MS step, as they will all elute in the first minutes and are not usually suitable for quantitation due to the ion suppression effect.

17. The desalting step can be replaced by evaporation, which will remove ammonium acetate. However, in our experience the C18 method is faster and more reliable.

Acknowledgments

This work was supported by the Ministry of Education, Youth and Sports of the Czech Republic under the project CEITEC 2020 (LQ1601) and TE02000177 (TACR), and by Brno PhD Talent 2017 (funded by Brno City Municipality) and IGA grant no. IP 15/2017 to H.H.

References

1. Milo R (2013) What is the total number of protein molecules per cell volume? A call to rethink some published values. BioEssays 35:1050–1055
2. Picotti P, Bodenmiller B, Mueller LN, Domon B, Aebersold R (2009) Full dynamic range proteome analysis of S. cerevisiae by targeted proteomics. Cell 138:795–806
3. Schubert OT, Gillet LC, Collins BC, Navarro P, Rosenberger G, Wolski WE et al (2015) Building high-quality assay libraries for targeted analysis of SWATH MS data. Nat Protoc 10(3):426–441
4. Acquadro A, Flavo S, Mila S, Albo AG, Comino C, Moglia A, Lanteri S (2009) Proteomics in globe artichoke: protein extraction and sample complexity reduction by PEG fractionation. Electrophoresis 30(9):1594–1602
5. Wang W-Q, Song B-Y, Deng Z-J, Wang Y, Liu S-J, Møller IM, Song S-Q (2015) Proteomic analysis of lettuce seed germination and thermoinhibition by sampling of individual seeds at germination and removal of storage proteins by polyethylene glycol fractionation. Plant Physiol 167(4):1332–1350
6. Cerna H, Černý M, Habánová H, Šafářová D, Abushamsiya K, Navrátil M, Brzobohatý B (2017) Proteomics offers insight to the mechanism behind Pisum sativum L. response to Pea seed-borne mosaic virus (PSbMV). J Proteomics 153:78–88
7. Baldrianová J, Černý M, Novák J, Jedelský PL, Divíšková E, Brzobohatý B (2015) Arabidopsis proteome responses to the smoke-derived growth regulator karrikin. J Proteomics 120:7–20
8. Batth TS, Francavilla C, Olsen JV (2014) Off-line high-pH reversed-phase fractionation for in-depth phosphoproteomics. J Proteome Res 13:6176–6186
9. Rappsilber J, Mann M, Ishihama Y (2007) Protocol for micro-purification, enrichment, pre-fractionation and storage of peptides for proteomics using StageTips. Nat Protoc 2:1896–1906
10. Mostovenko E, Hassan C, Rattke J, Deelder AM, van Veelen PA, Palmblad M (2013) Comparison of peptide and protein fractionation methods in proteomics. EuPA Open Proteom 1:30–37
11. Černý M, Skalák J, Kurková B, Babuliaková E, Brzobohatý BB (2011) Using a commercial method for rubisco immunodepletion in analysis of plant proteome. Chemické listy 105:640–642
12. Černý M, Jedelský PL, Novák J, Schlosser A, Brzobohatý B (2014) Cytokinin modulates proteomic, transcriptomic and growth responses to temperature shocks in Arabidopsis. Plant Cell Environ 37:1641–1655
13. Righetti PG, Boschetti E (2016) Global proteome analysis in plants by means of peptide libraries and applications. J Proteomics 143:3–14
14. Stevens R, Stevens L, Price N (1983) The stabilities of various thiol compounds used in protein purifications. Biochem Educ 11:70
15. Berka M, Luklová M (2017) Limited drying and its effect on peptide recovery rates. In: Polak O et al (eds) MendelNet 2017 Proceedings of 24th International PhD Students Conference. 24th International PhD Students Conference, Brno, November 2017. p 91
16. Nukarinen E, Tomanov K, Ziba I, Weckwerth W, Bachmair A (2017) Protein sumoylation and phosphorylation intersect in Arabidopsis signaling. Plant J 91:505–517

Chapter 7

A Systematic Analysis Workflow for High-Density Customized Protein Microarrays in Biomarker Screening

Rodrigo García-Valiente, Jonatan Fernández-García, Javier Carabias-Sánchez, Alicia Landeira-Viñuela, Rafael Góngora, María Gonzalez-Gonzalez, and Manuel Fuentes

Abstract

High-density protein microarrays constitute a promising high-throughput platform for the characterization of protein expression patterns, biomarker discovery, and validation. Different types of protein microarrays have been described according to several features (such as content, format, and detection system) presenting advantages and disadvantages which are relevant for the specific application and purposes. Therefore, an experimental design is key for any screening based on protein microarrays assays; in fact, the data analysis strategy is directly related to the experimental design, type of protein microarray and consequently the final outcome, the data and results interpretation, is also directly linked. Here, it is proposed a systematic workflow for biomarker discovery based on tailor-made protein microarrays platforms which obtain comprehensively info for the functional protein characterization in high-throughput format.

Key words Protein microarray, Analysis, Proteome, Antibodies, Fluorescence, Proteomics, Normalization, Biomarker, Protein microarray

1 Introduction

Despite advances in proteomics, deciphering the proteome in one single assay remains a challenge mainly because of the complexity, variety, and dynamics of the proteomes. Among others, the size of the proteome is high, as an example, the human transcriptome comprises more than 23,000 protein-coding genes, that generate more than 100,000 protein species, mainly derived after alternative splicing and posttranslational modifications (PTMs) . In addition, the wide dynamic range of the proteome is caused by huge quantitative variations at the protein levels; then, for systematic analysis

Rodrigo García-Valiente and Jonatan Fernández-García contributed equally to this work.

Xing Wang and Matthew Kuruc (eds.), *Functional Proteomics: Methods and Protocols*, Methods in Molecular Biology, vol. 1871, https://doi.org/10.1007/978-1-4939-8814-3_7,

extraction and enrichment methods (which are not always efficient) are required. Moreover, mass spectrometry only captures a glimpse of this complexity, and many biological and technical replicates are usually required. Overall, the comprehensive and exhaustive proteome characterization is still an amazing challenge. In order to avoid these challenges, protein microarrays have become one of the promising strategies for biomarker and drug discovery [1–3]. High-density protein microarrays allow to analyze hundreds to thousands of known proteins in a single experiment and in a high-throughput format [4].

Recently, thanks to the capacity of massive comprehensive and systematic analysis, proteomics have been considered an useful approach to look deeper into personalized medicine and biomarker discovery, via the five R criteria: right patient/target, right diagnosis, right treatment, right drug/target, and right dose/time [5]. As it was mentioned above, high-density tailor-made protein microarrays cover a wide number of applications in personalized medicine [6, 7]. Bearing this in mind, it is very relevant to choose the best options of the protein microarrays available strategies (i.e., content, format, detection system, experimental conditions) for the purpose of the protein assay [8].

Here, most of these relevant features in experimental design are briefly described:

1.1 Microarray Format

The majority of the protein microarrays are developed in two formats (depending of the surface) [6, 9–11]:

1.1.1 Planar Arrays

In this type of arrays, the content (protein, peptide, aptamer, tissue, or cell lysates) is immobilized in microspots arranged in a two-dimensions (2D) surface (around 250 μm of diameter and separated ~300 μm) over a solid matrix. In this 2D organized spatial distribution, spot density around 1000 spot per square centimeter is normally reached in most of the protein microarrays commercially available [4, 11].

In these arrays, several aspects are directly related with the robustness and reproducibility of the assay performance [12], such as spot (size, morphology, and reproducibility), the ligand (binding capacity), the sample, surface, and method (background signal), and the detection limits. In addition, other aspects are also related to physicochemical properties of the surface [13] or biomolecules, which have a consequence in the assay development, such as spotting buffer composition affecting the protein structure, printing method (contact or noncontact) [14], hygroscopy, and humidity. In addition, the ratio background/signal is important; then, the unspecific binding onto the array surface has to be evaluated and controlled in order to detect correctly the ligand-agent union [15]. For this purpose, classical blocking buffers as bovine serum albumin (BSA) or skimmed milk powder at different concentrations are typically used.

1.1.2 Bead Arrays

In this format, the ligand is bound to addressable beads (color-coded beads or quantum dots), whose diameter typically varies from 0.02 to 0.1 μm (nanoparticles) or from 0.1 μm (microspheres) [16]. Usually, to be able to distinguish them, beads with different ligands are previously labeled with different combination of fluorescent dyes.

Then, the color-coded beads are easily detected by flow cytometry (in which one or several lasers excite the internal dye), and a reporter fluorescence dye (which is directly linked to the identification or quantification of the target protein, and excitation/emission are far different from the ones used for the internal color-coded dyes). The detector captures the color profile and identifies the ligand and therefore the target protein (by an assigned intensity to the analyte).

1.1.3 Microarray Content

Protein microarrays offer a wide diversity of subclasses according to the biomolecules deposited or displayed onto the surface. To simplify the classification, they can be categorized as assembled arrays or self-assembled arrays [17].

Assembled Arrays

Composed by, typically, antibodies, purified proteins or other entities which are immobilized onto a functionalized surface. Some types of arrays included in this category are:

1. Capture arrays

   They are generated by printing analyte-specific reagents (ASRs), usually antibodies [18] (Fig. 1), but sometimes phages [19] or others, over the array surface. These ASRs serve to identify and quantify the presence of multiple entities simultaneously. Therefore, they are used to find biomarkers and detect molecular signatures. The quality of the results depends on the quality (specificity and affinity) of the ASRs, which, in the case of antibodies, is related to them being poly- or monoclonal.
   There are different detection methods for measuring the analyte-reagent union, being direct (e.g., fluorescent, Cy3 and/or Cy5 antibodies) or indirect (biotinylated sample revealed with streptavidine, or secondary antibodies marked with HRP) [17].
2. Reverse-phase arrays

   The concept is opposite to the capture arrays. On these arrays, the samples are deposited onto the surface (Fig. 2). The reverse-phase arrays are very useful to evaluate many samples against a single ASR. It is critical to ensure the affinity of the

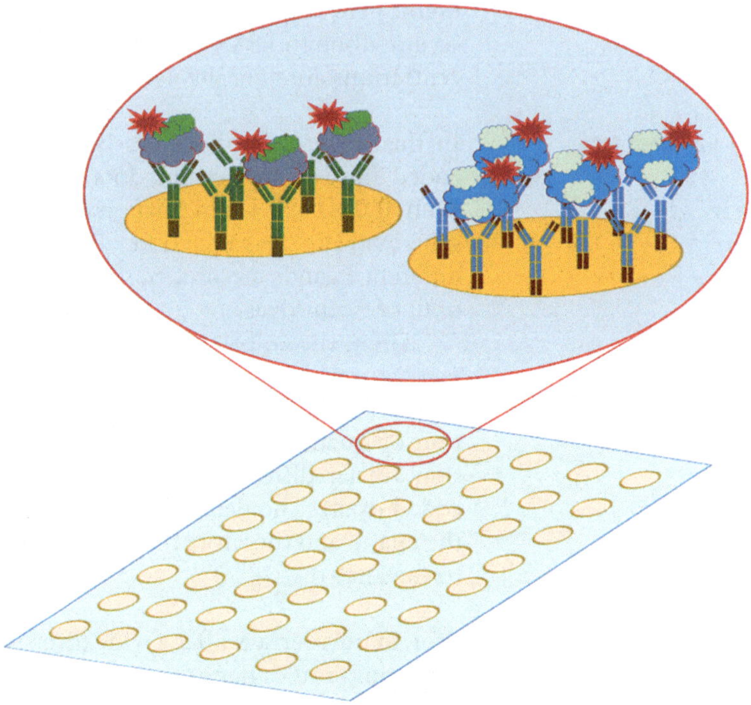

Fig. 1 Scheme of a capture microarray. In this particular example, it is labeled directly

ASR in order to avoid cross-reactivity. It is good to evaluate theoretical protein pathways in a high-throughput format [20]. However, it is highly time-consuming, and it, contrary to the capture arrays, may have difficulties detecting low abundance ligands in complex samples [10].

Self-Assembled Arrays

In these arrays, the protein is in situ expressed by an in vitro transcription-translation system from an immobilized cDNA encoding a protein of interest with a tag in carboxy or amino terminus (Fig. 3). It allows functional characterization of the in situ expressed proteins and also the identification of posttranslational modifications. Several types have been described so far: PISA [21], DAPA [22], PuCA [23], and NAPPA [24], the latter being optimized for high-throughput analysis.

For biomarker discovery, the more generalist and accessible platform is based on planar capture single-color microarrays printed by noncontact and incubated with chemically labeled samples (i.e., biotinylated), that allow to operate with a high number of samples independently. Therefore, this is the case that will be described in this chapter.

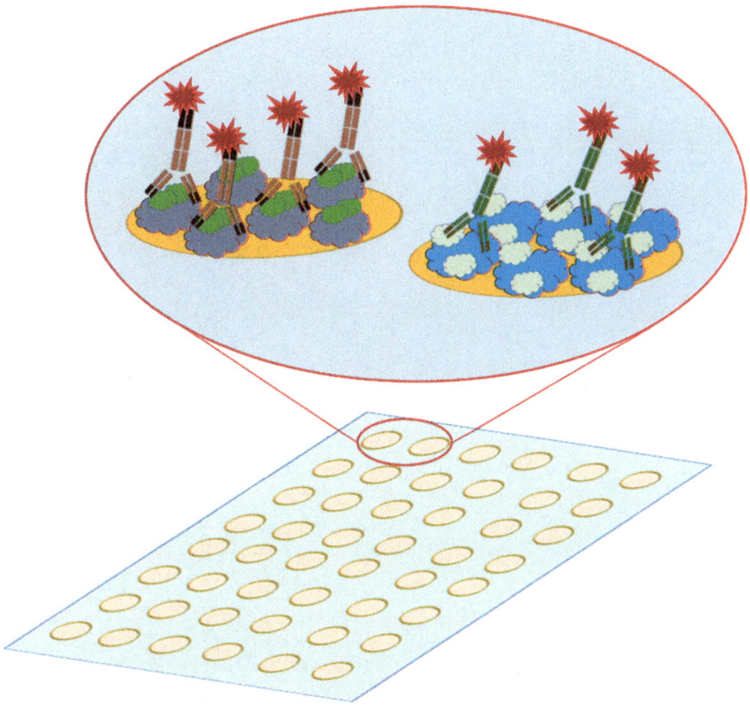

Fig. 2 Scheme of a reverse-phase protein microarray. A fluorescent tag is bound to each specific antibody

2 Materials

2.1 Experimental Materials

Prepare all materials at room temperature unless indicated otherwise.

2.1.1 Array Printing

Sample Preparation

- *General material*
  - JetStar™ Microarray-Specific 384 Microplates.
  - Micropipettes P10, P100 and correspondent tips.
  - Tube racks.
  - 1.5 mL tubes
  - Beaker of ≥200 mL.
- *Reactives*
  - PBS Na/K 1×.
  - 47% Glycerol.
  - BS3.
- *Benchtop instruments*
  - Thermoblock with agitator.
  - Centrifuge.
  - Vortex.

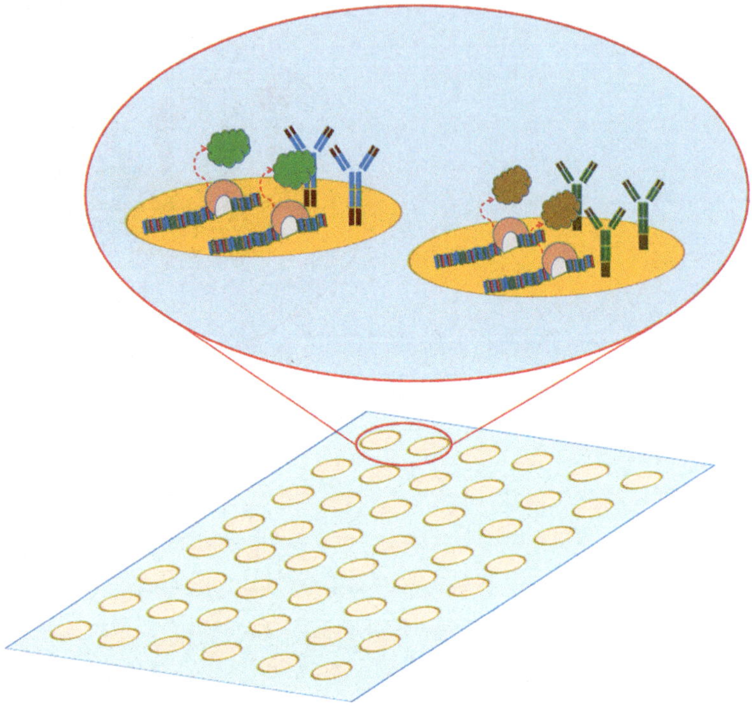

Fig. 3 Scheme of a self-assembled protein microarray. In this example, a nucleic acid programmable protein array—NAPPA

Array Printing

- *General material*
  - Bel-bulb pipettor.
- *Reactives*
  - Chemically active-surface slides.
- *Benchtop instruments*
  - Injection array printer, e.g., Arrayjet Marathon Argus.
  - Sonicator.

2.1.2 Array Assays

- *General material*
  - Microarray incubation chamber.
  - Microarray washing chamber.
  - Wet chamber.
  - 500 mL beaker.
  - Micropipettes P10, P100 and correspondent tips.
  - 1.5 mL tubes.
  - Cover slips.

- *Reactives*
  - Blocking buffer: PBS Na/K 1× + BSA 1% (w/v), 0.2% (w/v) Tween 20.
  - Streptavidine/fluor conjugate.
- *Benchtop instruments*
  - Agitator.
  - Orbital agitator.
  - Array washer platform.

2.1.3 Array Image Adquisition

- Benchtop instruments.
  - Microarray scanner, e.g., Sensovation's Fluorescent Array Imaging Reader.

2.2 Computational Resources

Recommended hardware requisites for the full computational workflow:

- IBM-compatible computer with an Intel Quad Core processor 1.8 GHz or faster.
- Microsoft Windows 7 64-bit edition operating system or superior.
- 8 GB RAM or more.
- Dedicated Video Card 512 MB or greater.
- 256 GB SSD unit (for image storage).
- 1280 × 1024 display system with 16 M colors.

2.2.1 Image Analysis

Recommended software:

- GenePix Pro software v. 7 or superior.
- Notepad++ v 7.5 or superior.

2.2.2 Data Analysis

Recommended software:

- R v. 3.0.1 or superior and RStudio v. 1.1.313 or superior, or, alternatively, Microsoft Excel 2010 or superior.

3 Methods

3.1 Customize Design

Customize arrays (Fig. 4) will be designed according to the particular needs of the specific screening. Some aspects have to be taken into account in an array-based assay.

1. According to the study, select appropriate Positive and Negative Controls and include them among the ligands. A standard negative control is the cleaning buffer. Including internal controls is an important tool to have an estimation of the behavior of the array and/or the sample. *"The more, the merrier."*

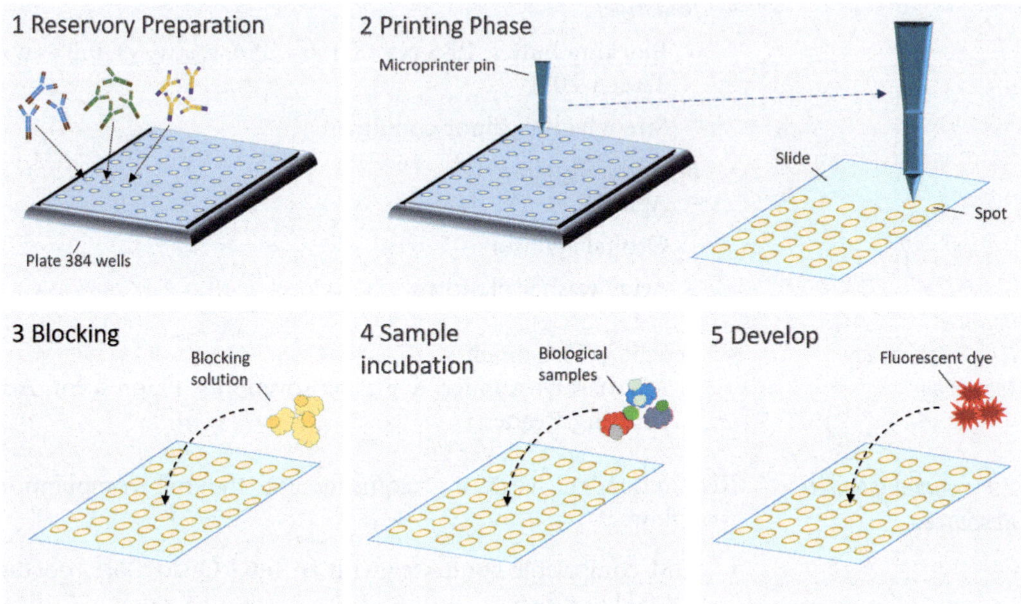

Fig. 4 Scheme of an antibody microarray printing followed by the assay. In the first phase (1), each specific antibody for each target protein is prepared in a plate, eluted in its corresponding master mix. The printing of the array (2) is carried out in a chamber with controlled temperature and humidity. Between each spot, the pin has to be cleaned with cleaning buffer to prevent cross contamination. After printing, the batch of arrays are dried and can be conserved for posterior use. Before using the microarray, it has to be blocked to prevent unspecific binding (3). After blocking and also after all the following steps (4 and 5) the microarray has to be thoroughly rinsed with distilled water. The array is incubated with the sample (4) and developed (5). After the last rinse, the array is dried and can be scanned. The microarrays can be preserved in darkness at r.t. in controlled humidity

2. Enough number of replicates have to be included. Technical replicates will give information about the quality of the biological and experimental aspects of the experiment, while biological replicates (at least three, more if possible) will provide information about the biological issue.
3. When working with planar microarrays, if there is a high number of samples and a moderate number of biomarkers to work with, an array may be divided in subsections called subarrays, each of one will have the same content as the others, but will be hybridized with a different sample. Therefore, the same sample in different subarrays will be considered technical replicates, while different samples of the same group in different subarrays will be considered biological replicates. On each subarray, all the content has to be displayed in, at least, triplicates, randomly distributed across the surface. Ideally, from 15 to 30 spots is a very suitable option (Fig. 4).

3.2 Experimental Workflow

Carry out all procedures at room temperature unless otherwise is specified on a protocol step.

In case of a high number of samples to be printed at once, then, technical replicates and biological replicates must be randomly and uniformly separated between the different experimental batches, to decrease the so-called batch effect.

It is important to optimize the amount of sample and ligand. For that, it is recommended to do previous smaller assays combining different dilutions of each, and to choose the combination that offers a less background/signal ratio.

3.2.1 Array Printing

Sample Preparation

1. Create an Excel file with the future and random distribution of the samples, negative controls and positive controls to print in the microplate (*see* **Note 1**).
2. Clean the laboratory work bench with 70% ethanol. Prepare required samples and reactives for the microplate preparation.
3. Protein samples shall be diluted in PBS Na/K 1×. As cross-linker BS3 at 50 mg/mL is used.
4. Microplates will be loaded according to the chosen distribution. Each sample will be loaded in a 1:1 dilution (v/v) with 47% glycerol (v/v).
5. After the microplates are set up, spin them in the centrifuge, using a microtiter adaptor.

Array Printing

1. Turn on the injection array printer, the associated control station, and linked computer.
2. Perform maintenance routine, specified by the manufacturer.
3. Introduce in the printer the microplate with the samples to print, and the functionalized slides, the microarrays, using the command to *Load microplates and slides*, ensuring that the microplates and the slides are correctly positioned.
4. In the computer, create a folder for the experiment. In said folder, save the execution parameters.
5. Start the printing, pressing the corresponding command. The machine will start printing each array automatically.
6. Once the printing process is over, save the quality results in the earlier mentioned folder.
7. Re-initialize the system (*see* **Note 2**).
8. Extract the microplate and store them in the conditions required by the samples.
9. Extract the printed microarrays. Label them (*see* **Note 3**).
10. Dry the printed microarrays in the stove at 37 °C along with an absorbent agent, silica gel.
11. Store the microarrays at room temperature.

3.2.2 Array Assays

1. Microarray blocking. Submerge the microarrays in the microarray washing chambers, in 6 mL of blocking solution on each one of them, during 1 h, in orbital agitation (*see* **Note 4**).
2. Microarray washing. Wash intensely during 10 min using milliQ water in the array washer platform. After this, wash three times, placing each array in the array washing chambers, filled with distilled water, in orbital agitation for 5 min each. They will be kept in distilled water until their processing.
3. Biologic sample processing. In this case, sample must be biotinilized, following the protocol described by Sierra A. et al., Journal Proteome Research 2016.
4. Microarray drying. Microarrays will be dried using centrifugation at 240 × *g* during 3 min, using the adaptors for 50 mL tubes. Once dried, they can be stored at 4 °C during a maximum of 15 days.
5. Incubate biological samples at the chosen dilution (*see* **Note 5**) overnight and at 4 °C at orbital agitation.
6. Microarray washing (*see* **Note 4**). Wash for 7 min using milliQ water in the array washer platform. After this, wash three times, placing each array in the array washing chambers, filled with distilled water, in orbital agitation for 5 min each. They will be kept in distilled water until their processing.
7. Sample incubation by indirect method given marked -biotinylated- samples.
   (a) Preparation of the streptavidine. Dilute 1:200 (v/v) of 0.1 mg of streptavidine/fluor conjugate in milliQ water.
   (b) Wet chamber preparation. Add distilled water to the wet chamber, enough to create humidity, but not enough to touch the arrays when they are added.
   (c) Incubation. Put the cover slips over the arrays, and the arrays in the wet chamber without touching the water. Incubate 200 μL of the reveal sample over the full microarray for 20 min in the wet chamber, at no light conditions.
8. Microarray washing (*see* **Note 4**). Wash for 7 min using milliQ water in the array washer platform. After this, wash three times, placing each array in the array washing chambers, filled with distilled water, in orbital agitation for 5 min each. They will be kept in distilled water until their processing.
9. Microarray drying. Microarrays will be dried using centrifugation at 240 × *g* during 3 min, using the adaptors for 50 mL tubes. Once dried, they have to be stored at no light exposure until scanned.

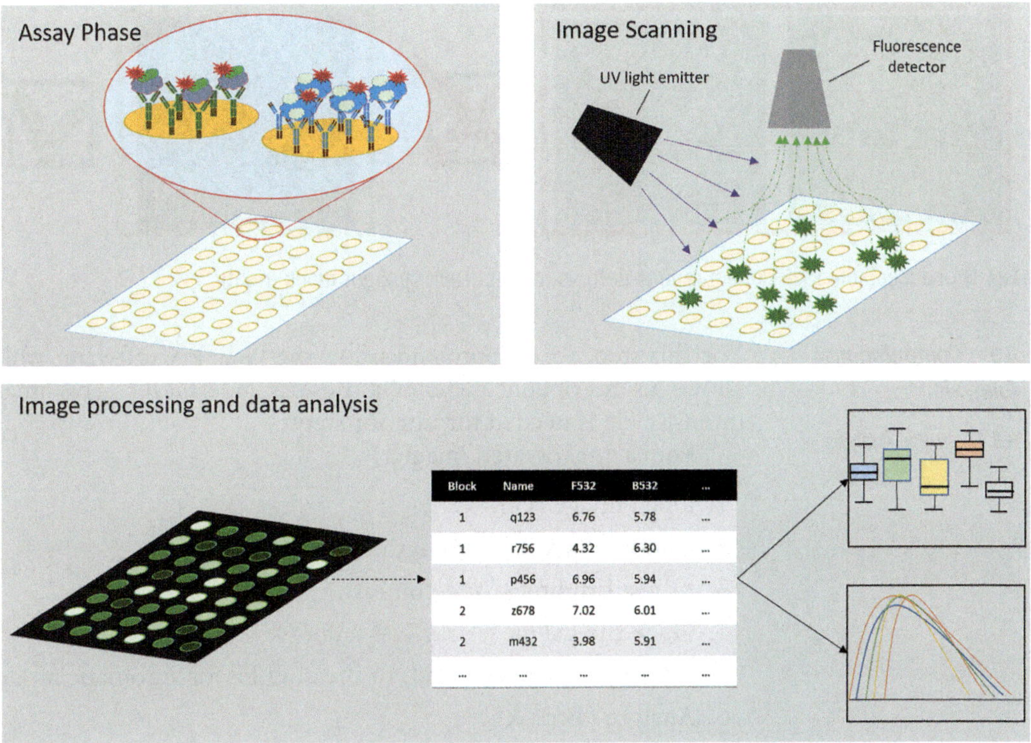

Fig. 5 General scheme of the analysis process after the assay. After the assay, the image is scanned in a fluorescent array imaging reader, which generates a file with the associated data for further statistical analysis

3.2.3 Array Image Adquisition

This protocol (Fig. 5) is designed to be performed in a Sensovation Fluorescent Array Imaging Reader. When operating with a different scanner, it should be adapted according to the characteristics of the instrument.

0. Turn on the equipment. Wait until everything is correctly loaded.
1. Open the Sensovation program.
2. Open the device to load as far as four arrays. **Steps 2–5** shall be repeated until all arrays have been scanned.
3. Click the Setup button, Rack Configuration to establish time exposure and focus parameters, to establish the correct spot visualization regarding the background signal (Fig. 6). Spots should be well delimited, and background noise should be homogeneous and well contrasted. Save parameters.
4. Click the Setup button, Assay Configuration to establish scanning options. Save them when correct.
5. Scan and save .tiff results (*see* **Note 6**).

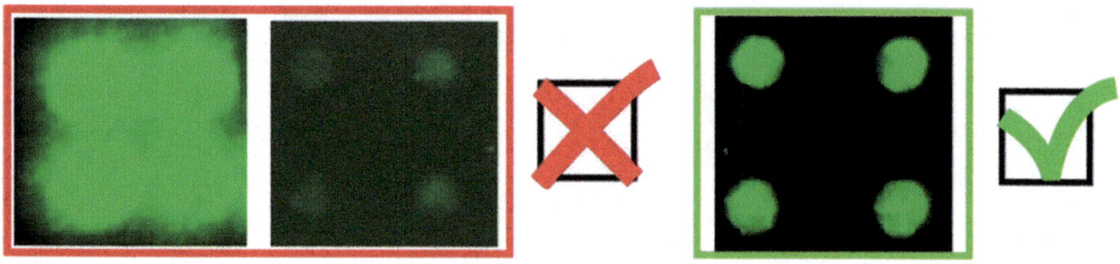

Fig. 6 Unacceptable rack configurations (left) vs. correct rack configurations (right)

3.3 Computational Analysis

3.3.1 Image Analysis

For this step, we recommend using the GenePix software, which allows us to obtain a GenePix Results file (GPR). This image intensity file is needed for ulterior steps.

For each generated image:

1. Open image (Ctrl+O).
2. Choose correct wavelength.
3. Adjust brightness and contrast.
4. Open the gal file/Array List (Alt+Y).
5. Adjust gal (manually and/or pressing F5 for automatic mode).
6. Analyze (Ctrl+A).
7. Configure. Background subtraction method: local.
8. Save results as .gpr.
9. Explore .gpr files with Notepad++, focusing on spot IDs, and checking they are correct.

3.3.2 Data Analysis

In order to obtain reliable and meaningful biological information from a microarray experiment, it has to be analyzed with a consistent statistical method (*see* **Note 7**). There are different available strategies to analyze microarrays, but not all of them, which were developed to analyze other microarray technologies such as DNA microarrays, are easily translatable to protein microarrays. The common steps to perform an analysis for this kind of assay are the following:

1. Import data sets. The Genepix output must be read into the software that will be used for the analysis. Suitable software choices for this task could be Python, R, or Matlab. GPR files contain, among other parameters, the mean, median, and standard deviation of the pixel intensity, as well as the total intensity for a spot at a given wavelength.
2. Background subtraction. A simple way to subtract the background for each intensity is using the default background measure from Genepix. Working with median values is recommended.

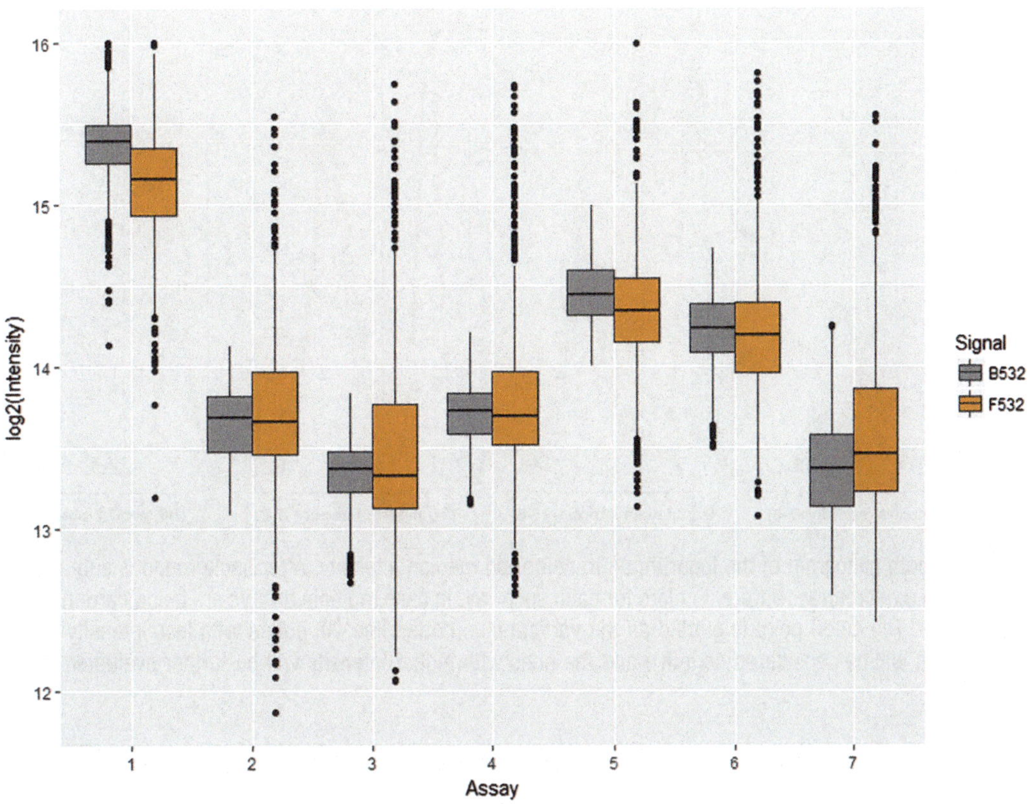

Fig. 7 Boxplots showing the distribution of the median intensities of the background (gray) and foreground (yellow) of every spot after logarithmic transformation. Seven assays are shown

3. A quality control check can be performed at this point, by detecting assays with an abnormally low overall signal (Fig. 7).
4. After subtracting the background, negative values are expected, which must be substituted by null values before **step 5**.
5. Apply a logarithmic transformation to the data sets.
6. Set a cutoff point to differentiate negative and positive spots (Fig. 8). This can be automatically achieved by determining the minimum between the positive and negative distribution, this can be achieved with a kernel density estimation of the data distribution.
7. Data normalization is required for establishing comparison across several assays. Various scaling methods are suited for this task (Fig. 9). Standard scoring is easy to implement, but it has to be applied to each assay individually with its respective mean and standard deviation. Quantile normalization could be used on demand [25] if some distributions display noisy patterns caused by technical variability but it could also subtract biological variability and trade off further statistical power.

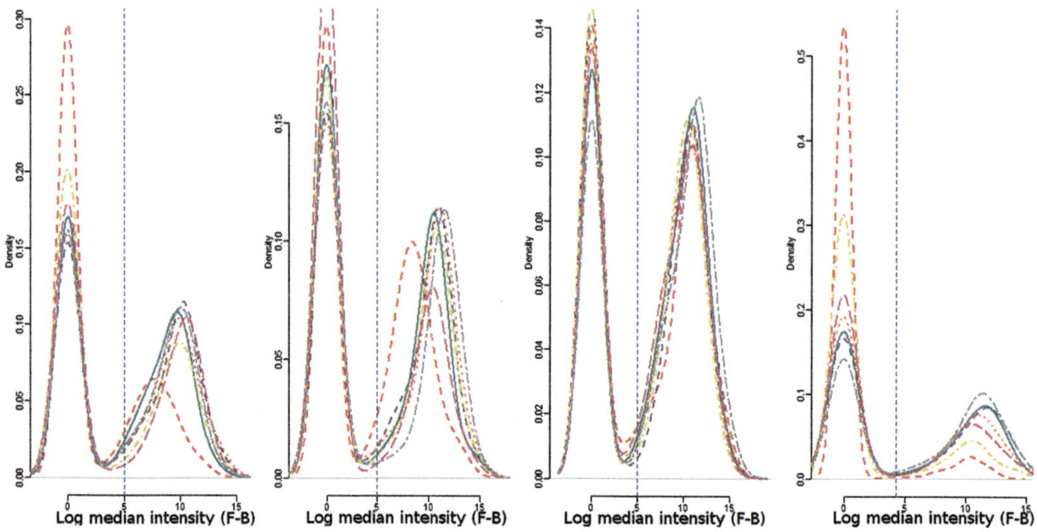

Fig. 8 Smooth histogram of the logarithmic transformed median intensity (after background is subtracted) of all spots in a microarray. Different colors for each subarray, in case multiple assays are being carried over the same slide. The cutoff point is plotted as the vertical blue dotted line. All points with less intensity than the cutoff point will be considered negative and the ones with higher intensity will be further evaluated

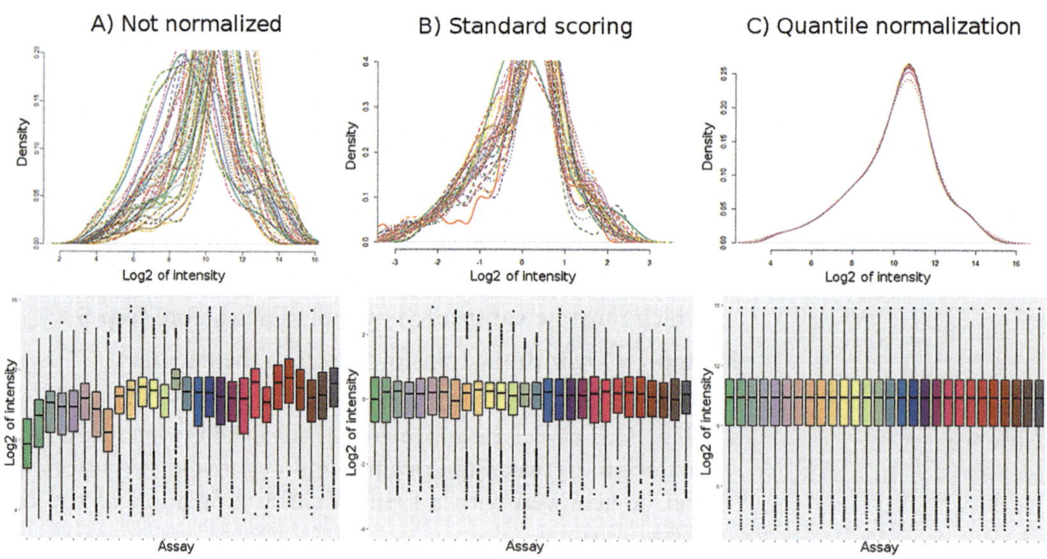

Fig. 9 Smooth histogram (top) and Boxplots (bottom) showing the median intensities of the positive spots (after logarithmic transformation) for every assay. Data not normalized (left), with standard scoring (center) and with a quantile normalization (right) are shown

8. Positive spots must be compared with negative control spots which act as a true nonarbitrary threshold for positivity on each individual assay.

9. To evaluate inter-array variability, same analyte spots that show true positive intensity (selected after the previous step) are counted. An index of confidence (IC), that ranges from 0 to 1, from *negative* to *positive detection in all replicates,* respectively, is assigned to every target protein (number of positive spots divided by total number of spots for that analyte).
10. Selection of differentially expressed proteins can be achieved by comparing the mean of the IC for each protein between groups of samples. This can be achieved with a standard *t*-test with a procedure to control FDR.

4 Notes

1. One of the recommended negative controls is 47% glycerol (v/v).
2. Between different printing essays, the device must be washed.
3. It is recommended that each label includes the batch, sample, and array information.
4. From this step on until the **step 8**, the slides/microarrays must be kept wet at all points.
5. To choose an optimal dilution for the biological samples doing previously the Subheading 3.2.2 with a reduced set of known control samples in different dilutions. However, typically 1:100 dilutions work well.
6. It is usually recommended to generate backup files, and it is essential to label correctly each file according to the scanned array.
7. These analysis can be executed in the online tool ProtArray (www.ProtArray.com).

Acknowledgments

We gratefully acknowledge financial support from the Spanish Health Institute Carlos III (ISCIII) for the grants: FIS PI14/01538, FIS PI17/01930, and CB16/12/00400. We also acknowledge Fondos FEDER (EU) and Junta Castilla-León (grant SA198A12-2). Fundación Solórzano FS38/2017. The Proteomics Unit belongs to ProteoRed, PRB3-ISCIII, supported by grant PT17/0019/0023, of the PE I + D + I 2017-2020, funded by ISCIII and FEDER.

References

1. Sierra-Sánchez Á, Garrido-Martín D, Lourido L, González-González M, Díez P, Ruiz-Romero C et al (2017) Screening and validation of novel biomarkers in osteoarticular pathologies by comprehensive combination of protein array technologies. J Proteome Res 16 (5):1890–1899
2. Zyuzin MV, Díez P, Goldsmith M, Carregal-Romero S, Teodosio C, Rejman J et al (2017) Comprehensive and systematic analysis of the immunocompatibility of polyelectrolyte capsules. Bioconjug Chem 28(2):556–564
3. Díez P, Ibarrola N, Dégano RM, Lécrevisse Q, Rodriguez-Caballero A, Criado I et al (2017) A systematic approach for peptide characterization of B-cell receptor in chronic lymphocytic leukemia cells. Oncotarget 8 (26):42836–42846
4. Merbl Y, Kirschner MW (2011) Protein microarrays for genome-wide posttranslational modification analysis. Wiley Interdiscip Rev Syst Biol Med 3(3):347–356
5. Dasgupta A (2008) Handbook of drug monitoring methods: therapeutics and drugs of abuse. Humana, Totowa, NJ, pp 1–445
6. Yu X, Schneiderhan-Marra N, Joos TO (2011) Protein microarrays and personalized medicine. Ann Biol Clin (Paris) 69(1):17–29
7. Yu X, Schneiderhan-Marra N, Joos TO (2010) Protein microarrays for personalized medicine. Clin Chem 56:376–387
8. Díez P, Dasilva N, González-González M, Matarraz S, Casado-Vela J, Orfao A et al (2012) Data analysis strategies for protein microarrays. Microarrays 1(3):64–83 http://www.mdpi.com/2076-3905/1/2/64/
9. Gonzalez-Gonzalez M, Jara-Acevedo R, Matarraz S, Jara-Acevedo M, Paradinas S, Sayagües JM et al (2012) Nanotechniques in proteomics: protein microarrays and novel detection platforms. Eur J Pharm Sci 45:499–506
10. Dasilva N, Díez P, Matarraz S, González-González M, Paradinas S, Orfao A et al (2012) Biomarker discovery by novel sensors based on nanoproteomics approaches. Sensors 12:2284–2308
11. Matarraz S, González-González M, Jara M, Orfao A, Fuentes M (2011) New technologies in cancer. Protein microarrays for biomarker discovery. Clin Transl Oncol 13:156–161
12. Ellington AA, Kullo IJ, Bailey KR, Klee GG (2010) Antibody-based protein multiplex platforms: technical and operational challenges. Clin Chem 56:186–193
13. Fuentes M, Díez P, Casado-Vela J (2016) Nanotechnology in the fabrication of protein microarrays. Methods Mol Biol 1368:197–208
14. Glökler J, Angenendt P (2003) Protein and antibody microarray technology. J Chromatogr B Anal Technol Biomed Life Sci 797:229–240
15. Kusnezow W, Jacob A, Walijew A, Diehl F, Hoheisel JD (2003) Antibody microarrays: an evaluation of production parameters. Proteomics 3(3):254–264
16. Casado-Vela J, González-González M, Matarraz S, Martínez-Esteso MJ, Vilella M, Sayagués JM et al (2013) Protein arrays: recent achievements and their application to study the human proteome. Curr Proteomics 10 (2):83–97. https://doi.org/10.2174/1570164611310020003
17. Lourido L, Diez P, Dasilva N, Gonzalez-Gonzalez M, Ruiz-Romero C, Blanco F, et al (2014) Protein microarrays: overview, applications and challenges. In: Genomics and proteomics for clinical discovery and development. Springer. p 147–173., https://doi.org/10.1007/978-94-017-9202-8_8
18. LaBaer J, Ramachandran N (2005) Protein microarrays as tools for functional proteomics. Curr Opin Chem Biol 9:14–19
19. Jara-Acevedo R, Díez P, González-González M, Dégano RM, Ibarrola N, Góngora R et al (2018) Screening phage-display antibody libraries using protein arrays. In: Phage display. Methods Mol Biol 1701:365–380
20. Spurrier B, Ramalingam S, Nishizuka S (2008) Reverse-phase protein lysate microarrays for cell signaling analysis. Nat Protoc 3 (11):1796–1808
21. He M, Taussig MJ (2001) Single step generation of protein arrays from DNA by cell-free expression and in situ immobilisation (PISA method). Nucleic Acids Res 29(15):E73–E73 http://www.ncbi.nlm.nih.gov/entrez/query.fcgi?cmd=Retrieve&db=PubMed&dopt=Citation&list_uids=11470888
22. He M, Stoevesandt O, Palmer EA, Khan F, Ericsson O, Taussig MJ (2008) Printing protein arrays from DNA arrays. Nat Methods 5 (2):175–177
23. Tao SC, Zhu H (2006) Protein chip fabrication by capture of nascent polypeptides. Nat Biotechnol 24(10):1253–1254
24. Ramachandran N, Raphael JV, Hainsworth E, Demirkan G, Fuentes MG, Rolfs A et al (2008) Next-generation high-density self-assembling functional protein arrays. Nat Methods 5 (6):535–538
25. Hicks SC, Irizarry RA (2014) When to use quantile normalization? bioRxiv. doi: https://doi.org/10.1101/012203. http://biorxiv.org/content/early/2014/12/04/012203.abstract

Chapter 8

Metaproteomics Study of the Gut Microbiome

Lisa A. Lai, Zachary Tong, Ru Chen, and Sheng Pan

Abstract

Proteomics is a widely used method for defining the protein composition of a complex sample. As this approach allows for identification and quantification of proteins across a broad dynamic range as well as detection of post-translational modifications, proteomics is an ideal platform to investigate the gut microbiome at a functional level. The gut microbiome is a dynamic environment which is crucial for overall health and fitness. Imbalances in the gut microbiome can influence nutrient absorption, pathogen resistance, inflammation, and various human diseases. Metaproteomic analysis of the gut microbiome is currently being performed on bacteria isolated from (1) fecal samples (2) colonic lavage, or (3) colon biopsies. Investigation of the gut microbiome has demonstrated that within the colon, there are distinct communities based on spatial location, and separable from the gut microbiomes isolated from stool. In addition to expanding our understanding of host–bacterial interactions for human health and disease, gut microbiome analysis is being utilized for biomarker development to discriminate normal individuals and diseased (i.e., inflammatory bowel disease or colon cancer) patients as well as to monitor disease activity and prognosis.

Key words Microbiome, Proteomics, Metaproteomics

1 The Gut Microbiota: An Introduction

The adult human gastrointestinal (GI) tract runs from the esophagus through the stomach and colon to the rectum. These organs are host to an enormous population of microorganisms, possibly upward of 100 trillion representing between 15,000 and 36,000 different species of bacteria [1–3]. Through host interaction, these bacteria, fungi, and viruses respond to stimuli within their microenvironment and impact a broad spectrum of essential functions including assisting with digestion of food, vitamin production/absorption, metabolism, nutrient extraction, immune response, and conferring resistance to pathogenic organisms [4, 5]. While the numbers and density of bacteria within the human gut are extremely high, the diversity is surprisingly low. Bacteria from four phyla *Bacteroides*, *Firmicutes*, *Proteobacteria*, and

Xing Wang and Matthew Kuruc (eds.), *Functional Proteomics: Methods and Protocols*, Methods in Molecular Biology, vol. 1871, https://doi.org/10.1007/978-1-4939-8814-3_8,

Actinobacteria constitute >98% of the microbes within the human GI tract [1], posing a unique challenge for metaproteomics analysis of gut microbiome, particularly with regard to species with low abundance.

The human gut microbiome begins to develop in utero as evidenced by microbiota detected in the placenta [6], amniotic fluid [7], and umbilical cord [8] and is fully colonized shortly after birth [9]. The development of the infant gut microbiota is impacted by the mode of delivery [10]. Infants delivered vaginally showed colonization by *Lactobacillus* and *Prevotella* whereas cesarean section delivered infants are first colonized by *Proteobacteria* and *Firmicutes.* This difference can persist up to the first 12 months of life. Rapid bacterial expansion and colonization of the gut occurs during early childhood such that by pre-adolescence, children have similar gut microbiomes to adults [11]. Once established, however, the gut microbiome does not stay static but rather is highly dynamic and can respond to environmental stimuli such as changes in diet and antibiotic treatment, among other factors.

Recently, the Human Microbiome Project (HMP) set out to characterize the microbiome in healthy individuals at multiple sites including the GI tract and feces [12]. Interestingly, differences in the microbial populations found in the lumen (stool) and in colon tissue samples from the same individual were reported [12, 13], suggesting that microbial communities may be spatially distinct. While the large intestine contained greater diversity than the small intestine, the microbial populations were similar, and dominated by bacteria of the phyla *Bacteriodetes* and *Firmicutes* [1]. The metagenomics data sets generated from the HMP has laid an important foundation for metaproteomics analysis of human gut microbiome. As the next step, the Integrative Human Microbiome Project (iHMP) focuses on the longitudinal investigation of microbiomes in various patient cohorts, including neonates, inflammatory bowel disease (IBD) patients, and diabetes patients [14]. Dysbiosis or an imbalance of the gut microbiota has been implicated in the pathogenesis of IBD [15], obesity [16, 17], metabolic disorders, and cardiovascular disease [18].

2 Modulation of the Gut Microbiome by Diet

There is much interest in the gut microbiome since it is a dynamic host-interaction which is responsive to environmental stimuli, including changes in diet. When mice were switched from a low fat plant-based diet to a high fat "Western" diet, microbiome gene expression changes were noted within a single day [19]. Diet studies in humans have shown changes in bacterial levels as needed for metabolism of plant polysaccharides in study participants eating a plant-based diet compared to increases in carbohydrate and protein fermentation for those on an animal-based diet [20]. Similar changes in carbohydrate metabolism were reported when comparing fecal microbiomes from rural vs. metropolitan communities showed differences in bacterial populations and profiles [21].

3 Changes in Gut Microbiota in Disease

There has been great interest in studying microbiome-dependent inflammation in diseases of the human gut, specifically, inflammatory bowel disease (i.e., ulcerative colitis (UC) or Crohn's disease (CD) [22]) and colorectal cancer [23, 24]. It has been shown that colorectal cancers are colonized by *Fusobacterium* and maintained in distal metastases [24]. In addition, treatment of mouse xenograft tumors with the antibiotic metronidazole reduced bacterial load and tumor growth [24]. In both feces and tissue samples, reduced diversity and imbalance of gut profiles have been reported in patients with GI disease.

In addition to studying causes of disease and disease pathogenesis, researchers are interested in the effects of reintroduction of bacteria to alleviate dysbiosis. Several studies have shown that fecal transplant is an effective treatment to reduce disease activity in IBD patients [25] or to treat recurrent diarrhea caused by *C. difficile* infection [26]. A recent study showed that fresh or frozen human feces could be successfully transplanted into gnotobiotic mice resulting in a gut microbiome which recapitulates the human donor within 1 week of transplant; the utility of this animal model for drug trials targeting gut microbiota was discussed [19].

4 Metaproteomics of Complex Samples Including Gut

While proteomics provides a comprehensive identification and quantitation of proteins within a sample, metaproteomics is the comprehensive characterization of expressed proteins within a microbiome community at a given point in time [27]. Metaproteomics has been used to investigate microbiomes in human and animals as well as in environmental communities such as soil [28], sludge [27], food [29], and the ocean [30]. This application is currently less common than metagenomics or metatranscriptomics studies due in part to the lack of consistent protocols for metaproteomic sample preparation, lack of efficient bioinformatics tools [31], and challenges of measuring low-abundance proteins within a complex sample [32]. Metaproteomics has been dependent upon databases and libraries generated from genome and metagenomics data for correct peptide identification and pathway analysis as well.

Historically, metaproteomics studies of the gut focused on bacterial populations isolated following short-term culture, but this approach was limited to strains that could be grown in vitro and oftentimes these cultured strains did not exhibit the same gut microbiota profiles. However, with advances in mass spectrometry, researchers have been increasingly successful at interrogating complex microbiomes. One study used two-dimensional differential gel electrophoresis (2D-DIGE) along with tandem mass spectrometry (MS/MS) and validation by SRM-based targeted proteomics to show that CD patients had overrepresentation of *Bacteroides* species and underrepresentation of *Clostridiales*, elevated expression of proteins involved in oxidative stress, energy saving, and IgA immunoglobulins and decreased GP2 (pancreatic glycoprotein 2 of zymogen granule membranes) which may promote inflammation [33].

Despite great advances in the field, this is still in the developmental stages due to the staggering complexity of the gut microbiome—more than 63 million unique proteins expressed from upward of 2100 different taxa [34]. Metaproteomic analysis requires enormous computing effort and power. In addition, the highly resistant cell walls of Gram-negative bacteria (which comprise but a portion of the strains within a microbiome) require additional mechanical cell wall disruption methods, such as bead beating or sonication, for optimal protein extraction [32, 35]. Also, there are challenges to correctly identifying peptides from homologous proteins, resulting in redundant protein identification and possibly skewing analysis since similar proteins from distinct species can have quite different functions.

5 Metaproteomics Using a Shotgun Proteomics Approach

Early metaproteomic analysis used two-dimensional gel electrophoresis (2D-GE) to separate proteins prior to mass spectrometry (MS). Then, shotgun metaproteomics using two-dimensional liquid chromatography (LC) coupled with nanospray tandem mass spectrometry (nano 2D LC-MS/MS) was used to interrogate infant fecal samples and demonstrated increased protein identification following enrichment of low-abundance microbial proteins from fecal samples [36]. This approach was used to show a significant decrease in proteins from the *Firmicutes* phylum in CD patients, higher abundance of inflammatory response proteins, and decreased expression of proteins involved in maintaining mucosal integrity, all of which may contribute to chronic inflammation [37]. An optimized LC-MS/MS workflow for analysis of the mouse fecal microbiome revealed identifications of 18,000 non-redundant tryptic peptides (93% of microbial origin), representing over 600 different microbial species and 250 protein families including members of the TonB-dependent receptor family which are involved in energy production [38].

6 Sample Collection to Investigate the Gut Microbiome

Analysis of the human gut microbiome has been approached using basically three types of clinically obtained materials—(1) stool, (2) colonic biopsies, and (3) colonic lavage. Of note, several studies have noted disparities between the microbiomes from paired fecal and rectal (colon) samples, even ones taken on the same day and without colonoscopy prep [39]. There are benefits and caveats to analysis of each type of clinical material, some of which are outlined below.

Many metaproteomics studies use fecal or stool samples since this material can be collected noninvasively and there is a large biomass of material to work with, considering that up to 30% of fecal biomass may be bacteria. Stool is a mixture of host cells, bacteria, food particles, and insoluble material, requiring enrichment by differential centrifugation [40], filtering to isolate microbial cells from larger insoluble particles [36], and/or precipitation. Protein extraction from feces yields a combination of bacterial and host secreted proteins, thus allowing interpretation of the interplay between host and bacteria. The first shotgun metaproteomics study of the human gut using fecal samples showed higher than expected expression of proteins related to translation, energy production, and carbohydrate metabolism, as well as proteins involved in novel microbial pathways and host immune response [40].

Colonic biopsies can be recovered during routine colonoscopy, and while they are minimally invasive, they do require disruption of the mucosal layer. While these biopsies sample a small area within the colon, they can be targeted for regions of dysplasia or specific regions within the colon (i.e., cecum, transverse colon, proximal colon, rectum). While the number of bacteria in a fecal sample can be up to 10^3 higher than in a biopsy sample, there is a difference in the bacterial communities isolated [41]. PCR-based techniques have been used to compare the profiles of bacterial subcommunities isolated from different parts of the colon as well as in feces. Although the predominant species detected from colon biopsies were consistently found independent of biopsy location, there was a distinct difference with the fecal samples, suggesting that fecal contamination during the colonoscopy was unlikely. The differences between colon and fecal microbiomes are less than between individuals [13].

Colonic lavage is a method of collecting bacteria following injection of a small amount of fluid into the colon during colonoscopy [42]. However, patients undergoing colonoscopy usually have completed a prep protocol, so much of the colon is cleansed before the procedure. Furthermore, analysis of mucosal lavage revealed a significant proportion of secreted human proteins (up to 63%) mixed in with the bacterial peptides (30%) [42]. By comparison, analysis of proteins isolated from fecal samples showed approximately 30% of identified proteins were of human origin [40]. Some studies suggest that mucosal lavage is a preferred sampling methodology since it would allow for niche-specific profiling as well as repeated sampling. Since the surface microbiota are recovered by lavage, they may be better adapted for adherence, host resistance, and other mucosal trophic factors than bacteria isolated from fecal samples. One study observed that in-gel digestion of mucosal lavage samples greatly increased the efficiency of trypsin digestion, which they postulate was likely due to inactivation of the trypsin inhibitor A1AT [42].

7 Challenges with Gut Microbiome Analysis

Some of the major problems in recovering microbiota from fecal samples are that samples are extremely heterogeneous, high amounts of insoluble material could skew protein measurements, and improper storage conditions could contribute to bacterial lysis [43]. To get around the heterogeneity issue, some researchers have homogenized the samples prior to aliquotting. Researchers have also successfully stored fecal samples in RNAlater rather than dry freezing which may preserve the amount of bacteria recovered—although storage conditions can have significant impact on microbiome profiles [43]. Bacteria recovered by extensive washing and then separated from the supernatant contain both lysed or secreted bacterial proteins and secreted host proteins. Comparison of different protein extraction methods (including beat beating and ultrasonication) combined with various lysis buffers (SDS, B-Per and urea) showed that while peptide identification and protein yield were highest with SDS buffer and bead beating, only B-Per was able to extract proteins from *Bacteriodetes*, and *Actinobacteria* was detected only with lysis in urea buffer. Extraction using urea resulted in good detection of post-translational modifications but less so using other lysis buffers [32].

8 Specialized Software Tools for Metaproteomics Analysis

Most of the reported metaproteomics studies use shotgun proteomics for identification of bacterial proteins which may be influenced by complexity and sensitivity issues such that detection of low abundant proteins may be challenging. However, the complexity and dynamic range issues may be addressed using the emerging spectral library based methods, such as data-independent acquisition (DIA), which provides a library-based, broader coverage for peptide/protein detection [44–47]. While such an approach has been increasingly applied in quantitative analysis of the human proteome, its applications in metaproteomics has lagged behind, in part due to the complexity involved in bioinformatics. Reference databases are constantly integrating mass spec analysis of cultured bacterial and pathogen species to generate reference databases for improved species identification [48].

Recently developed software tools designed specifically for metaproteomic data analysis have become available as well. MetaLab uses spectral clustering to improve peptide identification speeds [49]. An example of classification of the peptides identified from the metaproteomics analysis of a human fecal specimen is presented in Fig. 1. Others have shown improved protein identification in metaproteomics by employing de Bruijn graph assembly to predict protein sequences from metagenomics sequence data and

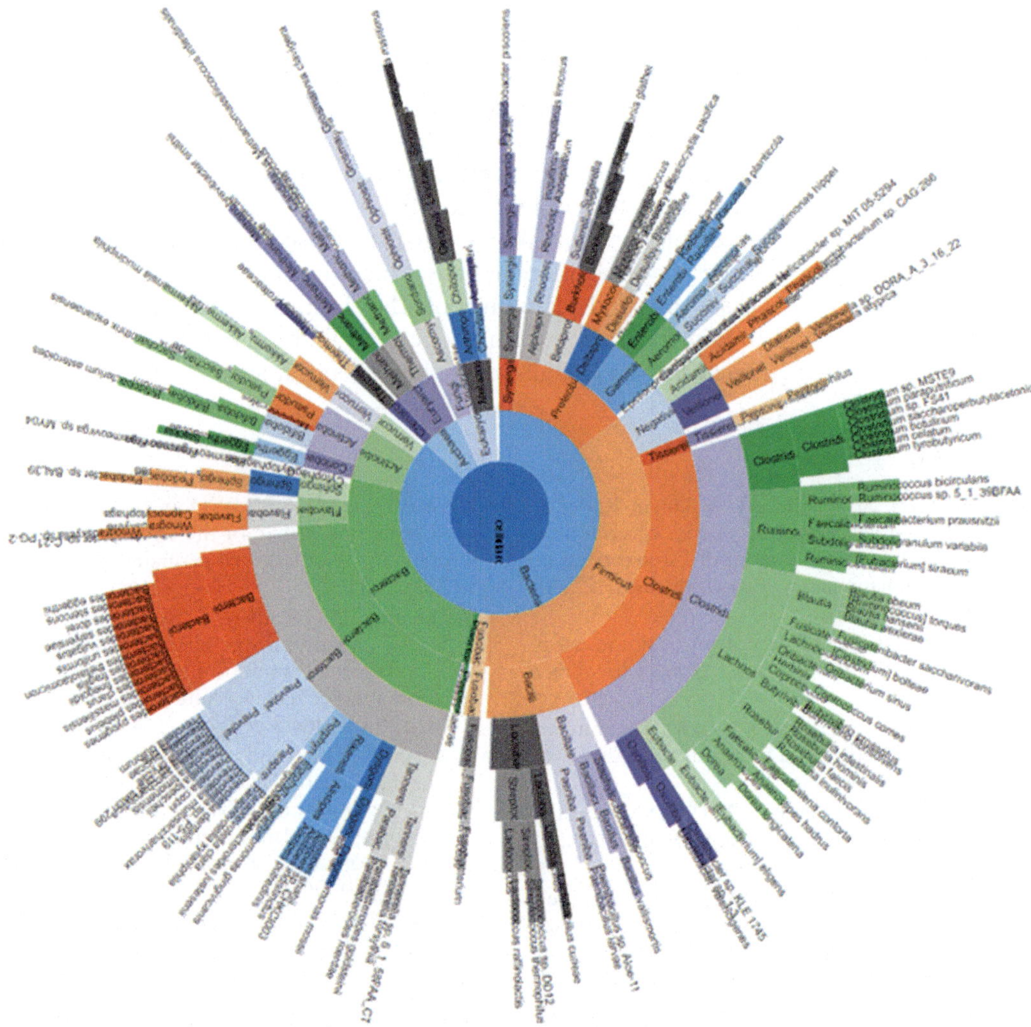

Fig. 1 Pie chart illustrates the taxonomy distribution from metaproteomics analysis of a human fecal sample. The center circle represents the organism with each concentric circle moving outward depicting taxonomies (i.e., domain, kingdom, phylum, class, order, family, genus, and species)

generate a reference database [50]. Taxon-specific classification of peptide sequences can be performed using UniPept (http://unipept.ugent.be) which uses shotgun proteomics data from UniProt KB with identification noise filtering to provide enhanced biodiversity analysis [51, 52]. Another option for analysis of metaproteomics data is MetaPro-IQ, which is ideal for fecal samples since the gut microbiome gene catalog was curated from fecal studies which negates the need for matched metagenomics data but makes it less applicable for other types of microbiome samples [53].

9 Conclusions and Perspectives

With recent advances in mass spectrometry instrumentation and data analysis, the comprehensive characterization of complex gut microbiome samples is becoming increasingly feasible. While previous efforts have focused on detection and identification of bacterial species, it is the hope that these efforts will blossom into analyses which can integrate metagenomics and metatranscriptomics data to provide better insight into the very complex and dynamic gut microbiome. The ultimate goal of these studies is to define localized and global interactions which impact human health and disease. By understanding of the intimate relationship between microbiota and gut, the hope is to find interventions which positively impact overall human health and fitness.

References

1. Frank DN, St Amand AL, Feldman RA et al (2007) Molecular-phylogenetic characterization of microbial community imbalances in human inflammatory bowel diseases. PNAS 104:13780–13785
2. Cresci GA (2015) The gut microbiome: what we do and don't know. Nutr Clin Pract 30:734–746
3. Lynch SV, Pederson O (2016) The human intestinal microbiome in health and disease. N Engl J Med 375:2369–2379
4. Tuddenham S, Sears CL (2015) The intestinal microbiome and health. Curr Opin Infect Dis 28:464–470
5. Rakoff-Nahoum S, Paglino J, Eslami-Varzaneh F et al (2014) Recognition of commensal microflora by toll-like receptors is required for intestinal homeostasis. Cell 118:229–241
6. Aagaard K, Ma J, Anthony K et al (2014) The placenta harbors a unique microbiome. Sci Transl Med 6:237ra65
7. Urushiyama D, Suda W, Ohnishi E et al (2017) Microbiome profile of the amniotic fluid as a predictive biomarker of perinatal outcome. Sci Rep 7:12171
8. DiGiulio DB, Romero R, Amogan HP et al (2008) Microbial prevalence, diversity and abundance in amniotic fluid during preterm labor: a molecular and culture-based investigation. PLoS One 3:e3056
9. Milani C, Duranti S, Bottacini F et al (2017) The first microbial colonizers of the human gut: composition, activities, and health implications of the infant gut microbiota. Microbiol Mol Biol Rev. https://doi.org/10.1128/MMBR.00036-17
10. Biasucci G, Rubini M, Riboni S et al (2010) Mode of delivery affects the bacterial community in the newborn gut. Early Hum Dev 86:13–15
11. Hollister EB, Riehle K, Luna RA et al (2015) Structure and function of the healthy

preadolescent pediatric gut microbiome. Microbiome 3:36

12. Human Microbiome Project Consortium (2012) Structure, function and diversity of the healthy human microbiome. Nature 486:207–212
13. Eckburg PB, Bik EM, Bernstein CN et al (2005) Diversity of the human intestinal microbial flora. Science 308:1635–1638
14. Proctor LM, Sechi S, DiGiacomo ND et al (2014) The integrative human microbiome project: dynamic analysis of microbiome-host omics profiles during periods of human health and disease. Cell Host Microbe 16:276–289
15. Tamboli CP, Neut C, Desreumaux P et al (2015) Dysbiosis in inflammatory bowel disease. Gut 53:1–4
16. Ley RE, Turnbaugh PJ, Klein S et al (2006) Microbial ecology: human gut microbes associated with obesity. Nature 444:1022–1023
17. Kolmeder CA, Ritari J, Verdam FJ et al (2015) Colonic metaproteomic signatures of active bacteria and the host in obesity. Proteomics 15:3544–3522
18. Kang Y, Cai Y (2017) Gut microbiota and hypertension: from pathogenesis to new therapeutic strategies. Clin Res Hepatol Gastroenterol. https://doi.org/10.1016/j.clinre.2017.09.006
19. Turnbaugh PJ, Ridaura VK, Faith JJ et al (2009) The effect of diet on the human gut microbiome: a metagenomic analysis in humanized gnotobiotic mice. Sci Transl Med 1:6ra14
20. David LA, Maurice CF, Carmody RN et al (2014) Diet rapidly and reproducibly alters the human gut microbiome. Nature 505:559–563
21. Yatsunenko T, Rey FE, Manary MJ et al (2012) Human gut microbiome viewed across age and geography. Nature 486:222–227
22. Wright EK, Kamm MA, Teo SM et al (2015) Recent advances in characterizing the gastrointestinal microbiome in Crohn's disease: a systematic review. Inflamm Bowel Dis 21:1219–1228
23. Arthur JC, Perez-Chanona E, Myhlbauer M et al (2012) Intestinal inflammation targets cancer-inducing activity of the microbiota. Science 338:120–123
24. Bullman S, Pedamallu CS, Sicinska E et al (2017) Analysis of Fusobacterium persistence and antibiotic response in colorectal cancer. Science. https://doi.org/10.1126/science.aal5240
25. Karolewska-Bochenek K, Grzesiowski P, Banaszkiewicz A et al (2017) A two-week fecal microbiota transplantation course in pediatric patients with inflammatory bowel disease. Springer, Heidelberg, Boston, MA, pp 1–7
26. Fischer M, Sipe B, Cheng YW et al (2017) Fecal microbiota transplant in severe and severe-complicated Clostridium difficile: a promising treatment approach. Gut Microbes 8:289–302
27. Wilmes P, Bond PL (2004) The application of two-dimensional polyacrylamide gel electrophoresis and downstream analyses to a mixed community of prokaryotic microorganisms. Env Microbiol 6:911–920
28. Bastida F, Hernandez T, Garcia C (2014) Metaproteomics of soils from semiarid environment: functional and phylogenetic information obtained with different protein extraction methods. J Proteomics 101:31–42
29. Maier TV, Lucio M, Lee H et al (2017) Impact of dietary resistant starch on the human gut microbiome, metaproteome, and metabolome. MBio 8:e01343–e01347
30. Williams TJ, Cavicchioli R (2014) Marine metaproteomics: deciphering the microbial metabolic food web. Trends Microbiol 22:248–260
31. Heyer R, Schallert K, Zoun R et al (2017) Challenges and perspectives of metaproteomic data analysis. J Biotechnol 261:24–36
32. Zhang X, Li L, Mayne J et al (2017) Assessing the impact of protein extraction methods for human gut metaproteomics. J Proteome. https://doi.org/10.1016/j.jprot.2017.07.001
33. Juste C, Kreil DP, Beauvallet C et al (2014) Bacterial protein signals are associated with Crohn's disease. Gut 63:1566–1577
34. Wilmes P, Heintz-Buschart A, Bond PL (2015) A decade of metaproteomics: where we stand and what the future holds. Proteomics 15:3409–3417
35. Glatter T, Ahrne E, Schmidt A (2015) Comparison of different sample preparation protocols reveals lysis buffer-specific extraction biases in gram-negative bacteria and human cells. J Proteome Res 14:4472–4485
36. Xiong W, Giannone RJ, Morowitz MJ et al (2015) Development of an enhanced metaproteomic approach for deepening the microbiome characterization of the human infant gut. J Proteome Res 14:133–141
37. Erickson AR, Cantarel BL, Lamendella R et al (2012) Integrated metagenomics/metaproteomics reveals human host-microbiota signatures of Crohn's disease. PLoS One 7:e49138
38. Tanca A, Palomba A, Pisanu S et al (2014) A straightforward and efficieint analytical pipeline

for metaproteome characterization. Microbiome 2:49
39. Durban A, Abellan JJ, Jimenez-Hernandez N et al (2011) Assessing gut microbial diversity from feces and rectal mucosa. Microb Ecol 61:123–133
40. Verberkmoes NC, Russell AL, Shah M et al (2009) Shotgun metaproteomics of the human distal gut microbiota. ISME J 3:179–189
41. Zoetendal EG, von Wright A, Vilpponen-Salmela T et al (2002) Mucosa-associated bacteria in the human gastrointestinal tract are uniformly distributed along the colon and differ from the community recovered from feces. Appl Environ Microbiol 68:3401–3407
42. Li X, LeBlanc J, Truong A et al (2011) A metaproteomic approach to study human-microbial ecosystems at the mucosal luminal interface. PLoS One 6:e26542
43. Choo JM, Leong LEX, Rogers GB (2015) Sample storage conditions significantly influence faecal microbiome profiles. Sci Rep 5:16350
44. Chapman JD, Goodlett DR, Masselon CD (2014) Mulitplexed and data-independent tandem mass spectrometry for global proteome profiling. Mass Spec Rev 33:452–470
45. Nigjeh EN, Chen R, Allen-Tamura Y et al (2017) Spectral library-based glycopeptide analysis--detection of circulating galectin-3 binding protein in pancreatic cancer. Proteomics Clin Appl 11:1700064
46. Rosenberger G, Liu Y, Rost HL et al (2017) Inference and quantification of peptidoforms in large sample cohorts by SWATH-MS. Nat Biotechnol 35:781–788
47. Rost HL, Aebersold R, Schubert OT (2017) Automated SWATH data analysis using targeted extraction of ion chromatograms. Methods Mol Biol 1550:289–307
48. Alispahic M, Hummel K, Jandreski-Cvetkovic D et al (2010) Species-specific identification and differentiation of Arcobacter, Helicobacter, and Campylobacter by full-spectral matrix-associated laser desorption/ionization time of flight mass spectrometry analysis. J Med Microbiol 59:295–301
49. Cheng K, Ning Z, Zhang X et al (2017) MetaLab: an automated pipeline for metaproteomic data analysis. Microbiome 5:157
50. Tang H, Li S, Ye Y (2016) A graph-centric approach for metagenome-guided peptide and protein identification in metaproteomics. PLoS Comput Biol 12:e1005224
51. Mesuere B, Devreese B, Debyser G et al (2012) Unipept: tryptic peptide-based biodiversity analysis of metaproteome samples. J Proteome Res 11:5773–5780
52. Mesuere B, Van der Jeugt F, Willems T et al (2018) High-throughput metaprotomics data analysis with Unipept: a tutorial. J Proteome 171:11–22
53. Zhang X, Ning Z, Moore JI et al (2016) MetaPro-IQ: a universal metaproteomic approach to studying human and mouse gut microbiota. Microbiome 4:31

Chapter 9

Double One-Dimensional Electrophoresis (D1-DE) Adapted for Immunoproteomics

Youcef Shahali, Hélène Sénéchal, and Pascal Poncet

Abstract

The classical proteomics approach for the identification of allergen candidates consists on the separation of proteins by high-resolution two-dimensional electrophoresis (2-DE) with subsequent IgE immunoblotting and further analysis of IgE-reactive protein spots with mass spectrometry. In this approach at least two gels most be run. One gel is used for staining and the other is for immunoblotting by antibodies labeled with specific immunostains. Additional functional characterizations require either protein purification or 2-DE replicates and appear to be time- and reagent-consuming. Here we described a modified double one-dimensional electrophoresis (D1-DE) allowing the conversion of a protein spot previously visualized by 2-DE into an extended protein band. In D1-DE, the purity of the protein of interest is similar to 2-DE spots, but its abundance is many times higher than what can be found in a 2-DE single spot allowing many other functional analyses from a single D1-DE separation.

Key words Double one-dimensional electrophoresis, D1-DE, 2-DE, Allergens, Proteomics

1 Introduction

Two-DE combined with immunoblotting and mass spectrometry analyses is now routinely used in immunoproteomics to isolate, identify, and characterize IgE-binding proteins of complex extracts from allergenic sources [1–4]. To date, this comprehensive approach led to the characterization of more than 850 allergens, improving the diagnosis and therapy of allergic diseases [1]. Despite its numerous advantages, 2-DE presents some limitations for functional studies on allergens. One of the main drawbacks of this technique is that only a single supplementary experiment can be performed on each single 2-DE protein spot (e.g., mass spectrometry or immunoreactivity after Western blotting) [5]. Additional functional or immunological characterizations will require either protein purification or 2-DE replicates. Another limitation of this technique is in the detection of low-abundance protein spots that constitute an important part of any proteome [6, 7]. In 2-DE

Xing Wang and Matthew Kuruc (eds.), *Functional Proteomics: Methods and Protocols*, Methods in Molecular Biology, vol. 1871, https://doi.org/10.1007/978-1-4939-8814-3_9,

immunoblotting, although disease-associated proteins often cover a very small percentage of the whole 2-DE separation, the entire gel must be transferred onto a membrane and individually incubated with a sizeable volume of serum (at least 250–300 μL for a membrane of 100 cm^2) [5–8]. In addition, with regard to allergomic studies, since the 2-DE has not yet been miniaturized, this method appeared to be tedious, time- and reagent-consuming when several allergenic sources have to be studied [9]. To overcome these drawbacks, we propose the use of a modified double one-dimensional electrophoresis (D1-DE) allowing the simultaneous screening of at least 30 allergic patient sera on the same blotted protein separated according to the isoelectric point (pI) and molecular mass (Mr), just as a 2-DE spot [10]. It is an extension of the first D1-DE concept originally reported by Atland et al. (1981) to study protein genetic variants [11]. This method consists of the sequential one-dimensional combination of two electrophoretic separations while the migration axis remains unchanged. The main advantage of this technique was obviously in the simultaneous separation and high resolution of proteins of interest in dual sorting parameters. For allergomics application, we designed and developed a D1-DE for the conversion of a protein spot previously visualized by 2-DE into an extended protein band. Therefore, the purity of the protein of interest is similar to 2-DE spots, but its abundance is many times higher than what can be found in a 2-DE single spot allowing many other functional analyses from a unique D1-DE separation.

2 Materials

2.1 Equipment

1. Horizontal flatbed electrophoresis apparatus (e.g., Multiphor II, GE Healthcare, Uppsala, Sweden).
2. Electrophoresis power supply apparatus of minimum 1000 V (e.g., EPL 3501 XL, GE Healthcare).
3. Thermostatic circulator (e.g., Multitemp II, GE Healthcare).
4. Semidry blotting apparatus (e.g., Multiphor II electro-semidry transfer apparatus, GE Healthcare).
5. Gel rehydration pool (e.g., GelPool, GE Healthcare).
6. Rehydration tray for IEF strips (e.g., rehydration trays, Serva, Heidelberg, Germany).
7. Flat Forceps.
8. A pair of scissors.
9. Magnetic stirrer.
10. Water bath.

2.2 First Dimension IEF

1. Polyacrylamide gel 4%T, 3%C (CleanGel, GE Healthcare BioSciences AB, Uppsala, Sweden).
2. Carrier ampholytes for IEF (Servalyt pH 2–11) were from Serva, Heidelberg, Germany.
3. Cathode and anode buffers (Serva IEF buffers, Serva, Heidelberg, Germany).
4. Electrode GF/B glass fiber strips (Whatman).

2.3 Second Dimension Separation Using SDS-PAGE

1. Chemicals composing the equilibration buffer: 114 mmol/L Tris pH 6.8 and 12% w/v sodium dodecyl sulfate (SDS) were all supplied by Sigma-Aldrich (St Louis, MO, USA).
2. Polyacrylamide 8–18% gradient gel (ExcelGel gradient 8–18%) were from GE Healthcare.
3. ExcelGel SDS buffer strips (anode and cathode; GE Healthcare).
4. Molecular masses (Mr) standard protein mixture was from Bio-Rad (Hercules, CA, USA).
5. Whatman N° 1 and 3 paper (GE Healthcare).

3 Methods

3.1 Principles of D1-DE

Like standard 2-DE, the D1-DE reported here consists of IEF as a first dimension followed by SDS-PAGE. The difference between the two techniques mainly resides in the migration axis of the second separation. This means that an acidic, neutral, or basic horizontal band from the first dimension (IEF) is transferred to SDS-PAGE (Fig. 1). Therefore, proteins are isolated as a long continuous band with purity comparable to 2-DE spots. In the present protocol, the sequential one-dimensional combination of IEF and SDS-PAGE followed by immunoblotting is described for the IgE screening of patients sera under equal conditions. For a better correlation between 2-DE and D1-DE patterns, the same IEF gel could be used as the first dimension of both 2-DE and D1-DE immunoblotting.

3.2 Preparation of IEF

1. A polyacrylamide gel 4%T, 3%C (CleanGel, GE Healthcare BioSciences AB, Uppsala, Sweden) is first hydrated in a solution containing carrier ampholytes (CleanGel Dry IEF-12.5 × 26.0 cm).
2. For this purpose, first prepare the rehydration buffer by adding 5% v/v Servalytes pH 2–11, (40% carrier w/v solution in water) (Serva, Heidelberg, Germany) in distilled water.
3. To rehydrate a complete CleanGel, pipet 25 mL rehydration solution into a Gelpool (tray for rehydration).

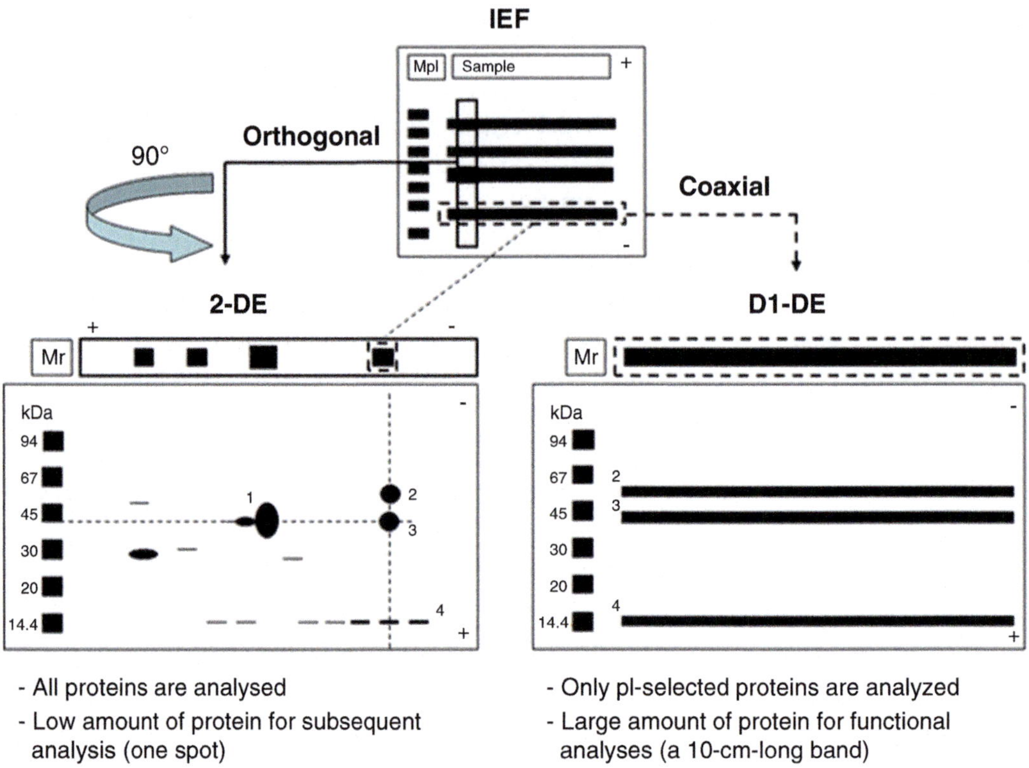

Fig. 1 Schematic illustration of a D1-DE adapted from Shahali et al. (2012) [10]. Both 2-DE and D1-DE consist of the sequential combination of IEF and SDS-PAGE. The difference between the two techniques resides in the axis of migration of the second separation

4. Place the CleanGel IEF precast gel into the pool for rehydration by laying the edge of the CleanGel, with the gel surface facing down into the rehydration buffer, slowly lower it in order to avoid air bubbles.
5. Using forceps, lift the film at the edges up to the middle and lower it again without catching air bubbles to distribute evenly the liquid. Very even rehydration is also obtained by shaking the GelPool at a slow rotation rate.
6. Meanwhile, switch on the cooling system (thermostatic circulator or Peltier cooling plate) of the horizontal flatbed apparatus.
7. Remove the gel from the Gelpool after 1 h making sure the rehydration has been evenly achieved.
8. Remove the excess buffer from the gel surface using the edges of a Whatman filter paper N° 1, until the gel surface is completely dry (*see* **Note 1**).
9. The rehydrated gel is now ready for IEF run.
10. Pipet a low amount of kerosene (about 2.5 mL) onto the horizontal cooling plate (e.g., Multhiphor II apparatus) to improve the cooling contact.

11. Place the gel (gel surface up/gel film or GelBond down) onto the center of the cooling plate.
12. Cut the electrode strips at the gel length (26 cm).
13. Align strips on the cathodal and anodal edges of the gels, the electrode strips overlapping the gel-edge with about 3 mm.
14. Apply evenly the cathode and anode buffers (Serva IEF buffers, Heidelberg, Germany) on the cathode and anode strips, respectively.
15. Cut two 10-cm-long and up to 0.7-cm-wide strips of a dry Whatmans N°1 paper (GE Healthcare) for sample application on anode side.
16. Cut a small piece of Whatman N° 1 paper (0.5 × 0.5 cm) for IEF protein standards (pI marker) and place it directly onto the upper center of the gel (anode side).
17. Use forceps to align the strips on both sides of the pI marker at the same distance from the paper piece (minimum 1 cm).
18. Pipet 2 μL of pI marker (Bio-Rad) for wide or short range IEF as comparative references.
19. Load samples (55–60 μg of proteins from the allergen extract) using a micropipette on each 10-cm-long sample paper piece.
20. Clean platinum electrode wires with a wet tissue paper.
21. Then, move electrodes on anode/cathode strips to ensure a complete overlapping between buffer strips, gel, and electrode wires.
22. For Multiphor, connect the cables of the electrodes to the electrophoresis power supply apparatus (EPS 3501 XL, GE Healthcare). Finally lower and close the safety lid.
23. Select the running program and start the IEF migration (*see* **Note 2**).
24. Stop the migration and wash the cathode and anode borders of the gel with PBS pH 7.5.
25. Lift the film at the edges up to the middle to form a U shape and pour promptly PBS on the middle of the gel.

3.3 Preparation of D1-DE

1. After IEF running, cut out the gel at the pI marker along with a small part of the sample separation for Coomassie blue or silver staining.
2. Lay the gel face down on a clean plastic film or a glass plate. Mark the area that should be cut out for transferring. The size of this area should be 10 cm long and up to 7 mm wide.
3. Cut out horizontally the IEF strip in the selected narrow pI range containing the allergenic fraction of interest by reference to pI markers and sample stained.

4. At this step, the IEF strips could be transferred to the SDS-PAGE separation or placed in a sealed bag until use, conserve the gel at −20 °C until use (up to 1 years).
5. Otherwise, incubate the IEF strips in the equilibration buffer containing 114 mmol/L Tris pH 6.8 and 12% w/v SDS for 3 × 10 min.
6. Meanwhile, switch on the thermostatic circulator adjusted to 12 °C.
7. Pipet a low amount of kerosene (about 2.5 mL) onto the horizontal cooling plate (e.g., flatbed electrophoretic chamber, Multiphor II, GE Healthcare) to improve the cooling contact.
8. During the equilibration procedure, place the thin 8–18% gradient polyacrylamide gel (ExcelGel; GE Healthcare) surface up/gel film down onto the center of the cooling plate.
9. Align the ExcelGel SDS buffer strips (anode and cathode; GE Healthcare) on the cathodal and anodal edges of the gels, the electrode strips overlapping the gel-edge with at least 5 mm.
10. Cut a small piece of Whatman N°1 paper (0.5 × 0.5 cm) for molecular masses protein standards (Mr) and place it directly onto the upper center of the gel (anode side).
11. Remove the 10-cm IEF strips from the equilibration trays.
12. Remove the excess buffer using the edges of a Whatman filter.
13. Using two forceps, place each IEF strip gel-side down into the top of the SDS ExcelGel and push them carefully toward the anode.
14. Pipet 2–5 μL of Mr (Bio-Rad) onto the small piece of Whatman paper.
15. Clean platinum electrode wires with a wet tissue paper.
16. Then, move electrodes on anode/cathode strips to ensure a complete overlapping between buffer strips, gel, and electrode wires.
17. For Multiphor, connect the cables of the electrodes to the basic electrophoresis power supply apparatus. Finally lower and close the safety lid.
18. Select the running program and start the SDS-PAGE migration (*see* **Note 3**).
19. Stop the migration and wash carefully the entire SDS gel with PBS pH 7.5. At this step, D1-DE separations could either be silver-stained according to Blum et al. [12], Coomassie blue-stained or blotted onto a cyanogen bromide-activated nitrocellulose (Fig. 2) (Optitrans BA-S 83, Schleicher and Schuell, Dassel, Germany) sheet [13] for further functional analyses (*see* **Note 4**) [10, 14].

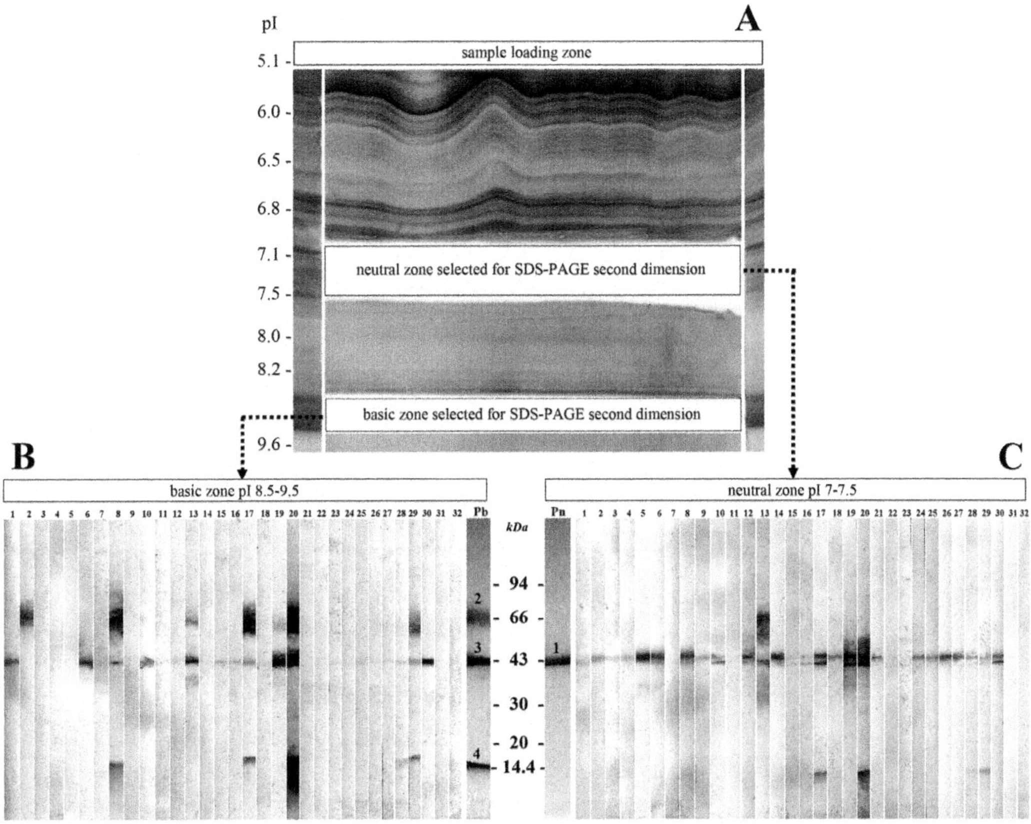

Fig. 2 D1-DE combined with IgE immunoblots (adapted from Shahali et al. 2012) [10]. (**a**) Represents the initial IEF separation. The silver staining was performed after excision of the basic and neutral bands. IgE immunoblots of the basic (**b**) and neutral (**c**) proteins of cypress pollen (CP) extracts probed with sera of 30 CP allergic patients: lanes 1–30; lane 31: healthy donor serum; lane 32: no serum (negative control). D1-DE allowed the MS/MS characterization of an allergenic polygalacturonase (PG) of 43 kDa which overlapped (in previous 1-DE experiments) with another CP major allergen belonging to the pectate lyase (PL) family referenced as Cup s 1 (*see* **Note 6**)

20. After staining, the protein band of interest could be cut out, digested, and submitted to mass spectrometry analyses (*see* **Note 5**) [10, 14].

4 Notes

1. Remove excess buffer from the gel surface with a piece of Whatman N° 1 paper until you can hear a "squeaking."
2. Running conditions and parameters for IEF experiment:

   Running temperature: 12 °C; total running time: 2 h 30 min. Phase 1: Set the voltage as constant at 50 V during 60 min. Remove the strips at the end of this phase.

Phase 2: Set the voltage as constant at 200 V during 60 min.
Phase 3: Set the voltage as constant at 150 V during 40 min.
Phase 4: Set the watt as constant at 1 W during 90 min.
Phase 5: Set the watt as constant at 2 W during 50 min.
Phase 6: Set the watt as constant at 3 W during 120 min.

3. Stop the IEF when cytochrome C (red color) marker is near cathode side
Running conditions and parameters for SDS-PAGE. Underlined values are constant.

Time (min)	Voltage (V)	Current (mA)	Power (W)	Temperature (°C)
75	100	40	40	12
110	750	20	15	12
120	900	25	15	12

4. Up to 40 blotted strips could be tested with various antibodies, with or without inhibitors, which gave rise to fundamental results concerning the specificity of the allergen recognition among cypress pollen allergic patients [14].
5. Besides multiplexing for immunoblotting, the D1-DE circumvents the problem of low protein amounts often encountered in a single 2-DE spot, which generally makes MS and microsequencing experiments difficult. The whole protein band can be excised, digested, and processed for analysis. A novel low-abundant allergen has been recently identified and characterized in cypress pollen using this approach [15].
6. D1-DE IgE screenings on the neutral and basic IEF fractions of CP extracts demonstrate that among 30 tested patients, 21 (70%) showed a positive IgE response to the novel 43 kDa basic allergen (PG), while 22 (~73%) were sensitized to Cup s 1 (PL). This novel CP major allergen has been recently indexed as Cup s 2 by the WHO/IUIS allergen nomenclature [14].

References

1. Nony E, Le Mignon M, Brier S, Martelet A, Moingeon P (2016) Proteomics for allergy: from proteins to the patients. Curr Allergy Asthma Rep 16:64
2. Mousavi F, Majd A, Shahali Y, Ghahremaninejad F, Shoormasti RS, Pourpak Z (2017) Immunoproteomics of tree of heaven (*Ailanthus atltissima*) pollen allergens. J Proteome 154:94–101
3. Hoffmann-Sommergruber K (2016) Proteomics and its impact on food allergy diagnosis. EuPA Open Proteom 12:10–12
4. Tiotiu A, Brazdova A, Longé C, Gallet P, Morisset M, Leduc V et al (2016) *Urtica dioica* pollen allergy: clinical, biological, and allergomics analysis. Ann Allergy Asthma Immunol 117:527–534

5. Poncet P, Sénéchal H, Clement G, Purohit A, Sutra JP, Desvaux FX et al (2010) Evaluation of ash pollen sensitization pattern using proteomic approach with individual sera from allergic patients. Allergy 65:571–580
6. D'Amato A, Bachi A, Fasoli E, Boschetti E, Peltre G, Sénéchal H et al (2010) In-depth exploration of *Hevea brasiliensis* latex proteome and "hidden allergens" via combinatorial peptide ligand libraries. J Proteome 73:1368–1380
7. Shahali Y, Sutra JP, Fasoli E, D'Amato A, Righetti PG, Futamura N et al (2012) Allergomic study of cypress pollen via combinatorial peptide ligand libraries. J Proteome 77:101–110
8. Shahali Y, Sutra JP, Peltre G, Charpin D, Sénéchal H, Poncet P (2010) IgE reactivity to common cypress (C. Sempervirens) pollen extracts: evidence for novel allergens. W Allergy Organ J 3:229–234
9. Shahali Y, Nicaise P, Brazdova A, Charpin D, Scala E, Mari A et al (2014) Complementarity between microarray and immunoblot for the comparative evaluation of IgE repertoire of French and Italian cypress pollen allergic patients. Folia Biol 60:192
10. Shahali Y, Sutra JP, Haddad I, Vinh J, Guilloux L, Peltre G et al (2012) Proteomics of cypress pollen allergens using double and triple one-dimensional electrophoresis. Electrophoresis 33:462–469
11. Altland K, Silke R, Hackler R (1981) Demonstration of human prealbumin by double one-dimensional slab gel electrophoresis. Electrophoresis 2:148–155
12. Blum H, Beier H, Gross HJ (1987) Improved silver staining of plant proteins, RNA and DNA in polyacrylamide gels. Electrophoresis 8:93–99
13. Demeulemester C, Peltre G, Laurent M, Panheleux D, David B (1987) Cyanogen bromide-activated nitrocellulose membranes: a new tool for immunoprint techniques. Electrophoresis 8:71–73
14. Shahali Y, Sutra JP, Hilger C, Swiontek K, Haddad I, Vinh J et al (2017) Identification of a polygalacturonase (Cup s 2) as the major CCD-bearing allergen in *Cupressus sempervirens* pollen. Allergy 72:1806–1810
15. Sénéchal H, Šantrůček J, Melčová M, Svoboda P, Zídková J, Charpin D et al (2018) A new allergen family involved in pollen food-associated syndrome: Snakin/gibberellin-regulated proteins. J Allergy Clin Immunol 141:411–414

Chapter 10

BioID: A Proximity-Dependent Labeling Approach in Proteomics Study

Peipei Li, Yuan Meng, Li Wang, and Li-jun Di

Abstract

Biological activities are mainly executed by proteins and in most of the occasions these activities are accomplished by protein complexes or through protein–protein interactions (PPI). So it is critical to reveal how the protein complexes are organized and demonstrate the PPIs involved in the biological processes. In addition to the traditional biochemical approaches, proximity-dependent labeling (PDL) has recently been proposed to identify the interacting partners of a given protein. PDL requires the fusion expression of the target protein with an enzyme which catalyzes the attachment of a reactive molecule to the interacting partners in a distance-dependent manner. Further analysis of all the proteins that are modified by the reactive molecule discloses the identity of these proteins which are presumed to be interacting partners of the target protein. BioID is one of those representative PDL methods with the most widely applications. The enzyme used in BioID is a biotin ligase BirA which catalyzes the biotinylation of target protein with the presence of biotin. Through streptavidin-mediated pull-down and mass spectrometry analysis, the interacting protein candidates of a given protein can be obtained.

Key words BioID, Protein–protein interactions, Proximity-dependent labeling

1 Introduction

Either stably or transiently, the proteins need to form complexes via PPIs to perform important biological functions [1, 2]. So the PPIs are the most fundamental biological activity that are critical to accomplish the biological processes. Thus the depicting of protein complex relies on the recognition of PPIs. Traditionally, the identification of PPI relies on the biochemical approaches, i.e., the interacting proteins can be captured by immunoprecipitating the target protein [3]. The co-immunoprecipitation requires the lysis of cells and known as in vitro technology. Some technologies are able to capturing PPIs in vivo such as two/three yeast hybridization system [4]. Comparing to the technologies in vitro, the in vivo determination of PPIs in live cell is preferred because it discloses the truly interacting proteins at the physiological condition. However, the

Xing Wang and Matthew Kuruc (eds.), *Functional Proteomics: Methods and Protocols*, Methods in Molecular Biology, vol. 1871, https://doi.org/10.1007/978-1-4939-8814-3_10,

specificity of the antibody used in these in vivo methods has always a big problem.

Fluorescence resonance emission transfer (FRET) technology has been applied to directly demonstrate the PPI in live cells. This technology is based on the fact that when the suspected interacting pair of proteins are both fused to different fluorescent proteins, the fused fluorophore will be close enough to transfer energy from the fluorescent donor to the fluorescent receptor that will alternate the wavelength of the emitted fluorescent signal of the energy receptor. Through analyzing the change of the fluorescent signal emitted by the energy receptor, the PPI can be quantitated. Another technology is known as bimolecular complementation technology. The proteins that are expected to interact are fused to complementary fragments of a fluorescent protein and the PPI will bring the complementary fragments in proximity and the fluorescent signal recovers [5, 6]. However, both of these methods can only validate the suspected interacting pairs of proteins but not discover novel ones.

The PDL methods are recently proposed and provide much more flexibility and reliability in identifying novel PPIs in vivo. The advantage of PDL methods is that both the direct interacting proteins, but also the proteins in close proximity can be identified. Roughly, the frequency or interacting affinity of PPIs can be quantitated through analyzing the hit rate of the peptide in mass analysis. The development of PDL methods relies on the enzymes that are capable of modifying the nearby proteins in vivo by catalyzing the attachment of reactive groups to the target amino acids of any proteins in proximity. Several PDL approaches have been reported such as peroxidase-based techniques including selective proteomic proximity labeling assay using tyramide (SPPLAT), enzyme-mediated activation of radical sources (EMARS), and Ascorbate Peroxidase (APEX) [7–9]. Among these methods, BioID is the earliest and most widely applied PDL method [10].

The prototype of BioID is based on the discovery that the biotin ligase BirA, isolated from *Escherichia coli,* can biotinylate the target proteins in mammalian cells. The limitation of the prototype of BioID is that the target protein has to have a consensus sequence which is recognized by BirA [11, 12]. The reason is that two continuous steps are required in BirA catalyzed biotinylation. Firstly, BirA catalyzes the formation of the reactive molecule 5-′-AMP-Biotin; this molecule cannot be released but stay at the reaction center. At the second step, the ε-amino groups of lysine residues of a target protein binds BirA reactive center, attacks the anhydride of the 5′-AMP-biotin to generate an amide bond and the biotin is transferred to the target protein [13, 14]. By mutating one amino acid (R118G; also known as BirA*), Roux et al. demonstrated that BirA* can be engineered to release 5′-AMP-biotin and biotinylate almost any proteins even when the required consensus sequence is absent. The upgraded technology is now known as BioID [15]. Since the instability of the reactive molecule, only the proteins that are close to the BirA reaction center have the chance to be modified. The biotinylated proteins can be captured by streptavidin-coated agarose beads and further analyzed by untargeted mass spectrometry or other targeted technologies [16].

BioID has been successfully applied in determining the composition of insoluble protein complexes including lamina and centrosomes [17, 18]; the components of cytoplasm membrane-bound protein complexes such as the tight junctions [19]; the Hippo pathway PPIs [20]; and the composition of protein complexes from infectious pathogens such as *Toxoplasma gondii*, HIV, and EBV [21–23]. Since BioID is an extremely sensitive labeling method in vivo, the nonspecific modifying of proteins is inevitable. To overcome this problem, some studies tried multi-bait in applying BioID and the real PPIs can be identified more efficiently [24].

In this manual, we introduce an application example of BioID technology and the detailed protocol is also described.

2 Materials

NOTE: All the reagents should be analytical grade; preparing the solutions by using 18 MΩ-cm of ultrapure water; all the reagents and solutions should be stored at indicated temperature. Only the specially required reagents are listed here and other regular required reagents are omitted.

2.1 BioID Vector

1. Expression vectors for BirA* can be obtained from Addgene (pcDNA3.1 MCS-BirA(R118G)-HA #36047 or pcDNA3.1 mycBioID #35700).
2. Cloned coding sequence of the interested protein.

2.2 Validation of BioID Fusion Protein

1. 293T or other appropriate cell lines and the appropriate cell culture medium.
2. 1 mM Biotin: 12.2 mg biotin (Sigma) dissolved in 50 mL H_2O, sterilize by passing through a 0.22-μm syringe-driven filter, store at 4 °C.
3. HRP-streptavidin (Sigma).
4. Primary antibodies for BioID fusion protein (e.g., anti-HA/MYC, anti-interest protein).
5. Streptavidin-Alexa Fluor.
6. DNA labeling reagent (e.g., DAPI).

2.3 BioID Pull-Down

1. Streptavidin conjugated beads (MyOne Streptavidin C1, Invitrogen).
2. Wash buffer 1: 2% SDS in H_2O, store at room temperature.
3. Wash buffer 2: 0.1% deoxycholate, 1% Triton X-100, 500 mM NaCl, 1 mM EDTA, and 50 mM Hepes, pH 7.5, store at room temperature.
4. Wash buffer 3: 250 mM LiCl, 0.5% NP-40, 0.5% deoxycholate, 1 mM EDTA, and 10 mM Tris, pH 8.0, store at room temperature.
5. Wash buffer 4: 50 mM Tris, pH 7.4, store at room temperature.
6. ACN buffer: 50 mM ammonium bicarbonate, store at room temperature.

3 Methods

NOTE: Conducting all procedures at room temperature unless otherwise specified.

3.1 Generation of BioID Construct

In order to study the interacting proteins of a give protein, this protein needs to be expressed in fusion with BirA*. The expression vectors of BirA* have provided multiple cloning site and the target protein coding sequence can be cloned into this site. The gene clone strategy is not given in this protocol. The fusion protein (now named as BioID vector) shouldn't affect the function and localization of the original protein, gene fragment inserted into either the N- or C-terminus of biotin ligase should be carefully considered. Additionally, the fusion protein may be observed to ensure the proper intracellular localization and showing the expected function.

NOTE: The BioID vector should contain HA tag or Myc tag for easy detection of the fusion protein. Also, it is important to select N-terminal or C-terminal for insertion of the target protein because the fusion protein may be influenced functionally if the insertion site hasn't been carefully tested.

3.2 Validation of BioID Fusion Protein

1. Preparing BioID vector by using QIAGEN Plasmid preparation kits.
2. Preparing a six-well plate cell culture (and/or an eight-well chamber cell culture) ready for transfection by the following vectors: empty vector with or without biotin, BioID vector with or without biotin.

NOTE: The empty vector transfection is important negative control for validation of BioID fusion protein and subsequent LC-mass analysis.

3. Transfecting the cells with the above conditions using Lipofectamine 3000, add 50 μM biotin into the culture medium 3 h post-transfection.

NOTE: The selection of transfection reagent should depend on cell lines. Excess biotin enhances protein biotinylation, but the concentration of 50uM should be sufficient for experiment.

4. After transfection for 24 h, the six-well plate cultured cells are ready for immunoblot assay and the eight-well chamber cultured cells are ready for immunofluorescence assay.
5. For immunoblot analysis, wash the cells with PBS prior to lysis, apply 100 μL cell lysis buffer into the cells on ice for 30 min with the presence of protease inhibitors, sonicate the sample, and centrifuge at 15,000 × *g* for 10 min at 4 °C. If the target protein is nuclear protein, nuclear extract is recommended.
6. Adding 5× SDS-PAGE sample loading buffer and heat to 98 °C for 5 min to denature proteins.
7. Performing SDS-PAGE electrophoresis and protein transfer, incubate membrane in 1% BSA blocking buffer for 30 min at room temperature.
8. Incubating membrane in primary antibodies for BioID fusion protein (e.g., anti-HA/MYC antibodies or anti-target protein antibody) for 1 h at room temperature.
9. Washing the membrane five times with PBST, 5 min each time.
10. Incubating membrane in secondary antibodies to detect primary antibody for 1 h at room temperature.
11. Washing the membrane five times with PBST, 5 min each time.
12. Taking images of the fusion protein by ECL. A representative western blotting of CtBP2-BirA* is provided as Fig. 1a.
13. Stripping the membrane in stripping buffer for 20 min, and rinsing membrane with PBST several times to remove the striping buffer.
14. Blocking the cells with 1% BSA for 0.5 h, followed by incubation of streptavidin-HRP (1:20,000) for 1 h at room temperature.
15. Washing the membrane five times with PBST, 5 min each time.
16. Observing the biotinylated proteins using ECL. The representative western blotting is provided as Fig. 1b. Be noted that the appearance of the spear indicates the BioID pull-down is successful.

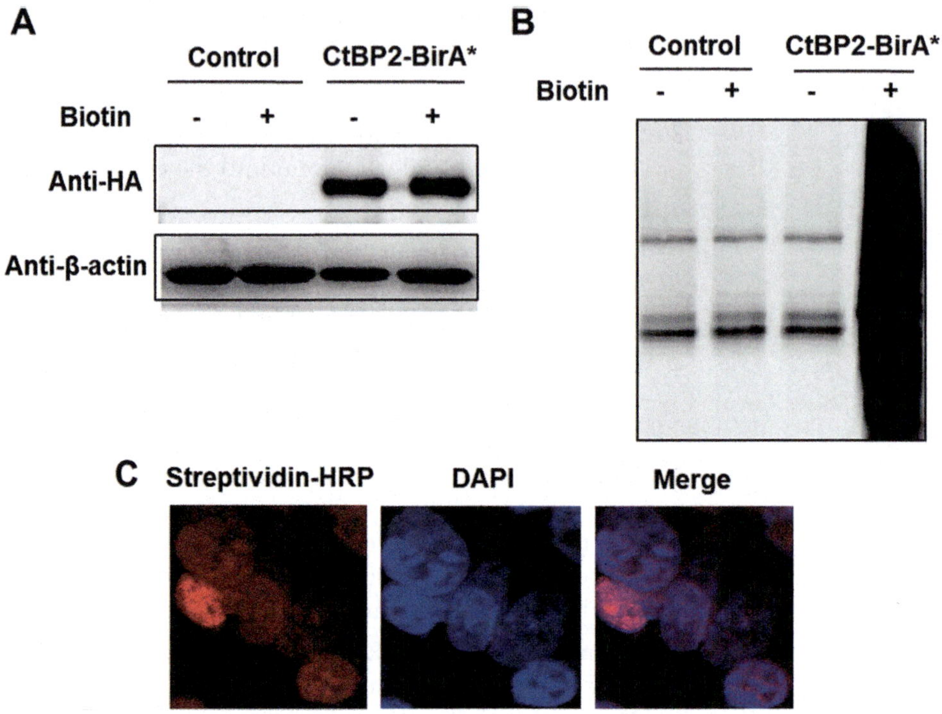

Fig. 1 Detection of biotinylated proteins by BioID in HEK293 cells expressing CtBP2-BirA*. (**a**) Cells expressing CtBP2-BirA* and control cells were incubated for 24 h with and without 50 μM biotin, the expression of CtBP2-BirA* was detected by anti-HA, and β-actin served as loading control. (**b**) Biotinylated proteins were identified by HRP-streptavidin in different conditions. (**c**) Fluorescence microscopy was used to observe the biotinylated proteins in CtBP2-BirA* overexpressed cells incubating for 24 h with 50 μM biotin, biotinylated proteins were detected by streptavidin-Alexa Fluor 594 (red), DNA was detected with DAPI (blue)

17. For immunofluorescence assay, fix the cell with 4% PFA for 10 min at room temperature, then permeate the cells with 0.2% TritonX-100 for 10 min.
18. Blocking the cells with 1% BSA for 0.5 h, followed by incubation of streptavidin-Alexa Fluor (1:1000) and DAPI, observe the fluorescence by microscopy (Fig. 1c).

NOTE: BSA has better effect in getting rid of free biotin than milk.

3.3 BioID Pull-Down Assay

This protocol describes the cells transiently expressing BioID fusion protein to process large-scale (6×10^7 cells) BioID pull-down assay and LC-mass (Liquid chromatograph mass spectrometry) analysis. Cells stably expressing BioID fusion protein are also subjected to BioID pull-down as well.

1. Plating two 15-cm dishes for each experimental condition (cells expressing BioID construct and control cells).

2. Transfecting empty vector or BioID vector into the cells using Lipofectamine or other appropriate transfection reagents and supplement 50 μM biotin into the culture medium 3 h post-transfection.

3. Incubating cells for 24 h.

NOTE: 24 h is enough for protein biotinylation, and extending incubation time will reduce the amount of biotinylated protein.

4. Washing the cells twice with PBS to remove the free biotin.

5. Adding 1.2 mL cell lysis buffer per dish into the cells, scrape and collect the cells.

6. Placing the tube on ice for 30 min, sonicate to shear DNA and centrifuge the samples at 15,000 × *g* for 10 min at 4 °C.

NOTE: Mix the sample well by vortex, and pipet the lysates up and down every 10 min during the incubation. Sonication is to break down the DNA and solubilize the proteins.

7. The supernatant of cell lysates are gently transferred to 2-mL tubes and are diluted to 2.5-fold with prechilled 50 mM Tris·Cl, pH 7.4. Subsequently the lysates are aliquoted to 1.5 mL per tube.

8. The magnetic streptavidin beads need to be equilibrium in 1:1 lysis buffer and 50 mM Tris·Cl, pH 7.4. Using the magnetic separation stand to collect the magnetic beads and remove the buffer after equilibrium.

9. The supernatant in **step** 7 are transferred to the tube in **step 8**, mix the samples and beads gently, and incubate the mixture on the rotator over night at 4 °C.

10. The tubes are placed on magnetic separation stand for 3 min until the beads accumulate at one side of tube, remove the supernatant by pipet. Avoid disturbing the beads on the tube wall.

11. Washing the magnetic beads once sequentially by Wash Buffer 1 to 3 once and twice by Wash Buffer 4.

12. Resuspending the beads with 200 μL of 50 mM Tris·Cl, pH 7.4.

13. 10% of the sample is reserved for SDS-PAGE, other 90% sample is washed twice by 200 μL of 50 mM ammonium bicarbonate. Sample volume is adjusted to 50 μL by 50 mM ammonium bicarbonate for LC-mass preparation or storage at −80 °C.

NOTE: The protein identification by LC-mass can be performed by specialist or commercial service providers and the details are omitted here.

4 Summary

We have successfully applied BioID to identify the interacting proteins of the nuclear transcriptional factors for several cases. Of noting is that BioID is extremely sensitive and produces unneglectable background. So for each experiment, several experimental repeats are strongly recommended and the empty vector control should also be included for the LC-mass analysis. To our experience, only the proteins that are repeatedly identified throughout the experiment repeats but not in the negative control groups are potential candidates, given their score in LC-mass analysis is among the top.

We also notice that after BioID pull-down, the western blotting is sensitive enough to validate the suspected PPIs. So BioID can be a reliable tool to validate the PPIs besides being an explorative tool to identify novel PPIs.

Acknowledgments

This work is supported by the Science and Technology Development Fund (FDCT) of Macao SAR (FDCT/0014/2018/A1), the Multi-Year Research Grant from the University of Macau (MYRG2018-00158-FHS), and the National Natural Science Foundation of China (NSFC 81772980) to LD. This work is also supported by the Multi-Year Research Grant from the University of Macau to LW (MYRG2016-00251-FHS).

References

1. Nooren IM, Thornton JM (2003) Diversity of protein-protein interactions. EMBO J 22 (14):3486–3492
2. Ngounou Wetie AG, Sokolowska I, Woods AG, Roy U, Loo JA, Darie CC (2013) Investigation of stable and transient protein-protein interactions: past, present, and future. Proteomics 13(3–4):538–557. https://doi.org/10.1002/pmic.201200328
3. Vermeulen M, Hubner NC, Mann M (2008) High confidence determination of specific protein-protein interactions using quantitative mass spectrometry. Curr Opin Biotechnol 19 (4):331–337. https://doi.org/10.1016/j.copbio.2008.06.001
4. Stynen B, Tournu H, Tavernier J, Van Dijck P (2012) Diversity in genetic in vivo methods for protein-protein interaction studies: from the yeast two-hybrid system to the mammalian split-luciferase system. Microbiol Mol Biol Rev 76(2):331–382. https://doi.org/10.1128/MMBR.05021-11
5. Kenworthy AK (2001) Imaging protein-protein interactions using fluorescence resonance energy transfer microscopy. Methods 24(3):289–296. https://doi.org/10.1006/meth.2001.1189
6. Sjohamn J, Bath P, Neutze R, Hedfalk K (2016) Applying bimolecular fluorescence complementation to screen and purify aquaporin protein:protein complexes. Protein Sci 25(12):2196–2208. https://doi.org/10.1002/pro.3046
7. Li XW, Funk PE, Rees JS, Farndale RW, Xue P, Lilley KS et al (2014) New insights into the DT40 B cell receptor cluster using a proteomic proximity labeling assay. J Biol Chem 289 (6):14434–14447. https://doi.org/10.1074/jbc.M113.529578

8. Miyagawa-Yamaguchi A, Kotani N, Honke K (2014) Expressed glycosylphosphatidylinositol-anchored horseradish peroxidase identifies co-clustering molecules in individual lipid raft domains. PLoS One 9(3):e93054. https://doi.org/10.1371/journal.pone.0093054.g001
9. Rhee HW, Zou P, Udeshi ND, Martell JD, Mootha VK, Carr SA et al (2013) Proteomic mapping of mitochondria in living cells via spatially restricted enzymatic tagging. Science 339:1328–1331
10. Li P, Li J, Wang L, Di LJ (2017) Proximity labeling of interacting proteins: application of BioID as a discovery tool. Proteomics 17(20). https://doi.org/10.1002/pmic.201700002
11. Parrott MB, Barry MA (2000) Metabolic biotinylation of recombinant proteins in mammalian cells and in mice. Mol Ther 1(1):96–104. https://doi.org/10.1006/mthe.1999.0011
12. Parrott MB, Barry MA (2001) Metabolic biotinylation of secreted and cell surface proteins from mammalian cells. Biochem Biophys Res Commun 281(4):993–1000. https://doi.org/10.1006/bbrc.2001.4437
13. Chapman-Smith A, Morris TW, Wallace JC, Cronan JE (1999) Molecular recognition in a post-translational modification of exceptional specificity. J Biol Chem 274(3):1449–1457
14. Prakash O, Eisenberg MA (1979) Biotinyl 5-′-adenylate corepressor role in the regulation of the biotin genes of Escherichia coli K-12. Proc Natl Acad Sci U S A 76:5592–5595
15. Roux KJ, Kim DI, Raida M, Burke B (2012) A promiscuous biotin ligase fusion protein identifies proximal and interacting proteins in mammalian cells. J Cell Biol 196(6):801–810. https://doi.org/10.1083/jcb.201112098
16. Kuroishi T, Rios-Avila L, Pestinger V, Wijeratne SS, Zempleni J (2011) Biotinylation is a natural, albeit rare, modification of human histones. Mol Genet Metab 104(4):537–545. https://doi.org/10.1016/j.ymgme.2011.08.030
17. Xie W, Chojnowski A, Boudier T, Lim JS, Ahmed S, Ser Z et al (2016) A-type lamins form distinct filamentous networks with differential nuclear pore complex associations. Curr Biol 26(19):2651–2658. https://doi.org/10.1016/j.cub.2016.07.049
18. Firat-Karalar EN, Stearns T (2015) Probing mammalian centrosome structure using BioID proximity-dependent biotinylation. Methods Cell Biol 129:153–170. https://doi.org/10.1016/bs.mcb.2015.03.016
19. Van Itallie CM, Aponte A, Tietgens AJ, Gucek M, Fredriksson K, Anderson JM (2013) The N and C termini of ZO-1 are surrounded by distinct proteins and functional protein networks. J Biol Chem 288 (19):13775–13788. https://doi.org/10.1074/jbc.M113.466193
20. Couzens AL, Knight JD, Kean MJ, Teo G (2013) Protein interaction network of the mammalian Hippo pathway reveals mechanisms of kinase-phosphatase interactions. Sci Signal 6:rs15
21. Nourani E, Khunjush F, Durmus S (2015) Computational approaches for prediction of pathogen-host protein-protein interactions. Front Microbiol 6:94. https://doi.org/10.3389/fmicb.2015.00094
22. Le Sage V, Cinti A, Valiente-Echeverria F, Mouland AJ (2015) Proteomic analysis of HIV-1 Gag interacting partners using proximity-dependent biotinylation. Virol J 12:138. https://doi.org/10.1186/s12985-015-0365-6
23. Holthusen K, Talaty P, Everly DN Jr (2015) Regulation of latent membrane protein 1 signaling through interaction with cytoskeletal proteins. J Virol 89(14):7277–7290. https://doi.org/10.1128/JVI.00321-15
24. Gupta GD, Coyaud É, Gonçalves J, Mojarad BA, Liu Y, Wu Q et al (2016) A dynamic protein interaction landscape of the human centrosome-cilium interface. Cell 163 (6):1484–1499. https://doi.org/10.1016/j.cell.2015.10.065

Chapter 11

Functional Application of Snake Venom Proteomics in In Vivo Antivenom Assessment

Choo Hock Tan and Kae Yi Tan

Abstract

Reverse-phase high-performance liquid chromatography is commonly employed as a decomplexing strategy in snake venom proteomics. The chromatographic fractions often contain relatively pure toxins that can be assessed functionally for toxicity level through the determination of their median lethal doses (LD_{50}). Further, antivenom efficacy can be evaluated specifically against these venom fractions to understand the limitation of the antivenom as the treatment for snake envenomation. However, methods of toxicity assessment and antivenom evaluation vary across laboratories; hence there is a need to standardize the protocols and parameters, in particular those related to the neutralizing efficacy of antivenom. This chapter outlines the important in vivo techniques and data interpretation that can be applied in the functional study of snake venom proteomes.

Key words Immuno-neutralization, Toxin-specific neutralization, Median lethal dose (LD_{50}), Median effective dose (ED_{50}), Potency (P)

1 Introduction

The complexity of snake venom can be greatly resolved through various proteomic techniques (for instance: decomplexing proteomics, Chapter 5) [1]. The identification and quantitative estimation of proteins in a venom has become relatively more achievable and time-saving [2, 3]. In decomplexing venom proteomics, high-resolutional reverse-phase high-performance liquid chromatography separates venom into pure/partially pure fractions that can be subjected to mass spectrometric analysis [4, 5]. In addition, the HPLC fractionation allows functional assessment of the venom components from a bottom-up approach, where the toxicity of the individual venom fractions and neutralization by antivenom can be determined [4, 6]. The approach for the functional correlation of venom proteome typically involves laboratory animal models to elucidate the in vivo pathophysiology of snake envenomation,

Xing Wang and Matthew Kuruc (eds.), *Functional Proteomics: Methods and Protocols*, Methods in Molecular Biology, vol. 1871, https://doi.org/10.1007/978-1-4939-8814-3_11,

and is essential for the robust preclinical evaluation of antivenom efficacy in neutralizing the venom toxicity [7].

To interpret the systemic toxicity of a snake venom, the venom fractions are usually administered through intravenous route to ensure a full systemic access of the venom components into the animal. The use of in vivo model is favored as it involves the whole biological system in which the snake venom can act. Using the in vivo model, the evolution of clinical syndrome upon envenomation can also be closely monitored, for instance the development of neuromuscular paralysis induced by the venom [8]. Through the animal model, the lethal effect of the various venom fractions can be determined. By correlating the data with the venom proteome, it is possible to determine the principal toxins responsible for venom-induced lethality. Of note, the lethality parameter is essential in antivenom studies as the neutralization of lethality has been regarded as the gold standard for antivenom assessment by WHO [9, 10]. Hence, functional venom proteomics can provide valuable information on the strength and weaknesses of an antivenom in neutralizing a venom as well as its principal toxins [11, 12]. However, the determination of the efficacy parameters of an antivenom often varies among different laboratories, and the interpretation (as well as comparison) of the data across different studies has been challenging. This chapter illustrates the common protocol that can be adopted to study the lethality of venom and toxin components, and recommends the standardization of parameters that are essential to gauge the efficacy or potency of antivenom.

2 Materials

2.1 Snake Venom/Toxin Fractions

Lyophilize the snake venom or its toxin fractions (obtained from protein decomplexation approach, Chapter 5). Store at −20 °C until use.

2.2 Snake Antivenoms

Lyophilized antivenom: Reconstitute the lyophilized antivenom according to manufacturer's instruction. Aliquot and store at −20 °C until use.

Liquid form antivenom: Dilute the liquid antivenom accordingly for the neutralization study. Store at 2–8 °C until use.

2.3 Chemical and Solution

Reconstitution buffer for antivenom: normal saline.

Protein concentration estimation: Bicinchoninic acid (BCA) assay/Lowry assay.

3 Methods

3.1 Lethality Determination of Venom/Toxin Fraction

1. Divide ICR mice into four groups ($n = 5$ per group, 20–25 g body weight). Each group of mice receives venom or toxin fraction at different concentrations (treated as "doses," see below). Estimate the median lethal dose using at least four doses of the venom or individual toxin fraction (*see* **Note 1**).
2. *Sample preparation:* Estimate the protein concentration of venom/toxin fraction using bicinchoninic acid assay or Lowry assay. Dilute the venom/toxin fraction with normal saline to serial concentrations. Fix the volume of injection at 100 μL per mouse.
3. Hold the mice using a rodent restrainer. Inject the appropriately diluted venom/toxin fraction intravenously into the mice via caudal vein (*see* **Note 2**).
4. Allow the mice access to food and water *ad libitum*. Monitor and record the survival ratio of the mice in each group for 48 h (*see* **Note 3**).
5. Determine the median lethal dose (LD_{50}) of the venom/toxin fraction using Probit analysis (*see* **Note 4**).
   **Figure 1 shows the schematic drawing of the assessment of median lethal dose (LD_{50}) of venom/toxin fraction.*

3.2 Lethality Neutralization of Venom/Toxin Fraction

1. Divide ICR mice into four groups ($n = 5$ per group, 20–25 g body weight). Each group of mice receives venom or toxin fraction that has been preincubated with varying doses of antivenom (see below). Estimate the antivenom efficacy and

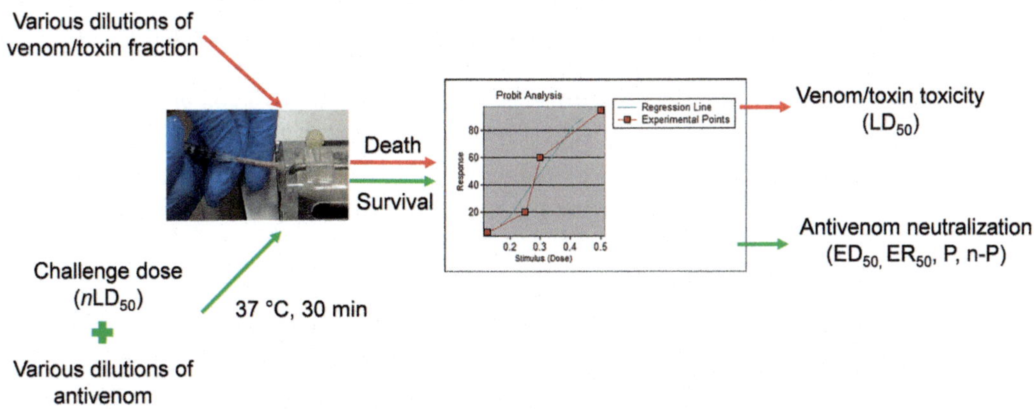

Fig. 1 A schematic diagram shows the assessment of median lethal dose (LD_{50}) of venom or toxin fraction (flow following red arrows) and neutralization efficacy as well as potency of antivenom (flow following green arrows). The results are determined using Probit analysis based on the recorded survival ratio throughout the experiment. The test samples (venom/toxin fraction and antivenom mixtures) are intravenously injected into mice to ensure a full systemic access under a control titration

potency using at least four antivenom doses for the neutralization of each venom or individual toxin fraction (*see* **Note 5**).

2. *Antivenom protein concentration*: Estimate the protein concentration of antivenom using bicinchoninic acid assay or Lowry assay.
3. *Antivenom incubation*: Prepare a challenge dose of venom/toxin fraction (5× LD_{50}) in 50 μL normal saline. Mix this challenge dose with various dilutions of antivenom in normal saline to give a total volume of 250 μL per injection. Incubate the venom/toxin fraction-antivenom mixture at 37 °C for 30 min.
4. Hold the mice using a rodent restrainer. Inject the 250 μL preincubated mixture intravenously into the mice via caudal vein.
5. Allow the mice access to food and water *ad libitum*. Monitor and record the survival ratio of the mice for 48 h.
6. Determine the efficacy and potency of antivenom neutralization through the following parameters:
   (a) Median effective dose, ED_{50}: Volume of reconstituted/liquid antivenom in μL at which 50% of mice survived (*see* **Note 6**).
   (b) Median effective ratio, ER_{50}: Ratio of the amount of venom/toxin (mg) to the volume dose of antivenom (mL) at which 50% of mice survived.
   (c) Potency, P: The amount of venom/toxin (mg) neutralized completely by 1 mL antivenom (*see* **Note 7**).
   (d) Normalized potency, n-P: The amount of venom/toxin (mg) that is completely neutralized by 1 g of antivenom proteins (*see* **Note 8**).
   **Figure 1 shows the schematic drawing of the assessment of antivenom neutralization against venom/toxin fractions.*
7. Adopt the following formulae for calculating the challenge dose of venom/toxin fraction and the various neutralization parameters for antivenom:

$$\text{Challenge dose } (\mu g) = n \times LD_{50}\ (\mu g/g) \times \text{mouse weight}(g)$$

*[As indicated in Subheading 3.2, **step 3**]*

$$\text{Median effective dose, } ER_{50} \text{ (standardize the unit as mg/mL)} = \frac{[n \times LD_{50}\ (\mu g/g) \times \text{mouse weight}(g)]}{ED_{50}(\mu L)}$$

*[As indicated in Subheading 3.2, **step 6b**]*

$$\text{Potency, P (standardize the unit as mg/mL)} = \frac{[(n-1) \times \text{LD}_{50}\,(\mu\text{g/g}) \times \text{mouse weight(g)}]}{\text{ED}_{50}\,(\mu\text{L})}$$

[As indicated in Subheading 3.2, ***step 6c****]*

$$\text{Normalized potency, n} - \text{P (standardize the unit as mg/g)} = \frac{\text{Potency, P (mg/mL)}}{\text{Antivenom protein concentration (mg/mL)}} \times 1000$$

[As indicated in Subheading 3.2, ***step 6d****].*

4 Notes

1. The initial starting dose for median lethal dose (LD50) determination can be estimated based on reported values for venoms or toxin derived from closely related species.
2. The route of administration is intravenous to ensure a full systemic access of the venom or toxin into the animal. This enables the assessment and interpretation of the systemic toxicity of venom/toxin that becomes fully bioavailable to the animal.
3. The survival ratios obtained from the lethality assay should contain an upper dose which shows 100% death of mice ($n = 5$), a lower dose with 100% survival ($n = 5$), and intermediate doses with a mix of death and survival.
4. Median lethal dose (LD_{50}) is determined with Probit analysis applying the Finney method.
5. If 200 μL of reconstituted antivenom failed to provide full protection to the mice, a lower challenge dose (2.5× or 1.5× LD_{50}) can be used. All challenge doses should be proven to be above the lethal dose 100% (LD_{100}) when injected intravenously into the mice. This can be assessed in an additional group of mice constituting the control.
6. The survival ratios obtained from the neutralization assay should contain an upper dose which shows 100% survival of mice ($n = 5$), a lower dose with 100% death ($n = 5$), and intermediate doses with a mix of death and survival. Median effective dose (ED_{50}) is determined using Probit analysis.
7. The neutralization potency (P) is an indicator of antivenom neutralizing capacity and is theoretically independent of the dosage of challenge dose. This is because it takes into consideration the dose of antivenom that is able to completely neutralize the lethal effect of venom/toxin by subtracting 1× LD_{50}

($n - 1$) from the total challenge dose as shown the formula under Subheading 3.2, **step 7**.

8. The normalized potency (n-P) takes into consideration the antivenom protein amount which could be varied between different products. By normalizing the P values of different antivenoms by their respective protein amounts, the n-P values can be used to compare the efficacy of neutralization across different antivenom products.

References

1. Calvete JJ (2013) Snake venomics: from the inventory of toxins to biology. Toxicon 75 (Suppl C):44–62 https://doi.org/10.1016/j.toxicon.2013.03.020
2. Tan KY, Tan CH, Fung SY, Tan NH (2015) Venomics, lethality and neutralization of *Naja kaouthia* (monocled cobra) venoms from three different geographical regions of Southeast Asia. J Proteome 120:105–125. https://doi.org/10.1016/j.jprot.2015.02.012
3. Wong KY, Tan CH, Tan KY, Naeem QH, Tan NH (2018) Elucidating the biogeographical variation of the venom of *Naja naja* (spectacled cobra) from Pakistan through a venom-decomplexing proteomic study. J Proteome 175:156–173. https://doi.org/10.1016/j.jprot.2017.12.012
4. Tan CH, Tan KY, Lim SE, Tan NH (2015) Venomics of the beaked sea snake, *Hydrophis schistosus*: a minimalist toxin arsenal and its cross-neutralization by heterologous antivenoms. J Proteome 126:121–130. https://doi.org/10.1016/j.jprot.2015.05.035
5. Oh AMF, Tan CH, Ariaranee GC, Quraishi N, Tan NH (2017) Venomics of *Bungarus caeruleus* (Indian krait): comparable venom profiles, variable immunoreactivities among specimens from Sri Lanka, India and Pakistan. J Proteome 164:1–18. https://doi.org/10.1016/j.jprot.2017.04.018
6. Tan CH, Tan KY, Yap MK, Tan NH (2017) Venomics of *Tropidolaemus wagleri*, the sexually dimorphic temple pit viper: unveiling a deeply conserved atypical toxin arsenal. Sci Rep 7:43237. https://doi.org/10.1038/srep43237
7. Tan CH, Wong KY, Tan KY, Tan NH (2017) Venom proteome of the yellow-lipped sea krait, *Laticauda colubrina* from Bali: insights into subvenomic diversity, venom antigenicity and cross-neutralization by antivenom. J Proteome 166:48–58. https://doi.org/10.1016/j.jprot.2017.07.002
8. Tan KY, Tan CH, Sim SM, Fung SY, Tan NH (2016) Geographical venom variations of the Southeast Asian monocled cobra (*Naja kaouthia*): venom-induced neuromuscular depression and antivenom neutralization. Comp Biochem Physiol C Toxicol Pharmacol 185–186:77–86 https://doi.org/10.1016/j.cbpc.2016.03.005
9. World Health Organization (2010) WHO Guidelines for the production control and regulation of snake antivenom immunoglobulins. WHO publication, 1–141
10. Faisal T, Tan KY, Sim SM, Quraishi N, Tan NH, Tan CH (2018) Proteomics, functional characterization and antivenom neutralization of the venom of Pakistani Russell's viper (Daboia russelii) from the wild. J Proteome 183:1–13. https://doi.org/10.1016/j.jprot.2018.05.003
11. Tan KY, Tan CH, Fung SY, Tan NH (2016) Neutralization of the principal toxins from the venoms of Thai *Naja kaouthia* and Malaysian *Hydrophis schistosus:* insights into toxin-specific neutralization by two different antivenoms. Toxins 8(4):86. https://doi.org/10.3390/toxins8040086
12. Wong KY, Tan CH, Tan NH (2016) Venom and purified toxins of the spectacled cobra (*Naja naja)* from Pakistan: insights into toxicity and antivenom neutralization. Am J Trop Med Hyg 94(6):1392–1399. https://doi.org/10.4269/ajtmh.15-0871

Chapter 12

Proteomic Detection of Carbohydrate-Active Enzymes (CAZymes) in Microbial Secretomes

Tina R. Tuveng, Vincent G. H. Eijsink, and Magnus Ø. Arntzen

Abstract

Secretomes from microorganisms growing on biomass contain carbohydrate-active enzymes (CAZymes) of potential biotechnological interest. By analyzing such secretomes, we may discover key enzymes involved in degradation processes and potentially infer the mode-of-action of biomass conversion. Some of these enzymes may have predicted functions in carbohydrate degradation, while others may not, while yet exhibiting a similar expression pattern; these latter enzymes constitute potential novel enzymes involved in the degradation process and provide a basis for further biochemical exploration. Hence, secretomes represent an important source for the study of both predicted and novel CAZymes. Here we describe a plate-based culturing technique that allows for collection of protein fractions that are highly enriched for secreted proteins, bound or unbound to the substrate, and which minimizes contamination by intracellular proteins trough unwanted cell lysis.

Key words Secretomics, Proteomics, Protein secretion, Carbohydrate-active enzymes, CAZymes

1 Introduction

Polysaccharides such as cellulose, hemicellulose, pectin, and chitin are abundantly produced in Nature but do not accumulate due to removal by the concerted action of specialized microbes and microbial consortia, including fungi and bacteria. These microorganisms exploit sophisticated enzyme systems to degrade biomass and the enzymes involved have potential in biotechnological applications, such as in biofuel production [1]. Microorganisms tend to degrade polysaccharides outside of the cell and then import the generated oligo- or monosaccharides for further intracellular metabolism. To do so, the microorganisms secrete a variety of carbohydrate-active enzymes (CAZymes) [2–5] depending on the growth substrate and the degrading strategy applied. The secretomes of microorganisms thus represent an important protein subfraction for the study of CAZymes.

Xing Wang and Matthew Kuruc (eds.), *Functional Proteomics: Methods and Protocols*, Methods in Molecular Biology, vol. 1871,
https://doi.org/10.1007/978-1-4939-8814-3_12,

Proteomics applied to secreted proteins is often referred to as secretomics or exoproteomics, and should ideally only concern proteins exported by the organisms. However, many seemingly cytosolic proteins (by prediction) have been reported in secretome studies [5–7], and this is commonly ascribed to cell lysis. To avoid that these cytosolic proteins mask the identification of truly secreted CAZymes involved in biomass conversion, it is of interest and importance to prepare secretome samples in a way that limits contamination by cell lysis to a minimum.

Recently we have developed a plate-based method for growing microorganisms on solid substrates that allows to selectively obtain fungal [8] and bacterial [9] protein subfractions that are highly enriched for secreted proteins. Here we present this method as a step-by-step protocol and provide detailed notes for successful identification of CAZymes in secretomes.

2 Materials

Prepare all solutions using ultrapure water (prepared by purifying deionized water, to attain a conductivity of 18 MΩ cm at 25 °C) and analytical grade reagents. Prepare and store all reagents at room temperature unless indicated otherwise. Follow all waste disposal regulations when disposing waste materials. There is no need to adjust pH of buffers, unless indicated otherwise.

2.1 Predicting CAZymes and Secreted Proteins

1. Computer with access to the internet.
2. FASTA file containing the predicted proteins encoded by the organism of interest. In most cases, this can be downloaded from UniProt (http://www.uniprot.org/) or NCBI (https://www.ncbi.nlm.nih.gov/), or in the case of an un-sequenced genome, be generated by genome sequencing, assembly and translation of the DNA sequences into protein sequences, to create a FASTA file of the proteins encoded by the organism of interest.

2.2 Culture Plates with a Membrane Filter

1. M9 minimal medium agarose plates:
   (a) Stock solution with 5× M9 salts (500 mL): Weigh in 17.04 g Na_2HPO_4, 7.5 g KH_2PO_4, 1.25 g NaCl, and 2.5 g NH_4Cl. Dissolve in approx. 400 mL ultrapure water and add ultrapure water to a total volume of 500 mL. Autoclave at 121 °C for 20 min.
   (b) 1 M $MgSO_4$ stock solution (25 mL): Weigh in 3.009 g $MgSO_4$. Dissolve in approx. 20 mL ultrapure water and add ultrapure water to total volume of 25 mL. Autoclave at 121 °C for 20 min.

(c) 0.1 M $CaCl_2$ stock solution (25 mL): Weigh in 0.227 g $CaCl_2$. Dissolve in approx. 20 mL ultrapure water and add ultrapure water to total volume of 25 mL. Autoclave at 121 °C for 20 min.

(d) Carbon source of interest:

- Insoluble carbon sources, such as microcrystalline cellulose, chitin, or filter paper, are available as powder, flakes, or similar and are used "as is" or after milling to a manageable particle size. Insoluble carbon sources will be sterilized later in the method, *see* Subheading 3.3.1.
- Soluble carbon sources, such as carboxymethyl cellulose or glucose, are to be prepared in a suitable sterile stock solution, e.g., 20% (w/v). Preparation of 100 mL 20% (w/v) glucose stock solution: Dissolve 20 g glucose in approx. 80 mL sterilized ultrapure water. Add sterilized ultrapure water to total volume of 100 mL. Carry out a final sterilization by passing the solution through a sterile 0.22 μm filter.

(e) Agarose (*see* **Note 1**).

2. Sterile grade QM-A Quartz Filters, circle, 47 mm diameter (GE Healthcare Life Sciences, Oslo, Norway).
3. Glass Petri dish with a diameter of 80 mm.
4. Heating/cooling cabinet capable of delivering a stable temperature suitable for the organism of interest.

2.3 Protein Extraction, Sample Preparation, and Analysis

1. Water bath that can reach 100 °C.
2. 50 mL polypropylene tubes, MS-friendly (e.g., Falcon from Fisher Scientific, New Hampshire, USA).
3. 400 mM Dithiothreitol (DTT) (1 mL) stock solution: Weigh in 0.062 g DTT and dissolve in 1 mL ultrapure water. Aliquot into volumes of 20 μL and store at −20 °C. Thaw just before use and do not refreeze.
4. 100 mM NH_4HCO_3 (100 mL) stock solution: Weigh in 0.791 g NH_4HCO_3. Dissolve in approx. 80 mL ultrapure water and add ultrapure water to total volume of 100 mL.
5. Trypsin, dissolved in liquid according to the manufacturer's recommendation, typically in 50 mM acetic acid, to a concentration of 500 ng/μL (*see* **Note 2**). Aliquot into volumes of 5 μL and store at −70 °C. Thaw just before use.
6. Disposable syringes, 2 mL.
7. Pipette tips of epT.I.P.S. quality (Eppendorf, Hamburg, Germany) or similar.

8. Protein LoBind tubes 2 mL (Eppendorf, Hamburg, Germany) or similar.
9. 10% (v/v) trifluoroacetic acid (TFA), LC-MS grade stock: In a fume hood, add 1 mL 100% TFA to 9 mL ultrapure water. Store at −20 °C.
10. C18 ZipTip pipette tips (Merck Millipore, Massachusetts, USA).
11. NanoHPLC-MS/MS system.

2.4 Data Integration and Heat Map Generation

1. Computer with a spreadsheet application (e.g., Excel from Microsoft, Washington, USA).
2. An installation of Perseus. Perseus is a free software package for the analysis of (prote)omics data and can be downloaded from http://www.coxdocs.org/doku.php?id=perseus:start. The version used in this chapter is 1.6.0.7.

3 Methods

3.1 Predicting Carbohydrate-Active Enzymes (CAZymes)

The CAZy database (http://www.cazy.org) is a database specialized in the display and analysis of genomic, structural, and biochemical information of CAZymes [10]. CAZy contains more than 300 families of catalytic and auxiliary modules, classified as glycoside hydrolases (GHs), glycosyltransferases (GTs), polysaccharide lyases (PLs), carbohydrate esterases (CEs), auxiliary activities (AAs), and carbohydrate-binding modules (CBMs). CAZy identifies families of evolutionary-related proteins using a classification based on significant amino acid sequence similarity with at least one biochemically characterized founding member [11]. Although the CAZy web page does not allow for automatic annotation of a given protein sequence query, it contains CAZy annotations of several sequenced genomes (CAZome, http://www.cazy.org/Genomes.html). If the CAZome of the organism under study is not found in the CAZy database, alternatives exist for automated CAZyme annotation, including dbCAN [12] and CAT [13]. Importantly, dbCAN has generated hidden Markov models (HMMs) representing the signature domains present in every CAZy family, and enables automatic annotation via these.

3.1.1 Predicting CAZymes Using the dbCAN Web Server

1. Access dbCAN at http://csbl.bmb.uga.edu/dbCAN/annotate.php.
2. Enter/paste one or more protein sequences in the submission form, or select a protein FASTA file (*see* Subheading 2.1, **step 2**) for upload of multiple protein sequences.
3. Leave your email address if many sequences are loaded; the results will be emailed when the job is done.

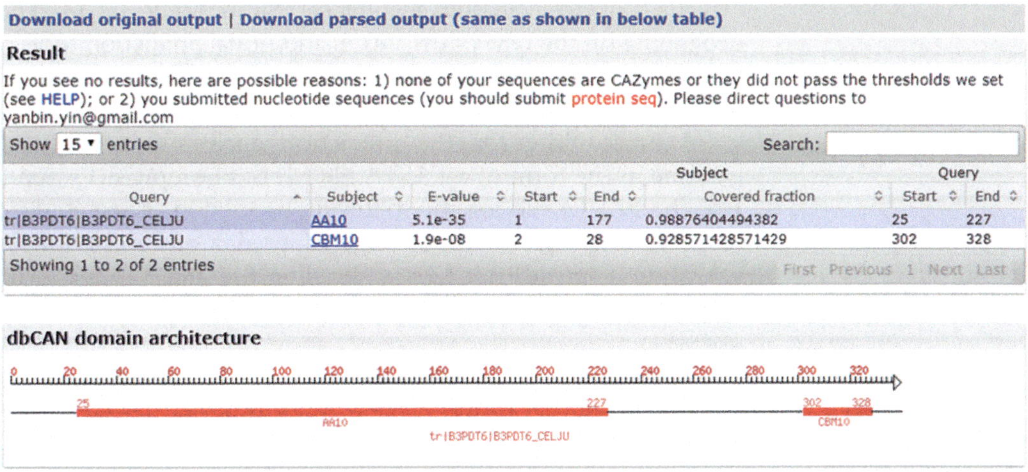

Fig. 1 The figure shows an example of CAZyme prediction using the dbCAN web server. The FASTA-formatted sequence of the chitin-binding protein (UniProt: B3PDT6) from *Cellvibrio japonicus* Ueda107 was used as input. A table shows the HMM hits to different parts of the protein sequence and a graphical representation shows the domain architecture within the protein

4. Click submit, and after some computation time, the results will be presented as a table and as a graphical representation of the domain architecture (Fig. 1). Note that several CAZy modules may be found within one protein sequence, leading to multiple rows in the table per protein.
5. Click "Download parsed output" to download a tab-separated file containing the results. This can be used for integration with proteomic expression values in a spreadsheet application, *see* Subheading 3.4.

3.2 Predicting Protein Secretion

Several algorithms have been developed for in silico prediction of different N-terminal signal peptides and, consequently, the subcellular locations of their cognate proteins. These prediction tools have since the 1980s developed from signal peptide prediction based on weight matrices and the presence of specific amino acid motifs to more sophisticated machine learning approaches [14], including also so-called leaderless proteins, i.e., secreted proteins without a typical signal peptide. Today, a combination of several prediction servers is often used to obtain the most reliable predictions possible [15]. It is important to remember that in silico prediction of the subcellular location of proteins is not trivial, and that the presence of a signal peptide is not necessarily a guarantee for secretion. There is a plethora of prediction servers available [16–18], some applicable to all types of organisms, while others are more "specialized" for, e.g., bacteria. In any case, the SignalP server (http://www.cbs.dtu.dk/services/SignalP/ [19]) offers a good starting point for prediction of signal peptides and is explained in detail in this chapter (*see* Subheading 3.2.1). Then, depending on the type of organism, additional prediction tools can be used to supplement the analysis to get an as good, and complete, prediction as possible.

A prerequisite for the prediction analysis is a sequenced genome in the form of a FASTA file (*see* Subheading 2.1, **step 2**). Use the FASTA file for the following prediction of secreted proteins. It is advisable to organize all predictions (secretion and CAZy) in a spreadsheet (*see* Subheading 3.4 and Table 1) where agreement between different prediction algorithms also can be assessed.

3.2.1 Step-by-Step Prediction of Secreted Proteins Using SignalP

1. Access SignalP at http://www.cbs.dtu.dk/services/SignalP/
2. Enter/paste one or more protein sequences in the submission form, or select a protein FASTA file for upload of multiple protein sequences. *See* **Note 3**.
3. Select organism group: Eukaryotes, Gram-negative bacteria, or Gram-positive bacteria.
4. One can choose between four output formats where "Standard" is the most applicable for a single/low number of protein sequences. This option provides a graphical representation of the N-terminal protein sequence(s) with highlighting of the predicted cleavage site, and several scores to aid the interpretation. The D-score (discrimination score) is the score that is used to discriminate signal peptides from non-signal peptides and this is accompanied by "YES" if a signal peptide is predicted. The output format "Short" is applicable to multiple protein entries such as FASTA files. This provides a single line per protein sequence including the different scores, and the letter "Y" is displayed when a signal peptide is predicted. This latter format is useful for copying into a spreadsheet application and to subsequentially integrate with proteomic abundance values, *see* Subheading 3.4. The two other output formats "Long" and "All" add extra information regarding each position in the sequence.
5. Choose "Short" output format and click "Submit."
6. After some computation time, the results will be presented as a table that can be selected and copied into a spreadsheet application for integration with proteomics expression values, *see* Subheading 3.4.

3.2.2 Prediction in Bacteria: Additional Considerations

- **Prediction of lipoproteins**. The LipoP web server (http://www.cbs.dtu.dk/services/LipoP/ [20]) discriminates between classical signal peptides (presence of signal peptidase I (SpI) cleavage site), signal peptides for lipoproteins (presence of signal peptidase II (SpII) cleavage site), transmembrane helices

(TMH), and cytosolic proteins (CYT). LipoP thus provides additional information compared to SignalP, which only predicts presence of SpI cleavage site, without further discrimination. LipoP was trained using Gram-negative bacteria, but also shows good performance for sequences from Gram-positive bacteria [20, 21]. Notably, dedicated prediction tools for lipoproteins in Gram-positive bacteria exist, such as PRED-LIPO (http://www.compgen.org/tools/PRED-LIPO [22]).

- **Prediction of twin-arginine signal peptides**. Translocation of completely folded proteins from the cytoplasm may occur via the twin-arginine translocation (TAT) pathway. The signal peptides involved in this secretion pathway are longer and less hydrophobic than normal signal peptides, and contain a distinctive pattern of two consecutive arginines in the N-terminal region. Several servers are available for prediction, such as PRED-TAT (http://www.compgen.org/tools/PRED-TAT) [23] and TatP (http://www.cbs.dtu.dk/services/TatP/) [24].
- **Prediction of transmembrane helices (TMHs)**. A large proportion of proteins contain membrane-spanning helices and knowledge of the presence and location of the TMHs is important for structural, as well as functional, annotation of the proteins. In general, proteins containing TMHs are not expected to be found in true secretomes unless they are part of vesicles produced by the microorganism. Albeit, LipoP occasionally reports proteins containing a TMH prediction, even in secretomes prepared according to the method described in this chapter, and these could be confused with a signal peptide, especially if the TMH is close to the N-terminus and only one TMH is predicted. These TMH predictions can be further validated using the TMHMM server, a very accurate predictor of TMHs (http://www.cbs.dtu.dk/services/TMHMM/) [25].

3.2.3 Prediction in Fungi: Additional Considerations

To get a reliable prediction of secreted proteins in fungi, it is possible to complement the SignalP prediction with other prediction servers where eukaryotic proteins can be analyzed. Two alternatives include Phobius (http://phobius.sbc.su.se/) [26], which predicts TMHs and signal peptides, and WoLF PSORT (https://wolfpsort.hgc.jp/) [27], which uses a combination of signal motifs and sequence-derived features, such as amino acid content, to predict secreted proteins. For a protein to be considered as secreted, it is advisable that two of three prediction servers agree on the prediction, such as applied in [8].

3.2.4 Nonclassically Secreted Proteins

Although most secreted proteins have a signal peptide that guides them to a secretory pathway, a limited number of proteins without such a signal peptide (also called leaderless proteins) are secreted

[15, 28]. For example, two chitinases encoded by *Serratia marcescens* lack a classical signal peptide, while they are known to be secreted via a process that remains unclear [29]. Such proteins are said to be subject to so-called nonclassical secretion. SecretomeP (http://www.cbs.dtu.dk/services/SecretomeP/) is a server that predicts nonclassically secreted proteins in either mammalian cells or bacteria by utilizing several sequence-derived features, including various posttranslational modifications and localization predictions, found in proteins known to be secreted [30, 31]. Proteins predicted as cytosolic proteins by other prediction servers could be candidates for further analysis using SecretomeP.

3.3 Culture Plates for Secretome Enrichment

Carry out all procedures at room temperature unless otherwise specified.

3.3.1 Casting of Plates with Membrane (Work in Sterile Hood)

1. The weights and volumes used below allow for preparation of 250 mL M9 agarose medium, which will give approx. 15 plates.
2. (a) Using an insoluble carbon source: To achieve a 1% (w/v) concentration of carbon source in the final plates (*see* **Note 4**): Weigh in 2.5 g insoluble carbon source and mix with 2.5 g agarose in 199.5 mL ultrapure water. Include a magnet for later homogenization (*see* **step 9** and **Note 5**).
   (b) Using a soluble carbon source: Weigh in 2.5 g agarose and mix with 187 mL ultrapure water. The soluble carbon source will be added to the medium in **step 7**.
3. Autoclave at 121 °C for 20 min and cool down to approx. 70 °C.
4. Add 250 μL 0.1 M $CaCl_2$ (to achieve a final concentration of 0.1 mM).
5. Add 250 μL 1 M $MgSO_4$ (to achieve a final concentration of 1 mM).
6. Add 50 mL 5× M9 salt (to achieve a final concentration of 1×).
7. If using a soluble carbon source: To achieve 1% (w/v) concentration of carbon source in the final plates (*see* **Note 4**): Add 12.5 mL of the sterile carbon source stock solution (given that the concentration of the stock is 20% (w/v), *see* Subheading 2.2).
8. Mix gently to ensure homogenization.
9. Poor 8 mL M9 minimal agarose medium (*see* **Note 5**) in an 80 mm glass Petri dish and let it solidify for 5–10 min. While waiting for the agarose to solidify, store the bottle with melted agarose at 60 °C.

10. Gently place a sterile QM-A Quartz Filter in the middle of the dish and poor another 8 mL (*see* **Note 5**) of the M9 minimal agarose medium over the filter. Let it solidify for 5–10 min.
11. Store plates at 4 °C until use. Always let the plates equilibrate to the temperature of interest before use.

3.3.2 Inoculation and Growth

Inoculation of Bacteria

1. Grow a liquid pre-culture in the desired medium.
2. Measure the OD_{600} of the pre-culture and plate an appropriate amount of cells (e.g., 100 μL of a culture with $OD_{600} \approx 0.5$). *See* **Note 6** for considerations regarding this step.
3. Incubate at the desired temperature for an appropriate amount of time (*see* **Note 7**).

Inoculation of Fungi

1. Collect an agar plug with a diameter of ~7 mm containing fungus grown on a M9 agarose plate with the carbon source of interest using a pipette tip (Fig. 2, **step 1A**).
2. Transfer the plug to a new plate with an embedded membrane. Position the plug in the center over the membrane (Fig. 2, **step 1B**).
3. Incubate at the desired temperature for an appropriate amount of time (*see* **Note 7** and Fig. 2, **step 2**).

3.3.3 Protein Extraction and Sample Preparation

As with all proteomics work, it is important to work clean to avoid contamination from fingers, hairs, and similar.

1. Weigh an empty 50 mL Falcon tube and note the weight.
2. Flip the agar disc and punch out the agar with the 50 mL Falcon tube directly under the filter (Fig. 2, **step 3**). Transfer the plug to the Falcon tube, weigh again, and calculate the sample net weight by subtracting the weight of the empty Falcon tube in **step 1**.
3. Add 10 μL of 400 mM DTT stock solution per g sample to achieve a final concentration of 4 mM DTT (by assuming 1 g sample = 1 mL sample).
4. Heat the sample until agar melts (i.e., by placing the tube in an 80–100 °C water bath) and vortex, then boil the sample for 30 min. In this step, proteins bound to the solid substrate are likely to be released.
5. Cool the sample to room temperature (the agar resolidifies).
6. Use a 2 mL disposable syringe (without needle), remove the plunger, and transfer the solid sample into the syringe from the top. Insert the plunger and crush the agar by pressing it through the syringe, back into the Falcon tube.
7. Add 1 mL of 100 mM NH_4HCO_3 stock solution per g sample to achieve a final concentration of 50 mM NH_4HCO_3. Mix by gentle vortexing.

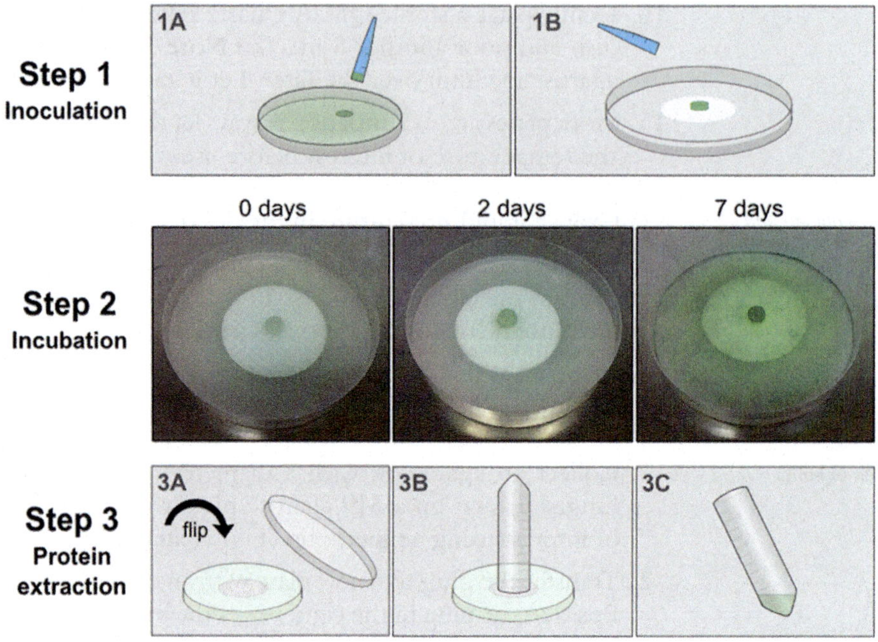

Fig. 2 Growth of microbes on membrane-containing plates and secretome sampling. **Step 1**: An agar plug containing fungus grown on a normal plate is collected using the back of a sterile pipette tip (**1A**) and transferred to a membrane-containing plate with the same carbon source and positioned in the center of the plate, i.e., over the membrane (**1B**). In the case of bacteria, one applies a cell suspension instead of an agar plug. **Step 2**: Incubate plates for the desired time. The pictures show growth of the fungus *Hypocrea jecorina*. **Step 3**: After incubation, the agar is flipped out of the Petri dish, thus exposing the cell free agar between the bottom of the dish and the agar-embedded membrane (**3A**). Use a sterile Falcon tube (or the back of a sterile pipette tip if a smaller sample is desirable) to punch out an agar disc containing secreted proteins that have passed through the filter (**3B–C**). This figure is reproduced from [8] with permission from Elsevier

8. Add 2 μg trypsin (e.g., 4 μL of a 500 ng/μL solution) and incubate overnight at 37 °C.
9. Freeze (−20 °C) and thaw the sample before centrifuging the Falcon tube briefly (4500 × *g* for 1 min); collect all liquid.
10. Transfer the liquid fraction to a 2 mL Eppendorf LoBind tube (*see* **Note 8**).
11. Centrifuge at 16,000 × *g* for 10 min.
12. Transfer the supernatant to a new 2 mL Eppendorf LoBind tube. Repeat the centrifugation if necessary (no agar pieces should be transferred).
13. Reduce the sample volume to 10–15 μL using a speed-vacuum centrifuge and add TFA to a final concentration of 0.1% (v/v) (*see* **Note 9**).
14. ZipTip the sample to concentrate the sample and remove buffers, using the supplier's recommendations (*see* **Note 10**).

15. Dry the sample using a speed-vacuum centrifuge and then dissolve in 10 μL of a solution that is appropriate for subsequent loading on the mass spectrometer (e.g. 0.1% TFA in ultrapure water).
16. Analyze the samples using a nanoHPLC-MS/MS system, e.g., as described previously [8, 9].

3.4 Integrating Quantitative Proteomics Data with Functional Annotation

Quantitative proteomics has become an indispensable analytical tool for microbial research. Quantitative data can be acquired using a vast number of analytical techniques including classical gel-based procedures, or modern MS-based quantitative techniques with metabolic or chemical labeling, or by using label-free approaches. A detailed description of these methods and their strengths and limitations is beyond the scope of this chapter and a recent review of current methods can be found in [32]. Independent of the quantification technique used, it is most practical to arrange the data in a tabular form with each row representing a protein (or protein group) with the quantitative data for the different conditions in separate columns. In addition, functional prediction and prediction of secretion should also be in separate columns. This can be achieved in a spreadsheet application. For an example, *see* Table 1. For final publication of proteomics data, it is highly advisable to deposit the data to the ProteomeXchange Consortium accessible at http://www.proteomexchange.org/. This ensures transparency and enables potential reuse of the data in the future.

3.4.1 Building a Heat Map Using Perseus

Perseus is a free software package for analyzing quantitative (prote) omics data and can be used with many different quantification techniques [33], *see* **Note 11**.

1. In Perseus, click on **Generic matrix upload**, a small green arrow in the upper-left corner.
2. Select the tabular file containing the quantitative proteomics data and the predicted functional data, similar to Table 1. Perseus supports tab- or comma-delimited text files.
3. Columns containing quantitative values should be selected as "Main" columns.
4. Select dbCAN and secretion predictions as "Categorical" columns.
5. Select accession number and protein name as "Text" columns. Click **OK**.
6. Click **Annotation of rows**, and select **Categorical annotation of rows**. Give biological replicates the same name, and keep default settings. Click **OK**.
7. If you have quantification values = 0, meaning that the protein was not detected in quantifiable amounts, click **Quality** and

Table 1
An example of how to structure quantitative proteomics data in a spreadsheet application

							α-Chitin			β-Chitin			Glucose		
Accn.	Protein name	SignalP	LipoP	SecretomeP	Secreted?	dbCAN	R1	R2	R3	R1	R2	R3	R1	R2	R3
B3PDT6	Chitin-binding protein, putative, cbp33/10B	Y	SpI	0.90	Y	AA10 CBM10	32.9	32.8	33.3	33.7	33.2	33.6	28.2	27.5	28.3
P14768	Endo-1,4-beta-xylanase A	Y	SpI	0.96	Y	CBM2 CBM10 GH10	26.8	26.8	27.5	33.4	32.6	34.0	25.8	ND	25.5
B3PDV8	Pullulanase, putative, pul13B	Y	SpII	0.94	Y	CBM48 GH13	29.4	30.3	31.5	29.6	29.3	28.7	25.3	26.0	25.7
B3PK74	Alpha-glucosidase, putative, adg97B	Y	SpI	0.58	Y	GH97	25.6	24.0	24.5	23.9	23.9	24.0	ND	ND	ND
B3PBG2	Pilin	N	TMH	0.93	N		31.0	30.5	30.9	30.2	30.2	30.7	30.6	30.4	30.8
B3PF53	Putative lipoprotein	N	SpII	0.90	Y		28.8	26.7	27.3	28.3	28.6	27.9	26.2	ND	25.8
B3PI93	SrpA-related protein	N	CYT	0.18	N		30.6	30.8	31.2	27.8	27.4	28.8	ND	ND	24.0

The data describes selected proteins detected in the secretome of *Cellvibrio japonicus* Ueda107 growing on α- or β-chitin or on glucose. The quantitative values are log2-transformed LFQ-values from the MaxQuant [37] software. The complete proteome FASTA file was used for predicting secreted proteins using different prediction servers (*see* Subheading 3.2) and for predicting CAZymes (*see* Subheading 3.1). The table shows results for four expressed CAZymes with varying expression levels on the different substrates and three non-CAZymes for comparison; the data is adapted from [9]. The column labeled "Secreted?" is based on the use of three prediction servers SignalP, LipoP, and SecretomeP and was marked with Y when at least two algorithms predicted secretion. *ND* not detected, *R1–3* replicate 1–3 select **Convert to NaN**. For "Invalid values should be," select **Less or equal**. The "Threshold" should be set to **0**. Click **OK**.

8. Click **Visualization, Histogram** and **OK** to see histograms for all samples. The graphs should look like bell-shaped distributions nicely spread across the intensity range (*x*-axis), but not necessarily normally distributed for all samples. If this is not what you observe, the quantitative data may need log-transformation. Click **Basic, Transform** and **OK** and then redo the histograms to reevaluate.
9. Assuming three biological replicates: Click **Filter rows**, select **Filter rows based on valid values**. "Min. valids" should be **Number** and set to **2**. "Mode" should be **In at least one group**. Other options are kept default. *See* **Note 12**. Click **OK**.
10. The generated matrix can now be used for further analysis of the data, e.g., calculation of the fraction of secreted proteins (*see* Subheading 3.4.2). Perseus offers a wide variety of statistical and visual tools (*see* **Note 11**); however, it is out of the scope of this chapter to go into the details of these options.
11. To filter the data to only contain CAZymes, click **Filter rows** and then **Filter rows based on categorical column**. Select the **dbCAN column**, and add all items in "Values" to the right container box. Change "Mode" to **Keep matching rows**. Then a new matrix will be generated where only CAZymes are visible, *see* **Note 13**. Click **OK**.
12. Click on **Clustering/PCA** and select **Hierarchical clustering**. Default cluster parameters are generally sufficient, but in some cases, it is more practical to deselect **Columns tree** and manually select the column order using the box labeled **Use for clustering**.
13. When clicking **OK**, the heat map will be generated as a new tab called **Clustering**. Therein it is possible to adjust the heat map color scale, the size of text headers, as well as the size and thickness of the dendrograms.
14. To visualize category columns in the heat map, click the button labeled **Configure row names** and choose the desired category as **Row color bar** (creates a color bar next to the heat map) and/or **Addtl. row names** (adds the category value as text).
15. Clusters can be defined manually by clicking on nodes in the dendrogram, or automatically by clicking on the button labeled **Define row clusters** and then either enter the number of desired clusters, or select clustering based on a distance threshold.
16. Remember to save your Perseus file, so you can continue to work with it at a later stage.
17. The heat map can be exported by clicking on the PDF button and select PDF as file type, *see* **Note 14**. An example of a heat map prepared for publication using Perseus can be found in Fig. 3.

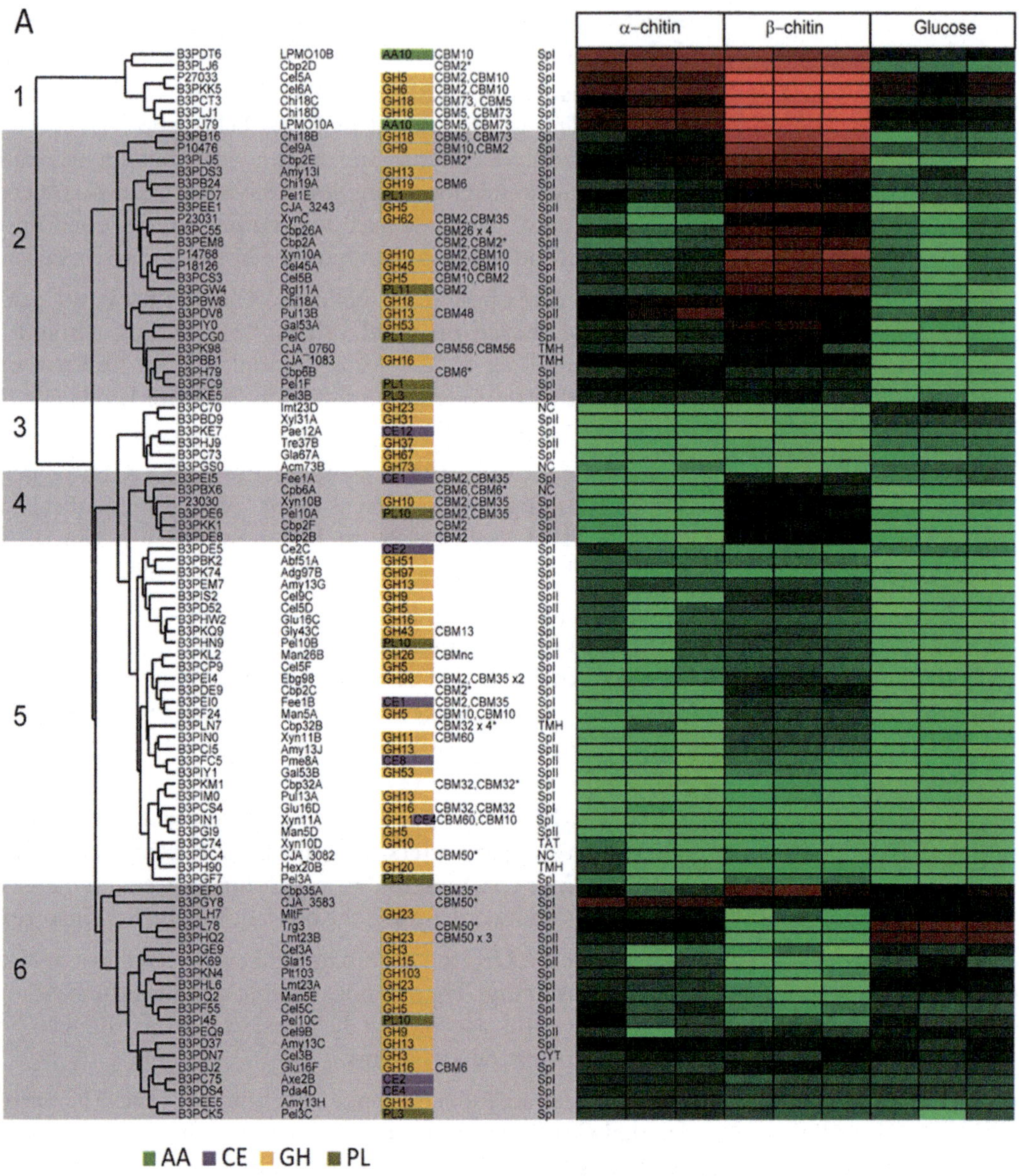

Fig. 3 A heat map representation of quantitative proteomics data where every protein (row) has a CAZyme annotation (*see* Subheading 3.1) and predicted cellular location (from LipoP, *see* Subheading 3.2). The figure is a filtered subset of the complete data set, showing only proteins with a CAZy annotation. The CAZy annotations are colored as indicated below the figure; *GH* glycoside hydrolase, *CE* carbohydrate esterase, *PL* polysaccharide lyase, *AA* auxiliary activity, *CBM* carbohydrate-binding module. The heat map generated by Perseus was further manually sectioned into six clusters based on similar expression patterns. The colors in the heat map indicate protein abundance, ranging from high (red color, MaxQuant LFQ 5×10^{10}) to low (green color, MaxQuant LFQ 7×10^{6}). Table 1 shows an example of the data used for generating this figure. The figure is reproduced from [9] with permission from John Wiley & Sons

Table 2
Table for calculating the secretome enrichment

	Protein count	Predicted secreted	NOT predicted secreted	Secreted fraction (%)
In experimental secretome	351	267	84	76
In complete proteome	3711	1076	2635	29

The secreted fraction in the experimental secretome is compared to the secreted fraction in the complete proteome. The data used in this example apply to *Cellvibrio japonicus* Ueda107 growing on α-chitin. The secretome was collected after 144 h growth. Data is adapted from [9]

3.4.2 Assessment of Secretome Enrichment in Samples

1. Calculate the percentage of predicted secreted proteins in the complete proteome according to Table 2 (*see* **Note 15**).
2. Calculate the percentage of predicted secreted proteins in the secretome sample using the same criteria as for the complete proteome. Comparing this percentage with the percentage calculated for the complete proteome is a good indication of the enrichment of secreted proteins in the secretome sample (*see* **Note 16**).

4 Notes

1. Agarose is used instead of agar due to its purity, hence reducing the risk of transferring contaminants to the mass spectrometer later on. In addition, compared to agar, agarose has less interactions with biomolecules such as proteins and DNA.
2. Trypsin must be of high quality and suitable for protein sequencing, e.g., Sequencing Grade Modified Trypsin from Promega (Wisconsin, USA).
3. Some prediction servers have limits when it comes to the number of sequences per submission and the number of amino acids per protein entry. In the case of complete proteomes, a FASTA file may often exceed these limits and splitting the FASTA file is necessary. This can readily be done using a text editor or online tools such as FaBox [34], available at http://users-birc.au.dk/biopv/php/fabox/ (click on "Fasta dataset splitter/divider," select your FASTA file and the desired number of sequences in each file).
4. The ideal concentration of carbon source might vary for different microorganisms and depends on the type of carbon source.
5. Homogenization is important before pouring out the 8 mL. We advise to use a magnet stirrer to mix the M9 agarose medium (include a magnet when sterilizing the agar and

carbon source). It can be useful to use a measuring cylinder to measure 8 mL before transferring the medium to the plate.

6. When plating bacteria from a liquid pre-culture you can consider harvesting cells, i.e., centrifuge and remove the medium, and resolubilize cells in a suitable medium before plating. Alternatively, streaking out directly from a fresh plate or from a −80 °C stock may be considered.
7. It is strongly recommended to do pre-experiments to establish how the microorganism behaves during growth on plates with the desired carbon source. Too much growth is not advisable, and since measuring growth is difficult on plates, sampling at different time-points is needed to find optimal conditions for secretome analysis. When comparison between different growth conditions (e.g., different carbon sources) is to be performed, and when, hence, comparison of "equal growth phases" is desirable, one can use the number of proteins detected in the secretome as a very rough estimate of growth. Note that prolonged growth inevitably leads to cell lysis, meaning that artificial high numbers of proteins will be detected and that the fraction of proteins predicted to be secreted will decrease. In our experience, it is important to invest time in finding optimal incubation times; optimal meaning that one obtains a high number of proteins and little contamination with cytosolic proteins.
8. When transferring the sample from the Falcon tube to the LoBind Eppendorf tube, it can be useful to cut the pipet tip to get a wider tip opening. Some agar pieces may be transferred, hence the centrifugation in the next step.
9. If the sample is accidentally dried completely during the speed-vacuum centrifugation process, dissolve the sample in 10–15 μL 0.1% TFA before the ZipTip procedure.
10. If you have many samples, it can be useful to elute the peptides bound to the ZipTip directly in the desired HPLC-vial (given that the speed-vacuum centrifugation unit is able to handle these tubes). This will reduce the number of LoBind Eppendorf tubes used and also potentially reduce sample loss.
11. In this chapter, we only explain how to generate a heat map representation of the quantitative data (*see* Subheading 3.4.1), but we recommend the reader to explore the other functionality of the software, such as data transformation, profile plots, statistical tests, and volcano plots that can aid the data analysis. Tutorials, user examples, etc. can be found at http://www.coxdocs.org/doku.php?id=perseus:start.
12. This filtering step removes proteins that are only identified in one out of three biological replicates. We recommend setting a threshold that a protein should be identified in at least two out

of three biological replicates, in at least one group (e.g., one carbon source), to be included in the analyses.

13. This filtering will remove all proteins without predicted CAZyme annotation and hence also remove (novel) CAZymes that have not yet been recognized as such, i.e., potentially novel enzymes involved in biomass conversion. Potential novel CAZymes or other enzyme possibly involved in conversion of the biomass in question are likely to be abundant proteins showing similar expression patterns as known CAZymes.
14. All plots in Perseus can be exported as PDF files, but typically, the figures need further improvement prior to publication. The PDF files can be imported into a vector graphics software, e.g., Inkscape (https://inkscape.org/en/) or Adobe Illustrator (http://www.adobe.com/products/illustrator.html), for generating figures suitable for publication.
15. When calculating the percentage of secreted proteins, include proteins predicted with a SpI and Tat signal peptide, and proteins predicted to be nonclassically secreted by SecretomeP. Lipoproteins (SpII) must be used with caution as many membrane-anchored proteins face the periplasm rather than the extracellular milieu.
16. In the majority of cases where we have applied the plate method for collecting secreted proteins, we have obtained protein fractions that are clearly enriched for secreted proteins ([8, 9] and in unpublished results). However, for *Serratia marcescens* growing on chitin, we observed larger fractions of cytosolic proteins than expected [35]. This could possibly be explained by nonclassical secretion systems used by this bacterium, as experimental data indicated that cell lysis was not a big problem (*see* discussion in [35]). Some bacteria transfer CAZymes to the external environment by expelling enzyme-loaded vesicles, which may lead to less good enrichment statistics, despite the secretome data being correct and relevant [36]. This emphasizes that the success of the method may vary between microorganisms, but also that increased knowledge of the organism under study enables better evaluation of the results.

References

1. Himmel ME, Xu Q, Luo Y, Ding S-Y, Lamed R, Bayer EA (2010) Microbial enzyme systems for biomass conversion: emerging paradigms. Biofuels 1(2):323–341. https://doi.org/10.4155/bfs.09.25
2. Payne CM, Knott BC, Mayes HB, Hansson H, Himmel ME, Sandgren M, Ståhlberg J, Beckham GT (2015) Fungal cellulases. Chem Rev 115(3):1308–1448. https://doi.org/10.1021/cr500351c

3. Benz JP, Chau BH, Zheng D, Bauer S, Glass NL, Somerville CR (2014) A comparative systems analysis of polysaccharide-elicited responses in *Neurospora crassa* reveals carbon source-specific cellular adaptations. Mol Microbiol 91(2):275–299. https://doi.org/10.1111/mmi.12459
4. Suzuki K, Suzuki M, Taiyoji M, Nikaidou N, Watanabe T (1998) Chitin binding protein (CBP21) in the culture supernatant of *Serratia marcescens* 2170. Biosci Biotechnol Biochem 62(1):128–135. https://doi.org/10.1271/bbb.62.128
5. Takasuka TE, Book AJ, Lewin GR, Currie CR, Fox BG (2013) Aerobic deconstruction of cellulosic biomass by an insect-associated *Streptomyces*. Sci Rep 3:1030. https://doi.org/10.1038/srep01030
6. Siljamäki P, Varmanen P, Kankainen M, Sukura A, Savijoki K, Nyman TA (2014) Comparative exoprotein profiling of different *Staphylococcus epidermidis* strains reveals potential link between nonclassical protein export and virulence. J Proteome Res 13(7):3249–3261. https://doi.org/10.1021/pr500075j
7. Adav SS, Cheow ESH, Ravindran A, Dutta B, Sze SK (2012) Label free quantitative proteomic analysis of secretome by *Thermobifida fusca* on different lignocellulosic biomass. J Proteome 75(12):3694–3706. https://doi.org/10.1016/j.jprot.2012.04.031
8. Bengtsson O, Arntzen MØ, Mathiesen G, Skaugen M, Eijsink VGH (2016) A novel proteomics sample preparation method for secretome analysis of *Hypocrea jecorina* growing on insoluble substrates. J Proteome 131:104–112. https://doi.org/10.1016/j.jprot.2015.10.017
9. Tuveng TR, Arntzen MØ, Bengtsson O, Gardner JG, Vaaje-Kolstad G, Eijsink VGH (2016) Proteomic investigation of the secretome of *Cellvibrio japonicus* during growth on chitin. Proteomics 16(13):1904–1914. https://doi.org/10.1002/pmic.201500419
10. Lombard V, Golaconda Ramulu H, Drula E, Coutinho PM, Henrissat B (2014) The carbohydrate-active enzymes database (CAZy) in 2013. Nucleic Acids Res 42(D1):D490–D495. https://doi.org/10.1093/nar/gkt1178
11. Cantarel BL, Coutinho PM, Rancurel C, Bernard T, Lombard V, Henrissat B (2009) The Carbohydrate-Active EnZymes database (CAZy): an expert resource for Glycogenomics. Nucleic Acids Res 37(Database):D233–D238. https://doi.org/10.1093/nar/gkn663
12. Yin Y, Mao X, Yang J, Chen X, Mao F, Xu Y (2012) dbCAN: a web resource for automated carbohydrate-active enzyme annotation. Nucleic Acids Res 40(Web Server issue):W445–W451. https://doi.org/10.1093/nar/gks479
13. Park BH, Karpinets TV, Syed MH, Leuze MR, Uberbacher EC (2010) CAZymes Analysis Toolkit (CAT): web service for searching and analyzing carbohydrate-active enzymes in a newly sequenced organism using CAZy database. Glycobiology 20(12):1574–1584. https://doi.org/10.1093/glycob/cwq106
14. Caccia D, Dugo M, Callari M, Bongarzone I (2013) Bioinformatics tools for secretome analysis. Biochim Biophys Acta Proteins Proteom 1834(11):2442–2453. https://doi.org/10.1016/j.bbapap.2013.01.039
15. Desvaux M, Hebraud M, Talon R, Henderson IR (2009) Secretion and subcellular localizations of bacterial proteins: a semantic awareness issue. Trends Microbiol 17(4):139–145. https://doi.org/10.1016/j.tim.2009.01.004
16. Nielsen H (2017) Predicting secretory proteins with SignalP. In: Kihara D (ed) Protein function prediction: methods and protocols. Springer, New York, pp 59–73. https://doi.org/10.1007/978-1-4939-7015-5_6
17. Nielsen H (2017) Protein sorting prediction. In: Journet L, Cascales E (eds) Bacterial protein secretion systems: methods and protocols. Springer, New York, pp 23–57. https://doi.org/10.1007/978-1-4939-7033-9_2
18. Nielsen H (2016) Predicting subcellular localization of proteins by Bioinformatic algorithms. In: Bagnoli F, Rappuoli R (eds) Protein and sugar export and assembly in gram-positive bacteria. Springer International Publishing, Cham, pp 129–158. https://doi.org/10.1007/82_2015_5006
19. Petersen TN, Brunak S, Heijne G, Nielsen H (2011) SignalP 4.0: discriminating signal peptides from transmembrane regions. Nat Methods 8:785. https://doi.org/10.1038/nmeth.1701
20. Juncker AS, Willenbrock H, Von Heijne G, Brunak S, Nielsen H, Krogh A (2003) Prediction of lipoprotein signal peptides in Gram-negative bacteria. Protein Sci 12(8):1652–1662. https://doi.org/10.1110/ps.0303703
21. Rahman O, Cummings SP, Harrington DJ, Sutcliffe IC (2008) Methods for the bioinformatic identification of bacterial lipoproteins encoded in the genomes of Gram-positive bacteria. World J Microbiol Biotechnol 24(11):2377. https://doi.org/10.1007/s11274-008-9795-2

22. Bagos PG, Tsirigos KD, Liakopoulos TD, Hamodrakas SJ (2008) Prediction of lipoprotein signal peptides in Gram-positive bacteria with a Hidden Markov Model. J Proteome Res 7(12):5082–5093. https://doi.org/10.1021/pr800162c

23. Bagos PG, Nikolaou EP, Liakopoulos TD, Tsirigos KD (2010) Combined prediction of Tat and Sec signal peptides with hidden Markov models. Bioinformatics 26 (22):2811–2817. https://doi.org/10.1093/bioinformatics/btq530

24. Bendtsen J, Nielsen H, Widdick D, Palmer T, Brunak S (2005) Prediction of twin-arginine signal peptides. BMC Bioinformatics 6:167. https://doi.org/10.1186/1471-2105-6-167

25. Krogh A, Larsson B, von Heijne G, Sonnhammer E (2001) Predicting transmembrane protein topology with a hidden Markov model: application to complete genomes. J Mol Biol 305:567–580. https://doi.org/10.1006/jmbi.2000.4315

26. Käll L, Krogh A, Sonnhammer EL (2007) Advantages of combined transmembrane topology and signal peptide prediction—the Phobius web server. Nucleic Acids Res 35 (Suppl 2):W429–W432. https://doi.org/10.1093/nar/gkm256

27. Horton P, Park K-J, Obayashi T, Fujita N, Harada H, Adams-Collier C, Nakai K (2007) WoLF PSORT: protein localization predictor. Nucleic Acids Res 35(suppl_2):W585–W587. https://doi.org/10.1093/nar/gkm259

28. Costa TR, Felisberto-Rodrigues C, Meir A, Prevost MS, Redzej A, Trokter M, Waksman G (2015) Secretion systems in Gram-negative bacteria: structural and mechanistic insights. Nat Rev Microbiol 13(6):343–359. https://doi.org/10.1038/nrmicro3456

29. Hamilton JJ, Marlow VL, Owen RA, Costa Mde A, Guo M, Buchanan G, Chandra G, Trost M, Coulthurst SJ, Palmer T, Stanley-Wall NR, Sargent F (2014) A holin and an endopeptidase are essential for chitinolytic protein secretion in *Serratia marcescens*. J Cell Biol 207(5):615–626. https://doi.org/10.1083/jcb.201404127

30. Bendtsen J, Kiemer L, Fausboll A, Brunak S (2005) Non-classical protein secretion in bacteria. BMC Microbiol 5(1):58. https://doi.org/10.1186/1471-2180-5-58

31. Bendtsen J, Jensen L, Blom N, von Heijne G, Brunak S (2004) Feature based prediction of non-classical protein secretion. Protein Eng Des Sel 17:349–356. https://doi.org/10.1093/protein/gzh037

32. Otto A, Becher D, Schmidt F (2014) Quantitative proteomics in the field of microbiology. Proteomics 14(4–5):547–565. https://doi.org/10.1002/pmic.201300403

33. Tyanova S, Temu T, Sinitcyn P, Carlson A, Hein MY, Geiger T, Mann M, Cox J (2016) The Perseus computational platform for comprehensive analysis of (prote)omics data. Nat Methods 13(9):731–740. https://doi.org/10.1038/nmeth.3901

34. Villesen P (2007) FaBox: an online toolbox for fasta sequences. Mol Ecol Resour 7 (6):965–968. https://doi.org/10.1111/j.1471-8286.2007.01821.x

35. Tuveng TR, Hagen LH, Mekasha S, Frank J, Arntzen MØ, Vaaje-Kolstad G, Eijsink VGH (2017) Genomic, proteomic and biochemical analysis of the chitinolytic machinery of *Serratia marcescens* BJL200. Biochim Biophys Acta Proteins Proteom 1865(4):414–421. https://doi.org/10.1016/j.bbapap.2017.01.007

36. Arntzen MO, Varnai A, Mackie RI, Eijsink VGH, Pope PB (2017) Outer membrane vesicles from *Fibrobacter succinogenes* S85 contain an array of carbohydrate-active enzymes with versatile polysaccharide-degrading capacity. Environ Microbiol 19(7):2701–2714. https://doi.org/10.1111/1462-2920.13770

37. Cox J, Mann M (2008) MaxQuant enables high peptide identification rates, individualized p.p.b.-range mass accuracies and proteome-wide protein quantification. Nat Biotechnol 26(12):1367–1372. https://doi.org/10.1038/nbt.1511

Chapter 13

An Overview of Mass Spectrometry-Based Methods for Functional Proteomics

J. Robert O'Neill

Abstract

The mechanism underlying many biological phenotypes remains unknown despite the increasing availability of whole genome and transcriptome sequencing. Direct measurement of changes in protein expression is an attractive alternative and has the potential to reveal novel processes. Mass spectrometry has become the standard method for proteomics, allowing both the confident identification and quantification of thousands of proteins from biological samples. In this review, mass spectrometry-based proteomic methods and their applications are described.

Key words Mass spectrometry, Proteomics, Quantitation, Label-free, Selective reaction monitoring, MALDI

1 The Challenge of Measuring the Proteome

The study of the entire protein content of an organism, tissue, or cell was first described as proteomics nearly 20 years ago [1]. Mass spectrometry has become the de facto standard method for proteomics, allowing the confident identification of proteins from complex mixtures [2].

Although the goal of measuring an entire eukaryotic proteome has been achieved [3], the human proteome has yet to be described in toto despite the publication of the complete human genome at the turn of the century [4]. The human proteome project has delivered progressive increments toward this goal [5, 6] yet as of the August 2017 data release of neXtProt, the most comprehensive human protein database available, no direct experimental evidence has been provided for 3031 (15%) of the predicted 20,199 proteins comprising the human proteome [7]. Several reasons can be proposed for this disparity.

The polymerase chain reaction (PCR) allows template nucleotide sequences to be copied with an increase in number of many orders of magnitude and very low error rates [8]. Complementary-

Xing Wang and Matthew Kuruc (eds.), *Functional Proteomics: Methods and Protocols*, Methods in Molecular Biology, vol. 1871, https://doi.org/10.1007/978-1-4939-8814-3_13, © Springer Science+Business Media, LLC, part of Springer Nature 2019

base pairing also allows cryptic nucleotide sequences to be rapidly deciphered [9]. The combination of these methods and advances in computational assembly of short sequence reads allows nucleotide sequencing to proceed in massively parallel configurations to sequence entire genomes within hours [10].

In contrast, the de novo identification of protein sequences contains greater intrinsic challenges. No method exists to amplify protein or peptide sequences and therefore proteomic methods are always restricted by the input mass. Similarly amino acids do not exhibit complementation and identification relies on mass measurement or, historically, chromatography or electrophoresis [11]. The proteome is also significantly larger than the genome with alternative splicing and alternative transcription start sites contributing to transcriptome and ultimately proteome diversity [12].

A further challenge is posed by greater combinatorial possibilities with up to 21 amino acids used interchangeably to generate peptides. This complexity is further increased by posttranslational modifications (PTMs) including the addition of biochemical groups such as a phosphate (phosphorylation), a carbohydrate (glycosylation), and at least 25 other distinct moieties or modifications [11].

A final compounding difficulty is the dynamic nature of the proteome. The genome sequence of an organism is constant across all cells in that organism and is relatively stable in the face of DNA extraction methods even allowing DNA sequences to be obtained from ancient specimens [13]. In contrast, the proteome varies from cell to cell [14] and is highly context-dependent with the post-translational state of a single protein varying across subcellular locations [15]. Extracting the proteome for quantification is also confounded by the rapid alterations in the PTM state induced by hypoxia and changes in intracellular pH with some phosphorylations reported to be lost within 60 min of tissue biopsy [16]. Many of these challenges have been addressed with technological advances, the most significant of which is the use of high scanning speed, high accuracy mass spectrometry [17].

2 Mass Spectrometry for Proteomics

The fundamental components of a mass spectrometer consist of an ion detector coupled to a mass analyzer that measures both the number and the mass-to-charge ratio (m/z) of ions generated into the gas phase from an ionization source. Variations on the instrumentation abound, however, each with their own strengths and weaknesses [18–22]. Despite this, all combine high sensitivity and high mass accuracy to finally bring the measurement of whole proteomes within reach.

Electrospray ionization (ESI) sources ionize analytes directly into the gas phase from liquid, commonly a polar volatile solvent eluted from a chromatography column [23]. These sources are most commonly used for the analysis of complex mixtures including cell lysates. Alternatives include matrix-assisted laser desorption ionization (MALDI) sources which use a laser to ionize analytes directly into the gas phase out of a solid matrix [24]. These sources are limited in the number of ions that can be generated and have previously been reserved for relatively homogenous analytes.

3 Protein Identification Using MS: Bottom-Up Approach

The de novo identification of proteins from a complex mixture can be achieved by several means. The most common method, termed "shotgun" or "bottom-up" proteomics relies on the identification of peptides generated by proteolytic digestion of the protein mixture. The presence of a protein in the original mixture is then inferred by interrogation of a protein sequence database with the identified peptide sequences. Matching a peptide sequence unique to a particular protein provides evidence of the protein in the original mixture [2]. An example overview of the workflow is illustrated in Fig. 1.

Shotgun proteomics relies on tandem mass spectrometry (MS/MS) where peptides are ionized to generate precursor ions, analyzed and separated according to their mass-to-charge ratio (m/z) in the primary mass spectrometry run (MS1) . Precursor ions are then fragmented, usually by collision ion dissociation, and the fragment ions are separated and analyzed in the second MS run (MS2) [25]. Multiple fragment species are generated from the same peptide and, with high-quality spectra and sufficient fragment ions, species differing by each individual amino acid in the peptide will be discernible as discrete ion peaks separated by a measured mass difference. As amino acids all have a fixed, defined mass, the measured difference can be used to identify the amino acid [26]. Thus the sequence of the peptide can be determined directly, defined as de novo peptide sequencing [27]. In practice, with complex peptide mixtures it is rarely possible to sequence all peptides directly and this labor-intensive approach is reserved for organisms with limited genome sequence information and therefore limited or absent potential protein databases.

More commonly, database searching is performed to generate peptide-spectrum matches. Several algorithms have been described but they generally follow the same principle; the measured precursor mass is used to filter a database of peptides generated by in silico digestion of a list of potentially identifiable proteins. Theoretical fragment-ion mass differences are generated for all the candidate peptides with a matching precursor mass [28]. These are compared

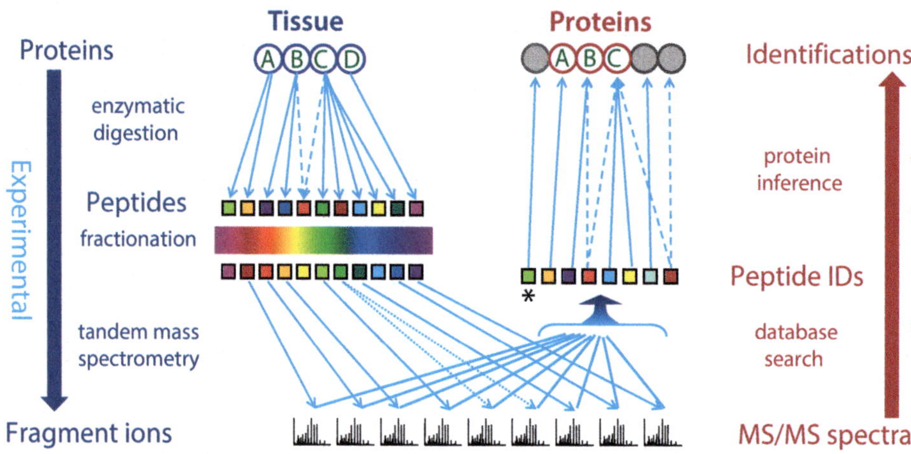

Fig. 1 An overview of protein identification by shotgun proteomics. A complex protein mixture, in this example a tissue sample containing proteins A–D, is proteolytically digested to yield peptides. Each peptide is illustrated as a colored box. To reduce the mixture complexity, peptides are fractionated by a common property such as isoelectric point. Peptide fractions are subjected to tandem mass spectrometry to yield fragment ion spectra. Peptide-spectrum matches (IDs) are made using a protein database, the peptide (precursor) ion masses, and a database search tool [91]. Not all fragment ion spectra result in a peptide match and some peptide matches are of low confidence (e.g., light green peptide; *). Using further statistical tools [28], proteins are identified with unique peptide matches confirming the presence of a protein in the original mixture. Each shotgun experiment only identifies a subset of the proteome from complex mixtures such as tissue lysate, so in this example protein "D" has not been identified

with the identified fragment-ion spectra and candidates ranked using a scoring algorithm, specific to the database search method [29].

These methods can identify peptides without the requirement for prior mass spectrometry. Organism-specific spectral libraries generated using stringent identification thresholds and evaluation of millions of published experimentally derived peptide spectra are now available [30, 31]. An alternative, or complementary, approach is to search identified spectra against these libraries, incorporating other spectral features such as relative ion intensity. This has been reported to enhance the number of peptide identifications compared to standard database search strategies [32, 33].

The sequences of identified peptides are then used to identify proteins using the original search database. A variety of statistical approaches are included in commonly used software packages to deal with protein inference problems such as repeated peptide sequencing events, peptides shared between multiple proteins and estimating the false-discovery rate [28].

3.1 Fractionation

A significant limitation of mass spectrometry is the throughput of ions that can be analyzed. Although this has improved with the current generation of instruments, the number of analytes that can be studied simultaneously is still often the limiting factor. Tissue lysates contain highly diverse mixtures of proteins. This diversity is further compounded by proteolytic digestion, presenting significant challenges for peptide spectrum matching. Samples are often fractionated to reduce this complexity. Approaches include strong cation exchange [34], subcellular fractionation [35], isoelectric focusing electrophoresis [36], high pH (basic) reversed phase [37], and other chromatography methods [38]. By delivering fractions with reduced numbers of unique peptides into the mass spectrometer, homogenous *m/z* fractions can be produced during the MS1 phase which can be accurately sequenced during the MS2 phase [39, 40].

4 Data-Dependent and -Independent Shotgun Proteomics

A key feature of the shotgun proteomic method as described previously is the selection of precursor ions for fragmentation in the MS2 phase. This is usually performed on the basis of precursor ion intensity and is referred to as the data-dependent approach [41]. This method has the limitation that a precursor ion must be detected to allow peptide sequencing and places an intrinsic bias toward abundant precursor species. An alternative strategy is to systematically fragment all precursor ions within windows of a defined *m/z* range regardless of whether a precursor ion was detected or not [21, 42]. In one iteration of this method, the precursor mass used for peptide spectrum matching is assigned as the center of the MS1 *m/z* window. When this method is applied, fragment ions yielding high-confidence peptide spectrum matches can be detected in up to 10% of cases in the absence of a precursor ion [42, 43] and this approach can enhance the dynamic range of detection by identifying low-abundance peptides. A disadvantage is the long data acquisition times required to obtain spectra across all *m/z* windows although faster instruments and optimized chromatography have reduced this [44].

5 Protein Identification Using MS: Top-Down Approach

An exciting recent development has been the ability to identify intact proteins by mass spectrometry, a so-called "top-down" approach [45]. Proof of concept studies have demonstrated the capacity to identify several thousand distinct protein isoforms (proteoforms) using cultured mammalian cells and extensive orthogonal fractionation in the liquid phase [46–48]. An advantage of this method is the direct identification of proteins, rather than inference from peptide identifications using the shotgun approach. This provides the potential to characterize the entire population of proteoforms generated from a single gene and identify dynamic changes in protein-processing, alternative-splicing or posttranslational modification often not possible from peptide-level data alone. Although not currently capable of proteome-scale analysis, with further developments in automated fractionation, instrumentation and data analysis methods, this may become feasible in the future [49]. Measuring dynamic changes in cellular states, the goal of most biological proteomic experiments, however, requires quantitation in addition to protein identification. Methods to undertake this using a hypothesis-free top-down approach are in early development and still lack the robustness of shotgun approaches [50].

6 Selective Reaction Monitoring

Mass spectrometry provides an ideal method to allow the hypothesis-free discovery of expressed proteins in biological samples using either the "bottom-up" or "top-down" approaches described. The development of high-quality comprehensive spectral libraries and the availability of synthetic peptides have allowed the development of robust mass spectrometry "assays" covering the entire human proteome and that of several model organisms [30, 51, 52]. These databases can be used for hypothesis-driven studies to quantify protein expression across samples. The commonest application of this method is selective reaction monitoring.

In this method, a peptide unique to the protein of interest and consistently identifiable by mass spectrometry is selected (proteotypic peptide) [53]. The spectral features of this peptide are then used to isolate precursor ions using defined mass windows. This significantly reduces the complexity of the ion mixture for subsequent fragmentation and peptide identification. This also significantly reduces the analysis time so that higher numbers of samples can be analyzed.

Many dozens of proteins can be assayed simultaneously by multiplexing this strategy (multiple reaction monitoring; MRM). By spiking-in isotopically labeled synthetic proteotypic peptides at a defined concentration, the absolute concentration of peptide, and by inference the protein, of interest can be determined with high accuracy. The higher throughput of MRM approaches means that they are commonly employed in the validation phase of biomarker development studies when shortlisted biomarker candidates determined in a discovery proteomic experiment in a small number of samples are assessed in several hundred further samples [54].

Improvements in instrument scanning speeds and the application of data analysis approaches from selective reaction monitoring have been employed in a further hybrid proteomic method. This technique, termed sequential window acquisition of all theoretical fragment-ion spectra or "SWATH MS" [21], has high technical reproducibility and quantitative accuracy [19, 55, 56]. In this method, data are acquired using a data-independent shotgun proteomic method, and peptide identifications are made on a candidate basis using the SRM approach. Proponents of this technique claim that a "digital" representation of the protein state of a biological sample is created and this can be assessed retrospectively as hypotheses are subsequently developed without the need for further mass spectrometry. Although compelling as a concept, complete proteome coverage is still not routinely achieved using current instruments and identifying biologically significant changes in protein modifications such as phosphorylation still requires careful experimental control and modification-specific sample preparation methods.

7 Quantitative Proteomics

A striking common finding of the increasing number of large-scale proteomic studies is that few proteins exhibit tissue-specific expression [57, 58]. In almost all cases, therefore, diverse phenotypes are manifest through changes in protein expression level, subcellular localization, or posttranslational modifications rather than the presence or absence of protein expression. If the experimental objective is to understand the mechanism governing an observed phenotype, then quantifying protein expression is of central importance.

7.1 Gel-Based Methods for Quantitative Proteomics

A typical proteomic experimental design is to compare a biological sample under two or more conditions and attempt to identify differentially expressed proteins. Historically, 2D gel electrophoresis would be used to separate the lysates from each condition according to protein mass and isoelectric point [59]. Gels could then be stained using a silver-based or other similar methods and differentially expressed proteins could be identified as spots of differing intensities [60]. A variation of this method minimized the gel-to-gel variability by labeling all the proteins in each sample with a different fluorophore and running all the samples together on the same gel [61]. By quantifying the relative emission from each fluorophore across the spots, the relative expression could be determined.

In both examples, differentially expressed spots containing proteins of unknown identify are excised, digested to peptides using proteolytic enzymes, and subjected to mass spectrometry for peptide and subsequent protein identification using similar strategies to shotgun proteomics. This method has the advantage of limiting the protein identifications to a small number of differentially expressed proteins, and providing a relatively homogenous sample for mass spectrometry. Unfortunately, despite advances in the automation of spot detection and quantification, these methods were only semiquantitative, labor-intensive, the data quality was highly user dependent, and protein identifications were limited to a few dozen per experiment.

7.2 Quantitative Shotgun Proteomics

Advances in sample processing and instrumentation have enabled the development of quantitative shotgun proteomic methods. These rely on lysis, digestion, and usually fractionation of samples prior to liquid chromatography (LC) and MS/MS. A labeling phase can be incorporated into the sample preparation stages prior to MS/MS or peptides can be quantified directly using label-free strategies [62].

7.3 Quantitative Shotgun Proteomics Using Labeling

7.3.1 Stable Isotope Labeling of Amino Acids in Culture (SILAC)

Chemical labeling can take place at the protein or peptide level. The use of stable carbon, hydrogen, and nitrogen isotopes allows differential labeling of amino acids such as Leucine, Lysine, and Arginine that will remain biochemically identical but through their mass differences are resolvable as discrete spectral peaks. This approach, termed Stable Isotopic Labeling of Amino Acids in Culture (SILAC), allows the proteins in mammalian cells in culture to be isotopically labeled by the use of medium containing only "heavy" amino acids [63]. A typical experiment would comprise one treated, "heavy"-labeled cell line and a control unlabeled, "light", cell line. Cell lysates are mixed in a 1:1 ratio and then subjected to standard LC-MS/MS workflow. Peptides are identified in the usual fashion and the relative expression between cell line conditions identified at the MS1 level by the ratio of heavy to light peptide ion intensities. This approach has been shown to be reproducible across the proteome with a coefficient of variation of ~30% [62]. By using both heavy lysine and heavy arginine combinations, three conditions can be compared simultaneously.

A disadvantage of SILAC approaches is the requirement for complete label uptake by cultured cells, which limits the application to cells which express stable phenotypes of interest across several passages. The requirement for prior labeling in the conventional SILAC method also precludes the study of human tissues samples although fully isotopically labeled organisms have been described which may have application in disease models [64–66].

7.3.2 Super-SILAC

A variation of the SILAC method, termed super-SILAC, has been applied to quantify the proteome of human cancer samples [67]. In this procedure a mixture of cell lines derived from the cancer tissue of interest and approximately covering the expression profile of the tissue of interest are heavy-labeled using the SILAC method. A mixture of lysates from these cell lines with a defined protein mass is spiked-in to each tissue lysate in a 1:1 ratio before digestion, fractionation, and LC-MS/MS using standard procedures. Peptide identification and quantitation then proceeds as for a standard SILAC experiment. The ratio of expression between heavy and light peptides is calculated for each tissue sample. The constant SILAC spike-in mass provides a method of normalizing between experimental runs and also, by calculating the ratio of ratios, allows the relative expression between tissue types to be calculated [68].

An advantage is the spike-in standard can be used in multiple experiments on multiple platforms and still allow normalization between experiments and, once the spike-in standard is generated, there is no further labeling steps or reagent costs. A disadvantage is that the accuracy of SILAC is highest at ratios <2 and therefore a relatively close match to the tissue expression profile is required [62]. SILAC media is not yet available for many primary cells and therefore primary human tissues or cancers with few available cell lines may be difficult to analyze with this technique. Similarly, proteins unique to a tissue sample will not be quantified.

7.3.3 Isotope-Coded Affinity Tags (ICAT)

In this method, the cysteine residues of reduced proteins are labeled with tags comprising a composite of a sulfhydryl reacting group, a deuterated linker, and a biotin affinity tag [69]. Proteins from discrete samples can be differentially labeled as both "light" and "heavy" isotopes of the linker are available. Labeled samples are then pooled and digested together. Cysteine-containing peptides are then enriched by avidin-affinity chromatography. Peptides can then be further fractionated or directly subjected to LC-MS/MS. The different isotopes of the deuterated linker provide discrete mass peaks during MS1 analysis to allow differential expression analysis.

Unfortunately only cysteine-containing proteins can be studied, limiting proteome-wide efforts and the bulky affinity group, biotin, introduces significant background into the MS/MS spectra [70]. Furthermore, deuterated labels are more hydrophobic and therefore are differentially eluted during reverse phase LC, complicating the MS analysis [71]. This technique still has a role, however, as the affinity enrichment step allows the study of low-abundance proteins, not easily accessible by other methods.

7.3.4 ^{18}O Labeling

An advantage of this strategy is that it can be applied to almost any sample. In an approach that predates the SILAC method, samples for comparison are either proteolytically digested in ^{18}O-containing water for the "heavy" sample or standard "light" water [72]. As the protease, in most cases trypsin, cleaves the peptide bonds, the heavy isotope is incorporated, so all tryptic peptides will be labeled. The subsequent data analysis is identical to SILAC methods. A disadvantage of this approach is the relative expense of $H_2{}^{18}O$.

7.3.5 Dimethyl Isotopic Labeling

A further method uses standard and deuterium isotopes of formaldehyde to label the amino-terminus of peptides or the amino group of Lysine residues [73]. The isotopes are subsequently resolved by their mass differences allowing peptide-level quantitation from the MS1 scan. A further limitation common to SILAC, ^{18}O, and Dimethyl labeling is that maximum of three samples can be compared per mass spectrometry analysis.

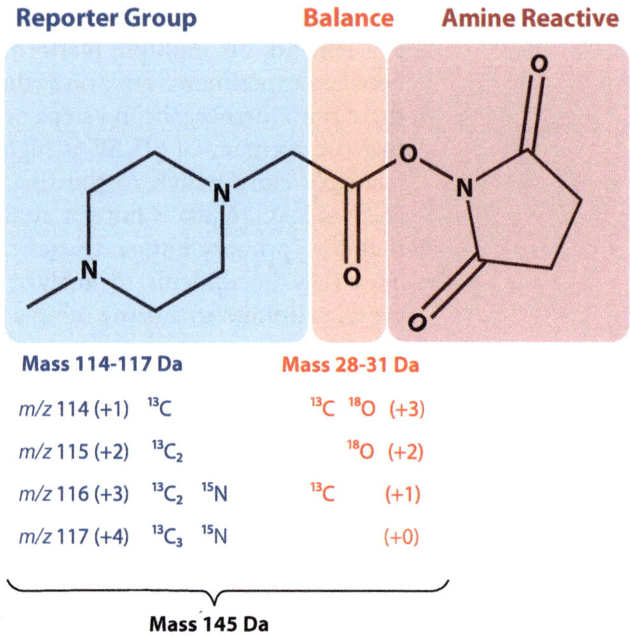

Fig. 2 Schematic of the four-plex iTRAQ peptide label. A reporter group with a defined mass between 114 and 117 Da is connected to a balancing linker. Together the reporter and linkers have a fixed mass of 145 Da and they are connected to an amine-reactive group which binds peptide amino-termini and lysine residues. The label is cleaved at the balancing linker during MS2 fragment ion generation to allow reporter ion detection. Figure adapted from [74]

7.3.6 Isobaric Peptide Labeling

Isobaric peptide labels offer greater multiplexing capabilities with 4-plex or 8-plex (Isobaric Tag for Relative and Absolute Quantification; iTRAQ) [74] or 6-plex, 10-plex or 11-plex (Tandem Mass Tags; TMT) commercial kits available [75]. These kits all rely on the same underlying principle.

Each label consists of an amine-reactive ester, a balancing carbonyl linker, and a reporter ion (Fig. 2). Tryptic peptides form amide linkages with the labels via N-termini or lysine residues. A label with a different reporter is used for each different sample and all the samples are mixed prior to fractionation and LC-MS/MS. Each label has the same total mass and chromatographic properties and therefore the LC retention time and mass/charge (*m/z*) separation of each sample are not differentially affected during the MS1 scan [74].

Precursor ions are then sampled for MS/MS analysis and the ionized-labeled peptides are fragmented with dissociation of the reporter ions from the balancing carbonyl linker. The peptide fragments are detected generating mass spectra in the usual manner. The reporter ions are also detected as peaks at a predefined *m/z*. For a four-plex iTRAQ experiment the reporters are detected at 114.1, 115.1, 116.1, and 117.1 *m/z* [74]. For a six-plex TMT experiment, the reporters are detected between 126.1 and 131.1 m/z [75].

Assuming complete peptide labeling of each sample, the more abundant peptides within each sample will have accumulated more label. When equal amounts of each sample are mixed and subjected to LC-MS/MS together, those samples with a greater original concentration of a peptide of interest will produce higher reporter ion peak intensities in the MS/MS scan. By comparing the relative reporter ion intensities, the relative peptide and therefore protein abundances in the original samples can be determined [76].

The multiplexing capabilities of isobaric labels are directly offset by the consequent dilution of each sample leading to challenges in identifying low-abundance peptides [44]. Samples also require to be lysed and digested separately which has the potential to introduce error. In contrast, cell populations can be mixed prior to lysis in SILAC experiments. The quantitative accuracy and dynamic range offered by isobaric labels are excellent, however, surpassing SILAC in a direct comparison [62].

7.4 Label-Free Quantitative Shotgun Proteomics

All sample manipulation steps during a proteomic workflow reduce the data yield due to loss of proteins [77] and are additional sources of variation [78]. Eliminating the sample processing steps to incorporate labels for quantitation is clearly an advantage and underlies the rationale to develop label-free methods of quantitation.

The total number of spectra matched to each peptide contributing to a protein identification, termed the spectral count, has been reported to correlate with absolute protein abundance [79]. Various methods have been proposed to refine the spectral count such as normalizing for protein length [80], or combination scores including peptide count and fragment-ion intensity [81]. For complex protein mixtures, spectral counts are still subject to significant between-run variability and are highly dependent on LC conditions and precursor ion selection. As a result, the quantitative reproducibility of spectral counting is inferior to isobaric labeling methods [82].

An alternative relies on the capture of precursor ion intensity as a function of time to produce an ion chromatogram. The area under the ion chromatogram curve is linearly proportional to the peptide concentration [83]. Challenges exist in applying this method across LC-MS/MS runs to allow differential analysis as the same peptide ion must be identified and quantified despite background noise, co-eluting peptides causing signal overlap, technical variations in retention time and total protein loading among other factors [84].

A simultaneous advantage and disadvantage of both label-free approaches is the requirement to analyze one sample per LC-MS/MS run. This prevents sample dilution, provides maximum potential coverage, and prevents the potential failure to identify dysregulated low-abundance proteins that occurs with multiplexed approaches. By comparing conditions across separate LC-MS/MS runs, however, the inherent changes in LC performance and the stochastic nature of protein identification by shotgun proteomics both contribute to data heterogeneity. Until these concerns are addressed, labeling strategies will still be widely employed.

8 MALDI-Imaging MS (MALDI-IMS)

A major disadvantage of lysing tissue biopsies for downstream mass spectrometry analysis is the loss of microscopic spatial information relating to protein expression. The local microenvironmental context of a cell is critical in determining behavior with cancer tissues being a well-recognized example [85]. Understanding the changes in protein expression that occur within a defined cellular niche may unveil novel insights not apparent from analysis of tissue biopsies in toto [86].

To preserve this heterogeneity, matrix-assisted laser desorption and ionization (MALDI) techniques have been adapted to allow direct ionization and mass spectrometry from tissue sections [87]. By co-registering spectra and histological images, patterns of protein expression can be interpreted within their biological and cellular context. The significant advantages of this imaging mass spectrometry (MALDI-IMS) method are offset by some of the limitations common to all MALDI approaches [88].

MALDI-IMS generates spectral features (m/z) which can be used to differentiate samples but does not provide protein identities. Hybrid approaches with downstream tandem mass spectrometry allow low mass proteins to be identified directly although proteome coverage has not yet reached parity with LC-MS/MS analysis of tissue lysates [89]. A further significant limitation is in the resolution of ionization sources. Current technologies allow a minimum resolvable area of 10 μm but most analyses are practically limited to areas of 100 μm [90]. This allows a granular expression map to be generated but the goal of identifying subcellular expression patterns, for example, at the tumor-stromal interface, remains elusive.

9 Conclusion

Biological phenotypes are governed by protein interactions and the correlation between RNA and protein expression is limited in many circumstances. The large-scale, direct measurement of protein expression is therefore an attractive prospect for the biologist. Mass spectrometry offers the potential to identify expressed proteins from biological samples in a hypothesis-free manner.

Technological advances have allowed the goal of proteome-wide measurement to become a reality in some model systems.

Monitoring dynamic changes in protein abundance has become feasible using both biochemical labeling strategies to provide highly accurate protein quantitation and label-free techniques. The evolution of selective reactive monitoring methods allows robust identification and quantitation across conditions. MALDI-IMS allows the spatial diversity of protein expression in biological tissues to be preserved and novel insights into disease processes such as cancer can be gleaned directly from tissue sections. By applying this array of proteomic techniques, scientists can address fundamental questions and begin to understand biological processes and disease pathophysiology.

References

1. Wasinger VC, Cordwell SJ, Cerpa-Poljak A, Yan JX, Gooley AA, Wilkins MR, Duncan MW, Harris R, Williams KL, Humphery-Smith I (1995) Progress with gene-product mapping of the Mollicutes: mycoplasma genitalium. Electrophoresis 16(7):1090–1094
2. Aebersold R, Mann M (2003) Mass spectrometry-based proteomics. Nature 422 (6928):198–207. https://doi.org/10.1038/nature01511
3. de Godoy LM, Olsen JV, Cox J, Nielsen ML, Hubner NC, Fröhlich F, Walther TC, Mann M (2008) Comprehensive mass-spectrometry-based proteome quantification of haploid versus diploid yeast. Nature 455 (7217):1251–1254. https://doi.org/10.1038/nature07341
4. Consortium IHGS (2004) Finishing the euchromatic sequence of the human genome. Nature 431(7011):931–945. https://doi.org/10.1038/nature03001
5. Legrain P, Aebersold R, Archakov A, Bairoch A, Bala K, Beretta L, Bergeron J, Borchers CH, Corthals GL, Costello CE, Deutsch EW, Domon B, Hancock W, He F, Hochstrasser D, Marko-Varga G, Salekdeh GH, Sechi S, Snyder M, Srivastava S, Uhlén M, Wu CH, Yamamoto T, Paik YK, Omenn GS (2011) The human proteome project: current state and future direction. Mol Cell Proteomics 10(7):M111.009993. https://doi.org/10.1074/mcp.M111.009993
6. Omenn GS, Lane L, Lundberg EK, Overall CM, Deutsch EW (2017) Progress on the HUPO Draft Human Proteome: 2017 Metrics of the Human Proteome Project. J Proteome Res 16:4281. https://doi.org/10.1021/acs.jproteome.7b00375
7. Gaudet P, Michel PA, Zahn-Zabal M, Britan A, Cusin I, Domagalski M, Duek PD, Gateau A, Gleizes A, Hinard V, Rech de Laval V, Lin J, Nikitin F, Schaeffer M, Teixeira D, Lane L, Bairoch A (2017) The neXtProt knowledgebase on human proteins: 2017 update. Nucleic Acids Res 45(D1):D177–D182. https://doi.org/10.1093/nar/gkw1062
8. Saiki RK, Gelfand DH, Stoffel S, Scharf SJ, Higuchi R, Horn GT, Mullis KB, Erlich HA (1988) Primer-directed enzymatic amplification of DNA with a thermostable DNA polymerase. Science 239(4839):487–491
9. Liu L, Li Y, Li S, Hu N, He Y, Pong R, Lin D, Lu L, Law M (2012) Comparison of next-generation sequencing systems. J Biomed Biotechnol 2012:251364. https://doi.org/10.1155/2012/251364
10. Pareek CS, Smoczynski R, Tretyn A (2011) Sequencing technologies and genome sequencing. J Appl Genet 52(4):413–435. https://doi.org/10.1007/s13353-011-0057-x
11. Niall HD (1973) Automated Edman degradation: the protein sequenator. Methods Enzymol 27:942–1010
12. Sperling J, Azubel M, Sperling R (2008) Structure and function of the Pre-mRNA splicing machine. Structure 16(11):1605–1615. https://doi.org/10.1016/j.str.2008.08.011
13. Ermini L, Olivieri C, Rizzi E, Corti G, Bonnal R, Soares P, Luciani S, Marota I, De Bellis G, Richards MB, Rollo F (2008) Complete mitochondrial genome sequence of the Tyrolean iceman. Curr Biol 18 (21):1687–1693. https://doi.org/10.1016/j.cub.2008.09.028

14. Elguoshy A, Hirao Y, Xu B, Saito S, Quadery AF, Yamamoto K, Mitsui T, Yamamoto T, JProS CXPTo (2017) Identification and validation of human missing proteins and peptides in public proteome databases: data mining strategy. J Proteome Res 16:4403. https://doi.org/10.1021/acs.jproteome.7b00423
15. Thul PJ, Åkesson L, Wiking M, Mahdessian D, Geladaki A, Ait Blal H, Alm T, Asplund A, Björk L, Breckels LM, Bäckström A, Danielsson F, Fagerberg L, Fall J, Gatto L, Gnann C, Hober S, Hjelmare M, Johansson F, Lee S, Lindskog C, Mulder J, Mulvey CM, Nilsson P, Oksvold P, Rockberg J, Schutten R, Schwenk JM, Sivertsson Å, Sjöstedt E, Skogs M, Stadler C, Sullivan DP, Tegel H, Winsnes C, Zhang C, Zwahlen M, Mardinoglu A, Pontén F, von Feilitzen K, Lilley KS, Uhlén M, Lundberg E (2017) A subcellular map of the human proteome. Science 356(6340):eaal3321. https://doi.org/10.1126/science.aal3321
16. Espina V, Edmiston KH, Heiby M, Pierobon M, Sciro M, Merritt B, Banks S, Deng J, VanMeter AJ, Geho DH, Pastore L, Sennesh J, Petricoin EF, Liotta LA (2008) A portrait of tissue phosphoprotein stability in the clinical tissue procurement process. Mol Cell Proteomics 7(10):1998–2018. https://doi.org/10.1074/mcp.M700596-MCP200
17. Aebersold R, Mann M (2016) Mass-spectrometric exploration of proteome structure and function. Nature 537 (7620):347–355. https://doi.org/10.1038/nature19949
18. Hu Q, Noll RJ, Li H, Makarov A, Hardman M, Graham Cooks R (2005) The Orbitrap: a new mass spectrometer. J Mass Spectrom 40 (4):430–443. https://doi.org/10.1002/jms.856
19. Collins BC, Hunter CL, Liu Y, Schilling B, Rosenberger G, Bader SL, Chan DW, Gibson BW, Gingras AC, Held JM, Hirayama-Kurogi-M, Hou G, Krisp C, Larsen B, Lin L, Liu S, Molloy MP, Moritz RL, Ohtsuki S, Schlapbach R, Selevsek N, Thomas SN, Tzeng SC, Zhang H, Aebersold R (2017) Multi-laboratory assessment of reproducibility, qualitative and quantitative performance of SWATH-mass spectrometry. Nat Commun 8 (1):291. https://doi.org/10.1038/s41467-017-00249-5
20. Michalski A, Damoc E, Lange O, Denisov E, Nolting D, Müller M, Viner R, Schwartz J, Remes P, Belford M, Dunyach JJ, Cox J, Horning S, Mann M, Makarov A (2012) Ultra high resolution linear ion trap Orbitrap mass spectrometer (Orbitrap Elite) facilitates top down LC MS/MS and versatile peptide fragmentation modes. Mol Cell Proteomics 11(3):O111.013698. https://doi.org/10.1074/mcp.O111.013698
21. Gillet LC, Navarro P, Tate S, Röst H, Selevsek N, Reiter L, Bonner R, Aebersold R (2012) Targeted data extraction of the MS/MS spectra generated by data-independent acquisition: a new concept for consistent and accurate proteome analysis. Mol Cell Proteomics 11(6):O111.016717. https://doi.org/10.1074/mcp.O111.016717
22. Souza GH, Guest PC, Martins-de-Souza D (2017) LC-MS(E), multiplex MS/MS, ion mobility, and label-free quantitation in clinical proteomics. Methods Mol Biol 1546:57–73. https://doi.org/10.1007/978-1-4939-6730-8_4
23. Hardman M, Makarov AA (2003) Interfacing the orbitrap mass analyzer to an electrospray ion source. Anal Chem 75(7):1699–1705
24. Krutchinsky AN, Kalkum M, Chait BT (2001) Automatic identification of proteins with a MALDI-quadrupole ion trap mass spectrometer. Anal Chem 73(21):5066–5077
25. Johnson RS, Martin SA, Biemann K, Stults JT, Watson JT (1987) Novel fragmentation process of peptides by collision-induced decomposition in a tandem mass spectrometer: differentiation of leucine and isoleucine. Anal Chem 59(21):2621–2625
26. Hughes C, Ma B, Lajoie GA (2010) De novo sequencing methods in proteomics. Methods Mol Biol 604:105–121. https://doi.org/10.1007/978-1-60761-444-9_8
27. Johnson RS, Taylor JA (2002) Searching sequence databases via de novo peptide sequencing by tandem mass spectrometry. Mol Biotechnol 22(3):301–315. https://doi.org/10.1385/MB:22:3:301
28. Nesvizhskii AI, Vitek O, Aebersold R (2007) Analysis and validation of proteomic data generated by tandem mass spectrometry. Nat Methods 4(10):787–797. https://doi.org/10.1038/nmeth1088
29. Sadygov RG, Cociorva D, Yates JR (2004) Large-scale database searching using tandem mass spectra: looking up the answer in the back of the book. Nat Methods 1 (3):195–202. https://doi.org/10.1038/nmeth725
30. Kusebauch U, Campbell DS, Deutsch EW, Chu CS, Spicer DA, Brusniak MY, Slagel J, Sun Z, Stevens J, Grimes B, Shteynberg D, Hoopmann MR, Blattmann P, Ratushny AV, Rinner O, Picotti P, Carapito C, Huang CY, Kapousouz M, Lam H, Tran T, Demir E,

Aitchison JD, Sander C, Hood L, Aebersold R, Moritz RL (2016) Human SRMAtlas: a resource of targeted assays to quantify the complete human proteome. Cell 166(3):766–778. https://doi.org/10.1016/j.cell.2016.06.041

31. Picotti P, Clément-Ziza M, Lam H, Campbell DS, Schmidt A, Deutsch EW, Röst H, Sun Z, Rinner O, Reiter L, Shen Q, Michaelson JJ, Frei A, Alberti S, Kusebauch U, Wollscheid B, Moritz RL, Beyer A, Aebersold R (2013) A complete mass-spectrometric map of the yeast proteome applied to quantitative trait analysis. Nature 494(7436):266–270. https://doi.org/10.1038/nature11835
32. Dasari S, Chambers MC, Martinez MA, Carpenter KL, Ham AJ, Vega-Montoto LJ, Tabb DL (2012) Pepitome: evaluating improved spectral library search for identification complementarity and quality assessment. J Proteome Res 11(3):1686–1695. https://doi.org/10.1021/pr200874e
33. Lam H (2011) Building and searching tandem mass spectral libraries for peptide identification. Mol Cell Proteomics 10(12): R111.008565. https://doi.org/10.1074/mcp.R111.008565
34. Jmeian Y, El Rassi Z (2009) Liquid-phase-based separation systems for depletion, prefractionation and enrichment of proteins in biological fluids for in-depth proteomics analysis. Electrophoresis 30(1):249–261. https://doi.org/10.1002/elps.200800639
35. Boisvert FM, Lam YW, Lamont D, Lamond AI (2010) A quantitative proteomics analysis of subcellular proteome localization and changes induced by DNA damage. Mol Cell Proteomics 9(3):457–470. https://doi.org/10.1074/mcp.M900429-MCP200
36. Chenau J, Michelland S, Sidibe J, Seve M (2008) Peptides OFFGEL electrophoresis: a suitable pre-analytical step for complex eukaryotic samples fractionation compatible with quantitative iTRAQ labeling. Proteome Sci 6:9. https://doi.org/10.1186/1477-5956-6-9
37. Batth TS, Olsen JV (2016) Offline high pH reversed-phase peptide fractionation for deep phosphoproteome coverage. Methods Mol Biol 1355:179–192. https://doi.org/10.1007/978-1-4939-3049-4_12
38. Boersema PJ, Mohammed S, Heck AJ (2008) Hydrophilic interaction liquid chromatography (HILIC) in proteomics. Anal Bioanal Chem 391(1):151–159. https://doi.org/10.1007/s00216-008-1865-7
39. Bantscheff M, Boesche M, Eberhard D, Matthieson T, Sweetman G, Kuster B (2008) Robust and sensitive iTRAQ quantification on an LTQ Orbitrap mass spectrometer. Mol Cell Proteomics 7(9):1702–1713. https://doi.org/10.1074/mcp.M800029-MCP200
40. Garbis SD, Roumeliotis TI, Tyritzis SI, Zorpas KM, Pavlakis K, Constantinides CA (2011) A novel multidimensional protein identification technology approach combining protein size exclusion prefractionation, peptide zwitterion-ion hydrophilic interaction chromatography, and nano-ultraperformance RP chromatography/nESI-MS2 for the in-depth analysis of the serum proteome and phosphoproteome: application to clinical sera derived from humans with benign prostate hyperplasia. Anal Chem 83(3):708–718. https://doi.org/10.1021/ac102075d
41. Michalski A, Cox J, Mann M (2011) More than 100,000 detectable peptide species elute in single shotgun proteomics runs but the majority is inaccessible to data-dependent LC-MS/MS. J Proteome Res 10(4):1785–1793. https://doi.org/10.1021/pr101060v
42. Panchaud A, Scherl A, Shaffer S, Haller P, Kulasekara H, Miller SI, Goodlett DR (2009) Precursor acquisition independent from ion count: how to dive deeper into the proteomics ocean. Anal Chem 81:6481–6488
43. Scherl A, Shaffer SA, Taylor GK, Kulasekara HD, Miller SI, Goodlett DR (2008) Genome-specific gas-phase fractionation strategy for improved shotgun proteomic profiling of proteotypic peptides. Anal Chem 80 (4):1182–1191. https://doi.org/10.1021/ac701680f
44. Panchaud A, Jung S, Shaffer SA, Aitchison JD, Goodlett DR (2011) Faster, quantitative, and accurate precursor acquisition independent from ion count. Anal Chem 83 (6):2250–2257. https://doi.org/10.1021/ac103079q
45. Savaryn JP, Catherman AD, Thomas PM, Abecassis MM, Kelleher NL (2013) The emergence of top-down proteomics in clinical research. Genome Med 5(6):53. https://doi.org/10.1186/gm457
46. Tran JC, Zamdborg L, Ahlf DR, Lee JE, Catherman AD, Durbin KR, Tipton JD, Vellaichamy A, Kellie JF, Li M, Wu C, Sweet SM, Early BP, Siuti N, LeDuc RD, Compton PD, Thomas PM, Kelleher NL (2011) Mapping intact protein isoforms in discovery mode using top-down proteomics. Nature 480 (7376):254–258. https://doi.org/10.1038/nature10575
47. Catherman AD, Durbin KR, Ahlf DR, Early BP, Fellers RT, Tran JC, Thomas PM, Kelleher NL (2013) Large-scale top-down proteomics of the human proteome: membrane proteins,

mitochondria, and senescence. Mol Cell Proteomics 12(12):3465–3473. https://doi.org/10.1074/mcp.M113.030114

48. Fornelli L, Durbin KR, Fellers RT, Early BP, Greer JB, LeDuc RD, Compton PD, Kelleher NL (2017) Advancing top-down analysis of the human proteome using a Benchtop Quadrupole-Orbitrap mass spectrometer. J Proteome Res 16(2):609–618. https://doi.org/10.1021/acs.jproteome.6b00698
49. Fornelli L, Toby TK, Schachner LF, Doubleday PF, Srzentić K, DeHart CJ, Kelleher NL (2017) Top-down proteomics: where we are, where we are going? J Proteome 175:3. https://doi.org/10.1016/j.jprot.2017.02.002
50. Cai W, Guner H, Gregorich ZR, Chen AJ, Ayaz-Guner S, Peng Y, Valeja SG, Liu X, Ge Y (2016) MASH suite pro: a comprehensive software tool for top-down proteomics. Mol Cell Proteomics 15(2):703–714. https://doi.org/10.1074/mcp.O115.054387
51. Desiere F, Deutsch EW, King NL, Nesvizhskii AI, Mallick P, Eng J, Chen S, Eddes J, Loevenich SN, Aebersold R (2006) The PeptideAtlas project. Nucleic Acids Res 34(Database issue):D655–D658. https://doi.org/10.1093/nar/gkj040
52. Zolg DP, Wilhelm M, Schnatbaum K, Zerweck J, Knaute T, Delanghe B, Bailey DJ, Gessulat S, Ehrlich HC, Weininger M, Yu P, Schlegl J, Kramer K, Schmidt T, Kusebauch U, Deutsch EW, Aebersold R, Moritz RL, Wenschuh H, Moehring T, Aiche S, Huhmer A, Reimer U, Kuster B (2017) Building ProteomeTools based on a complete synthetic human proteome. Nat Methods 14(3):259–262. https://doi.org/10.1038/nmeth.4153
53. Mallick P, Schirle M, Chen SS, Flory MR, Lee H, Martin D, Ranish J, Raught B, Schmitt R, Werner T, Kuster B, Aebersold R (2007) Computational prediction of proteotypic peptides for quantitative proteomics. Nat Biotechnol 25(1):125–131. https://doi.org/10.1038/nbt1275
54. Ebhardt HA, Root A, Sander C, Aebersold R (2015) Applications of targeted proteomics in systems biology and translational medicine. Proteomics 15(18):3193–3208. https://doi.org/10.1002/pmic.201500004
55. Navarro P, Kuharev J, Gillet LC, Bernhardt OM, MacLean B, Röst HL, Tate SA, Tsou CC, Reiter L, Distler U, Rosenberger G, Perez-Riverol Y, Nesvizhskii AI, Aebersold R, Tenzer S (2016) A multicenter study benchmarks software tools for label-free proteome quantification. Nat Biotechnol 34(11):1130–1136. https://doi.org/10.1038/nbt.3685
56. Röst HL, Aebersold R, Schubert OT (2017) Automated SWATH data analysis using targeted extraction of ion chromatograms. Methods Mol Biol 1550:289–307. https://doi.org/10.1007/978-1-4939-6747-6_20
57. Geiger T, Velic A, Macek B, Lundberg E, Kampf C, Nagaraj N, Uhlen M, Cox J, Mann M (2013) Initial quantitative proteomic map of 28 mouse tissues using the SILAC mouse. Mol Cell Proteomics 12(6):1709–1722. https://doi.org/10.1074/mcp.M112.024919
58. Uhlén M, Fagerberg L, Hallström BM, Lindskog C, Oksvold P, Mardinoglu A, Sivertsson Å, Kampf C, Sjöstedt E, Asplund A, Olsson I, Edlund K, Lundberg E, Navani S, Szigyarto CA, Odeberg J, Djureinovic D, Takanen JO, Hober S, Alm T, Edqvist PH, Berling H, Tegel H, Mulder J, Rockberg J, Nilsson P, Schwenk JM, Hamsten M, von Feilitzen K, Forsberg M, Persson L, Johansson F, Zwahlen M, von Heijne G, Nielsen J, Pontén F (2015) Proteomics. Tissue-based map of the human proteome. Science 347(6220):1260419. https://doi.org/10.1126/science.1260419
59. Görg A, Weiss W, Dunn MJ (2004) Current two-dimensional electrophoresis technology for proteomics. Proteomics 4(12):3665–3685. https://doi.org/10.1002/pmic.200401031
60. Granier F, de Vienne D (1986) Silver staining of proteins: standardized procedure for two-dimensional gels bound to polyester sheets. Anal Biochem 155(1):45–50
61. Dowsey AW, Dunn MJ, Yang GZ (2003) The role of bioinformatics in two-dimensional gel electrophoresis. Proteomics 3(8):1567–1596. https://doi.org/10.1002/pmic.200300459
62. Altelaar AF, Frese CK, Preisinger C, Hennrich ML, Schram AW, Timmers HT, Heck AJ, Mohammed S (2013) Benchmarking stable isotope labeling based quantitative proteomics. J Proteome 88:14–26. https://doi.org/10.1016/j.jprot.2012.10.009
63. Ong SE, Blagoev B, Kratchmarova I, Kristensen DB, Steen H, Pandey A, Mann M (2002) Stable isotope labeling by amino acids in cell culture, SILAC, as a simple and accurate approach to expression proteomics. Mol Cell Proteomics 1(5):376–386
64. Krüger M, Moser M, Ussar S, Thievessen I, Luber CA, Forner F, Schmidt S, Zanivan S, Fässler R, Mann M (2008) SILAC mouse for quantitative proteomics uncovers kindlin-3 as an essential factor for red blood cell function.

Cell 134(2):353–364. https://doi.org/10.1016/j.cell.2008.05.033

65. Sury MD, Chen JX, Selbach M (2010) The SILAC fly allows for accurate protein quantification in vivo. Mol Cell Proteomics 9 (10):2173–2183. https://doi.org/10.1074/mcp.M110.000323
66. Larance M, Bailly AP, Pourkarimi E, Hay RT, Buchanan G, Coulthurst S, Xirodimas DP, Gartner A, Lamond AI (2011) Stable-isotope labeling with amino acids in nematodes. Nat Methods 8(10):849–851. https://doi.org/10.1038/nmeth.1679
67. Geiger T, Cox J, Ostasiewicz P, Wisniewski JR, Mann M (2010) Super-SILAC mix for quantitative proteomics of human tumor tissue. Nat Methods 7(5):383–385. https://doi.org/10.1038/nmeth.1446
68. Geiger T, Wehner A, Schaab C, Cox J, Mann M (2012) Comparative proteomic analysis of eleven common cell lines reveals ubiquitous but varying expression of most proteins. Mol Cell Proteomics 11(3):M111.014050. https://doi.org/10.1074/mcp.M111.014050
69. Gygi SP, Rist B, Gerber SA, Turecek F, Gelb MH, Aebersold R (1999) Quantitative analysis of complex protein mixtures using isotope-coded affinity tags. Nat Biotechnol 17 (10):994–999. https://doi.org/10.1038/13690
70. Zhou H, Ranish JA, Watts JD, Aebersold R (2002) Quantitative proteome analysis by solid-phase isotope tagging and mass spectrometry. Nat Biotechnol 20(5):512–515. https://doi.org/10.1038/nbt0502-512
71. Zhang R, Sioma CS, Wang S, Regnier FE (2001) Fractionation of isotopically labeled peptides in quantitative proteomics. Anal Chem 73(21):5142–5149
72. Antonov VK, Ginodman LM, Rumsh LD, Kapitannikov YV, Barshevskaya TN, Yavashev LP, Gurova AG, Volkova LI (1981) Studies on the mechanisms of action of proteolytic enzymes using heavy oxygen exchange. Eur J Biochem 117(1):195–200
73. Hsu JL, Huang SY, Chow NH, Chen SH (2003) Stable-isotope dimethyl labeling for quantitative proteomics. Anal Chem 75 (24):6843–6852. https://doi.org/10.1021/ac0348625
74. Ross PL, Huang YN, Marchese JN, Williamson B, Parker K, Hattan S, Khainovski N, Pillai S, Dey S, Daniels S, Purkayastha S, Juhasz P, Martin S, Bartlet-Jones M, He F, Jacobson A, Pappin DJ (2004) Multiplexed protein quantitation in Saccharomyces cerevisiae using amine-reactive isobaric tagging reagents. Mol Cell Proteomics 3(12):1154–1169. https://doi.org/10.1074/mcp.M400129-MCP200
75. Thompson A, Schäfer J, Kuhn K, Kienle S, Schwarz J, Schmidt G, Neumann T, Johnstone R, Mohammed AK, Hamon C (2003) Tandem mass tags: a novel quantification strategy for comparative analysis of complex protein mixtures by MS/MS. Anal Chem 75(8):1895–1904
76. Bouchal P, Roumeliotis T, Hrstka R, Nenutil R, Vojtesek B, Garbis SD (2009) Biomarker discovery in low-grade breast cancer using isobaric stable isotope tags and two-dimensional liquid chromatography-tandem mass spectrometry (iTRAQ-2DLC-MS/MS) based quantitative proteomic analysis. J Proteome Res 8(1):362–373. https://doi.org/10.1021/pr800622b
77. Luk VN, Wheeler AR (2009) A digital microfluidic approach to proteomic sample processing. Anal Chem 81(11):4524–4530. https://doi.org/10.1021/ac900522a
78. Rai AJ, Gelfand CA, Haywood BC, Warunek DJ, Yi J, Schuchard MD, Mehigh RJ, Cockrill SL, Scott GB, Tammen H, Schulz-Knappe P, Speicher DW, Vitzthum F, Haab BB, Siest G, Chan DW (2005) HUPO Plasma Proteome Project specimen collection and handling: towards the standardization of parameters for plasma proteome samples. Proteomics 5 (13):3262–3277. https://doi.org/10.1002/pmic.200401245
79. Lundgren DH, Hwang SI, Wu L, Han DK (2010) Role of spectral counting in quantitative proteomics. Expert Rev Proteomics 7 (1):39–53. https://doi.org/10.1586/epr.09.69
80. Carvalho PC, Hewel J, Barbosa VC, Yates JR (2008) Identifying differences in protein expression levels by spectral counting and feature selection. Genet Mol Res 7(2):342–356
81. Griffin NM, Yu J, Long F, Oh P, Shore S, Li Y, Koziol JA, Schnitzer JE (2010) Label-free, normalized quantification of complex mass spectrometry data for proteomic analysis. Nat Biotechnol 28(1):83–89. https://doi.org/10.1038/nbt.1592
82. Li Z, Adams RM, Chourey K, Hurst GB, Hettich RL, Pan C (2012) Systematic comparison of label-free, metabolic labeling, and isobaric chemical labeling for quantitative proteomics on LTQ Orbitrap Velos. J Proteome Res 11 (3):1582–1590. https://doi.org/10.1021/pr200748h
83. Chelius D, Bondarenko PV (2002) Quantitative profiling of proteins in complex mixtures

using liquid chromatography and mass spectrometry. J Proteome Res 1(4):317–323

84. Listgarten J, Emili A (2005) Statistical and computational methods for comparative proteomic profiling using liquid chromatography-tandem mass spectrometry. Mol Cell Proteomics 4(4):419–434. https://doi.org/10.1074/mcp.R500005-MCP200
85. Hanahan D, Weinberg RA (2011) Hallmarks of cancer: the next generation. Cell 144 (5):646–674. https://doi.org/10.1016/j.cell.2011.02.013
86. Gerlinger M, Rowan AJ, Horswell S, Larkin J, Endesfelder D, Gronroos E, Martinez P, Matthews N, Stewart A, Tarpey P, Varela I, Phillimore B, Begum S, McDonald NQ, Butler A, Jones D, Raine K, Latimer C, Santos CR, Nohadani M, Eklund AC, Spencer-Dene B, Clark G, Pickering L, Stamp G, Gore M, Szallasi Z, Downward J, Futreal PA, Swanton C (2012) Intratumor heterogeneity and branched evolution revealed by multiregion sequencing. N Engl J Med 366 (10):883–892. https://doi.org/10.1056/NEJMoa1113205
87. Caprioli RM, Farmer TB, Gile J (1997) Molecular imaging of biological samples: localization of peptides and proteins using MALDI-TOF MS. Anal Chem 69(23):4751–4760
88. Aichler M, Walch A (2015) MALDI Imaging mass spectrometry: current frontiers and perspectives in pathology research and practice. Lab Investig 95(4):422–431. https://doi.org/10.1038/labinvest.2014.156
89. Minerva L, Clerens S, Baggerman G, Arckens L (2008) Direct profiling and identification of peptide expression differences in the pancreas of control and ob/ob mice by imaging mass spectrometry. Proteomics 8(18):3763–3774. https://doi.org/10.1002/pmic.200800237
90. Balluff B, Schöne C, Höfler H, Walch A (2011) MALDI imaging mass spectrometry for direct tissue analysis: technological advancements and recent applications. Histochem Cell Biol 136 (3):227–244. https://doi.org/10.1007/s00418-011-0843-x
91. Perkins DN, Pappin DJ, Creasy DM, Cottrell JS (1999) Probability-based protein identification by searching sequence databases using mass spectrometry data. Electrophoresis 20 (18):3551–3567. https://doi.org/10.1002/(SICI)1522-2683(19991201)20:18<3551::AID-ELPS3551>3.0.CO;2-2

Chapter 14

Functional Proteomic Analysis to Characterize Signaling Crosstalk

Sneha M. Pinto, Yashwanth Subbannayya, and T. S. Keshava Prasad

Abstract

The biological activities of a cell are determined by its response to external stimuli. The signals are transduced from either intracellular or extracellular milieu through networks of multi-protein complexes and post-translational modifications of proteins (PTMs). Most PTMs including phosphorylation, acetylation, ubiquitination, and SUMOylation, among others, modulate activities of proteins and regulate biological processes such as proliferation, differentiation, as well as host pathogen interaction. Conventionally, reverse genetics analysis and single molecule-based studies were employed to identify and characterize the function of PTMs and enzyme-substrate networks regulated by them. With the advent of high-throughput technologies, it is now possible to identify and quantify thousands of PTM sites in a single experiment. Here, we discuss recent advances in enrichment strategies of various PTMs. We also describe a method for the identification and relative quantitation of proteins using a tandem mass tag labeling approach combined with serial enrichment of phosphorylation, acetylation and succinylation using antibody enrichment strategy.

Key words Signaling pathways, Mass spectrometry, Phosphorylation, Crosstalk, Quantitation

1 Introduction

The biological activities of a cell are governed by various cellular and physiological processes. These processes are mediated by proteins, the functional units that either act in toto or in concert with other proteins. Distinct post-translational modifications (PTMs) dynamically regulate protein function and thereby are involved in driving a majority of cellular processes. These include changes mediated by phosphorylation and activation of downstream substrates, ubiquitination or SUMOylation of target proteins resulting in their degradation, subcellular trafficking or regulation of transcriptional mechanisms. In most biological processes, the relay of information is mediated by multiple protein PTMs. The crosstalk provides cues to decide the fate of the proteins and consequently govern cellular activity. In order to understand cellular

Xing Wang and Matthew Kuruc (eds.), *Functional Proteomics: Methods and Protocols*, Methods in Molecular Biology, vol. 1871, https://doi.org/10.1007/978-1-4939-8814-3_14,

mechanisms, it is imperative to identify and characterize PTMs, thereby understanding its role in health and disease.

It is estimated that there are over 200 types of post-translation modifications and 300 types of chemical and biological modifications [1, 2]. These PTMs are mediated by enzymes including kinases, phosphatases, transferases, and ligases under the influence of various stimuli. Conventionally, PTMs, especially phosphorylation, have been identified using in vitro kinase assays or site-directed mutagenesis-based assays. In case of ubiquitination or acetylation modifications, the site of modification and its effect have been studied depending on the availability of the antibodies against the protein of interest using immunoblotting approaches. In recent years, the utility of protein/peptide array chip for the detection of post-translational modifications and PTM-dependent interactions has been on the rise [3–5]. However, a vast majority of these studies have focused on methods for interrogating protein phosphorylation-mediated signaling systems in various diseases ([6, 7], reviewed in [8]). The advent of mass spectrometry technique has not only enabled identification and quantification of proteins in a high-throughput fashion but has rapidly become popular for the discovery of PTMs, thereby revolutionizing the current understanding of signaling pathways and its role in regulating cellular processes. With the unveiling of the two drafts of the human proteome [9, 10], the PTMome represents the next frontier of grand challenges. In the forthcoming sections, we will briefly describe advances made in the strategies for PTM enrichment, mass spectrometry methods, data analysis, and data-mining approaches.

1.1 Strategies to Enrich Post-translational Modifications

Identification and characterization of PTMs and enzyme-substrate networks regulated by them were conventionally studied using reverse genetics or by single molecule-based studies. With the advent of high-throughput technologies, it is now possible to identify and quantify thousands of PTM sites in a single experiment. However, the low stoichiometry of PTM-modified peptides necessitates the use of enrichment strategies to enable their detection in biological samples. Toward this end, substantial progress has been made toward the development and optimization of enrichment and fractionation strategies for global studies on PTM (reviewed in [11]). Specific enrichment strategies have primarily been developed based on the properties of affinity chromatography either by utilizing the ionic charge properties of the modifying group or based on antibody recognition. The enrichment strategies are applied either at the protein level or at the peptide level. Peptide-based enrichment strategies are by far the more common and amenable method. A brief description of the various enrichment strategies has been provided.

1.1.1 Antibody-Based Enrichment methods

One of the widely employed methodology to assay PTMs of interest is the use of antibody-based enrichment strategies coupled with LC-MS/MS analysis. The most common as well as the most popular PTM assayed includes the use of antibodies that specifically enrich for tyrosine [12–14] or serine/threonine phosphorylated proteins [15]. A detailed review of this topic has been provided by Harsha et al. [16]. Antibody-based enrichment strategies have now been extended to profile and quantify other PTMs including ubiquitination, SUMOylation, lysine modifications such as acetylation, succinylation, and malonlylation, among others. Furthermore, Cell Signaling Technologies and PTM Biolabs have also developed antibodies targeting methylation, crotonylation [17], and modifications on lysine and arginine residues [18]. In addition to pan-PTM-specific antibodies, antibodies specific to PTM motifs have also been employed to identify and quantify substrates [19].

1.1.2 Metal-Ion-Based Enrichment Strategies

The alternative approach for enrichment of PTMs especially phosphorylation moieties are chromatographic methods based on metal affinity enrichment referred as immobilized metal affinity chromatography (IMAC) and titanium dioxide (TiO_2) based enrichment [20]. Metal ions such as Fe(3+), Zr(4+), and Ga(3+) are employed in IMAC, which is immobilized on beads and covalently bound to negatively charged phosphopeptides [21–23]. TiO_2 also works on a similar principle and both approaches are employed in combination with chromatography-based separation techniques such as strong cation exchange chromatography and high pH reverse phase chromatography [14, 16, 24]. Sequential enrichment using IMAC followed by TiO_2 has also been reported to predominantly enrich for monophosphorylated peptides [25].

1.1.3 Chemical Proteomics

With advances in chemical biology, PTM enrichment using chemical tagging or derivatization has enabled detection and enrichment of several PTMs including N-linked and O-linked glycosylation [26, 27], acylation [28], palmitoylation [29, 30], and S-nitrosylation. The tagging methods include both in vitro and in vivo labeling of small chemical tag at the site of the modification and subsequent enrichment using a capture reagent such as biotin. In case of enrichment and separation of N-linked glycopeptides, boronic acid chemistry, hydrazide chemistry, and alkyne click chemistry, among others, have been employed [31]. A detailed description of advances in chemical proteomics to study the PTMome has been described by Tate [32].

1.1.4 Serial Enrichment of PTMs

It is known that proteins can be simultaneously modified by multiple PTMs and these PTMs may form the basis of regulatory cross talk events [33]. Therefore, it is ideal to simultaneously study numerous PTMs in a single experiment. This can be achieved by carrying out PTM enrichment in a serial manner. Serial enrichment technique has been explored for the enrichment of phosphorylation and ubiquitylation in *Saccharomyces cerevisiae* to understand cross talk mechanisms [34]. Mertins and colleagues have also employed serial enrichments of different posttranslational modifications (SEPTM) and extensively studied the changes in the protein expression, phosphorylation, acetylation, and ubiquitination events in the same sample [35].

1.2 Mass Spectrometry to Characterize Multiple Post-translational Modifications

Over the years, mass spectrometry-based proteomic analysis has considerably increased our understanding of the occurrence and dynamics of protein PTMs. Following PTM enrichment, sequence identification and confident site localization is vital. In case of bottom-up approach which involves proteolytic digestion of proteins followed by enrichment of PTMs, information on MS2 fragmentation provides data that is being used to deduce the sequence as well as identify the site of modification. The fragmentation approaches include collision-induced dissociation (CID) and higher energy collisional dissociation (HCD), both of which generate predominantly b- and y-type ions and are universally employed for global protein profiling as well as for the analyses of stable PTMs. In the CID mode, fragmentation and data collection occurs in ion trap at a faster acquisition speed resulting in higher number of data points. On the contrary, HCD mode of fragmentation results in Fourier transform detection of the MS2 fragments in the Orbitrap mass analyzer with data being acquired at a higher resolution albeit lower scan rate. In case of labile PTMs such as acetylation and glycosylation, electron transfer dissociation (ETD) has been described as a preferred method for data acquisition [36]. The mass accuracy of mass spectrometers can affect the identification and accurate site localization. Mass accuracy can also significantly impact distinction of isobaric/indistinguishable mass PTMs when mass analyzers such as ion trap are employed rather than high-resolution mass analyzers such as Orbitrap. It is therefore essential to choose the right mass spectrometry method for data acquisition. To maximize the detection of PTM-modified peptides, a combination of fragmentation methods can also be employed.

1.3 Quantitative Proteomic Strategies to Assess Dynamic States of Proteins

In addition to detection of PTMs on thousands of proteins in a single experiment, recent advances in mass spectrometry-based techniques also allow quantitation of the relative changes of modification in multiple biological conditions. Several quantitative proteomic technologies are currently available to determine differences in protein abundances across biological states including metabolic labeling techniques such as SILAC [37] and chemical labeling techniques such as iTRAQ [38] and TMT (Tandem Mass Tags) [39]. These tools have enabled multiplexing and achieve quantitation of PTMs across multiple conditions. In addition, the development of label-free approaches such as iBAQ [40], NSAF [41],

SWATH [42], and APEX [43] have also enabled quantitation of PTMs without the need for chemical or metabolic labeling. Software algorithms such as ptmRS [44] and A-Score [45] now allow localization of modification sites, thereby enabling quantitation of the extent of PTMs.

1.4 Studies on Mass Spectrometry-Based PTM Analysis

A partial list of studies on mass spectrometry-based PTM analysis is provided in Table 1. For the purpose of this chapter, we have briefly described studies on phosphorylation, succinylation, acetylation, ubiquitination, glycosylation, and palmitoylation.

Reversible protein phosphorylation at serine, threonine, and tyrosine residues is one of the most common and widely studied PTM of proteins. Other residues such as histidine, asparagine, arginine, cysteine, and lysine are also known to be phosphorylated. However, the extent of phosphorylation is relatively less known owing to its occurrence in a low frequency or challenges in efficient enrichment and identification. A previous study on human renal cancer used a Ni-NTA-based enrichment method and identified 44,728 phosphosites on 6415 proteins [46]. Another study expanded the understanding of the IL-33 signaling pathway by identifying 7191 phosphorylation sites on 2746 proteins [14]. Phosphorylation events have also been shown to be widespread in prokaryotes using metal affinity-based enrichment techniques [24].

Succinylation of lysine was identified as a new posttranslational modification a few years ago [47]. Succinylation has been observed to occur widely across both prokaryotes and eukaryotes [48]. Enrichment of succinylated peptides has been typically carried out using antibody-based affinity enrichment [49]. These enriched peptides are then subjected to strong cation exchange chromatography and subjected to LC-MS/MS.

Acetylation is known to regulate diverse cellular processes and has several implications in the disease pathophysiology [50]. One of the early studies surveyed acetylation sites in HeLa cells and mouse liver mitochondria using an antibody-based enrichment approach identified 388 sites [51]. Lysine acetylation has also been found to be conserved in prokaryotes such as *Escherichia coli* [52] and *Saccharomyces cerevisiae* [53] and rat models [54] using antibody-based enrichment methods. Quantitative proteomic techniques such as TMT labeling are amenable to antibody-based acetyl lysine peptide enrichment and have been applied to study lysine acetylation sites on perturbation [55]. In addition, technologies such as SWATH are being increasingly used for quantitation of acetylated proteins [56].

The ubiquitination PTM drives the localization, stability, and activity of the protein by covalent attachment of the ubiquitin moiety to cellular proteins [57]. Protein ubiquitination usually is

Table 1
Details of studies on mass spectrometry-based PTM analysis

PTM type	Enrichment method	Quantification method	Number of protein and sites	Model used	Mass spectrometer type	Reference
Phosphorylation	Ni-NTA		44,728 phosphosites on 6415 proteins	Human renal cancer	LTQ Orbitrap Velos	Peng et al. [46]
	TiO_2-based phosphopeptide enrichment	TMT	512 phosphorylation sites derived from 257 proteins	*Mycobacterium tuberculosis*	Orbitrap Fusion Tribrid	Verma et al. [24]
	TiO_2-based phosphopeptide enrichment	SILAC	7191 phosphorylation sites on 2746 proteins	RAW264.7 cells	LTQ-Orbitrap Elite	Pinto et al. [14]
Succinylation	Anti-succinyl lysine antibody	SILAC	2004 lysine succinylation sites on 738 proteins	HeLa cells	Q-Exactive	Weinert et al. [48]
	Anti-succinyl lysine antibody		1931 lysine succinylated peptides from 642 proteins	*Vibrio parahaemolyticus*	Q Exactive Plus	Pan et al. [77]
	Anti-succinyl lysine antibody	TMT	815 succinylation sites on 407 proteins	Rat	Q-Exactive	Cheng et al. [78]
Acetylation	Anti-acetyl lysine antibody	SILAC	2803 lysine acetylation sites in 782 proteins	*Escherichia coli*	LTQ Orbitrap Velos	Colak et al. [79]
	Anti-acetyl lysine antibody		1206 lysine acetylation sites in 576 proteins	Human	LTQ-Orbitrap	Sun et al. [80]
	Anti-acetyl lysine antibody		1128 lysine acetylation sites in 658 proteins	*Mycobacterium tuberculosis*	Q-Exactive	Xie et al. [81]
Ubiquitination	Anti-lysine ubiquitination (Kub) antibody	SILAC	1067 ubiquitination sites on 613 proteins	A549 cells	Q Exactive Plus	Wu et al. [58]
	Anti-lysine ubiquitination antibody	TMT	1248 ubiquitinated peptides	Mouse J1 and E14 embryonic stem cells	Q Exactive HF	Karg et al. [82]
	Anti-lysine ubiquitination antibody	SILAC	2299 ubiquitination sites	*Saccharomyces cerevisiae*	Q-Exactive	Iesmantavicius et al. [59]

Glycosylation	Lectin affinity chromatography		720N-glycosylated sites on 372 proteins	Human seminal plasma	LTQ Orbitrap Velos	Yang et al. [60]
	Boronic acid-based chemical enrichment		816N-glycosylation sites on 332 proteins	*Saccharomyces cerevisiae*	LTQ Orbitrap elite	Chen et al. [62]
	Hydrazide chemistry method	Label-free	608 *N*- glycosylation sites on 317 proteins	Human Bronchial epithelial cells	LTQ-Orbitrap XL	Sudhir et al. [61]
Palmitoylation	17-octadecynoic acid (17-ODYA)-based click chemistry	SILAC	338 proteins	BW5147-derived mouse T-cell hybridoma cells	Orbitrap Velos	Martin et al. [65]
	Acyl-biotinyl exchange chemistry	SILAC	280 proteins	Primary CD4 + T cells, Jurkat cells	Orbitrap elite	Morrison et al. [29]
	Acyl-biotinyl exchange chemistry		401 proteins	*Toxoplasma gondii*	Linear quadrupole ion trap	Caballero et al. [83]

enriched using an antibody affinity-based approach. An antibody-based ubiquitination enrichment approach followed by mass spectrometry identified 1067 ubiquitination sites on 613 proteins in A549 lung cancer cells [58]. A similar approach identified 2299 ubiquitination sites in *Saccharomyces cerevisiae*, of which 581 sites were found to be differentially regulated in response to rapamycin [59].

Glycosylation, a PTM where a carbohydrate moiety is covalently attached to the proteins, is one of the most abundant post-translational modifications. Protein glycosylation events are enriched either using lectin affinity chromatography [60] or hydrazide chemistry [61]. A study on human seminal plasma identified 720 N-glycosylated sites on 372 proteins using lectin affinity-based enrichment. Hydrazide-based enrichment method identified 608 N- glycosylation sites on 317 proteins in human bronchial epithelial cells. An enrichment method using boronic acid-based chemical enrichment has also been described previously [62]. Using a yeast model, this method was used to identify 816 N-glycosylation sites on 332 proteins.

Palmitoylation is one of the most common mechanisms of lipid modifications besides prenylation and myristoylation. Palmitoylation is known to drive mechanisms of protein targeting toward the membrane and subcellular protein trafficking [63]. Several studies have used acyl-biotinyl exchange chemistry to enrich palmitoylated proteins. A previous study used this technique to identify 280 palmitoylated proteins in primary and Jurkat T cells [29]. Protein S-palmitoylation has also been enriched by using 17-octadecynoic acid (17-ODYA)-based click chemistry. 17-ODYA is metabolically incorporated into palmitoylation machinery. This can be combined with metabolic labeling techniques such as SILAC. The extent of palmitoylation modification can be assessed using click chemistry ligation to biotin-azide [64]. Using this method, global quantitative profiling of dynamic protein palmitoylation was carried out in mouse T-cell hybridoma cells [65].

1.5 Data-Mining Approaches for PTM Analysis

Advances in PTM enrichment techniques and mass spectrometry technologies have led to increasingly large amounts of studies on various PTMs. Researchers are cataloging these data into comprehensive public databases accessible to the scientific community. Several PTM-specific databases are currently available to researchers for comparison and analysis, some which are provided in Table 2. A few databases currently provide information pertaining to multiple PTMs. These include PhosphositePlus (https://www.phosphosite.org) [66], Human Protein Reference Database (HPRD) (http://hprd.org) [67], dbPTM (http://dbptm.mbc.nctu.edu.tw/) [68], and CPLM (Compendium of Protein Lysine Modifications) (http://cplm.biocuckoo.org/) [69]. However, several databases

Table 2
Databases pertaining to posttranslational modifications

Database	PTM type	Link	Reference
CPLM (Compendium of Protein Lysine Modifications)	Nε-lysine acetylation, ubiquitination, methylation, sumoylation, glycation, butyrylation, crotonylation, malonylation, propionylation, succinylation, phosphoglycerylation, pupylation	http://cplm.biocuckoo.org/	Liu et al. [69]
dbGSH	S-Glutathionylation	http://csb.cse.yzu.edu.tw/dbGSH/	Chen et al. [84]
dbPTM	20 types of PTM	http://dbptm.mbc.nctu.eDU.tw/	Huang et al. [68]
dbSNO	S-nitrosylation	http://140.138.144.145/~dbSNO/	Chen et al. [85]
DEPOD	Dephosphorylation	http://depod.bioss.uni-freiburg.de/	Duan et al. [86]
Human Protein Reference Database (HPRD)	27 types of PTM	http://hprd.org/index_html	Prasad et al. [67]
MYRbase	Myristoylation	http://mendel.imp.ac.at/myristate/myrbase/	Maurer-Stroh et al. [87]
O-GlycBase	O- and C-glycosylation	http://www.cbs.dtu.dk/databases/OGLYCBASE/	Gupta et al. [88]
PHOSIDA	phosphorylation, acetylation, and N-glycosylation	http://141.61.102.18/phosida/index.aspx	Gnad et al. [89]
Phospho.ELM	Phosphorylation	http://phospho.elm.eu.org/	Dinkel et al. [90]
PhosphositePlus	Acetylation, Caspase cleavage, Di-methylation, Methylation, Mono-methylation, O-Galnac, O-Glcnac, Phospho-Ser, Phospho-Thr, Phospho-Tyr, Succinylation, Sumoylation, Tri-methylation, Ubiquitylation	https://www.phosphosite.org/homeAction.action	Hornbeck et al. [66]
PRENbase	Farnesylation, Geranylgeranylation	http://mendel.imp.ac.at/PrePS/PRENbase/	Maurer-Stroh et al. [91]
PupDB	Pupylation	http://cwtung.kmu.edu.tw/pupdb/	Tung C. W [92].

(continued)

Table 2 (continued)

Database	PTM type	Link	Reference
SNObase	S-nitrosylation	http://www.nitrosation.org/index.html	Zhang et al. [93]
Succinsite	Succinylation	http://systbio.cau.edu.cn/SuccinSite/	Hasan et al. [94]
UbiProt	Ubiquitination	http://ubiprot.org.ru/	Chernorudskiy et al. [95]

specialize in one type of PTM and exclusively provide information pertaining to these. PTM-specific databases will enable the growth and development of specific assays for multiple reaction monitoring of modified peptides. Technological innovations such as IMAC-MRM (Immobilized Metal Affinity Chromatography coupled to Multiple Reaction Monitoring) now enable measurement of modified peptide levels [70]. Given that mass spectrometry and enrichment technologies are becoming increasingly accessible to researchers, the number of specialized PTM databases is expected to grow phenomenally.

Several data-mining approaches are also being developed to carry out analysis of mass spectrometry-derived PTM data. These methods have enabled the measurement of PTM abundances at the basal level. Several unassigned spectra in proteomics datasets have been attributed to the presence of unknown or unexpected post-translational modifications. Data from high-resolution mass spectrometers may be searched for potential post-translational modifications using wide-tolerance searches. One such approach using a large precursor ion tolerant search identified several post-translational modifications including phosphorylation, N-terminal acetylation, glycosylation, and monomethylation demethylation and rarer PTM forms including glycerol phosphorylethanolamine (GPE) and glutamylation in HEK293 cells [71]. This approach can be potentially used to mine for PTM data from existing proteomic datasets.

1.6 PTM Cross Talk in Biological Systems

PTM profiling studies have shown that numerous proteins can be modified by multiple PTMs, suggesting the possibility of PTM crosstalk. Several examples of complex PTM crosstalk events regulating biological processes exist in the literature. An example of a complex crosstalk involving PKCδ, Caspase-3, and p53 through multiple PTMs is known to regulate various biological functions

[72]. Co-occurrence of PTM sites has been used to measure PTM co-evolution across eukaryotes [73]. Minguez and colleagues collected data for 13 PTMs across 8 eukaryotes and derived a global network of functionally associated PTMs.

Complex PTM cross talk mechanisms are not fully understood due to nonavailability of enrichment methods for most types of PTMs. To date, ~200 PTM crosstalk pairs in 77 human proteins have been reported in the literature [13]. Recent advances in affinity enrichment of PTMs have enabled the analysis of a few of the multiple PTMs in a single sample. To a certain extent, PTM crosstalk events can be identified using serial enrichment techniques [33]. A previous study in murine synaptosomes identified crosstalk between O-GlcNAcylation and phosphorylation events using a mass spectrometry-based method [74]. The serial use of multiple enrichment techniques coupled with high-resolution mass spectrometry will help explore the dynamics of cross talk between various PTMs will help us in a better understanding of the cellular processes and underlying mechanisms of disease pathophysiology.

1.7 Challenges in Multi-PTM Analysis

As is evidenced by several studies, studying protein PTMs is vital as it provides a strong foundation for a better understanding of cellular networks and mechanisms of cell regulation. However, there are several challenges when it comes to identification and characterization of PTMs including sub-stoichiometric levels of expression, high dynamic range as protein modification is a reversible process. Furthermore, the existence of multiple PTM on the same site or same modifications on different sites can bring about further complexity. Lastly, suppression of modified peptide signal by non-modified peptides at MS level can severely deter their identification. This necessitates strategies that will enable efficient identification of the PTM proteome. Advance in enrichment techniques such as chemical methods, affinity enrichment, and chromatographic approaches together with improvements in MS-based detection has dramaticalyly influenced comprehensive characterization of PTM repertoire. PTMs can be erroneously assigned in the mass spectrometry-based analysis due to several factors discussed in detail by Kim et al. [75]. Some of these factors may be attributed to isobaric PTMs and chemical modifications, amino acid substitutions, SNPs, shared peptides, and poor MS/MS fragmentation. Most of these errors can be resolved with countermeasures including use of high-resolution analyzers such as Orbitrap, use of multiple proteolytic enzymes, and employment of statistical measures for site localization.

2 Materials

2.1 Protein Extraction and Estimation

1. SDS-Lysis buffer: 2% (w/v) SDS, 50 mM TEABC, pH 8.0, 1 mM sodium fluoride, 1 mM sodium orthovanadate, 2.5 mM sodium pyrophosphate, 1 mM beta glycerophosphate).
2. Sonicator—Branson Sonifier 250.
3. Benchtop centrifuge.
4. Pierce BCA protein assay kit (Thermo Scientific, Catalog # 23225).

2.2 In-Solution Digestion of Proteins

1. Reducing reagent (100 mM Dithiothreitol (DTT) in 50 MM TEABC buffer.
2. Alkylating reagent (200 mM Iodoacetamide (IAA) in 50 MM TEABC buffer (*see* **Note 1**).
3. Ice-cold acetone.
4. Triethylammonium bicarbonate (TEABC) (Sigma-Aldrich, Catalog # T7408-500ML).
5. Sequencing grade modified trypsin (Promega, Catalog #V5111).

2.3 Tandem Mass Tag Labeling of Peptides

1. TMT 6plex or 10plex isobaric label reagent set.
2. Anhydrous acetonitrile.
3. 5% hydroxylamine (Quenching Reagent).
4. Ultrasonic water bath.
5. Benchtop centrifuge and vortex mixer.

2.4 Basic pH Reverse Phase Chromatography (bRPLC)

1. TEABC 1.0 M, pH8.4–8.6.
2. Acetonitrile.
3. Waters XBridge column (Waters Corporation, Milford, MA; 130 Å, 5 μm, 250 × 9.4 mm).
4. HPLC system.
5. Solvents:
   (a) Solvent A: 10 mM TEABC buffer, pH 9.5.
   (b) Solvent B: 10 mM TEABC buffer, 90% acetonitrile, pH 9.5.
6. 96-well plates for fraction collection.
7. Speedvac concentrator with cold trap.

2.5 Enrichment of Phosphorylated Peptides

1. Titansphere 10 μm or 5 μm, GL Sciences Inc., Japan (P/N 5020-75010).
2. 2,5-Dihydrobenzoic acid (DHB), (Sigma Catalog #. 149357-100G).

3. Ammonium hydroxide (J.T. Baker Catalog #9721-01).
4. Trifluoroacetic acid (TFA) (Fisher Scientific Catalog # A116-50).
5. Empore C_8 disks.
6. Solvents:
   (a) Wash solution 1: 80% acetonitrile, 3% TFA.
   (b) Wash solution 2: 80% acetonitrile, 1% TFA.
   (c) Wash solution 3: 80% acetonitrile, 0.1% TFA.
   (d) DHB solution (5% 2,5-dihydroxybenzoic acid (DHB), i.e. 50 mg of DHB in 1 mL of Wash solution 1).
7. Ammonia solution (4% NH_4OH, pH ~10.5, i.e. 40 μL of NH_4OH in 1 mL of 40% acetonitrile).

2.6 SepPak-Based Sample Cleanup

1. Sep-Pak C18 columns (Waters Corporation, Milford, MA, USA).
2. 5-mL syringes.
3. Solvents:
   (a) 100% acetonitrile.
   (b) Solvent A: 0.1% formic acid.
   (c) Solvent B: 40% acetonitrile, 0.1% formic acid.

2.7 Immunoaffinity Enrichment of Acetylated and Succinylated Lysine Residues

1. Anti-Acetyl-Lysine Motif [Ac-K] Immunoaffinity Beads (Cell Signaling Technology, Catalog # 13416) or Pan anti-acetyllysine antibody agarose conjugated beads (PTM Bio PTM-104).
2. Pan anti-succinyllysine antibody agarose conjugated beads (PTM Bio PTM-402).
3. Universal pH Indicator strips.
4. Gel loading tips.
5. Benchtop centrifuge.
6. Tube rotator.
7. Solvents:
   (a) Immunoaffinity purification buffer (1×): 50 mM MOPS pH 7.2, 10 mM sodium phosphate, 50 mM NaCl (pH 7.0–7.5) (*see* **Note 2**).
   (b) 1 M Tris base pH ~10 (1.21 mg in 1 mL water).
   (c) MilliQ water.
   (d) 0.15% TFA.

2.8 Sample Cleanup of Enriched Peptides

1. Empore C_{18} disks.
2. 200 μL pipette tips.

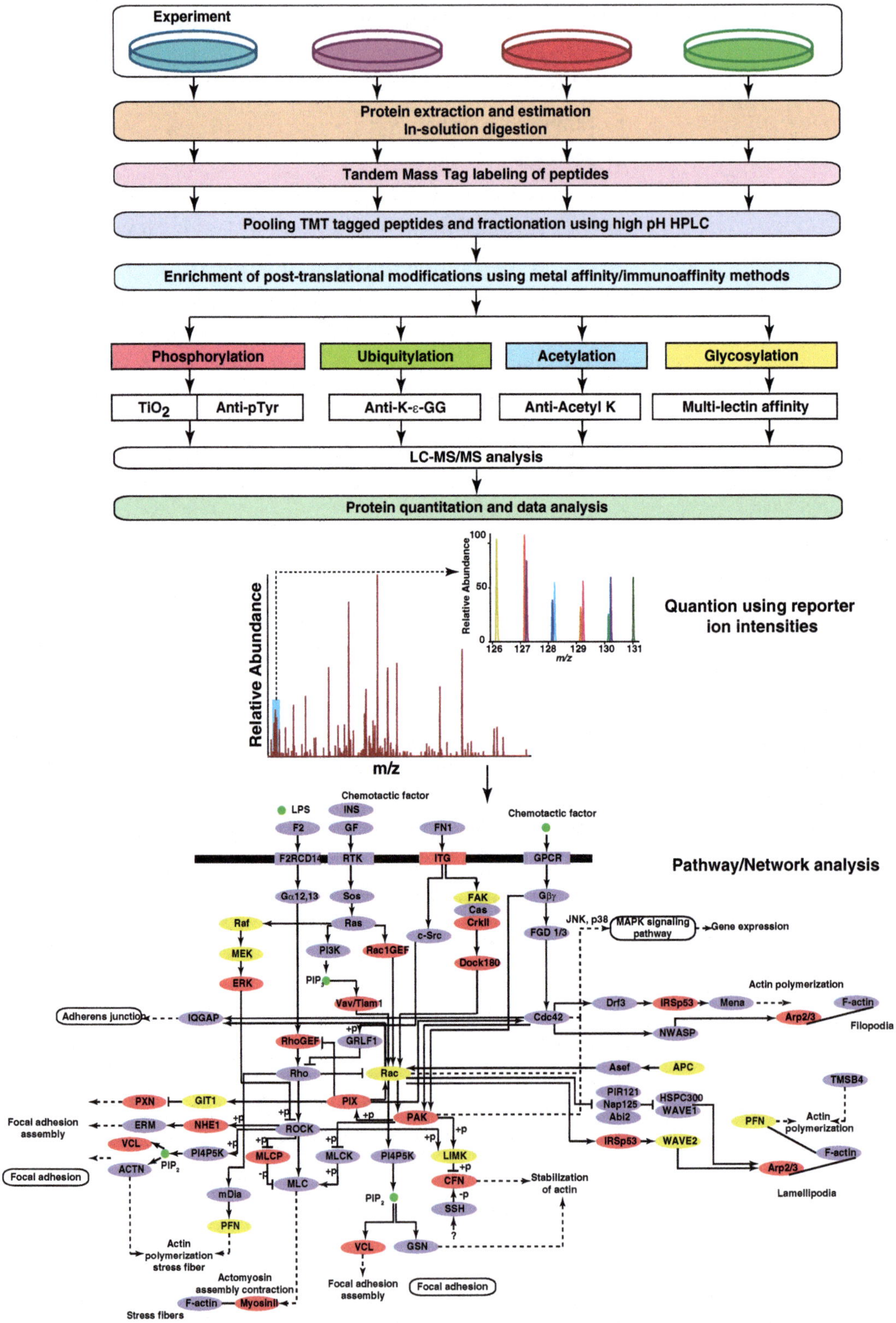

Fig. 1 Workflow for the identification and quantification of PTM-modified peptides using LC-MS/MS analysis. The cells are cultured in appropriate growth media and stimulated with growth factors/cytokines or treated with inhibitors to study their effect. Proteins from each condition are harvested by cell lysis, enzymatically

3. Syringes.
4. Blunt end needle.
5. Benchtop centrifuge for microfuge tubes.
6. Vacuum concentrator.
7. Solvents.
   (a) Acetonitrile (LC-MS grade).
   (b) 0.1% formic acid.
   (c) 80% acetonitrile, 0.5% acetic acid.
   (d) 50% acetonitrile, 0.1% formic acid.

2.9 LC-MS Analysis and Data Analysis

1. High-resolution mass spectrometer (Orbitrap Fusion Tribrid mass spectrometer, Thermo Electron, Bremen, Germany).
2. Nanoflow liquid chromatography system such as Easy-nLC1200 or Ultimate 3000 Quaternary RSLC (Thermo Electron, Bremen, Germany).
3. Data analysis software such as Proteome Discoverer software suite, MaxQuant, ProteinPilot.

3 Methods

A brief overview of the methodology employed for the identification and quantitation of modified peptides using high-resolution LC-MS/MS analysis is provided in Fig. 1.

3.1 Protein Extraction and Estimation

1. Culture cells in appropriate growth media. Prior to stimulation with growth factors/cytokines or treatment with drugs to study their effect, it is recommended to suspend the cells in media free of serum/growth factor supplements.
2. Once the cells have reached the desired confluency and density, aspirate media from the culture plates/flasks. In case of adherent cells, aspirate media, wash the cells with chilled 1× PBS thrice (*see* **Note 3**).

Fig. 1 (continued) digested and subsequently labeled with Tandem mass tags (TMT). The TMT-labeled samples are pooled, fractionated using high pH reverse phase chromatography. The fractionated samples are dried and subjected to the serial enrichment of posttranslational modifications using metal affinity/immunoaffinity methods. It is preferable to first enrich for phosphorylated residues using metal affinity chromatography as the solvents used are volatile and the flowthrough can be then subjected to immunoaffinity purification. The enriched peptides as well as a fraction of the total peptides are analyzed using high-resolution nano-LC-MS/MS. The raw data derived from the data acquisition are processed using software suites to identify and quantify PTM-modified peptides and proteins. The reporter ions are used for the relative quantitation of the abundance of peptides

3. Add 0.5–1 mL SDS lysis buffer (2% SDS, 50 mM TEABC, pH 8.0, 1 mM sodium fluoride, 1 mM sodium orthovanadate, 2.5 mM sodium pyrophosphate, 1 mM beta glycerophosphate) to dish/cell pellet (*see* **Note 4**).
4. Gently scrape cells into the lysis buffer using cell scraper and transfer to 1 mL microfuge tubes, keep on ice. In case of cell pellet obtained from suspension, add lysis buffer to the cell pellet and vortex to disrupt cell pellet.
5. Sonicate the lysed samples using a probe sonicator-12 pulses of 10 s each at an amplitude of 30–40%. Heat the lysate at 95 °C for 5 min to ensure complete denaturation. Cool the tubes to room temperature.
6. Centrifuge the samples at 14,000 × *g* for 10 min at 4 °C. Transfer the supernatant to a new microfuge tube (*see* **Note 5**).
7. Estimate the protein concentration of the cleared lysates using BCA assay.

3.2 Protein Digestion

1. Transfer 600–800 μg per condition into a new tube. Adjust the final volume such that it is equal in all tubes.
2. Add reducing agent to a final concentration of 10 mM and incubate at 60 °C for 20 min. Cool the tubes to room temperature.
3. Add alkylating agent to the final concentration of 20 mM and incubate for 10 min in dark at room temperature.
4. Add six volumes of ice-cold acetone to each tube and incubate at −20 °C overnight (*see* **Note 6**).
5. Centrifuge the samples at 14,000 × *g* for 10 min at 4 °C. Discard the supernatant and air-dry the pellet. Resuspend the protein pellet in 200 μL of 50 mM TEABC buffer (*see* **Note 7**).
6. Prepare trypsin solution at a concentration of 1 mg/mL in 50 mM TEABC buffer. Add Trypsin solution at 1:20 (Enzyme to substrate ratio) to the peptide digest and incubate the microfuge tubes overnight at 37 °C (*see* **Note 8**).
7. After the confirmation of digestion efficiency evaporate the sample to complete dryness.

3.3 Tandem Mass Tag (TMT) Labeling of Peptides

1. Bring down the TMT reagents to room temperature and add 41 μL of anhydrous acetonitrile to each of the TMT reagent (0.8 mg) tube.
2. Briefly vortex the tubes to ensure complete dissolution.
3. Centrifuge the tubes and keep for 5 min at room temperature. Check the pH (7.5–8).
4. Dissolve the vacuum-dried sample in 100 μL of 50 mM TEABC (*see* **Note 9**). Centrifuge the tubes at 12,000 × *g* for 10 min.

5. Transfer the supernatant directly to TMT reagent vial. Briefly vortex the samples and centrifuge the tubes to bring down the solution. Incubate at room temperature for 1 h.
6. Add 8 μL of 5% hydroxylamine to quench the reaction and incubate at room temperature for about 15 min.
7. Labeling efficiency check: Pipette out volume equivalent to 2 μg from each TMT channel tube and pool the sample. Vortex briefly and centrifuge at 14,000 × *g* for 10 min. Transfer the supernatant and desalt using C_{18} StageTip.
8. Carry out LC-MS analysis of the desalted sample to check the labeling efficiency. If labeling is efficient, pool the samples from all TMT channels and evaporate the sample to dryness.

3.4 Fractionation

1. Fractionate the peptide digest using basic reverse phase chromatography. Reconstitute the TMT-labeled pooled peptide digest in 1 mL Solvent A (10 mM TEABC buffer, pH 9.5).
2. Load the sample onto XBridge C18 column. Resolve the peptides using a gradient of solvent A and B for 120 min with a flow rate of 1 mL/min.
3. Collect the fractions in 96-well plate using a fraction collector programmed to collect fractions from the start of the gradient (*see* **Note 10**).
4. Reduce the volume of fractions at freezing temperature. Concatenate the fractions to obtain a total of 6 fractions. Dry the samples using vacuum concentrator equipped with refrigerated vapor trap.

3.5 Enrichment of Post-translational Modifications

3.5.1 Enrichment of Phosphorylated Peptides Using Metal Affinity Chromatography

1. Reconstitute the fractions with 500 μL of 5% DHB solution. Ensure the peptides are entirely dissolved (*see* **Note 11**).
2. For enrichment of phosphorylated peptides, the beads-to-peptide ratio is 1:2. Weigh required amount of TiO_2 beads. Pre-warm the TiO_2 beads by keeping the tube in the dry bath for 15–20 min.
3. Suspend TiO_2 beads in 5% DHB solution and incubate for 15 min on the rotator at room temperature.
4. Add beads to each tube at a ratio of 1:2 (TiO_2: peptide) and incubate the peptide-DHB-TiO_2 mix on a rotator for 30 min at room temperature.
5. Centrifuge the tubes at 1500 × *g* for 1 min. Transfer supernatant to another microfuge tube.
6. Wash the peptide-bound beads with 500 μL wash solution 1 (80% ACN, 3% TFA). Centrifuge the tubes at 1500 × *g* for 1 min. Discard supernatant.

7. Wash the beads with 500 μL wash solution 2 (80% ACN, 1% TFA). Centrifuge the tubes at 1500 × *g* for 1 min. Discard supernatant.
8. Prepare StageTips with C_8 plug. Resuspend TiO_2 beads in 200 μL wash solution 3 and transfer the entire content to C_8 StageTip.
9. Connect the StageTips to 5-mL syringes with the plunger pulled out. Slowly push the plunger down to remove the wash solution. The peptide-bound-TiO_2 beads remain in the tip.
10. In separate collection tubes, add 40 μL of 3% TFA and place it on ice. Elute phosphopeptides from TiO_2 beads by adding 4% NH_4OH solution. Prepare the elution buffer shortly before use.
11. Repeat elution twice. Dry the pooled samples in vacuum concentrator with cold trap attached.
12. Resuspend the dried peptides in 0.1% formic acid and desalt them by C_{18} StageTip. Store the enriched phosphopeptides in −80 °C until LC-MS/MS analysis.
13. The supernatants obtained following incubation with TiO_2 beads from each fraction can be further used to enrich other PTMs. For this, the samples must first be desalted to remove DHB.

SepPak C18 cleaning:
All the steps for SepPak purification of peptides should be carried out at room temperature.

1. Connect Sep-Pak C 18 columns to the shorter end of 5-mL syringes. Remove the plunger as the body of the syringe serves as a reservoir to hold the solvents as well as the sample.
2. Activate the column using 5 mL of 100% acetonitrile. Connect the plunger and apply slight pressure such that the solvent starts to elute. Remove the plunger and allow the solvent to flow due to gravity.
3. Equilibrate the columns with 7 mL of Solvent A (0.1% Formic acid) twice. Ensure the column does not dry up at any time during the course of the sample cleanup.
4. Load the supernatant/peptide sample and follow the procedure described in **step 2**. Reload the eluate.
5. Wash the column with 12 mL of Solvent A added in two batches of 6 mL each.
6. Elute the bound peptides using 2 mL of Solvent B (40% acetonitrile, 0.1% formic acid).
7. Evaporate the purified peptides to dryness using vacuum concentrator. Proceed for downstream processing.

3.5.2 Enrichment of PTM-Modified Peptides Using Immunoaffinity Beads

All steps involving enrichment using anti-phosphotyrosine, anti-acetyllysine, or anti-succinyllysine antibody should be carried out on ice or at 4 °C.

1. Dissolve the lyophilized peptide digest with 1.4 mL of 1× immunoaffinity purification (IAP) buffer. Sonicate the digest until a clear solution is obtained with no turbidity observed. Centrifuge the sample at 3000 × *g* for 30 s.
2. Pipette a small volume (1–2 μL) onto a pH indicator strip. If the pH is less than 7.0 adjust to the desired pH range (7.0–7.5) using 1 M Tris base. The volume of 1 M Tris solution should not exceed 30 μL.
3. Centrifuge the sample at 10,000 × *g* for 5 min. Transfer the supernatant to a fresh tube.
4. Wash the antibody-bead slurry with 1 mL of 1× PBS buffer. Care should be taken not to introduce air bubbles. Gently invert the tube 3–4 times to obtain a homogenous suspension (*see* **Note 12**).
5. Centrifuge at 2000 × *g* for 1 min. Aspirate the buffer leaving behind ~50 μL volume as there are chances of losing beads when complete aspiration is performed. Repeat **step 4** twice. Finally wash the beads with 1× IAP buffer.
6. Transfer the peptide solution to the antibody beads containing vial. Pipette the sample directly on top of the beads. Gently invert the tube 3–4 times to ensure homogenous mixing. Care should be taken not to introduce any air bubbles.
7. Incubate the peptide-antibody beads on a rotator for 2 h at 4 °C.
8. Centrifuge at 1500 × *g* for 1 min. Keep tubes on ice for 1–2 min for the beads to settle down completely. Carefully transfer supernatant using a fresh microfuge tube without disturbing the beads (*see* **Note 13**).
9. Wash the peptide-bound antibody beads twice with 1 mL of 1× IAP buffer added to the beads each time. Gently mix the contents by inverting the tubes 4–5 times. Ensure that no air bubbles are introduced. Centrifuge at 1500 × *g* for 30 s and remove supernatant.
10. Wash the beads to remove unbound/nonspecific bound peptides using 1 mL chilled HPLC grade water. Mix well to create a homogenous suspension. Centrifuge at 1500 × *g* for 1 min and remove supernatant. Repeat the steps for a total of three washes.
11. Following the last washing step, carefully remove the entire solvent using a gel loading tip.

12. Add 55 μL of 0.15% TFA to elute the enriched PTM peptides. Gently tap the bottom of the tube several times such that a homogenous slurry is formed. Incubate for 10 min at room temperature (*see* **Note 14**).
13. Centrifuge at 1500 × *g* for 30 s. Transfer motif-enriched peptides to fresh microfuge tubes.
14. Repeat **step 13** twice and pool the eluates. Use gel loader tips to recover the enriched peptides without disturbing the beads.

C_{18} StageTip cleanup:

1. Prepare C_{18}StageTip using Empore C_{18} disk.
2. Equilibrate the C18 disk with 120 μl of 80% ACN, 0.5% acetic acid solution followed by 120 μL of 50% ACN, 0.1% Formic acid, and 120 μL of 0.15% TFA.
3. Load the pooled eluate onto the equilibrated C_{18} StageTip connected to the end of a 2-mL syringe. Elute the sample into a microfuge tube and reload the eluate.
4. Add 40 μL of 0.15% TFA to wash the bound peptides. Ensure that no beads are present.
5. Elute the bound peptides with 10 μL of 50% ACN, 0.1% Formic acid. Repeat the process thrice. Combine the eluate and evaporate to dryness using vacuum concentrator.
6. Store the enriched modified peptides in −80 °C until LC-MS/MS analysis.

3.6 Mass Spectrometry Data Acquisition of PTM Experiments

To characterize dynamic states of protein complexes in signaling pathways, the TMT-labeled fractions enriched for PTM-modified peptides can be analyzed using high-resolution nanoflow HPLC-MS/MS setup that is being mostly employed for shotgun proteomics analysis (*see* **Note 15**). The PTM enriched sample from each fraction can be reconstituted in 0.1% formic acid and loaded onto trap column that is typically of 2 cm length and is packed with 3 μm C_{18} material. The enriched peptides are resolved on 15–50 cm C_{18} analytical columns with 75 μm diameter. The typical flow rate employed for resolving the peptides is 250–300 nL/min with a column oven temperature set at 40 °C. A linear gradient of 5–35% solvent B (80% acetonitrile in 0.1% formic acid) over 100 min is generally used in our setting to resolve the peptide mixture with a total run time of 120 min. Data dependent acquisition with full scans in 400–1600 m/z range is typically carried out using an Orbitrap mass analyzer at a mass resolution of 120,000 at 200 m/z in the positive ion mode. From the precursor survey scan, we generally select the most intense precursor ions for an MS2 scan using TopSpeed mode. The selected precursor ions were fragmented using HCD mode with 40–42% normalized collision energy. The scan range is typically set as 100–1600 m/z and detected at a mass resolution of 60,000 at 200 *m/z*. Studies have reported that for most isobaric-based quantification methods, co-isolating precursor species result in distorted reporter ion intensities, thereby providing an inaccurate estimate of its abundance in any given sample. One approach would be to narrow down the precursor isolation width; however, this may hamper the peak picking of the isotopic cluster. This issue has been partly circumvented with the advent of Orbitrap Fusion mass spectrometer which enables synchronous precursor selection (SPS) [76]. Availability of three mass analyzers enables increased scan rates as the MS2 events occur in the IonTrap and the MS3 events occur in parallel in the Orbitrap mass analyzer. Enabling SPS to include 10 MS2 fragment ions with an increased collision energy of up to 55% and a narrow *m/z* range provides TMT reporter ion abundance. SPS increase the sensitivity and quantitation accuracy by isolating multiple fragment ions in the MS2 spectrum. The data obtained can be searched against reference databases for the organism under the study. The search parameters should include dynamic modification of the PTMs under study. If quantitative methodology is employed, then the appropriate quantitation node with TMT/iTRAQ should be enabled. Statistical assessment of site localization can be carried out using algorithms such as ptmRS [44] and A-Score [45].

4 Notes

1. Prepare IAA shortly before use and since it is light sensitive, it is recommended to be stored in the dark until use.
2. Prepare a stock of 10× IAP buffer. This solution can be stored at 4 °C for 2 months. Prior to use, dilute the requisite volume with MilliQ water to 1× concentration before use. Store for up to 1 month at 4 °C.
3. In case of suspension cells, ensure the cells are suspended homogenously, centrifuge and pellet down the cells. Aspirate the media and resuspend the cell pellet in 15 mL chilled 1× PBS buffer. Make a homogenous suspension. Repeat centrifugation and completely aspirate PBS. Repeat the process thrice.
4. The volume of lysis buffer can be adjusted such that the final protein concentration does not exceed 5 mg/mL. For acetyl or succinyl enrichment, appropriate inhibitors such as sodium butyrate (final concentration 5 mM) should be added.
5. In case of microorganisms or infectious agents employed to study host-pathogen interactions, it is recommended to filter the lysate using 0.22 μm filter to ensure no viable microbes are retained.

6. Ensure the centrifuge tubes are made of acetone compatible material else the plasticizers may leach and interfere with LC-MS analysis. Acetone precipitates the protein in solution and based on our experience, the protein loss is minimal when the samples are reduced and alkylated prior to precipitation. In addition to effective removal of SDS, acetone also enables removal of excess DTT and IAA added to the lysates.
7. Acetone-precipitated pellet may not dissolve completely. It is recommended to use sonicator water bath to disrupt the pellet and dissolve the samples. In case the pellet does not dissolve, an additional 100 μL of 50 mM TEABC buffer can be added.
8. It is important to check the digestion efficiency before proceeding to TMT labeling of the samples. Pre-digest samples can be taken prior to addition of chilled acetone and the post-digest sample (10 μg/condition) can be taken post 14–16 h incubation with trypsin.
9. Use a water bath sonicator to ensure complete dissolution of peptide digest.
10. Add 50 μL of 1% Formic acid to 96 wells prior to fraction collection to neutralize peptides.
11. The tubes can be kept in the thermomixer set at 1150 rpm for 20 min to ensure complete dissolution. Alternatively, the tubes can be placed on vortex mixer set at low speed for 20 min.
12. Pipette the desired volume of IAP buffer directly onto the beads such that the beads are dislodged and enter into the solution. Forceful ejection of the solvent must be avoided as it can introduce air bubbles. Care should also be taken to ensure that the beads are not agitated which can result in the breaking of the beads further leading to nonspecific binding.
13. The supernatant obtained at this step can be subjected to enrichment using other PTM affinity antibody beads. For instance, after phosphopeptide enrichment using TiO_2 beads which predominantly enriches phosphorylation at serine and threonine residues, the supernatant can be subjected to enrichment using anti-phosphotyrosine antibody beads. The supernatant thereby obtained can be further subjected to enrichment using anti-acetyllysine or anti-succinyllysine antibody.
14. Tap the tubes gently to ensure that the beads do not settle at the bottom and remain in suspension every 2–3 min.
15. In case of a quantitative PTMomic experiment, it is ideal to carry out a corresponding quantitative proteomic profiling experiment using the same sample. This will help in ascertaining that changes of PTM abundance are not due to changes in the protein abundance.

Acknowledgements

The authors acknowledge Yenepoya (Deemed to be University) for access to mass spectrometry instrumentation facility. We also thank Karnataka Biotechnology and Information Technology Services (KBITS), Government of Karnataka, for the support to the Center for Systems Biology and Molecular Medicine at Yenepoya University under the Biotechnology Skill Enhancement Programme in Multiomics Technology (BiSEP GO ITD 02 MDA 2017). SMP is a recipient of INSPIRE Faculty Award from Department of Science and Technology (DST), Government of India.

References

1. UniProt C (2015) UniProt: a hub for protein information. Nucleic Acids Res 43(Database issue):D204–D212. https://doi.org/10.1093/nar/gku989
2. Creasy DM, Cottrell JS (2004) Unimod: protein modifications for mass spectrometry. Proteomics 4(6):1534–1536. https://doi.org/10.1002/pmic.200300744
3. Zhang H, Shi X, Pelech S (2016) Monitoring protein kinase expression and phosphorylation in cell lysates with antibody microarrays. Methods Mol Biol 1360:107–122. https://doi.org/10.1007/978-1-4939-3073-9_9
4. Shi J, Sharif S, Ruijtenbeek R, Pieters RJ (2016) Activity based high-throughput screening for novel O-GlcNAc transferase substrates using a dynamic peptide microarray. PLoS One 11(3):e0151085. https://doi.org/10.1371/journal.pone.0151085
5. Zhu B, Farris TR, Milligan SL, Chen H, Zhu R, Hong A, Zhou X, Gao X, McBride JW (2016) Rapid identification of ubiquitination and SUMOylation target sites by microfluidic peptide array. Biochem Biophys Rep 5:430–438. https://doi.org/10.1016/j.bbrep.2016.02.003
6. Al-Ejeh F, Miranda M, Shi W, Simpson PT, Song S, Vargas AC, Saunus JM, Smart CE, Mariasegaram M, Wiegmans AP, Chenevix-Trench G, Lakhani SR, Khanna KK (2014) Kinome profiling reveals breast cancer heterogeneity and identifies targeted therapeutic opportunities for triple negative breast cancer. Oncotarget 5(10):3145–3158. https://doi.org/10.18632/oncotarget.1865
7. Scholma J, Fuhler GM, Joore J, Hulsman M, Schivo S, List AF, Reinders MJ, Peppelenbosch MP, Post JN (2016) Improved intra-array and interarray normalization of peptide microarray phosphorylation for phosphorylome and kinome profiling by rational selection of relevant spots. Sci Rep 6:26695. https://doi.org/10.1038/srep26695
8. Baharani A, Trost B, Kusalik A, Napper S (2017) Technological advances for interrogating the human kinome. Biochem Soc Trans 45(1):65–77. https://doi.org/10.1042/BST20160163
9. Kim MS, Pinto SM, Getnet D, Nirujogi RS, Manda SS, Chaerkady R, Madugundu AK, Kelkar DS, Isserlin R, Jain S, Thomas JK, Muthusamy B, Leal-Rojas P, Kumar P, Sahasrabuddhe NA, Balakrishnan L, Advani J, George B, Renuse S, Selvan LD, Patil AH, Nanjappa V, Radhakrishnan A, Prasad S, Subbannayya T, Raju R, Kumar M, Sreenivasamurthy SK, Marimuthu A, Sathe GJ, Chavan S, Datta KK, Subbannayya Y, Sahu A, Yelamanchi SD, Jayaram S, Rajagopalan P, Sharma J, Murthy KR, Syed N, Goel R, Khan AA, Ahmad S, Dey G, Mudgal K, Chatterjee A, Huang TC, Zhong J, Wu X, Shaw PG, Freed D, Zahari MS, Mukherjee KK, Shankar S, Mahadevan A, Lam H, Mitchell CJ, Shankar SK, Satishchandra P, Schroeder JT, Sirdeshmukh R, Maitra A, Leach SD, Drake CG, Halushka MK, Prasad TS, Hruban RH, Kerr CL, Bader GD, Iacobuzio-Donahue CA, Gowda H, Pandey A (2014) A draft map of the human proteome. Nature 509(7502):575–581. https://doi.org/10.1038/nature13302
10. Wilhelm M, Schlegl J, Hahne H, Gholami AM, Lieberenz M, Savitski MM, Ziegler E, Butzmann L, Gessulat S, Marx H, Mathieson T, Lemeer S, Schnatbaum K, Reimer U, Wenschuh H, Mollenhauer M, Slotta-Huspenina J, Boese JH, Bantscheff M, Gerstmair A, Faerber F, Kuster B (2014) Mass-spectrometry-based draft of the human

proteome. Nature 509(7502):582–587. https://doi.org/10.1038/nature13319

11. Zhao Y, Jensen ON (2009) Modification-specific proteomics: strategies for characterization of post-translational modifications using enrichment techniques. Proteomics 9 (20):4632–4641. https://doi.org/10.1002/pmic.200900398
12. Sathe G, Pinto SM, Syed N, Nanjappa V, Solanki HS, Renuse S, Chavan S, Khan AA, Patil AH, Nirujogi RS, Nair B, Mathur PP, Prasad TSK, Gowda H, Chatterjee A (2016) Phosphotyrosine profiling of curcumin-induced signaling. Clin Proteomics 13:13. https://doi.org/10.1186/s12014-016-9114-0
13. Yu Y, Gaillard S, Phillip JM, Huang TC, Pinto SM, Tessarollo NG, Zhang Z, Pandey A, Wirtz D, Ayhan A, Davidson B, Wang TL, Shih Ie M (2015) Inhibition of spleen tyrosine kinase potentiates paclitaxel-induced cytotoxicity in ovarian cancer cells by stabilizing microtubules. Cancer Cell 28(1):82–96. https://doi.org/10.1016/j.ccell.2015.05.009
14. Pinto SM, Nirujogi RS, Rojas PL, Patil AH, Manda SS, Subbannayya Y, Roa JC, Chatterjee A, Prasad TS, Pandey A (2015) Quantitative phosphoproteomic analysis of IL-33-mediated signaling. Proteomics 15 (2–3):532–544. https://doi.org/10.1002/pmic.201400303
15. Zahari MS, Wu X, Pinto SM, Nirujogi RS, Kim MS, Fetics B, Philip M, Barnes SR, Godfrey B, Gabrielson E, Nevo E, Pandey A (2015) Phosphoproteomic profiling of tumor tissues identifies HSP27 Ser82 phosphorylation as a robust marker of early ischemia. Sci Rep 5:13660. https://doi.org/10.1038/srep13660
16. Harsha HC, Pinto SM, Pandey A (2013) Proteomic strategies to characterize signaling pathways. Methods Mol Biol 1007:359–377. https://doi.org/10.1007/978-1-62703-392-3_16
17. Bao X, Wang Y, Li X, Li XM, Liu Z, Yang T, Wong CF, Zhang J, Hao Q, Li XD (2014) Identification of 'erasers' for lysine crotonylated histone marks using a chemical proteomics approach. elife 3. https://doi.org/10.7554/eLife.02999
18. Gu H, Ren JM, Jia X, Levy T, Rikova K, Yang V, Lee KA, Stokes MP, Silva JC (2016) Quantitative profiling of post-translational modifications by immunoaffinity enrichment and LC-MS/MS in cancer serum without immunodepletion. Mol Cell Proteomics 15 (2):692–702. https://doi.org/10.1074/mcp.O115.052266
19. Matsuoka S, Ballif BA, Smogorzewska A, McDonald ER 3rd, Hurov KE, Luo J, Bakalarski CE, Zhao Z, Solimini N, Lerenthal Y, Shiloh Y, Gygi SP, Elledge SJ (2007) ATM and ATR substrate analysis reveals extensive protein networks responsive to DNA damage. Science 316(5828):1160–1166. https://doi.org/10.1126/science.1140321
20. Pinkse MW, Uitto PM, Hilhorst MJ, Ooms B, Heck AJ (2004) Selective isolation at the femtomole level of phosphopeptides from proteolytic digests using 2D-NanoLC-ESI-MS/MS and titanium oxide precolumns. Anal Chem 76(14):3935–3943. https://doi.org/10.1021/ac0498617
21. Li Y, Xu X, Qi D, Deng C, Yang P, Zhang X (2008) Novel Fe3O4@TiO2 core-shell microspheres for selective enrichment of phosphopeptides in phosphoproteome analysis. J Proteome Res 7(6):2526–2538. https://doi.org/10.1021/pr700582z
22. Feng S, Ye M, Zhou H, Jiang X, Jiang X, Zou H, Gong B (2007) Immobilized zirconium ion affinity chromatography for specific enrichment of phosphopeptides in phosphoproteome analysis. Mol Cell Proteomics 6 (9):1656–1665. https://doi.org/10.1074/mcp.T600071-MCP200
23. Thingholm TE, Jensen ON (2009) Enrichment and characterization of phosphopeptides by immobilized metal affinity chromatography (IMAC) and mass spectrometry. Methods Mol Biol 527:47–56, xi. https://doi.org/10.1007/978-1-60327-834-8_4
24. Verma R, Pinto SM, Patil AH, Advani J, Subba P, Kumar M, Sharma J, Dey G, Ravikumar R, Buggi S, Satishchandra P, Sharma K, Suar M, Tripathy SP, Chauhan DS, Gowda H, Pandey A, Gandotra S, Prasad TS (2017) Quantitative proteomic and phosphoproteomic analysis of H37Ra and H37Rv strains of mycobacterium tuberculosis. J Proteome Res 16(4):1632–1645. https://doi.org/10.1021/acs.jproteome.6b00983
25. Thingholm TE, Jensen ON, Robinson PJ, Larsen MR (2008) SIMAC (sequential elution from IMAC), a phosphoproteomics strategy for the rapid separation of monophosphorylated from multiply phosphorylated peptides. Mol Cell Proteomics 7(4):661–671. https://doi.org/10.1074/mcp.M700362-MCP200
26. Bertozzi CR, Kiessling LL (2001) Chemical glycobiology. Science 291(5512):2357–2364
27. Vocadlo DJ, Hang HC, Kim EJ, Hanover JA, Bertozzi CR (2003) A chemical approach for identifying O-GlcNAc-modified proteins in cells. Proc Natl Acad Sci U S A 100

(16):9116–9121. https://doi.org/10.1073/pnas.1632821100
28. Lanyon-Hogg T, Faronato M, Serwa RA, Tate EW (2017) Dynamic protein acylation: new substrates, mechanisms, and drug targets. Trends Biochem Sci 42(7):566–581. https://doi.org/10.1016/j.tibs.2017.04.004
29. Morrison E, Kuropka B, Kliche S, Brugger B, Krause E, Freund C (2015) Quantitative analysis of the human T cell palmitome. Sci Rep 5:11598. https://doi.org/10.1038/srep11598
30. Roth AF, Wan J, Bailey AO, Sun B, Kuchar JA, Green WN, Phinney BS, Yates JR 3rd, Davis NG (2006) Global analysis of protein palmitoylation in yeast. Cell 125(5):1003–1013. https://doi.org/10.1016/j.cell.2006.03.042
31. Zhang Y, Zhang C, Jiang H, Yang P, Lu H (2015) Fishing the PTM proteome with chemical approaches using functional solid phases. Chem Soc Rev 44(22):8260–8287. https://doi.org/10.1039/c4cs00529e
32. Tate EW (2008) Recent advances in chemical proteomics: exploring the post-translational proteome. J Chem Biol 1(1–4):17–26. https://doi.org/10.1007/s12154-008-0002-6
33. Webb K, Bennett EJ (2013) Eavesdropping on PTM cross-talk through serial enrichment. Nat Methods 10(7):620–621. https://doi.org/10.1038/nmeth.2526
34. Swaney DL, Beltrao P, Starita L, Guo A, Rush J, Fields S, Krogan NJ, Villen J (2013) Global analysis of phosphorylation and ubiquitylation cross-talk in protein degradation. Nat Methods 10(7):676–682. https://doi.org/10.1038/nmeth.2519
35. Mertins P, Qiao JW, Patel J, Udeshi ND, Clauser KR, Mani DR, Burgess MW, Gillette MA, Jaffe JD, Carr SA (2013) Integrated proteomic analysis of post-translational modifications by serial enrichment. Nat Methods 10(7):634–637. https://doi.org/10.1038/nmeth.2518
36. Kim MS, Zhong J, Kandasamy K, Delanghe B, Pandey A (2011) Systematic evaluation of alternating CID and ETD fragmentation for phosphorylated peptides. Proteomics 11(12):2568–2572. https://doi.org/10.1002/pmic.201000547
37. Ong SE, Blagoev B, Kratchmarova I, Kristensen DB, Steen H, Pandey A, Mann M (2002) Stable isotope labeling by amino acids in cell culture, SILAC, as a simple and accurate approach to expression proteomics. Mol Cell Proteomics 1(5):376–386
38. Ross PL, Huang YN, Marchese JN, Williamson B, Parker K, Hattan S, Khainovski N, Pillai S, Dey S, Daniels S, Purkayastha S, Juhasz P, Martin S, Bartlet-Jones M, He F, Jacobson A, Pappin DJ (2004) Multiplexed protein quantitation in Saccharomyces cerevisiae using amine-reactive isobaric tagging reagents. Mol Cell Proteomics 3(12):1154–1169. https://doi.org/10.1074/mcp.M400129-MCP200
39. Thompson A, Schafer J, Kuhn K, Kienle S, Schwarz J, Schmidt G, Neumann T, Johnstone R, Mohammed AK, Hamon C (2003) Tandem mass tags: a novel quantification strategy for comparative analysis of complex protein mixtures by MS/MS. Anal Chem 75(8):1895–1904
40. Schwanhausser B, Busse D, Li N, Dittmar G, Schuchhardt J, Wolf J, Chen W, Selbach M (2011) Global quantification of mammalian gene expression control. Nature 473(7347):337–342. https://doi.org/10.1038/nature10098
41. Zybailov B, Mosley AL, Sardiu ME, Coleman MK, Florens L, Washburn MP (2006) Statistical analysis of membrane proteome expression changes in Saccharomyces cerevisiae. J Proteome Res 5(9):2339–2347. https://doi.org/10.1021/pr060161n
42. Keller A, Bader SL, Kusebauch U, Shteynberg D, Hood L, Moritz RL (2016) Opening a SWATH window on posttranslational modifications: automated pursuit of modified peptides. Mol Cell Proteomics 15(3):1151–1163. https://doi.org/10.1074/mcp.M115.054478
43. Lu P, Vogel C, Wang R, Yao X, Marcotte EM (2007) Absolute protein expression profiling estimates the relative contributions of transcriptional and translational regulation. Nat Biotechnol 25(1):117–124. https://doi.org/10.1038/nbt1270
44. Taus T, Kocher T, Pichler P, Paschke C, Schmidt A, Henrich C, Mechtler K (2011) Universal and confident phosphorylation site localization using phosphoRS. J Proteome Res 10(12):5354–5362. https://doi.org/10.1021/pr200611n
45. Beausoleil SA, Villen J, Gerber SA, Rush J, Gygi SP (2006) A probability-based approach for high-throughput protein phosphorylation analysis and site localization. Nat Biotechnol 24(10):1285–1292. https://doi.org/10.1038/nbt1240
46. Peng X, Xu F, Liu S, Li S, Huang Q, Chang L, Wang L, Ma X, He F, Xu P (2017) Identification of missing proteins in the phosphoproteome of kidney cancer. J Proteome Res 16(12):4364–4373. https://doi.org/10.1021/acs.jproteome.7b00332

47. Zhang Z, Tan M, Xie Z, Dai L, Chen Y, Zhao Y (2011) Identification of lysine succinylation as a new post-translational modification. Nat Chem Biol 7(1):58–63. https://doi.org/10.1038/nchembio.495
48. Weinert BT, Scholz C, Wagner SA, Iesmantavicius V, Su D, Daniel JA, Choudhary C (2013) Lysine succinylation is a frequently occurring modification in prokaryotes and eukaryotes and extensively overlaps with acetylation. Cell Rep 4(4):842–851. https://doi.org/10.1016/j.celrep.2013.07.024
49. Xie Z, Dai J, Dai L, Tan M, Cheng Z, Wu Y, Boeke JD, Zhao Y (2012) Lysine succinylation and lysine malonylation in histones. Mol Cell Proteomics 11(5):100–107. https://doi.org/10.1074/mcp.M111.015875
50. Drazic A, Myklebust LM, Ree R, Arnesen T (2016) The world of protein acetylation. Biochim Biophys Acta 1864(10):1372–1401. https://doi.org/10.1016/j.bbapap.2016.06.007
51. Kim SC, Sprung R, Chen Y, Xu Y, Ball H, Pei J, Cheng T, Kho Y, Xiao H, Xiao L, Grishin NV, White M, Yang XJ, Zhao Y (2006) Substrate and functional diversity of lysine acetylation revealed by a proteomics survey. Mol Cell 23 (4):607–618. https://doi.org/10.1016/j.molcel.2006.06.026
52. Zhang J, Sprung R, Pei J, Tan X, Kim S, Zhu H, Liu CF, Grishin NV, Zhao Y (2009) Lysine acetylation is a highly abundant and evolutionarily conserved modification in Escherichia coli. Mol Cell Proteomics 8 (2):215–225. https://doi.org/10.1074/mcp.M800187-MCP200
53. Henriksen P, Wagner SA, Weinert BT, Sharma S, Bacinskaja G, Rehman M, Juffer AH, Walther TC, Lisby M, Choudhary C (2012) Proteome-wide analysis of lysine acetylation suggests its broad regulatory scope in Saccharomyces cerevisiae. Mol Cell Proteomics 11(11):1510–1522. https://doi.org/10.1074/mcp.M112.017251
54. Lundby A, Lage K, Weinert BT, Bekker-Jensen DB, Secher A, Skovgaard T, Kelstrup CD, Dmytriyev A, Choudhary C, Lundby C, Olsen JV (2012) Proteomic analysis of lysine acetylation sites in rat tissues reveals organ specificity and subcellular patterns. Cell Rep 2 (2):419–431. https://doi.org/10.1016/j.celrep.2012.07.006
55. Xie C, Shen H, Zhang H, Yan J, Liu Y, Yao F, Wang X, Cheng Z, Tang TS, Guo C (2017) Quantitative proteomics analysis reveals alterations of lysine acetylation in mouse testis in response to heat shock and X-ray exposure. Biochim Biophys Acta 1866:464. https://doi.org/10.1016/j.bbapap.2017.11.011
56. Meyer JG, D'Souza AK, Sorensen DJ, Rardin MJ, Wolfe AJ, Gibson BW, Schilling B (2016) Quantification of lysine acetylation and Succinylation stoichiometry in proteins using mass spectrometric data-independent acquisitions (SWATH). J Am Soc Mass Spectrom 27 (11):1758–1771. https://doi.org/10.1007/s13361-016-1476-z
57. Pickart CM, Eddins MJ (2004) Ubiquitin: structures, functions, mechanisms. Biochim Biophys Acta 1695(1–3):55–72. https://doi.org/10.1016/j.bbamcr.2004.09.019
58. Wu Q, Cheng Z, Zhu J, Xu W, Peng X, Chen C, Li W, Wang F, Cao L, Yi X, Wu Z, Li J, Fan P (2015) Suberoylanilide hydroxamic acid treatment reveals crosstalks among proteome, ubiquitylome and acetylome in non-small cell lung cancer A549 cell line. Sci Rep 5:9520. https://doi.org/10.1038/srep09520
59. Iesmantavicius V, Weinert BT, Choudhary C (2014) Convergence of ubiquitylation and phosphorylation signaling in rapamycin-treated yeast cells. Mol Cell Proteomics 13 (8):1979–1992. https://doi.org/10.1074/mcp.O113.035683
60. Yang X, Liu F, Yan Y, Zhou T, Guo Y, Sun G, Zhou Z, Zhang W, Guo X, Sha J (2015) Proteomic analysis of N-glycosylation of human seminal plasma. Proteomics 15 (7):1255–1258. https://doi.org/10.1002/pmic.201400203
61. Sudhir PR, Chen CH, Pavana Kumari M, Wang MJ, Tsou CC, Sung TY, Chen JY, Chen CH (2012) Label-free quantitative proteomics and N-glycoproteomics analysis of KRAS-activated human bronchial epithelial cells. Mol Cell Proteomics 11(10):901–915. https://doi.org/10.1074/mcp.M112.020875
62. Chen W, Smeekens JM, Wu R (2014) A universal chemical enrichment method for mapping the yeast N-glycoproteome by mass spectrometry (MS). Mol Cell Proteomics 13 (6):1563–1572. https://doi.org/10.1074/mcp.M113.036251
63. Guan X, Fierke CA (2011) Understanding protein palmitoylation: biological significance and enzymology. Sci China Chem 54 (12):1888–1897. https://doi.org/10.1007/s11426-011-4428-2
64. Martin BR (2013) Nonradioactive analysis of dynamic protein palmitoylation. Curr Protoc Protein Sci 73(Unit 14):15. https://doi.org/10.1002/0471140864.ps1415s73
65. Martin BR, Wang C, Adibekian A, Tully SE, Cravatt BF (2011) Global profiling of dynamic

protein palmitoylation. Nat Methods 9 (1):84–89. https://doi.org/10.1038/nmeth.1769

66. Hornbeck PV, Zhang B, Murray B, Kornhauser JM, Latham V, Skrzypek E (2015) PhosphoSitePlus, 2014: mutations, PTMs and recalibrations. Nucleic Acids Res 43(Database issue): D512–D520. https://doi.org/10.1093/nar/gku1267
67. Keshava Prasad TS, Goel R, Kandasamy K, Keerthikumar S, Kumar S, Mathivanan S, Telikicherla D, Raju R, Shafreen B, Venugopal A, Balakrishnan L, Marimuthu A, Banerjee S, Somanathan DS, Sebastian A, Rani S, Ray S, Harrys Kishore CJ, Kanth S, Ahmed M, Kashyap MK, Mohmood R, Ramachandra YL, Krishna V, Rahiman BA, Mohan S, Ranganathan P, Ramabadran S, Chaerkady R, Pandey A (2009) Human protein reference database—2009 update. Nucleic Acids Res 37 (Database):D767–D772. https://doi.org/10.1093/nar/gkn892
68. Huang KY, Su MG, Kao HJ, Hsieh YC, Jhong JH, Cheng KH, Huang HD, Lee TY (2016) dbPTM 2016: 10-year anniversary of a resource for post-translational modification of proteins. Nucleic Acids Res 44(D1): D435–D446. https://doi.org/10.1093/nar/gkv1240
69. Liu Z, Wang Y, Gao T, Pan Z, Cheng H, Yang Q, Cheng Z, Guo A, Ren J, Xue Y (2014) CPLM: a database of protein lysine modifications. Nucleic Acids Res 42(Database issue):D531–D536. https://doi.org/10.1093/nar/gkt1093
70. Kennedy JJ, Yan P, Zhao L, Ivey RG, Voytovich UJ, Moore HD, Lin C, Pogosova-Agadjanyan EL, Stirewalt DL, Reding KW, Whiteaker JR, Paulovich AG (2016) Immobilized metal affinity chromatography coupled to multiple reaction monitoring enables reproducible quantification of phospho-signaling. Mol Cell Proteomics 15(2):726–739. https://doi.org/10.1074/mcp.O115.054940
71. Chick JM, Kolippakkam D, Nusinow DP, Zhai B, Rad R, Huttlin EL, Gygi SP (2015) A mass-tolerant database search identifies a large proportion of unassigned spectra in shotgun proteomics as modified peptides. Nat Biotechnol 33(7):743–749. https://doi.org/10.1038/nbt.3267
72. Pan Z, Liu Z, Cheng H, Wang Y, Gao T, Ullah S, Ren J, Xue Y (2014) Systematic analysis of the in situ crosstalk of tyrosine modifications reveals no additional natural selection on multiply modified residues. Sci Rep 4:7331. https://doi.org/10.1038/srep07331
73. Minguez P, Parca L, Diella F, Mende DR, Kumar R, Helmer-Citterich M, Gavin AC, van Noort V, Bork P (2012) Deciphering a global network of functionally associated post-translational modifications. Mol Syst Biol 8:599. https://doi.org/10.1038/msb.2012.31
74. Trinidad JC, Barkan DT, Gulledge BF, Thalhammer A, Sali A, Schoepfer R, Burlingame AL (2012) Global identification and characterization of both O-GlcNAcylation and phosphorylation at the murine synapse. Mol Cell Proteomics 11(8):215–229. https://doi.org/10.1074/mcp.O112.018366
75. Kim MS, Zhong J, Pandey A (2016) Common errors in mass spectrometry-based analysis of post-translational modifications. Proteomics 16(5):700–714. https://doi.org/10.1002/pmic.201500355
76. Hughes CS, Spicer V, Krokhin OV, Morin GB (2017) Investigating acquisition performance on the Orbitrap fusion when using tandem MS/MS/MS scanning with isobaric tags. J Proteome Res 16(5):1839–1846. https://doi.org/10.1021/acs.jproteome.7b00091
77. Pan J, Chen R, Li C, Li W, Ye Z (2015) Global analysis of protein lysine succinylation profiles and their overlap with lysine acetylation in the marine bacterium vibrio parahemolyticus. J Proteome Res 14(10):4309–4318. https://doi.org/10.1021/acs.jproteome.5b00485
78. Cheng Y, Hou T, Ping J, Chen G, Chen J (2016) Quantitative succinylome analysis in the liver of non-alcoholic fatty liver disease rat model. Proteome Sci 14:3. https://doi.org/10.1186/s12953-016-0092-y
79. Colak G, Xie Z, Zhu AY, Dai L, Lu Z, Zhang Y, Wan X, Chen Y, Cha YH, Lin H, Zhao Y, Tan M (2013) Identification of lysine succinylation substrates and the succinylation regulatory enzyme CobB in Escherichia coli. Mol Cell Proteomics 12(12):3509–3520. https://doi.org/10.1074/mcp.M113.031567
80. Sun G, Jiang M, Zhou T, Guo Y, Cui Y, Guo X, Sha J (2014) Insights into the lysine acetylproteome of human sperm. J Proteome 109:199–211. https://doi.org/10.1016/j.jprot.2014.07.002
81. Xie L, Wang X, Zeng J, Zhou M, Duan X, Li Q, Zhang Z, Luo H, Pang L, Li W, Liao G, Yu X, Li Y, Huang H, Xie J (2015) Proteome-wide lysine acetylation profiling of the human pathogen Mycobacterium tuberculosis. Int J Biochem Cell Biol 59:193–202. https://doi.org/10.1016/j.biocel.2014.11.010
82. Karg E, Smets M, Ryan J, Forne I, Qin W, Mulholland CB, Kalideris G, Imhof A, Bultmann S, Leonhardt H (2017) Ubiquitome

analysis reveals PCNA-associated factor 15 (PAF15) as a specific ubiquitination target of UHRF1 in embryonic stem cells. J Mol Biol 429(24):3814–3824. https://doi.org/10.1016/j.jmb.2017.10.014

83. Caballero MC, Alonso AM, Deng B, Attias M, de Souza W, Corvi MM (2016) Identification of new palmitoylated proteins in toxoplasma gondii. Biochim Biophys Acta 1864(4):400–408. https://doi.org/10.1016/j.bbapap.2016.01.010
84. Chen YJ, Lu CT, Lee TY, Chen YJ (2014) dbGSH: a database of S-glutathionylation. Bioinformatics 30(16):2386–2388. https://doi.org/10.1093/bioinformatics/btu301
85. Chen YJ, Lu CT, Su MG, Huang KY, Ching WC, Yang HH, Liao YC, Chen YJ, Lee TY (2015) dbSNO 2.0: a resource for exploring structural environment, functional and disease association and regulatory network of protein S-nitrosylation. Nucleic Acids Res 43(Database issue):D503–D511. https://doi.org/10.1093/nar/gku1176
86. Duan G, Li X, Kohn M (2015) The human DEPhOsphorylation database DEPOD: a 2015 update. Nucleic Acids Res 43(Database issue):D531–D535. https://doi.org/10.1093/nar/gku1009
87. Maurer-Stroh S, Gouda M, Novatchkova M, Schleiffer A, Schneider G, Sirota FL, Wildpaner M, Hayashi N, Eisenhaber F (2004) MYRbase: analysis of genome-wide glycine myristoylation enlarges the functional spectrum of eukaryotic myristoylated proteins. Genome Biol 5(3):R21. https://doi.org/10.1186/gb-2004-5-3-r21
88. Gupta R, Birch H, Rapacki K, Brunak S, Hansen JE (1999) O-GLYCBASE version 4.0: a revised database of O-glycosylated proteins. Nucleic Acids Res 27(1):370–372
89. Gnad F, Gunawardena J, Mann M (2011) PHOSIDA 2011: the posttranslational modification database. Nucleic Acids Res 39(Database issue):D253–D260. https://doi.org/10.1093/nar/gkq1159
90. Dinkel H, Chica C, Via A, Gould CM, Jensen LJ, Gibson TJ, Diella F (2011) Phospho.ELM: a database of phosphorylation sites—update 2011. Nucleic Acids Res 39(Database issue):D261–D267. https://doi.org/10.1093/nar/gkq1104
91. Maurer-Stroh S, Koranda M, Benetka W, Schneider G, Sirota FL, Eisenhaber F (2007) Towards complete sets of farnesylated and geranylgeranylated proteins. PLoS Comput Biol 3(4):e66. https://doi.org/10.1371/journal.pcbi.0030066
92. Tung CW (2012) PupDB: a database of pupylated proteins. BMC Bioinformatics 13:40. https://doi.org/10.1186/1471-2105-13-40
93. Zhang X, Huang B, Zhang L, Zhang Y, Zhao Y, Guo X, Qiao X, Chen C (2012) SNObase, a database for S-nitrosation modification. Protein Cell 3(12):929–933. https://doi.org/10.1007/s13238-012-2094-6
94. Hasan MM, Yang S, Zhou Y, Mollah MN (2016) SuccinSite: a computational tool for the prediction of protein succinylation sites by exploiting the amino acid patterns and properties. Mol BioSyst 12(3):786–795. https://doi.org/10.1039/c5mb00853k
95. Chernorudskiy AL, Garcia A, Eremin EV, Shorina AS, Kondratieva EV, Gainullin MR (2007) UbiProt: a database of ubiquitylated proteins. BMC Bioinformatics 8:126. https://doi.org/10.1186/1471-2105-8-126

Chapter 15

Identification of Unexpected Protein Modifications by Mass Spectrometry-Based Proteomics

Shiva Ahmadi and Dominic Winter

Abstract

Peptide identification relies in the majority of mass spectrometry-based proteomics experiments on matching of experimental data against peptide and fragment ion masses derived from in silico digests of protein databases. One of the main drawbacks of this approach is that modifications have to be defined for database searching and therefore no unexpected modifications can be identified in a standard setup. Consequently, in many bottom-up proteomics experiments, unexpected modifications are not identified, even if high-quality fragment ion spectra of the modified peptides were acquired. It is therefore often not straightforward to identify unexpected modifications. In this protocol, we describe a stepwise procedure to identify unexpected modifications at peptides using the database search algorithm Mascot. The workflow includes parallel searches for the identification of known modifications at unexpected amino acids, error tolerant searches for modifications unexpected in the sample but known to the community, and mass tolerant searches for entirely unknown modifications. Furthermore, we suggest a follow-up strategy consisting of (1) verification of identified modifications in the initial dataset and (2) targeted experiments using synthetic peptides.

Key words Mass spectrometry, Unexpected modifications, Posttranslational modifications, Bottom-up proteomics, Data analysis, Mascot, Error tolerant search, Mass tolerant search

1 Introduction

Mass spectrometry has become the most versatile and powerful technique for the identification, quantification, and characterization of proteins. In the large majority of studies, the so-called bottom-up approach is used which employs proteolytic digestion of proteins followed by analysis of the resulting peptides using MALDI-MS or MS/MS, as well as LC-ESI-MS/MS [1]. In this method, MS/MS data are most often collected in the data-dependent acquisition (DDA) mode. In DDA, the instrument is set up to select the most abundant (multiply charged) precursor ions for fragmentation [2]. The resulting MS/MS spectra are then usually identified by matching against theoretical spectra originating from in silico digests of protein databases [3]. Despite the

Xing Wang and Matthew Kuruc (eds.), *Functional Proteomics: Methods and Protocols*, Methods in Molecular Biology, vol. 1871, https://doi.org/10.1007/978-1-4939-8814-3_15, © Springer Science+Business Media, LLC, part of Springer Nature 2019

widespread use of this approach, a large number of spectra (on average 75%) remain unidentified in a standard proteomics experiment [4]. One major factor preventing the assignment of spectra is unexpected modifications of peptides, either due to post-translational modifications (PTMs [5]) or chemical modifications, which are introduced during sample preparation [6]. Up to date, several hundreds of these modifications are known and can be found in specific databases like Unimod (www.unimod.org [7]). Unexpected modifications prohibit straightforward identification of the peptides as they alter their precursor mass and therefore prevent them from being matched to the theoretical masses calculated for the in silico digest [8]. Dependent on the experiment, it may be of interest to identify these unknown modifications. Unexpected PTMs, for example, could result in the discovery of cellular regulatory mechanisms, which may be of high importance with respect to the investigated biological question. Unintended chemical modifications, on the other hand, can affect the performance of the experiment, and are therefore usually undesired. Their identification could allow adapting the parameters for sample preparation and/or database searching and accordingly increasing the number of identified peptides. Furthermore, once a modification is identified, characterization of the modified peptides can be improved using a range of well-established approaches for the fractionation of peptides due to their physicochemical properties. These include for example affinity capture using antibodies, columns, or beads [9], and fractionation by chromatography techniques [10].

While expected modifications are often easily identified by defining them as variable modification for database searching, the identification of unknown modifications is usually not straightforward and requires further data analysis and validation. In this chapter, we provide a strategy for the discovery of unexpected modifications in complex samples by bottom-up LC-ESI-MS/MS. We suggest points to be considered for the preparation and analysis of such samples and exemplify the identification of unexpected modifications by different strategies using the commonly used peptide search engine Mascot (www.matrixscience.com). We demonstrate the workflow based on a recent study from our group addressing unspecific modifications resulting from reduction and alkylation of proteins in proteomics experiments [11].

2 Materials

2.1 Cell Culture

Cell culture medium: Dulbecco's modified Eagle medium (DMEM) supplemented with 10% FCS, 100 U/mL penicillin, and 100 μg/mL streptomycin.

2.2 Cell Harvest, Lysis, and Preparation of Cytosolic Fraction

1. Phosphate Buffered Saline (PBS, 1×).
2. Sucrose buffer: 250 mM sucrose, 15 mM KCl, 1.5 mM MgAc, 10 mM HEPES, 1× Protease Inhibitor Cocktail (*see* **Note 1**).
3. Acetone.
4. Lysis buffer: 0.1 M Tris HCl/ 4% SDS.
5. A kit for the determination of the protein concentration of a sample.
6. Dounce homogenizer.
7. Cell scraper.
8. Tabletop centrifuge.
9. Ultracentrifuge.

2.3 In-Gel Reduction, Alkylation, and Digestion

1. Thermomixer.
2. Sample loading buffer (modified from [12]): 0.25 M Tris–HCl pH 6.8, 8% (w/v) SDS, 40% (v/v) glycerol, and 0.004% (w/v) bromophenol blue.
3. Reduction solutions: 20 mM dithiothreitol (DTT), 20 mM tris-(2-chlorethyl)-Phosphate (TCEP), 40 mM β-mercaptoethanol (BME) in 0.1 M NH_4HCO_3.
4. Alkylation solutions: 55 mM iodoacetamide (IAA), iodoacetic acid (IAC), acrylamide (AA), chloroacetic acid CAA in 0.1 M NH_4HCO_3.
5. 10% SDS-PAGE gel.
6. Coomassie brilliant blue.
7. Gel destaining solution: 30% acetonitrile (ACN), 70 mM NH_4HCO_3.
8. Wash solution: 0.1 M NH_4HCO_3.
9. Trypsin solution: 0.1 μg/μL sequencing grade trypsin in 0.1 M NH_4HCO_3.
10. Peptide extraction solutions.
    (a) 0.1% trifluoroacetic acid (TFA), 50% ACN.
    (b) 0.1 M NH_4HCO_3.
    (c) 100% ACN.
11. Resuspension solution: 0.01% acetic acid 3% ACN.
12. MS sample buffer: 5% ACN, 5% formic acid (FA).
13. Vortex.

2.4 Dimethyl Labeling

1. Dimethylation solution 1: Light label: 4% (v/v) CH_2O, medium label: 4% (v/v) CD_2O, heavy label: 4% (v/v) $^{13}CD_2O$ (*see* **Note 2**, [13]).

2. Dimethylation solution 2: Light label: 0.6 M $NaBH_3CN$, intermediate label: 0.6 M $NaBH_3CN$, heavy label: 0.6 M $NaBD_3CN$ (*see* **Note 2**, [13]).
3. Quenching solution: 1% (v/v) NH_4OH.
4. Formic acid (FA).
5. Vacuum Centrifuge.
6. Fume hood.
7. Thermomixer.

2.5 LC-MS/MS Measurements

1. C_{18} analytical column: ESI spray tip produced in house with a Sutter Instruments P2000 laser puller from a 360 μm outer diameter, 100 μm inner diameter fused silica capillary and packed with 5 μm particles [Dr. Maisch, Reprosil C-18 AQ]. Alternatively, any type of commercially available nanoflow C_{18} column in combination with an appropriate emitter can be used.
2. Thermo Scientific EASY-nLC 1000 or similar nanoflow high/ or ultrahigh-performance liquid chromatography system.
3. Thermo Scientific Orbitrap Velos Mass Spectrometer or any other high-resolution / high accuracy mass spectrometer.
4. Solvent A: Water with 0.1% FA.
5. Solvent B: ACN with 0.1% FA.

2.6 Software

1. Mascot (www.matrixscience.com).
2. Proteome Discoverer (Thermo Scientific).

3 Methods

To be able to identify unexpected modifications, it is essential that the modified peptides are fragmented by the mass spectrometer. Peptide signals are usually selected for fragmentation in the DDA mode based on their intensity in the survey spectrum. Since it is unlikely that the modified peptides are among the most abundant peptides in the sample, it is advisable to use samples of medium to low complexity. This results in over- rather than under-sampling and is therefore more likely to yield fragment ion spectra from modified peptides. When analyzing single protein bands from SDS-gels or purified proteins, the complexity is usually sufficiently low. If, however, cells or tissues are to be analyzed at a proteome-wide scale, the sample should be divided into several fractions either at the protein or peptide level. Methods commonly employed for fractionation before proteolytic digestion include SDS-PAGE (in combination with in-gel digestion) [14], as well as size exclusion [15] or ion exchange chromatography [16], followed by proteolytic digestion of the resulting fractions. On the peptide level, basic reversed-phase high-performance liquid chromatography (RP-HPLC) [17], strong anion exchange chromatography (SAX) [18], strong cation exchange chromatography (SCX) [19], or isoelectric focusing (IEF) [20] are the available techniques for fractionation of digested peptides. Dependent on the desired depth of analysis, protein and peptide fractionation can be combined. Furthermore, subcellular fractionation (e.g., by differential centrifugation or enrichment of organelles) is another approach to generate samples of reduced complexity [21].

In the current study, we used a cytosolic fraction of HeLa cells, separated the proteins by SDS-PAGE according to their molecular weight, and performed in-gel digestion of a subset of proteins (ranging from 30 to 50 kDa). Additionally, we performed in-solution digestion of the whole sample followed by fractionation using SAX stop and go extraction (STAGE) tips [22]. As the strategy for LC-MS/MS measurement and data analysis is similar for both types of sample processing, we will restrict our descriptions in this chapter to the analysis of the in-gel digested samples. The details concerning the in-solution digested samples can be found in [11].

3.1 Sample Preparation: In-Gel Digestion of a Cytosolic HeLa Fraction

If the choice of the specific subcellular fraction is not influenced by a biological question and it serves only as a source of proteins, the cytosolic fraction is the easiest to generate, as the other parts of the cell can be removed by centrifugation. Furthermore, the cytosolic fraction is of adequate complexity in order to allow a sufficient coverage of the proteins present in the sample and to provide a reasonable number of proteins in order to resemble the complexity observed in medium to large scale proteomics studies. When planning the experiment, care should be taken that the differential treatment of the samples is not influencing the subsequent steps, possibly resulting in secondary effects. For this reason, we decided to perform reduction and alkylation after performing SDS-PAGE as differential modification of the proteins may influence their running behavior in the gel electrophoresis.

1. Place the cell culture dishes on ice, aspirate the medium, and wash the cells 3× with 10 mL ice-cold PBS (*see* **Note 3**).
2. Add 1 mL sucrose buffer, detach the cells using a cell scraper, and transfer the cell suspension to a dounce homogenizer (*see* **Note 4**).
3. Homogenize cells on ice using the dounce homogenizer (30 strokes) and transfer the suspension to a 2 mL microtube (for higher volumes 15 or 50 mL conical tubes can be used).
4. Centrifuge at 1000 × *g*, 4 °C for 10 min in order to pellet intact cells and nuclei. Transfer the supernatant to a new 2 mL microtube.

5. Resuspend the pellet from the first tube in 1 mL sucrose buffer and repeat **steps 3** and **4**.
6. Combine the supernatants from both homogenization steps, transfer them to an appropriate ultracentrifugation tube, and centrifuge at 100,000 × *g*, 4 °C for 1 h to pellet cell organelles and membranes. The clear supernatant represents the cytosolic fraction.
7. In order to precipitate proteins, mix the supernatant (cytosolic fraction) with acetone (−20 °C) in a ratio of 1–4 (v/v), vortex for 30 s, and incubate at −20 °C overnight (*see* **Notes 5** and **6**).
8. Centrifuge samples at 20,000 × *g*, 4 °C for 30 min, discard the supernatant, and air-dry the pellet at 23 °C.
9. Resuspend the pellet in the lysis buffer by heating at 95 °C for 5 min, centrifuge at 20,000 × *g*, RT for 30 min, and transfer the clear supernatant to a new tube (*see* **Note 7**).
10. Determine the protein concentration using a protein assay.
11. Perform SDS-PAGE, stain the gel using Coomassie brilliant blue (several hours to overnight, dependent on sample amount) and destain the gel using MilliQ water.
12. Excise the region to be analyzed from the gel and cut it into small pieces of ~1 mm^2. If PTMs in general are to be analyzed and no specific protein is targeted, cut the whole gel lane to pieces of similar complexity (as determined by the number of Coomassie stained bands). If unspecific modifications due to chemical sample treatments are to be analyzed, a single section of the gel is usually sufficient. In our example study we selected a region between 30 and 50 kDa.
13. Destain the gel using 500 μL gel destaining solution at 25 °C, 800 rpm for 30 min in a thermomixer and discard the liquid. Repeat this step until the gel pieces are colorless.
14. Reduce disulfide bonds using 100 μL of reduction solution for 45 min at 56 °C, 800 rpm.
15. Remove the reducing reagent and alkylate thiol groups by adding 100 μL of alkylation solution. Incubate for 30 min at RT in the dark.
16. Wash the gel pieces using 500 μL wash solution for 15 min at 800 rpm. Discard the liquid.
17. Dehydrate the gel pieces with 100% ACN for 15 min at 800 rpm. Discard the liquid and dry the gel pieces using a vacuum centrifuge.
18. Add 10 μL of trypsin solution, incubate for 15 min, and then add a sufficient volume of wash solution to cover the gel pieces. Control after 30 min if all gel pieces are covered and add more wash solution if needed. Incubate overnight at 37 °C.

19. Extract peptides consecutively using 50 μL of peptide extraction solutions A, B, and C for 15 min each at 25 °C, 800 rpm and pool the peptide extracts (*see* **Note 8**). Dry the sample using a vacuum centrifuge.
20. Reconstitute the samples in 100 μL resuspension solution and desalt, e.g., using STAGE tips [22].
21. Dry the eluate fractions using a vacuum centrifuge and reconstitute in 20 μL MS sample buffer.

3.2 LC-MS/MS Data Acquisition

For the identification of unexpected modifications, it is beneficial to measure the samples with high mass accuracy. This allows in the subsequent steps to analyze the data with a small mass tolerance window reducing the analysis time and probability of false positive assignments. We therefore advice, if possible, to use Orbitrap or QTOF instruments which provide high resolution and mass accuracy at a good sensitivity and scan speed. If only ion trap instruments are available, it should be considered to use (at least for the survey scan of the intact peptide ions), the zoom or ultra-zoom scan mode and to apply longer LC gradients to compensate for the lower survey scan speed. In our example study, we used a LTQ Orbitrap Velos mass spectrometer acquiring the survey scans for the intact precursor ions in the Orbitrap part of the instrument and the fragment ion spectra by collision induced dissociation (CID) in the ion trap. We chose this option over fragmentation by higher energy collisional dissociation (HCD), followed by measurement of the fragment ions in the Orbitrap, since in our experience the increased mass accuracy of the HCD spectra cannot compensate for the reduction in sensitivity and scan speed compared to the LTQ in this instrument.

1. Set up the gradient for the reversed-phase (RP) chromatography system in a way that the samples are rather over-sampled. This means, the instrument should be able to acquire more MS/MS spectra than necessary for the complexity of the sample which in turn increases the chance of identifying low-abundant modified peptides (*see* **Note 9**).
2. If the sample is too complex, an increase in the number of MS/MS spectra performed for each survey scan may allow for a more efficient identification of low-abundant peaks. This will come at a cost of deterioration of chromatographic peak shapes. Additionally, the identification rate of low-abundant peptides can be increased by performing technical replicates and/or definition of exclusion lists, while the latter was in our hands not very efficient.
3. Set the capillary voltage to the lowest possible value to prevent extensive in-source fragmentation as this may result in an increase of spectrum complexity. We use, if the spray tip allows, 1.6 kV in the positive ion mode.

4. Set the repeat count to one and the dynamic exclusion window to a value larger than the average chromatographic peak width at the basis in order to prevent repetitive fragmentation of highly abundant ions.
5. If certain neutral losses are expected (e.g., −98 for phosphorylation) define them in the MSA scan setup (in case an Orbitrap mass spectrometer is used).
6. If available, activate the lock mass option in order to achieve maximal mass accuracy.
7. If possible, use either small RP particles or avoid the use of a trap column to achieve maximal chromatographic resolution. This results in increased signal intensities due to narrower peak shapes aiding the identification of low-abundant species.
8. Perform nano(U)HPLC-MS/MS analysis of the sample.

3.3 Data Analysis

Dependent on the availability of sample material and controls, different strategies can be taken. Figure 1 shows the workflow for identification of unexpected modifications, including possible steps and the order of their execution. There are several algorithms available to perform such analyses; we used Mascot (www.matrixscience.org) either in combination with Proteome Discoverer (Thermo Scientific) or individually by manual submission of mascot generic files (mgf). Alternatively, dependent on their features other programs like MaxQuant, SEQUEST, PEAKS DB, ProteinPilot, pFind, Byonic, or X! Tandem can be used [23].

3.3.1 Initial Analyses Defining Known Modifications at All Amino Acids

In case of modifications due to a chemical treatment (in our case unspecific alkylation of amino acid side chains), whose composition is known but not the amino acid residue(s) modified, the possible modification(s) should initially be searched at all possible amino acids. As the number of variable modifications which can be defined in a Mascot search is limited to 9 at a time, three individual searches have to be performed. This results in redundant assignments of MS/MS spectra for the unmodified peptides in the single searches. Furthermore, if the fragment ions are not conclusive for assigning the modification to a distinct amino acid, the same spectrum may be annotated with different modification positions in the separate searches. Proteome discoverer allows dealing with these issues by combining the results of the individual searches in one FDR analysis.

1. Extend the specificity definition of the modification in the Mascot configuration editor to all possible amino acids and define them as active so they can be selected for searching. In our study, we specified carbamidomethyl (for IAA and CAA), carboxymethyl (for IAC), and propionamide (for AA) at all amino acids.

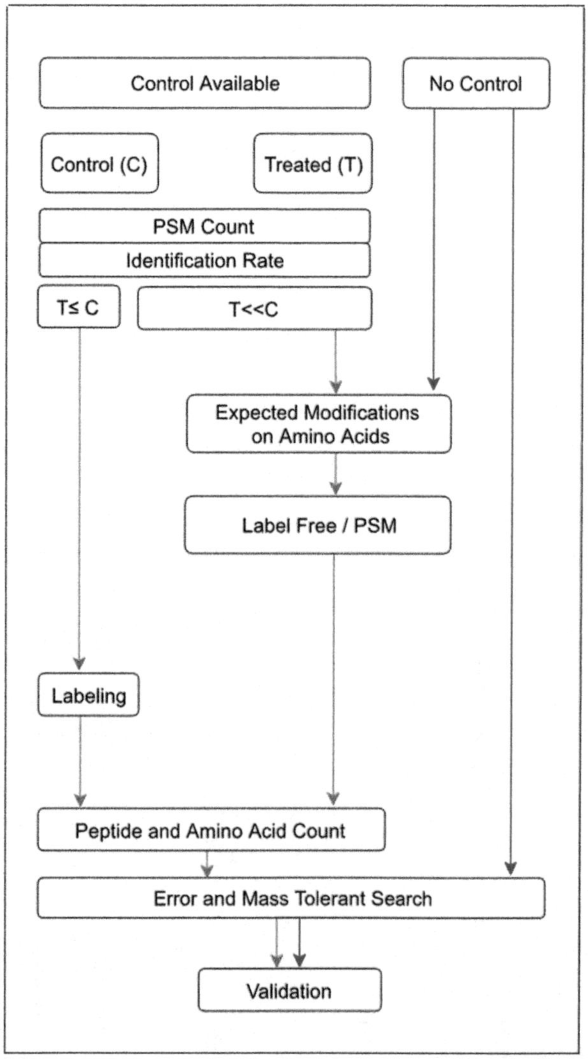

Fig. 1 Data analysis strategy for the identification of unexpected modifications. Dependent on the availability of a control sample, different possible routes can be taken. Identified novel modifications are finally validated by manual inspection of the datasets and synthetic peptide experiments. PSM: peptide spectral match; $T \leq C$: Number of PSMs in the treated sample is slightly lower or similar compared to the control sample; $T \ll C$: Number of PSMs in the treated sample is significantly lower compared to the control sample

2. Create a workflow with three parallel Mascot searches, each containing one subset of modifications.
3. Merge the results in a single percolator processing step in order to allow FDR calculation for the combined results and to remove redundantly assigned MS/MS spectra.

If Proteome Discoverer is not available, alternatively, the redundant assignments can be removed using MS Excel. Due to the high number of variable modifications, this search strategy will almost always result in the identification of modifications at every amino acid. In order to estimate the credibility of the modifications, manual analysis of the type of modified amino acid has to be performed. In our example study, for instance, we found alkylation at several amino acids with side chains which are highly unlikely to react with the alkylation reagents like alanine, leucine, or valine (Fig. 2a, b). Manual analysis revealed that all of these amino acids were located at the peptide N-terminus and that the observed modification was located at the N-terminal amino group and not at the peptide side chain. It became apparent that only seven amino acids as well as the peptide N-terminus were modified. We then

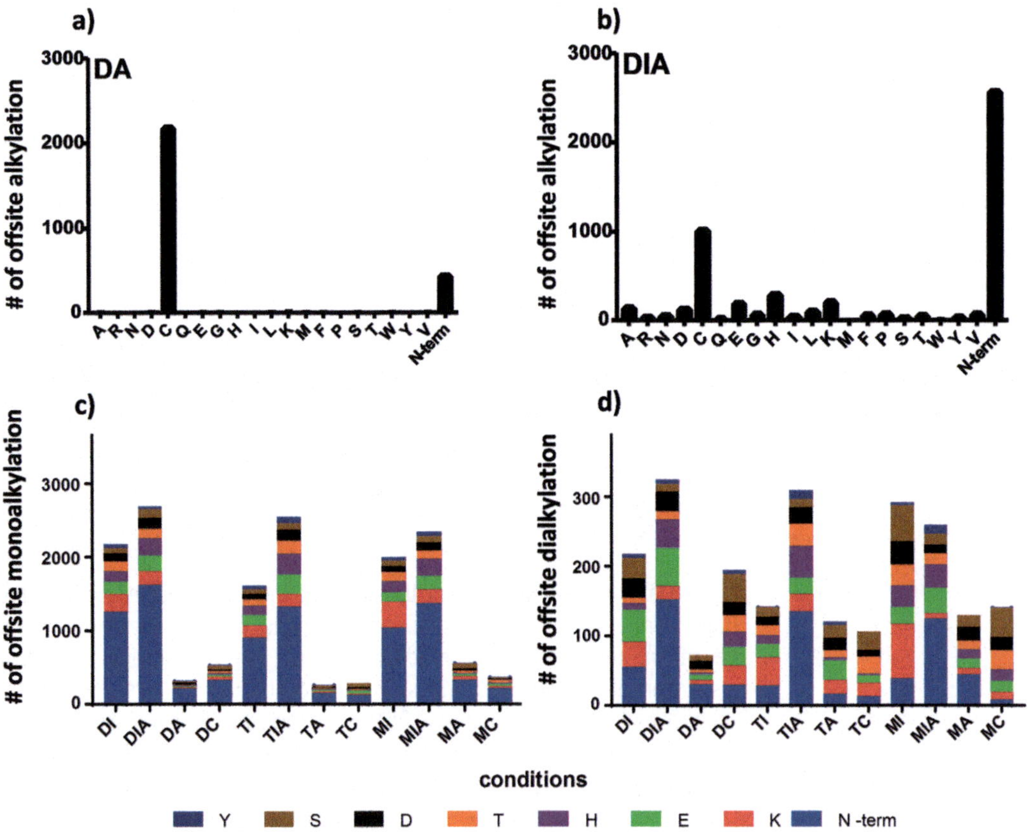

Fig. 2 Amino acids found to be modified by the reduction and alkylation procedure. Shown are the combined results from six replicates. (**a**, **b**) Investigation for off-site alkylation at all possible amino acids in samples reduced with DTT and alkylated with (**a**) AA (DA) and (**b**) IAA (DIA). Manual analysis of (**b**) revealed frequent alkylation of seven amino acids and the peptide N-terminus. (**c**, **d**) PSMs annotated to be off-site (**c**) mono-alkylated and (**d**) di-alkylated at Y, S, D, T, H, E, K, and the peptide N terminus (reproduced from [11]). *DI* DTT/IAA, *DIA* DTT/IAC, *DA* DTT/AA, *DC* DTT/CAA, *TI* TCEP/IAA, *TIA* TCEP/IAC, *TA* TCEP/AA, *TC* TCEP/CAA, *MI* BME/IAA, *MIA* BME/IAC, *MA* BME/AA, *MC* BME/CAA, *N-term* peptide N terminus

performed a second round of searches, this time only including such amino acids, which allowed us to identify a high number of unspecific offside alkylation by monomers and dimers of the alkylation reagents (Fig. 2c, d).

3.3.2 Estimating the Abundance of Unexpected Modifications

If the database searches with known modifications do not yield any conclusive results, a reasonable initial step is to investigate which unmodified peptides are depleted from the sample as a consequence of the investigated treatment. Therefore, if possible, the sample exposed to the biological stimulus or chemical treatment should be compared to an untreated control sample (*see* **Note 10**). This allows estimating the extent of the modification(s) either due to the lack of identification of unmodified peptides or their reduction in signal intensity. In our example study, we compared the samples treated with different alkylation reagents to an untreated control. Each sample was generated in two independent replicates and then each replicate measured three times by LC-MS/MS in order to exclude false conclusions due to DDA-dependent variation in peptide identification. This high degree of redundancy allowed us to assess if the observed differences are random or follow a systematic pattern.

1. If needed, convert MS raw files to an appropriate format (*see* **Note 11**).
2. Define parameters for the initial database search.
   (a) The precursor and fragment ion tolerance should be as narrow as possible, dependent on the performance of the mass spectrometer used. For most high-resolution mass spectrometers, a precursor mass tolerance of 10 ppm is reasonable (also applied for our dataset). For fragment ions, 50 mmU should be selected for high-resolution MS/MS spectra (Orbitrap and QTOF instruments) and 0.6–0.8 Da for low resolution fragment ion spectra acquired in the LTQ part of the instrument (*see* **Note 12**).
   (b) Choose a database and define the taxonomy based on the organism used. When searching for unexpected modifications, it is advisable to use rather small high confidence databases like, e.g., SwissProt (*see* **Note 13**).
   (c) Define a way of false discovery rate (FDR) estimation; we usually use Percolator [24] (which is included in Proteome Discoverer) and set the FDR to 1% on the peptide level.
   (d) Based on the alkylation reagent, define the respective fixed modification at cysteine, as well as the expected variable modifications (e.g., oxidation of methionine) (*see* **Note 14**).

3. Export the search results as *.csv file and import them in MS Excel (*see* **Note 15**). Calculate the percentage of MS/MS spectra assigned to peptide sequences passing the threshold for 1% FDR relative to the total number of acquired MS/MS spectra and compare the different conditions. There are three scenarios:
   (a) The number of acquired MS/MS spectra decreases but the identification rate does not change significantly. This implies that less peptides were generated and interference of the treatment with proteolytic digestion should be considered (*see* **Note 16**).
   (b) The number of acquired MS/MS spectra is constant and the identification rate is reduced. This indicates the presence of unassigned modifications.
   (c) The number of MS/MS spectra is constant and the identification rate does not vary considerably. In such a case the modification may only affect a small fraction of the peptides and high-abundance peptides are reduced in intensity, but their identification efficiency is not affected. In this case either sample complexity has to be reduced or relative quantification of peptide signal intensities has to be performed (*see* Subheading 3.3.3).

In our example study, we analyzed the identification rates of MS/MS spectra for the different alkylation reagents using standard database searching conditions. Figure 3 shows the resulting assigned peptide spectral matches (PSMs) and the relative identification rates. Dependent on the alkylation reagent, the percentage of identified spectra varied strongly. This implied that, dependent

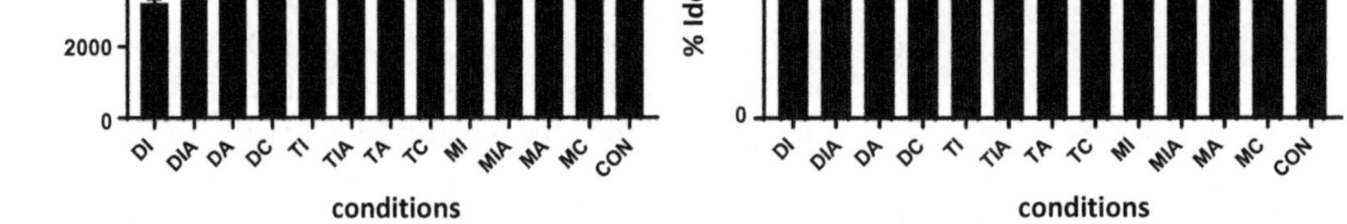

Fig. 3 Results of the analyses of six replicates of in-gel digested cytosolic fractions of HeLa cells reduced and alkylated with different combinations of reagents. (**a**) Number of peptide spectral matches (PSMs); (**b**) percentage of identified spectra (reproduced from [11]). *DI* DTT/IAA, *DIA* DTT/IAC, *DA* DTT/AA, *DC* DTT/CAA, *TI* TCEP/IAA, *TIA* TCEP/IAC, *TA* TCEP/AA, *TC* TCEP/CAA, *MI* BME/IAA, *MIA* BME/IAC, *MA* BME/AA, *MC* BME/CAA, *Con* control sample

on the treatment, different portions of unexpectedly modified peptides were present in the individual samples as these peptides were fragmented but not identified.

3.3.3 Relative Quantification of Signal Reduction

On the PSM level, the influence of a modification can only be identified if the investigated treatment results in a strong reduction of the peptide. If it, however, only results in the modification of a small percentage of the peptides, it may be difficult to realize the effect based on the PSM number; as also ions reduced in abundance are usually sufficient to result in MS/MS spectra allowing for peptide assignment. In such cases, relative quantification can be used in order to identify changes and to estimate the extent of modification. While the easiest approach is a label-free quantification by calculation of the area under the curve, a variety of stable isotope labeling-based methods is available which allows for more accurate quantification. They can be grouped in methods inducing stable isotope labels through metabolic incorporation ($^{15}N/^{14}N$ metabolic labeling and SILAC), chemical derivatization (ICAT, iTRAQ, TMT), or during proteolytic processing (18O labeling) [25]. As the labels are introduced at different steps of the protocol by the varying methods, the method of choice is dependent on the experiment. In our example study, we applied dimethyl labeling as it allowed us to use the same starting material for all conditions and to label the peptides after reduction and alkylation with the different reagents.

1. After sample treatment and proteolytic digestion, desalt the peptides, dry them using a vacuum centrifuge, and resuspend them at 1 μg/μL in 100 mM HEPES (pH 5–8.5) (*see* **Note 17**).
2. For 100 μL of sample volume, add 16 μL of light, medium, or heavy dimethylation solution 1 to the respective samples.
3. Add 16 μL of light, medium, or heavy dimethylation solution 2 to the samples followed by incubation in the fume hood for 1 h at RT while mixing at 800 rpm.
4. Quench the labeling with 64 μL of dimethyl quenching solution/100 μg sample.
5. Add 32 μL of FA for acidification of the sample.
6. Combine the different labeled samples, desalt them using STAGE tips [22], and proceed to MS analysis (*see* **Note 18**).
7. Process the raw files using the relative quantification by dimethyl labeling option in Proteome Discoverer. Exclude peptides containing the presumably modified amino acids (e.g., cysteine in this study) from the list of the results as they exhibit different masses and chromatographic retention times due to the different modified groups. Also exclude samples with missing values in one of the channels (*see* **Note 19**).

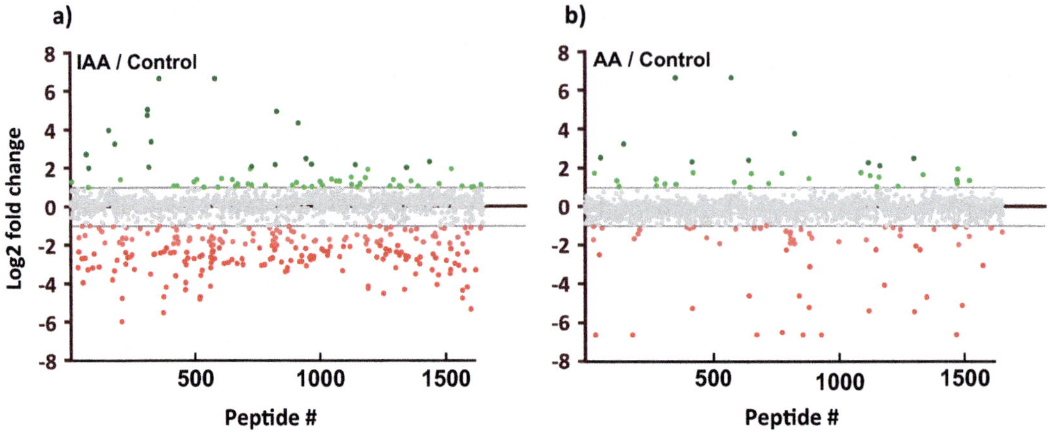

Fig. 4 Quantification of changes in peptide abundance due to differential alkylation. Shown are the log2 values of the average fold-change from two independent experiments. Proteins were reduced and alkylated in the gel, digested by trypsin, the individual peptide samples labeled by the dimethylation reagents (IAA: light, AA: medium, control: heavy) and combined. PSMs with a negative fold change between 1 and 2 are indicated in light red and those with >twofold in dark red. PSMs with a positive fold change between 1 and 2 are colored light green and >twofold dark green (reproduced from [11])

8. Investigate peptide abundance ratios between the different treatments and the control sample in order to identify peptides which decrease significantly in abundance.

In order to determine the significance cutoff, it is often sufficient to plot the normalized treated/control ratios in log2-scale if not enough replicates are available to calculate *p*-values. This allows to visually determine a reasonable cutoff as the majority of peptides should fall in one unregulated group. In our example study, where samples alkylated with IAA were labeled with the light, samples alkylated with AA with the medium, and the untreated control samples with the heavy label, respectively, we applied a >twofold change in abundance cutoff. The results are displayed in Fig. 4 (*see* **Note 20**). In these data, downregulation of the unmodified version of the peptide implies the upregulation of its modified form.

3.3.4 Identification of Affected Amino Acids or Functional Groups (at Peptide Termini or Specific Side Chains)

The MS/MS identification rate data (Subheading 3.3.2) or the relative peptide abundances (Subheading 3.3.3) can now be used to determine if the affected amino acids follow a certain pattern. For this purpose, all peptides which were either absent or significantly downregulated as a consequence of the treatment are analyzed for their amino acid composition. If the unknown modification is specific for a certain amino acid or functional group of the peptide/protein, as it usually is the case, the affected amino acid is expected to be overrepresented in the regulated population of peptides. This analysis can easily be performed in MS Excel:

1. Copy the peptide sequences (excluding any annotations for variable modifications) of all regulated peptides in one Excel column.
2. In the adjacent 20 columns, define the following function: =LEN (PS)-LEN (SUBSTITUTE (PS;"AmAc ";" ")), where PS is the cell containing the peptide sequence and AmAc is the amino acid of interest.
3. Summarize the abundance of each individual amino acid and normalize it to the total number of amino acids found in the whole data set.

In our example study, we performed this analysis for both the initial dataset and the dimethyl-based relative quantification dataset. The results are shown in Fig. 5. The analysis revealed that the only amino acid strongly affected by the alkylation procedure was methionine allowing us to narrow down our search window of the residues affected by the unexpected modification.

3.3.5 Database Search Strategies for Unexpected Modifications

A number of computational strategies have been developed to search in complex samples for unexpected modifications. In Mascot, there are two approaches available: error tolerant [26] and mass tolerant [7] searches. While error tolerant searches have been established more than a decade ago and can be selected in the Mascot search interface (not through Proteome Discoverer), mass tolerant searches have been introduced more recently. In error tolerant searches, Mascot generates a reduced database containing proteins which have been identified in a normal first pass search. This database is then used for the error tolerant search in which all modifications contained in the Unimod database (www.unimod.org) are searched successively. Furthermore, point mutations, changes in reading frames, and relaxed enzyme specificity are accounted for by the algorithm. In mass tolerant searches, no variable modifications are defined and therefore also modifications which are not listed in Unimod can be identified. Instead, a large mass tolerance window (up to ±250 Da) for the precursor ion is defined. This will then result in peptides which are identified as being unmodified along with a mass error which represents the mass of the unexpected modification. For both search strategies, the raw files have to be converted in a file format which is accepted by the Mascot MS/MS Ions Search interface. We usually use the mascot generic format (*.mgf) and generate for each condition one merged *.mgf file using Proteome Discoverer with a top N setting of 6 in 100 Da. If Proteome Discoverer is not available, several other algorithms can be used for this purpose (*see* **Note 21**).

Error Tolerant Search

Error tolerant searching can either be executed in the initial search by selecting the respective option in the MS/MS Ions Search interface or on the results page of any MS/MS ion search. The

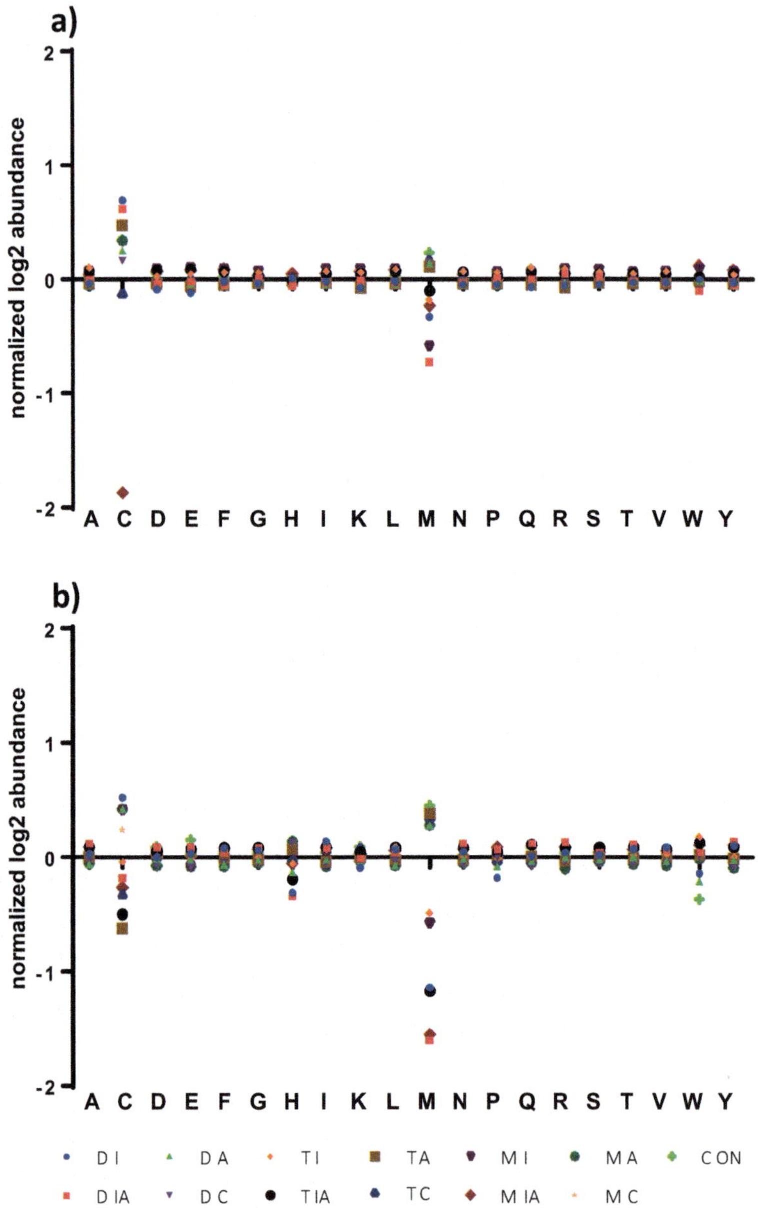

Fig. 5 Normalized amino acid abundance in (**a**) peptides regulated in the dimethylation experiment and (**b**) PSMs regulated in the whole initial dataset. Abundances are calculated based on the over-/underrepresentation of the indicated amino acid in the group of regulated peptides/PSMs with respect to the distribution of all amino acids in the corresponding whole dataset (reproduced from [11]). *DI* DTT/IAA, *DIA* DTT/IAC, *DA* DTT/AA, *DC* DTT/CAA, *TI* TCEP/IAA, *TIA* TCEP/IAC, *TA* TCEP/AA, *TC* TCEP/CAA, *MI* BME/IAA, *MIA* BME/IAC, *MA* BME/AA, *MC* BME/CAA, *Con* control sample

first option results in Mascot performing error tolerant searches for all proteins which have been identified in the first pass search. In our experience, this is only advisable if the samples are not of high complexity, as the search is otherwise likely to occupy the server for an extremely long time. If such searches are to be performed nevertheless, the server time-out (which is 1 day by default) should be adjusted accordingly to allow the search to finish. Therefore, often the second alternative is more practicable. In order to be able to select specific proteins from the results, they have to be displayed in the select summary format (*see* **Note 22**). In our example study, we selected the top 20 proteins for error tolerant searching.

Mass Tolerant Search

Mass tolerant searches have to be performed on the whole dataset. If they result in too much server occupancy, the *.mgf files can be generated from fewer *.raw files, or if the *.raw files are too large, they can be divided in smaller packages. This has to be determined based on the capacity of the server used. The wide-tolerance window for the precursor mass defined in the mass tolerant search (250 Da in our example study) cannot be defined in the Mascot MS/MS Ions Search interface. Therefore, the *.mgf file has to be modified manually.

1. Create the *.mgf file and open it using an appropriate software (e.g., Wordpad or Notepad++).
2. Modify the header of the *.mgf file to include the wide mass tolerance of your choice. In our example study, we changed the header to: peptide mass tolerance: 250 Da; fragment ion tolerance: 0.8 Da; search type: SQ; fixed modifications: carbamidomethyl, carboxymethyl, or propionamide, respectively, at cysteine residues depending on the alkylation reagent.
3. Perform Mascot MS/MS ion searches using the modified *.mgf file. The mass tolerance settings in the *.mgf file overwrite the settings selected in the MS/MS Ions Search interface; therefore the default settings can be kept.

Interpretation of Error Tolerant and Mass Tolerant Search Results

Both the error tolerant and the mass tolerant searches result in high numbers of false positive identifications and are not compatible with decoy database searching. Therefore, great care has to be taken which peptide matches are accepted. In such case, an arbitrary cutoff can be selected; in our experience an ion score of 30 is a reasonable compromise between selectivity and sensitivity. Alternatively, a cutoff to achieve 1% FDR at the peptide level can be determined in a first target/decoy database search. One should be, however, careful as the value determined by this approach will definitely be too low. Therefore, it should only be used as a tendency rather than a strict cutoff.

1. Export the search results with a reasonable ion score cutoff. Open the results in MS Excel or any other software accepting tab-delimited files.
2. Error tolerant searches:
   (a) Count the frequency with which the modifications have been annotated. A straightforward way to do this is using Pivot Tables in Excel. Specific modifications should be more abundant in comparison to the control sample.
   (b) Correlate the presumably specifically modified peptides with the list of affected peptides/amino acids found in Subheading 3.3.4 (if applicable). This can either be achieved by matching the peptide sequences themselves or by filtering for peptides containing the amino acid (s) found to be affected.
3. Mass tolerant searches:
   (a) Transfer the peptide sequences and mass errors in Dalton to an Excel sheet. Use the "frequency" function of Excel to count the number of peptides falling within specific mass ranges (5 Da is a reasonable value); count the number of peptides in these bins.
   (b) Plot the results of the treated and the control sample and compare the distribution of mass errors.
   (c) If a bin with a certain mass error is identified to be overrepresented in the treated sample, extract the peptides in this bin.
   (d) Correlate these peptides with the data obtained in Subheading 3.3.4.

For both the error and mass tolerant searches, the amino acids found to be affected in Subheading 3.3.4 are expected to be present in the identified modified peptides. If this is not the case, alternatively the identification/quantification information for these peptides can be extracted from the previous datasets to assess if they were regulated. While the error tolerant search provides a suggestion for the observed modification, the mass tolerant search only gives a mass value which does not provide any explanation. In this case, it would be ideal if it is possible to suggest modifications based on the treatment investigated. If this is not applicable, a reasonable first step is the determination of the modifications' elemental composition. This can be achieved by calculation of its accurate mass through the mean value of the mass error for all peptides found with the respective modification. This then allows to suggest possible sum formulas using the involved atoms (usually C, H, O, N, and S).

In our example study, error tolerant searches did not result in the identification of any unusual modifications. In mass tolerant

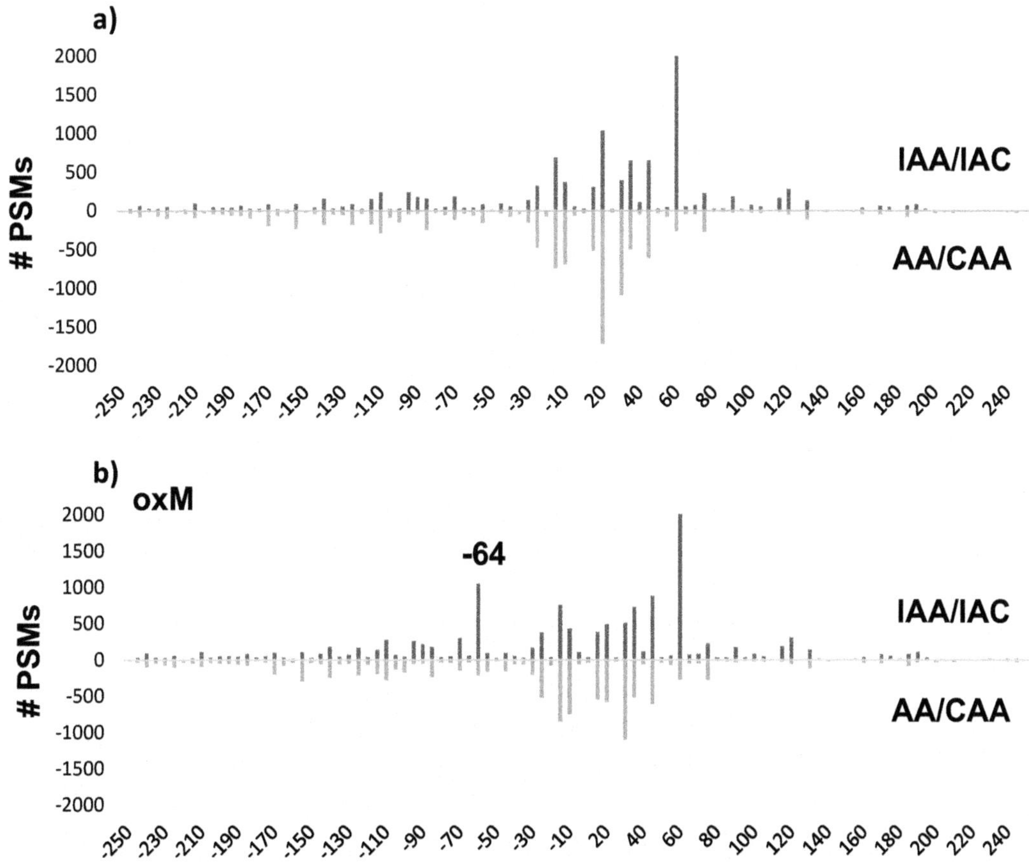

Fig. 6 Combined results of the mass tolerant searches for two independent replicates of all reduction and alkylation reagents. Shown are the summed abundances for peptide mass errors binned in groups of 5 Da for all reduction reagents in combination with the iodine-containing alkylation reagents (IAA/IAC, upper panel), as well as the non-iodine-containing reagents (AA/CAA, lower panel). (**a**) Searches performed without variable modifications did not identify any differences between the treatments except of the already identified offside alkylation (+57 Da); (**b**) searches including oxidation at Methionine (oxM) as variable modification resulted in the assignment of ~1000 PSMs with a mass error of −64 Da. All of these PSMs contained methionine (reproduced from [11])

searches, we initially also did not observe any differences except an increased number of peptides with a mass error of 57 Da for the iodine-containing alkylation reagents, which was due to the off-site alkylation already detected in the previous searches (Fig. 6a). We then repeated the mass tolerant searches with oxidation at methionine as variable modification. This resulted in the identification of ~1000 PSMs annotated with a mass error of −64 Da in the samples treated with iodine-containing alkylation reagents (Fig. 6b). Intriguingly, almost all of these peptides contained methionine, which was identified earlier (*see* Subheading 3.3.4) to be the main affected amino acid. This implied that alkylation of peptides with iodine-containing reagents results in a loss of the side chain of methionine (64 Da equals the molecular weight of the side chain of oxidized methionine).

Theoretically, one would have expected to identify this modification already in the first mass tolerant search, which was performed without any variable modifications. Manual analysis revealed that this was not the case, as the fragment ions including the modified methionine residue were not identified by Mascot—their difference in mass compared to the unmodified fragments interfered with their identification. This resulted in a lower ion score and prevented the respective peptides to pass our threshold for acceptance. It may be therefore beneficial to perform several rounds of mass tolerant searches including known modifications. This exemplifies a general drawback of mass tolerant searches: The mass error observed is solely determined from the precursor ion mass. As only unmodified fragment ions are matched, the number of assigned b/y-ions is reduced, which in turn decreases the peptide score. Additionally, the MS/MS spectrum does not provide any confirmation of the identified mass error. Therefore, results obtained from mass tolerant searches are more vulnerable toward false positive identifications resulting from, e.g., erroneously determined charge states or co-isolation of another precursor ion in MS/MS data acquisition. We therefore advise to only consider modifications identified in mass tolerant searches which have been detected at significant numbers and to also take lower scoring peptides into consideration as the reduced number of matched fragment ions may be due to the modification rather than the quality of the spectrum.

3.3.6 Manual Data Validation for Identified Modifications

The identified modifications have to be validated in order to prevent false conclusions. Due to the lack of an FDR analysis, this is especially of importance in the case of error and mass tolerant searches. Therefore, the initial database search should be repeated with the newly discovered modification(s) as variable modification at the respective amino acid(s). The modification can be defined in the Mascot configuration editor. If the modifications' elemental composition could not be determined, one can alternatively use a sum formula matching the detected mass error and increase the search tolerance window to compensate for possible minor differences in the molecular weight of the modification. The amino acids to be specified as modified are indicated in case of the error tolerant searches; for the mass tolerant searches, however, as the mass errors are only determined from the precursor ion, no suggestions can be provided by Mascot. If no amino acid abundance analysis (*see* Subheading 3.3.4) was performed or if it was inconclusive, another set of parallel searches (*see* Subheading 3.3.1) can be performed in order to narrow down the list of possibly affected amino acids. Subsequently, several spectra with high ion scores should be evaluated manually by comparison of the fragment ion masses calculated by Mascot and the annotated spectrum. Alternatively, manual de novo sequencing [27] can be performed by calculating the theoretical fragment ion masses. If the assignment is not conclusive, a MS/MS spectrum of the matching unmodified peptide fragmented in the control sample can be used for comparison—the fragment ion series not carrying the identified modification should match between both spectra. This allows to determine if the suggested modification is indeed likely to be present at the identified amino acid(s). In our example study, we suspected methionine residues to be alkylated resulting in a labile side chain and a neutral loss from the modified peptides. After defining the loss of the methionine side chain as variable modification in Mascot, we identified several hundred peptides with this modification (data not shown, for more information *see* ref. 11). Manual analysis of a number of peptide hits revealed, that the neutral loss of the alkylated side chain is occurring either in the ion source of the mass spectrometer (Fig. 7), or during MS/MS fragmentation (data not shown, for more information *see* ref. 11).

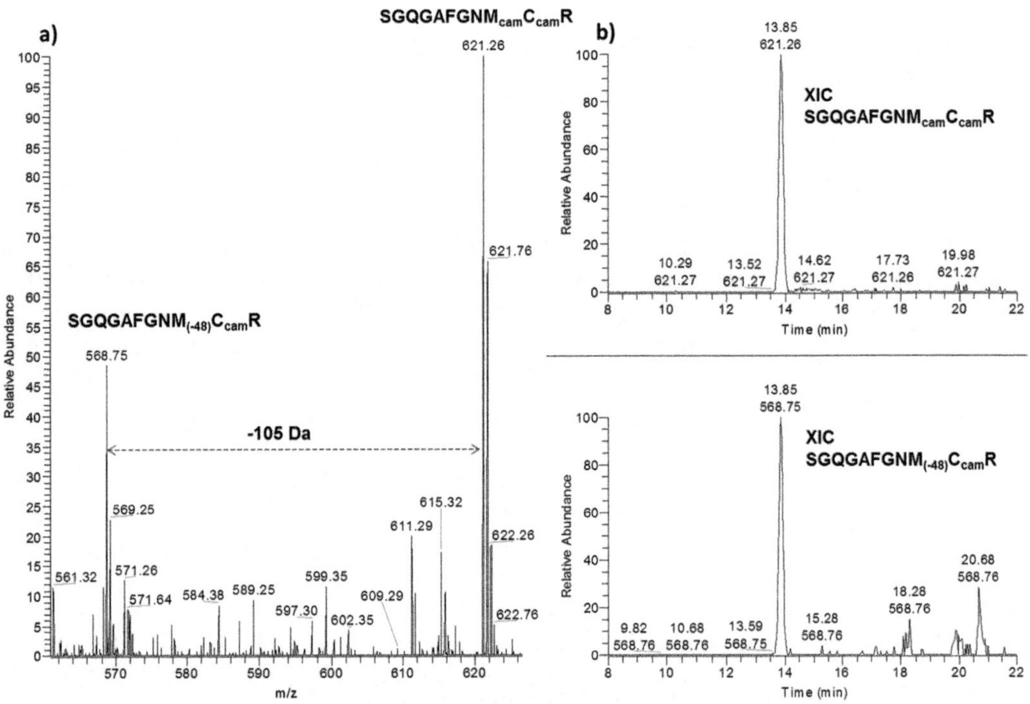

Fig. 7 Carbamylated methionine results in a neutral loss of the amino acid side chain due to in-source fragmentation. (**a**) Survey spectrum showing the neutral loss of the side chain from the peptide with alkylated methionine; (**b**) extracted ion chromatogram (XICs) of the peptide with alkylated methionine, and its version which lost the side chain due to in-source fragmentation, showing perfect co-elution (reproduced from [11])

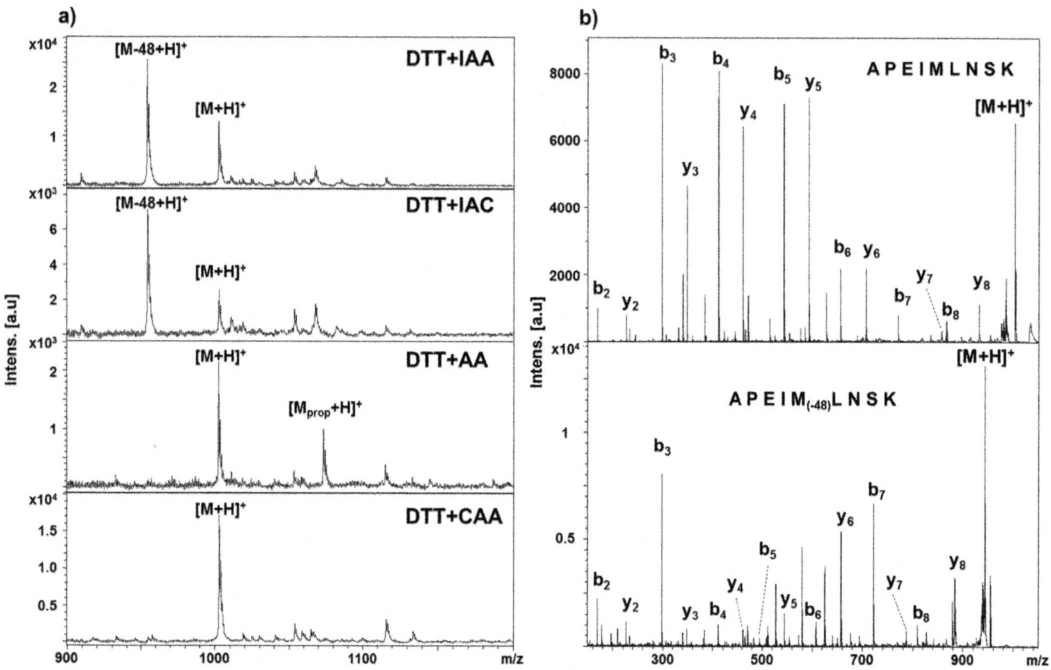

Fig. 8 (**a**) MALDI-MS spectra of the synthetic peptide APEIMLNSK, reduced with DTT and alkylated with IAA, IAC, AA, and CAA, respectively, showing a loss of the methionine side chain (−48 Da) in peptides alkylated using iodine-containing reagents. (**b**) MALDI-MS/MS spectra of the unmodified peptide (upper panel) and the signal at M-48 Da (lower panel). The fragment ion series confirms that the mass difference of 48 Da results from a loss of the methionine side chain (reproduced from [11])

3.3.7 Validation of Identified Modification (s) Using Synthetic Peptides

To provide final proof that the identified spectra indeed originate from the proposed modification(s), experiments with synthetic peptides should be performed. In the case of chemical artifacts, this procedure is straightforward as any synthetic peptide containing the amino acid(s) susceptible to the modification can be used (in our example study methionine). In case of biological modifications, ideally one or several of the peptides found to be modified should be chemically synthesized and should be measured using the same mass spectrometer in order to generate a reference spectrum, to which the initial samples can be compared to. If the identified modification is indeed correct, the fragment ion patterns are expected to match perfectly. In our example study, we used a peptide containing methionine, incubated it with the different alkylation reagents and measured the sample by MALDI-MS/MS (Fig. 8). This showed clearly that only iodine-containing reagents are resulting in an unspecific alkylation of methionine followed by the loss of its side chain (*see* **Note 23**).

3.4 Follow-Up Studies

Once a novel modification is unambiguously identified, it should be documented in online databases like Unimod to make it available to the community. In the case of biological modifications, protocols for its enrichment can be established in order to determine the extent of the modification in biological systems. Ultimately, mutation experiments are needed to prove the biological significance. For unintended chemical modifications, the protocols should be modified in a way to prevent these modifications. Based on the type of modification identified, it is often possible to recognize the reagents responsible for it. If it is not possible to prevent the modification, the strategy for database searching can be adapted including it as a variable modification. This will at least allow identification of a higher percentage of peptides in the samples. In our example study, it was revealed that iodine-containing alkylation reagents resulted in massive offside alkylation significantly reducing numbers of identified peptides. We therefore changed our protocols toward the use of the non-iodine-containing alkylation reagent acrylamide. The identified loss of the alkylated methionine side chain was documented in the Unimod database.

4 Notes

1. If protease inhibitor tablets are used, they can be dissolved in water to generate a 100× stock solution. The stock can be stored for several months at −20 °C. This solution should be added to the lysis buffer directly before performing the experiment.
2. Dimethyl labeling reagents (all cyano- and formaldehyde solutions) are highly toxic upon skin contact. Therefore, they should be handled with extreme care.
3. It is necessary to put plates on ice and to use ice-cold PBS to suppress any biological reactions.
4. To obtain a sufficiently pure cytosolic fraction, the cells have to be lyzed in a gentle way keeping the organelles intact. This allows to remove the organelles by centrifugation and prevents excessive contamination with proteins from them. We used a sucrose-containing buffer in combination with a dounce homogenizer. Alternative gentle lysis methods can be used based on the type of biological sample.
5. Protein precipitation is an optional step. If the lysis buffer is compatible with proteolytic digestion it can be omitted.
6. If the protein pellet is not suspended in SDS, Chloroform methanol precipitation usually leads to better results as the precipitated proteins are more efficiently digested [28].
7. As we performed in-gel digestion, the pellet was reconstituted in SDS buffer. In case of direct in-solution digestion, other buffers such as Rapigest [29] can be used. Urea, which is also

commonly used for in-solution digestion, should be avoided as it may introduce unspecific carbamylation [30].

8. The volume of peptide extraction solutions is dependent on the amount of the gel pieces. The volume should be sufficient to cover the gel pieces.
9. In a top N experiment, if the majority of survey scans are followed by the maximum number of MS/MS scans, the sample contains more peaks than the mass spectrometer can fragment. This can be investigated using software like Raw Meat (Vast Scientific): the part of the gradient in which the majority of peptides elute should be investigated by plotting how many MS/MS are triggered from each survey scan. If the number of MS/MS events is constantly at the maximum value, the sample complexity is too high for the mass spectrometer resulting in under-sampling. In this case, low-abundant peptides are not identified. On the other hand, if the MS/MS scans are below the maximum number per survey scan, the sample is over-sampled which is favorable for the identification of low-abundant modified peptides.
10. In case no comparison to a control sample is considered in your experimental setting, proceed to Subheading 3.3.5.
11. For a MS/MS ion search in Mascot in addition to *.mgf files the following formats are supported: Finnigan (*.ASC), Waters/Micromass (*.PKL), Sequest (*.DTA), PerSeptive (*.PKS), Sciex API III, Bruker (*.XML), mzData (*.XML), mzML (*.mzML).
12. While the right choice of precursor mass tolerance is of high importance, in our experience the MS/MS ion tolerance is less critical.
13. When datasets are searched against large databases, the search will take longer and the ion score cutoff at 1% FDR will be higher. Therefore, if only "normal" proteins from well-studied organisms like human and mouse are identified, a smaller database often results in better performance.
14. It is advisable to select the smallest number of variable modifications possible at this point. Higher numbers of modifications will result in higher score cutoffs due to an increased number of decoy hits. This in turn results in lower numbers of peptide identifications.
15. Proteome Discoverer includes a wide variety of analysis tools which allow the visualization of data in different ways. These tools may be used for some of these analyses as well.
16. In case of reduced enzymatic activity, it is advisable to perform a step to remove the interfering chemicals before proteolytic

digestion. This could be achieved, e.g., through protein precipitation, SDS-PAGE, or molecular weight cutoff spin filters.

17. For resuspension of peptides, avoid using buffers containing primary amines such as Tris or ammonium bicarbonate, since the substances used for dimethyl labeling are amine reactive and such buffers would prevent the efficient labeling of peptides. In case of in-solution digestion setups, the tryptic digestion can be carried out directly in a compatible buffer such as HEPES and the desalting step omitted. Before dimethyl labeling, a protein assay can be performed to make sure that the correct amount of peptide is used for the labeling reaction.
18. If samples with protein amounts exceeding 15–20 μg are generated, high capacity cartridges such as Oasis or Sep-Pak (both from Waters) can be used. In case of increased sample complexity, samples can be further fractionated (using, e.g., SAX STAGE tips) and the resulting fractions desalted by individual STAGE tips.
19. Mascot itself does not allow for quantification based on precursor ion intensities as this information is not contained in the *.mgf files used for database searching. Therefore, quantification has to be performed using Proteome Discoverer or other algorithms (for instance Mascot Distiller). Also other protein identification algorithms frequently include such quantification options.
20. The >twofold change cutoff applied in our example study is a frequently used value in proteomic studies. It is, however, rather arbitrary. Another possibility, which is widely accepted, is to compute the significance cutoff ratio based on a 95% confidence interval in the reference group (control study).
21. The *.mgf files can be created using several tools like the Trans-Proteomic Pipeline (TPP, [31]) or the Proteowizard MSConvert GUI [32].
22. For MS/MS searches with several hundreds to a few thousands of spectra, the "peptide summary" report can easily be accessed on a HTML page. However, in case of large and complex MS/MS searches, it is not practical to simply open the results as the file may become too big. There are a number of switches which can be modified for the individual report to allow opening large MS/MS searches as a "peptide summary" report. More information can be found in the Mascot help page (http://www.matrixscience.com/help/results_help.html).
23. In this study we used a MALDI instrument for the synthetic peptide experiments due to its ease of use. In principle, however, any mass spectrometer which is capable of performing MS/MS experiments can be used.

References

1. Aebersold R, Mann M (2016) Mass-spectrometric exploration of proteome structure and function. Nature 537:347–355. https://doi.org/10.1038/nature19949
2. Kalli A, Smith GT, Sweredoski MJ et al (2013) Evaluation and optimization of mass spectrometric settings during data-dependent acquisition mode: focus on LTQ-orbitrap mass analyzers. J Proteome Res 12:3071–3086. https://doi.org/10.1021/pr3011588
3. Eng JK, McCormack AL, Yates JR (1994) An approach to correlate tandem mass spectral data of peptides with amino acid sequences in a protein database. J Am Soc Mass Spectrom 5:976–989. https://doi.org/10.1016/1044-0305(94)80016-2
4. Griss J, Perez-Riverol Y, Lewis S et al (2016) Recognizing millions of consistently unidentified spectra across hundreds of shotgun proteomics datasets. Nat Methods 13:651–656. https://doi.org/10.1038/nmeth.3902
5. Nielsen ML, Savitski MM, Ra Z (2006) Extent of modifications in human proteome samples and their effect on dynamic range of analysis in shotgun proteomics. Mol Cell Proteomics 5:2384–2391. https://doi.org/10.1074/mcp.M600248-MCP200
6. Nesvizhskii AI, Roos FF, Grossmann J et al (2006) Dynamic spectrum quality assessment and iterative computational analysis of shotgun proteomic data: toward more efficient identification of post-translational modifications, sequence polymorphisms, and novel peptides. Mol Cell Proteomics 5:652–670. https://doi.org/10.1074/mcp.M500319-MCP200
7. Chick JM, Kolippakkam D, Nusinow DP et al (2015) A mass-tolerant database search identifies a large proportion of unassigned spectra in shotgun proteomics as modified peptides. Nat Biotechnol 33:743–749. https://doi.org/10.1038/nbt.3267
8. Tanner S, Shu H, Frank A et al (2005) InsPecT: identification of posttranslationally modified peptides from tandem mass spectra. Anal Chem 77:4626–4639. https://doi.org/10.1021/ac050102d
9. Jensen ON (2004) Modification-specific proteomics: characterization of post-translational modifications by mass spectrometry. Curr Opin Chem Biol 8:33–41. https://doi.org/10.1016/j.cbpa.2003.12.009
10. Zhao Y, Jensen ON (2009) Modification-specific proteomics: strategies for characterization of post-translational modifications using enrichment techniques. Proteomics 9:4632–4641. https://doi.org/10.1002/pmic.200900398
11. Müller T, Winter D (2017) Systematic evaluation of protein reduction and alkylation reveals massive unspecific side effects by iodine-containing reagents. Mol Cell Proteomics 16:1173–1187. https://doi.org/10.1074/mcp.M116.064048
12. Laemmli UK (1970) Cleavage of structural proteins during the assembly of the head of bacteriophage T4. Nature 227:680–685. https://doi.org/10.1038/227680a0
13. Boersema PJ, Raijmakers R, Lemeer S et al (2009) Multiplex peptide stable isotope dimethyl labeling for quantitative proteomics. Nat Protoc 4:484–494. https://doi.org/10.1038/nprot.2009.21
14. Shevchenko A, Wilm M, Vorm O et al (1996) Mass spectrometric sequencing of proteins from silver-stained polyacrylamide gels. Anal Chem 68:850–858. https://doi.org/10.1021/ac950914h
15. Chin Y, Aiken GR, O'Loughlin E (1994) Molecular weight, polydispersity, and spectroscopic properties of aquatic humic substances. Environ Sci 28:1853–1858. https://doi.org/10.1021/es00060a015
16. Williams A, Frasca V (2001) Ion-exchange chromatography. Curr Protoc Protein Sci 15:8.2.1–8.2.30
17. Chen J, Lee CS, Shen Y et al (2002) Integration of capillary isoelectric focusing with capillary reversed-phase liquid chromatography for two-dimensional proteomics separation. Electrophoresis 23:3143–3148. https://doi.org/10.1002/1522-2683(200209)23:18<3143::AID-ELPS3143>3.0.CO;2-7
18. Nühse TS, Stensballe A, Jensen ON et al (2003) Large-scale analysis of in vivo phosphorylated membrane proteins by immobilized metal ion affinity chromatography and mass spectrometry. Mol Cell Proteomics 2:1234–1243. https://doi.org/10.1074/mcp.T300006-MCP200
19. Beausoleil SA, Jedrychowski M, Schwartz D et al (2004) Large-scale characterization of HeLa cell nuclear phosphoproteins. Proc Natl Acad Sci 101:12130–12135. https://doi.org/10.1073/pnas.0404720101
20. Michel PE, Reymond F, Arnaud IL et al (2003) Protein fractionation in a multicompartment device using Off-Gel™ isoelectric focusing. Electrophoresis 24:3–11. https://doi.org/10.1002/elps.200390030

21. Huber LA, Pfaller K, Vietor I (2003) Organelle proteomics: implications for subcellular fractionation in proteomics. Circ Res 92:962–968. https://doi.org/10.1161/01.RES.0000071748.48338.25

22. Rappsilber J, Ishihama Y, Mann M (2003) Stop and go extraction tips for matrix-assisted laser desorption/ionization, nanoelectrospray, and LC/MS sample pretreatment in proteomics. Anal Chem 75:663–670. https://doi.org/10.1021/ac026117i

23. Verheggen K, Raeder H, Berven FS et al (2017) Anatomy and evolution of database search engines-a central component of mass spectrometry based proteomic workflows. Mass Spectrom Rev. https://doi.org/10.1002/mas.21543

24. Brosch M, Yu L, Hubbard T et al (2009) Accurate and sensitive peptide identification with mascot percolator. J Proteome Res 8:3176–3181. https://doi.org/10.1021/pr800982s

25. Bantscheff M, Schirle M, Sweetman G et al (2007) Quantitative mass spectrometry in proteomics: a critical review. Anal Bioanal Chem 389:1017–1031. https://doi.org/10.1007/s00216-007-1486-6

26. Creasy DM, Cottrell JS (2002) Error tolerant searching of uninterpreted tandem mass spectrometry data. Proteomics 2:1426–1434. https://doi.org/10.1002/1615-9861(200210)2:10<1426::AID-PROT1426>3.0.CO;2-5

27. Seidler J, Zinn N, Boehm ME et al (2010) De novo sequencing of peptides by MS/MS. Proteomics 10:634–649. https://doi.org/10.1002/pmic.200900459

28. Winter D, Steen H (2011) Optimization of cell lysis and protein digestion protocols for the analysis of HeLa S3 cells by LC-MS/MS. Proteomics 11:4726–4730. https://doi.org/10.1002/pmic.201100162

29. Yu YQ, Gilar M, Lee PJ et al (2003) Enzyme-friendly, mass spectrometry compatible surfactant for in-solution enzymatic digestion of proteins. Anal Chem 75:6023–6028. https://doi.org/10.1021/ac0346196

30. Kollipara L, Zahedi RP (2013) Protein carbamylation: in vivo modification or in vitro artefact? Proteomics 13:941–944. https://doi.org/10.1002/pmic.201200452

31. Deutsch EW, Mendoza L, Shteynberg D et al (2015) Trans-Proteomic Pipeline, a standardized data processing pipeline for large-scale reproducible proteomics informatics. Proteomics Clin Appl 9:745–754. https://doi.org/10.1002/prca.201400164

32. Holman JD, Tabb DL, Mallick P (2014) Employing ProteoWizard to convert raw mass spectrometry data. Curr Protoc Bioinformatics 46:13.24.1–13.24.9. https://doi.org/10.1002/0471250953.bi1324s46

Chapter 16

Label-Free LC-MS/MS Strategy for Comprehensive Proteomic Profiling of Human Islets Collected Using Laser Capture Microdissection from Frozen Pancreata

Lina Zhang, Giacomo Lanzoni, Matteo Battarra, Luca Inverardi, and Qibin Zhang

Abstract

Diabetes mellitus is caused by either loss of pancreatic islets β-cells (Type 1 Diabetes, T1D), insufficient insulin release in the islet β-cells coupled with insulin resistance in target tissues (Type 2 Diabetes, T2D), or impaired insulin release (genetic forms of diabetes and, possibly, T1D subtypes). The investigation of the islet proteome could elucidate facets of the pathogenesis of diabetes. Enzymatically isolated and cultured (EIC) islets are frequently used to investigate biochemical signaling pathways that could trigger β-cell changes and death in diabetes. However, they cannot fully reflect the natural protein composition and disease process of in vivo islets due to the stress from isolation procedures and in vitro culture. The laser capture microdissection method employs a high-energy laser source to separate the desired cells from the remaining tissue section in an environment which is well conserved and close to the natural condition. Here, we describe a label-free proteomic workflow of laser capture microdissected (LCM) human islets from fresh-frozen pancreas sections of cadaveric donors to obtain an accurate and unbiased profile of the pancreatic islet proteome. The workflow includes preparation of frozen tissue section, staining and dehydration, LCM islets collection, islet protein digestion, label-free Liquid Chromatography-Tandem Mass Spectrometry (LC-MS/MS), database search, and statistical analysis.

Key words LCM human pancreatic islets, Label-free proteomics, LC-MS/MS, MaxQuant, Perseus

1 Introduction

Diabetes mellitus is caused by either loss of pancreatic islets β-cells (Type 1 Diabetes, T1D), insufficient insulin release in the islet β-cells coupled with insulin resistance in target tissues (Type 2 Diabetes, T2D), or impaired insulin release (genetic forms of diabetes and, possibly, T1D subtypes). The investigation of the islet proteome could elucidate facets of the pathogenesis of diabetes. Enzymatically isolated and cultured (EIC) islets [1–3] have been frequently used to investigate biochemical signaling pathways that could trigger β-cell changes and death. However, such in vitro

Xing Wang and Matthew Kuruc (eds.), *Functional Proteomics: Methods and Protocols*, Methods in Molecular Biology, vol. 1871, https://doi.org/10.1007/978-1-4939-8814-3_16,

models have some limitations: they do not fully reflect what happens in vivo due to a lack of the natural environment where islets exist and due to the changes in cell physiology induced by isolation and culture. The procedure of enzymatic isolation of pancreatic islets causes major structural changes and induces upregulation of stress-related genes in islets [4]. Furthermore, EIC islets frequently contain a significant percentage of contaminating acinar cells and duct cells [5]. Alternatively, human pancreatic tissue can be collected from cadaveric individuals and preserved frozen for further laser-capture microdissected (LCM) isolation. LCM employs a high-energy laser source to separate the desired cells from the remaining tissue section [6], a strategy that can minimize the contamination of surrounding tissue. LCM isolation also enables the extraction of samples from an environment which is well conserved and close to the natural condition, to better investigate cell physiology [7], cell biology [8], cell transcriptome [4], and proteome [9]. The exploration of the proteome signature of LCM islets with an unbiased method may provide information on the changes of protein composition occurring in dysfunctional islets, even with limited sample amounts, which may facilitate understanding of the pathogenesis of diabetes.

Here we describe a workflow for label-free proteomic analysis of LCM islets obtained from sections of fresh-frozen human pancreas. This method enables accurate and unbiased profiling of the pancreatic islet proteome. The strategy avoids enzymatic treatment for cell dissociation and in vitro culture, and is designed to maintain protein composition close to that of the original tissue. The method can be easily adapted to other tissues, organs, and species. The workflow covers preparation of frozen tissue sections, immunohistochemical staining of reference sections, staining and dehydration for LCM, LCM of pancreatic islets and acinar tissue, preparation of samples for proteomic analysis, label-free Liquid Chromatography-Tandem Mass Spectrometry (LC-MS/MS) data acquisition, database search for protein identification, quantification, and statistical analysis for the determination of the proteins differentially expressed between LCM islets and LCM acinar tissue.

2 Materials

Common solvents and reagents [acetic acid, dithiothreitol (DTT), iodoacetamide (IAA), Ammonium Bicarbonate (NH_4HCO_3), Hydrochloric acid (HCl), Formic Acid (FA), Acetonitrile (CH_3CN)] were purchased from Sigma-Aldrich (St. Louis, MO). Tissue-Tek O.C.T. compound and 100% ethanol were purchased from VWR. Toluidine blue O and Anti-insulin antibody clone K36aC10 were purchased from Sigma-Aldrich. Leica polyethylene naphthalate (PEN) Membrane slides were purchased from Leica.

PBS was purchased from Gibco Life Technologies. Kimwipes and Drierite were purchased from VWR. Peroxo-Block™ was purchased from Thermofisher Scientific. Histostain Plus Broadspectrum AEC kit was purchased from Invitrogen. Elite Mini PAP Pen was purchased from Diagnostic BioSystems (Pleasanton, CA). PPS Silent Surfactant was purchased from Expedeon (San Diego, CA). The BCA protein assay was obtained from ThermoFisher Scientific (Rockford, IL), the sequencing-grade Trypsin was purchased from Promega (Madison WI). All solvents used are HPLC-grade.

Instrumentation: cryotome (Leica Cryotome CM3050 S, Leica), histology slide scanner (PathScan Enabler IV, with workstation and PathScan Enabler software, Meyer Instruments), laser microdissection system (Leica Microscope LS LMD, with workstation and Leica LMD software, Leica), liquid chromatography and mass spectrometry system (UltiMate 3000 RSLCnano system and a Q Exactive HF mass spectrometer coupled with an EASY-Spray ion source, ThermoFisher Scientific). Details of reagents and materials used in each step are listed below.

2.1 For Frozen Tissue Sections

1. Tissue-Tek Cryomold Standard.
2. Tissue-Tek O.C.T. compound.
3. 100% ethanol.
4. Leica PEN Membrane slides.

2.2 Immunohistochemical Staining of Reference Sections

1. Elite Mini PAP Pen.
2. 10 mM phosphate-buffered saline (PBS) pH 7.4.
3. Anti-insulin antibody clone K36aC10 1:1000 dilution (*see* **Note 1**).
4. Ready-to-use biotinylated secondary antibody (from Histostain Plus Broadspectrum AEC kit).
5. HRP Substrate/Chromogen reagents: AEC Single Solution.

2.3 For LCM Staining and Dehydration

1. 100% ethanol.
2. 70% ethanol: combine 70 mL ethanol and 30 mL H_2O.
3. 90% ethanol: combine 90 mL ethanol and 10 mL H_2O.
4. 0.5% w/v Toluidine blue O staining solution was prepared in 70% ethanol (*see* **Note 2**).
5. Drierite desiccant.

2.4 For LCM of Pancreatic Islets and Acinar Tissue

1. 50 mM NH_4HCO_3 pH 8: add 0.40 g of NH_4HCO_3 to 100 mL of H_2O.

2.5 For Protein Digestion

1. Prepare 1% PPS by adding 100 μL 50 mM NH_4HCO_3 to 1mg PPS bottle (*see* **Note 3**).
2. 50 mM dithiothreitol (DTT): weigh 0.77 mg of DTT in a microcentrifuge tube, and add 100 μL DI water (*see* **Note 4**).
3. 50 mM iodoacetamide (IAA): weigh 0.925 mg of IAA in a microcentrifuge tube, and add 100 μL DI water (*see* **Note 5**).
4. Trypsin stock solution: prepare 1 μg/μL in 50 mM acetic acid, and store at −20 °C before use.
5. 2 M HCl: 16.52 mL 37% HCl and add H_2O to 100 mL.

2.6 For LC-MS/MS

Buffer A: 0.1% FA; Buffer B: 0.1% FA in CH_3CN.

3 Methods

The major procedures involved in analysis of LCM human islets proteome is shown in Fig. 1.

The human pancreas tissue from three donors was used in this study. Three technical replicates of LCM islets were collected from each donor and six islet equivalent human pancreatic islets were collected from each replication. Meanwhile, the same equivalent acinar tissue was collected from surrounding of the same islets. Two technique replicates were collected for each donor. LCM acinar tissue was used to confirm no contamination in LCM islets.

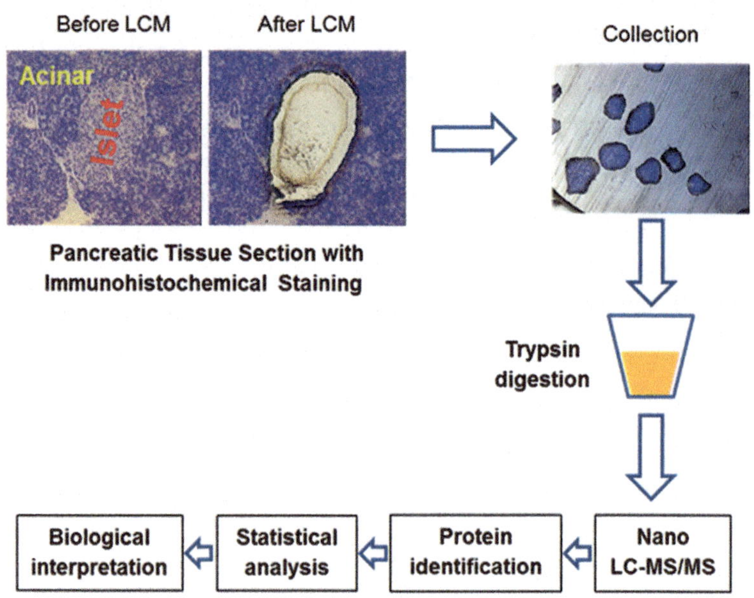

Fig. 1 Schematic representation of the experimental work flow

3.1 Preparation of Frozen Tissue Sections

1. Exercise care while processing pancreatic tissue: avoid squeezing or stretching the tissue, use a scalpel to obtain blunt cuts.
2. Resect pancreatic tissue blocks from the neck region of cadaveric pancreata from organ donors.
3. Collect tissue fragments of approximately 1 cm × 0.5 cm and position fragments in the center of cryomolds.
4. Embed tissue in Tissue-Tek O.C.T. compound and immediately freeze at −80 °C by placing the mold holding the tissue on top of dry ice (*see* **Note 6**).
5. Cut the blocks into sections of 10 μm thickness with a cryotome, with temperature set at −20 °C (*see* **Note** 7).
5. Transfer three pancreas sections onto each of 10 Leica PEN Membrane slides.
6. Prepare extra reference sections on a regular glass slide for insulin immunohistochemistry and mapping (*see* **Note 8**).

3.2 Immuno-histochemical Staining of Reference Sections

1. Fix the reference sections in 10% formalin for 15 min (*see* **Note 9**).
2. Wash four times in PBS.
3. Leave a drop of PBS on each tissue section and draw a circle with a PAP pen to surround each section.
4. Remove the PBS.
5. Add Peroxo-Block™ for 45 s. Wash immediately.
6. Add 100 μL of anti-insulin antibody clone K36aC10 solution (dilution 1:300) to each section to completely cover tissue.
7. Incubate in a humidified chamber at room temperature for 60 min.
8. Rinse with PBS for 5 min, three times.
9. Add 100 μL of secondary antibody to each section to completely cover tissue and incubate for 10 min.
10. Rinse with PBS for 5 min, three times.
11. Add enough enzyme conjugate solution (from the Histostain Plus Broadspectrum AEC kit) to each section to completely cover tissue and incubate for 10 min.
12. Rinse with PBS for 2 min, three times.
13. Add chromogen AEC Single Solution and incubate 5–10 min.
14. Scan reference sections with a PathScan Enabler IV instrument to obtain maps of the entire sections and identify stained insulin-containing islets.

3.3 Staining and Dehydration for LCM

1. The staining and dehydration protocol is performed with 8 clean Coplin jars, prefilled with 50 mL of 70% (jar # 1–5), 90% (jar #6), and 100% (jar #7 and 8) ethanol.

2. Jars # 1–5 are maintained chilled on ice during the staining, jars # 6–8 are maintained at room temperature to avoid condensation after dehydration.
3. The PEN membrane slides with tissue sections are dipped for 30 s in each jar from #1 to #3 of the ethanol series.
4. After jar #3, each slide is drained by gently placing the side edge of the glass on a Kimwipe, then placed horizontally, with the tissue sections on top.
5. The sections are stained for 90 s by adding 200 μL of Toluidine blue O staining solution, then drained and transferred to jar # 4, to continue dehydration (*see* **Note 10**).
6. The slides are dipped for 30 s in each jar following the numerical order (jar # 4–8), to obtain dehydration.
7. The stained and dehydrated slide is drained with a Kimwipe and placed under a laminar flow hood for 4 min to enable ethanol evaporation.
8. The slide is placed in a slide box containing the desiccant Drierite wrapped in Kimwipes and closed with tape (*see* **Note 11**).

3.4 LCM of Pancreatic Islets and Acinar Tissue (see Note 12)

1. The stage of the Leica Microscope LS LMD system is positioned in a clear acrylic box (microdissection chamber) where the atmosphere can be controlled.
2. 1 h before microdissection, the workplace is cleaned and the microdissection chamber is dehydrated using 2 kg of fresh Drierite to minimize the humidity and enable membrane microdissection.
3. The Leica LMD software is used to set up the laser (*see* Table 1), to initialize the instrument and to control the movements of the laser on the tissue section.
4. The PEN membrane slide with the stained tissue is positioned on the stage, with the tissue facing down.
5. Empty sterile collection tubes are placed under the cutting area (RNAse-free Eppendorf tubes, 500 μL volume, flat cap).
6. The scans of the reference sections are used to map the insulin-containing islets.
7. Pancreatic islets are identified in the toluidine-stained tissue by visualizing in bright field and in phase-contrast with ×10 magnification. In bright field, pancreatic islets appear as clusters of cells with lightly colored cytoplasm, whereas the surrounding acinar tissue is composed by cells with darker cytoplasm (*see* Fig. 1). In phase-contrast, islet cells appear finely granulated. Visualize islet borders at ×20 magnification (*see* **Note 13**).

Table 1
System configuration for laser capture microdissection with the Leica LMD instrument

Parameters	×10 magnification	×20 magnification
Aperture	10	13
Intensity	40	35
Speed	4	7
Offset	26	34
Ap Diff	8	8
Option	Med	Med

8. Pancreatic islets and pancreatic exocrine tissue are collected in separate tubes. The area of each microdissection is annotated.
9. The volume of microdissected tissue is calculated by multiplying the total area collected by 10 μm (thickness of the section).
10. Microdissected islets are collected into the cap of 500 μL sterile tubes (*see* **Note 14,** Fig. 1).
11. Acinar tissue is microdissected from neighboring areas and collected into separate tubes.
12. Additional tissue is microdissected until the total volume for each sample corresponds to 1.06×10^7 μm^3, six Islet Equivalents (*see* next paragraph).
13. Microdissection session should not last more than 60 min (stain additional tissue sections every 60 min).
14. Carefully remove the collection tubes from the microdissection chamber.
15. Resuspend the microdissected tissue with 50 μL of 50 mM NH_4HCO_3.
16. Close the tube, centrifuge the resuspended tissue for 2 min at 13,000 rpm (*see* **Note 15**).
17. Freeze by placing on dry ice, maintain frozen at −80 °C.

3.5 Conversion of LCM Areas to Volumes and to IEQ

The total volume of isolated islets can be expressed as number of islet equivalents (IEQ) [10]. An IEQ corresponds to the volume of a "standard" islet, a sphere with a diameter $d = 150$ μm and a volume of $V_{IEQ} = 1.77 \times 10^6$ μm^3. 1 IEQ contains approximately 1560 islet cells [11]. The area of laser-captured tissue is recorded, and the volume is calculated by multiplying the total area collected for the thickness of the tissue section (10 μm). The target total volume of each microdissected sample is 1.06×10^7 μm^3, corresponding to 6 IEQ ($6 \times 1.77 \times 10^6\ \mu m^3$). At any time during the collection, the total volume of laser-captured tissue can be divided by the standard volume of 1 IEQ ($V_{IEQ} = 1.77 \times 10^6\ \mu m^3$) to obtain the corresponding number of laser-captured IEQ.

3.6 Protein Digestion

1. Add 6 μL of 1% pps silent surfactant (PPS) to extract and solubilize hydrophobic proteins.
2. Add 1.5 μL of 50 mM DTT and incubate at 95 °C for 6 min.
3. Sonicate sample for 3 min.
4. Alkylate with 7.5 μL 50 mM iodoacetamide for 25 min at 45 °C in the dark.
5. Add 1 μg stock trypsin at 37 °C overnight.
6. Hydrolyze PPS by adding 12 μL 2 M HCl at room temperature for 2 h.
7. Centrifuge samples at 16,000*g* for 12 min and separate the supernatant for LC-MS/MS analysis (*see* **Note 16**).

3.7 LC-MS/MS Analysis

Protocols for LC-MS/MS analysis can vary because of diversity of LC systems (manufacturer, column, solvent composition, gradient, flow rate, etc.) and MS instruments (manufacturer, electrospray condition, fragmentation, MS parameters, analyzer, etc.). The following is the practice routinely used in our laboratory.

1. The LC-MS/MS platform consists of an UltiMate 3000 RSLCnano system and a Q Exactive HF mass spectrometer coupled with an EASY-Spray ion source (ThermoFisher Scientific).
2. Peptide separation is performed on a PepMap C18 analytical column (2 μm particle, 50 cm × 75 μm, ThermoFisher Scientific). Injection volume is 2.5 μL (0.5 μg peptide amount loaded into column) per sample (*see* **Note 17**).
3. A binary solvent system consisting of 0.1% FA in water (solvent A) and 0.1% FA in CH_3CN (solvent B) is used at a flow rate of 250 nL min^{-1} (*see* **Note 18**).
4. LC separation is performed using the following gradient setting: hold at 4% B for 3 min (for desalting), from 4 to 8% B in 0.1 min, 8 to 40% B in 90 min (effective gradient), 40 to 90% B in 0.1% min, hold at 90% B for 10 min (for washing column), 90% to 4% B in 0.1 min, and hold at 4% B for 17 min for re-equilibrating column (*see* **Note 19**).
5. MS data are acquired in profile mode and resolution for full scan (400–2000 *m/z*) is set to 120,000 (at *m/z* 200) with maximum ion injection time of 50 ms, and automatic gain control (AGC) target of $1e^6$.

6. MS/MS data are acquired with data-dependent method of top 15. An isolation window of 1.4 *m/z* is used to isolate precursor ions for fragmentation by higher-energy collisional dissociation (HCD) at normalized collision energy of 28. Resolution for MS/MS spectrum is set to 15,000 (at *m/z* 200) with maximum ion injection time of 100 ms. AGC target for MS/MS scans is $1e^5$.
7. Precursor ions with single, seven, and higher charge states are excluded from fragmentation, and dynamic exclusion time is set to 20 s.

3.8 Database Search for Protein Identification and Quantification

Many database search software packages are available for this purpose. MaxQuant is demonstrated here [12].

1. The acquired datasets (.raw files) are analyzed using MaxQuant and the built-in Andromeda search engine against a UniProt human database (*see* **Note 20**).
2. Variable modifications include protein N-terminal acetylation and methionine oxidation.
3. Fixed modifications contain cysteine carbamidomethylation.
4. A maximum of two missed cleavages are allowed for the search.
5. Trypsin/P is selected as the specific proteolytic enzyme (*see* **Note 21**).
6. For label-free quantification, "match between runs" is selected (*see* **Note 22**).
7. The false discovery rate (FDR) cutoff used for both peptides and proteins is 0.01 (1%) using decoy database.
8. Only the razor/unique peptides are used for quantitative calculations.
9. The other parameters are the default settings in MaxQuant software for processing orbitrap-type data.

3.9 Statistical Analysis

The search results in ProteinGroups.txt generated by MaxQuant are directly processed by Perseus software [13]. The differentially expressed proteins are identified by statistical analysis tools built in Perseus.

1. Import the quantitative data from ProteinGroups.txt into Perseus.
2. The potential contaminants, reverse hits and proteins only identified by modification site are excluded.
3. Filter out the protein with unique peptides less than 1.
4. The protein intensities are $\log_2$-transformed.
5. Categorize the samples into two groups: LCM islets and LCM acinar tissue.

6. Filter out the proteins not quantified in all the samples
7. Two-samples tests coupled with Benjamini-Hochberg (FDR cut off of 0.05) correction are performed to identify the differentially expressed proteins [14].

3.10 Additional Resources for Data Analysis and Biological Interpretation

- ***http://string-db.org***: a database of known and predicted protein interactions. The interactions include direct (physical) and indirect (functional) associations; they are derived from four sources: Genomic context, High-throughput experiments, Coexpression, Previous knowledge. This tool can be used to interpolate proteins in functional and interaction networks. The participation of proteins in networks was established by references in literature.
- ***http://www.proteinatlas.org***: a database of Antibody-based Proteomics. This tool enables the analysis of gene and protein expression in various human tissues. The database presents data related to the binding specificity of commercially available antibodies.
- ***http://compartments.jensenlab.org***: a subcellular localization database. The database integrates evidence on protein subcellular localization from manually curated literature, high-throughput analyses, automatic text mining, and sequence-based prediction methods.

4 Notes

1. The dilution of the anti-insulin antibody should be prepared freshly.
2. The 0.5% w/v Toluidine blue O staining solution in 70% ethanol should be prepared freshly.
3. PPS solution should be prepared freshly. Once the package is opened to air, the contents should be immediately reconstituted in aqueous buffer (pH 7–8), protected from elevated temperatures, and used within 12 h.
4. Stock DTT solution should be freshly prepared before use.
5. Keep the IAA solution in the dark.
6. Wipe the inner chamber and the stage of the cryotome with 100% ethanol.
7. Change the blade, wipe the inner chamber and the stage of the cryotome with 100% ethanol after each sample in order to avoid contamination.
8. Sections are maintained frozen and stored at −80 °C.

9. Immunohistochemical staining of reference slides with insulin antibodies enables the identification and mapping of islets with β-cells.
10. Toluidine blue O staining of frozen pancreas sections enables good discrimination of islets and acinar tissue: islets appear lightly colored compared to the surrounding acinar tissue; moreover, endocrine cells have a characteristic granulated or "rugged" aspect in phase-contrast illumination. If the humidity in the microdissection chamber is too high (as indicated by a pink drierite), the tissue section may rehydrate in one hour or less: this determines visible tissue degradation and hampers further laser-capture microdissection.
11. Insulin-containing islets were mapped via conventional immunohistochemical staining of reference sections, and Toluidine Blue O staining was used to guide the laser-capture microdissection in ethanol-dehydrated sections.
12. Each LCM session lasted a total of 60 min, to avoid tissue rehydration and degradation.
13. The setup for Laser capture microdissection is optimized by the operator and adjusted to the nature of the samples.
14. The collection tubes should be sterile and RNase/DNase/Protease free.
15. Centrifugation of the tissue at this stage enabled us to avoid loss of tissue.
16. A rough estimation of solvent volume used for reconstitution of peptides in each fraction can be determined by the amount of peptides loaded onto column for fractionation and the number of final fractions. For instance, 100 μg divided by 24 fractions yields 4.2 μg per fraction, and the preparation of samples at 0.2 μg/μL requires the addition of 21.0 μL of solvent for reconstitution.
17. The injection volume depends on the sample loop of autosampler, column loading capacity, and MS detector, therefore injection volume need to be adjusted based on the actual setup.
18. A flow rate of 250 nL/min for C18 50 cm × 75 μm i.d. column results in around 550–600 bar column pressure when heating column at 35 °C.
19. The gradient used for peptide separation can be modified depending on separation performance. However, all samples must be run under the same condition to limit variations between samples.
20. The database information needs to include the type, sequence entry number, and releasing date of database.

21. The selection of enzyme used for search is based on the enzyme that is chosen for protein digestion in Subheading 3.6.
22. "Match between runs" should be selected because it can improve the search results for less missing values.

Acknowledgements

This work was supported by the National Institutes of Health (R01 DK114345) and by the Diabetes Research Institute Foundation.

References

1. Schrimpe-Rutledge AC, Fontès G, Gritsenko MA, Norbeck AD, Anderson DJ, Waters M, Adkins JN, Smith RD, Poitout V, Metz TO (2012) Discovery of novel glucose-regulated proteins in isolated human pancreatic islets using LC–MS/MS-based proteomics. J Proteome Res 11(7):3520–3532
2. Waanders LF, Chwalek K, Monetti M, Kumar C, Lammert E, Mann M (2009) Quantitative proteomic analysis of single pancreatic islets. Proc Natl Acad Sci U S A 106 (45):18902–18907
3. Eizirik DL, Sammeth M, Bouckenooghe T, Bottu G, Sisino G, Igoillo-Esteve M, Ortis F, Santin I, Colli ML, Barthson J, Bouwens L, Hughes L, Gregory L, Lunter G, Marselli L, Marchetti P, McCarthy MI, Cnop M (2012) The human pancreatic islet transcriptome: expression of candidate genes for type 1 diabetes and the impact of pro-inflammatory cytokines. PLoS Genet 8(3):e1002552
4. Marselli L, Thorne J, Ahn YB, Omer A, Sgroi DC, Libermann T, Otu HH, Sharma A, Bonner-Weir S, Weir GC (2008) Gene expression of purified beta-cell tissue obtained from human pancreas with laser capture microdissection. J Clin Endocrinol Metab 93 (3):1046–1053
5. Marselli L, Thorne J, Dahiya S, Sgroi DC, Sharma A, Bonner-Weir S, Marchetti P, Weir GC (2010) Gene expression profiles of Beta-cell enriched tissue obtained by laser capture microdissection from subjects with type 2 diabetes. PLoS One 5(7):e11499
6. Bonner RF, Emmert-Buck M, Cole K, Pohida T, Chuaqui R, Goldstein S, Liotta LA (1997) Laser capture microdissection: molecular analysis of tissue. Science 278(5342):1481, 1483.
7. Sturm D, Marselli L, Ehehalt F, Richter D, Distler M, Kersting S, Grutzmann R, Bokvist K, Froguel P, Liechti R, Jorns A, Meda P, Baretton GB, Saeger HD, Schulte AM, Marchetti P, Solimena M (2013) Improved protocol for laser microdissection of human pancreatic islets from surgical specimens. J Vis Exp 71:50231
8. Marciniak A, Cohrs CM, Tsata V, Chouinard JA, Selck C, Stertmann J, Reichelt S, Rose T, Ehehalt F, Weitz J, Solimena M, Slak Rupnik M, Speier S (2014) Using pancreas tissue slices for in situ studies of islet of Langerhans and acinar cell biology. Nat Protoc 9 (12):2809–2822
9. Nishida Y, Aida K, Kihara M, Kobayashi T (2014) Antibody-validated proteins in inflamed islets of fulminant type 1 diabetes profiled by laser-capture microdissection followed by mass spectrometry. PLoS One 9(10): e107664
10. Ricordi C, Gray DW, Hering BJ, Kaufman DB, Warnock GL, Kneteman NM, Lake SP, London NJ, Socci C, Alejandro R et al (1990) Islet isolation assessment in man and large animals. Acta Diabetol Lat 27(3):185–195
11. Pisania A, Weir GC, O'Neil JJ, Omer A, Tchipashvili V, Lei J, Colton CK, Bonner-Weir S (2010) Quantitative analysis of cell composition and purity of human pancreatic islet preparations. Lab Invest 90 (11):1661–1675
12. Tyanova S, Temu T, Cox J (2016) The MaxQuant computational platform for mass spectrometry-based shotgun proteomics. Nat Protoc 11(12):2301–2319
13. Tyanova S, Temu T, Sinitcyn P, Carlson A, Hein MY, Geiger T, Mann M, Cox J (2016) The Perseus computational platform for comprehensive analysis of (prote)omics data. Nat Methods 13(9):731–740
14. Tusher VG, Tibshirani R, Chu G (2001) Significance analysis of microarrays applied to the ionizing radiation response. Proc Natl Acad Sci U S A 98(9):5116–5121

Chapter 17

Targeted Proteomics

Yun Chen and Liang Liu

Abstract

Targeted proteomics detects proteins of interest with high sensitivity, quantitative accuracy, and reproducibility. In a targeted proteomics assay, surrogate peptides are generated by proteolytic digestion of target proteins and selected reaction monitoring (SRM) assays are developed to quantify these peptides using liquid chromatography-tandem mass spectrometry (LC-MS/MS). In this report, we describe the details of quantitative analysis of target protein in cells and tissue samples.

Key words Targeted proteomics, Liquid chromatography-tandem mass spectrometry, Protein quantification, Cells and tissue samples

1 Introduction

With a growing demand for protein quantification across multiple samples, liquid chromatography-tandem mass spectrometry (LC-MS/MS)-based targeted proteomics has emerged as a powerful tool in systems biology, biomedical research, and clinical proteomics because of its high sensitivity, quantitative accuracy, and reproducibility [1–4]. Targeted proteomics was selected as a method to watch in both 2009 and 2010 and also as method of the year in 2012 by Nature Methods [5–7]. In a targeted analysis, target proteins are first digested into peptides using a proteolytic enzyme, commonly trypsin. Then, surrogate peptides that can uniquely represent the target proteins are selectively analyzed by selected/multiple reaction monitoring (SRM/MRM), which is typically performed on a triple quadruple mass spectrometer. Therein, the ion mass of the precursor peptide of interest is set in the first mass analyzer (Q1), while peptide product ions, which is generated by collision-induced dissociation in Q2, is predefined in the third mass analyzer (Q3). Precursor ion/product ion *m/z* pairs, referred to as SRM/MRM transitions, are used to yield LC-MS/MS chromatogram. The area under the curve of the chromatogram

Xing Wang and Matthew Kuruc (eds.), *Functional Proteomics: Methods and Protocols*, Methods in Molecular Biology, vol. 1871, https://doi.org/10.1007/978-1-4939-8814-3_17, © Springer Science+Business Media, LLC, part of Springer Nature 2019

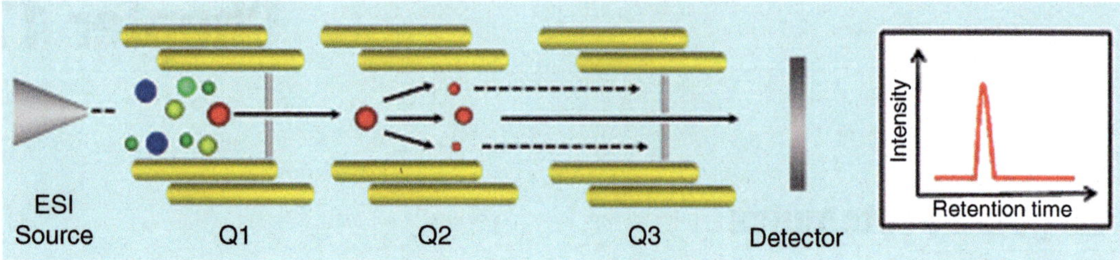

Fig. 1 A schematic of targeted proteomics strategy using a triple quadrupole mass spectrometer operating in SRM mode

provides a quantitative measurement for each desired peptide and target protein (Fig. 1).

To date, targeted proteomics has become increasingly popular for different research area and applications. In our lab, we have regularly applied this technique for protein quantification, such as p-glycoprotein (P-gp) [8, 9], transferrin (TRF) [10], transferrin receptor (TfR) [11], folate receptor (FR) [12], extracellular regulated protein kinase (ERK) [13], murine double minute 2 (MDM2) [14], and heat shock proteins 27 (HSP27) [15] in cells and tissue samples. The typical pipeline consists of (1) sample pretreatment and protein extraction, (2) protein digestion, (3) selection of appropriate surrogate peptides of target proteins, (4) chemical synthesis of internal standard peptides by incorporation of heavy stable isotopes, (5) assay development and validation, and (6) sample analysis (Fig. 2).

2 Materials

2.1 Cell Viability Test

Hemocytometer, cover slip, and counter.

10 mM phosphate-buffered saline (PBS).

Trypan blue (0.4%).

2.2 Tissue Homogenization

Scissor and homogenizer.

Homogenization buffer: 50 mM Tris–HCl, pH 7.4, 2 mM Ethylenediaminetetraacetic acid (EDTA), 1 mM DL-dithiothreitol (DTT), 150 mM NaCl, and 1% protease inhibitor (*see* **Note 1**).

2.3 Protein Extraction Buffers

2.3.1 Cytosolic Proteins

RIPA lysis buffer: 50 mM Tris–HCl, pH 7.4, 150 mM NaCl, 1% Triton X-100, 1% sodium deoxycholate, 0.1% sodium dodecyl sulfate (SDS) (*see* **Note 2**).

2.3.2 Membrane Proteins

Extraction buffer: 50 mM Tris–HCl, pH 7.4, 1 mM DTT, 2 mM EDTA, 1% protease inhibitor, 1% Triton X-114.

Washing buffer: 50 mM Tris–HCl, pH 7.4, 1 mM DTT, 2 mM EDTA, 0.06% Triton X-114.

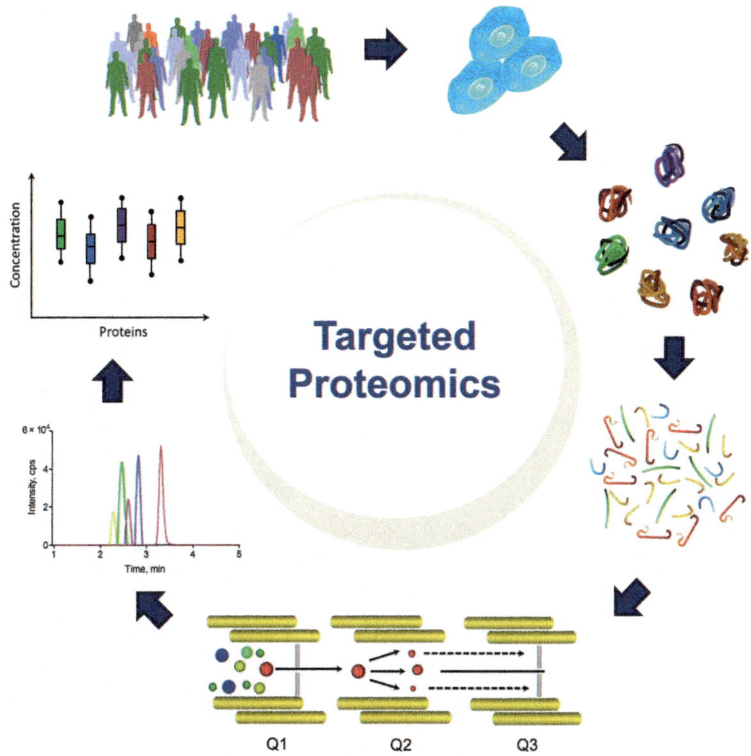

Fig. 2 A typical pipeline of targeted proteomics

2.4 Tryptic Digestion Buffer

50 mM DTT, 400 mM iodoacetamide (IAA), and 50 mM ammonium bicarbonate (NH_4HCO_3).

2.5 LC-MS/MS Instrument

An Agilent Series 1200 HPLC system and a 6410 Triple Quad LC/MS mass spectrometer (*see* **Note 3**).

3 Methods

3.1 Sample Preparation

3.1.1 Cell Pretreatment

1. Carefully remove culture medium from cells.
2. Wash cells twice with cold PBS.
3. Assess cell viability using trypan blue (0.4%) exclusion. Mix cell suspension, PBS, and trypan blue in a 2:3:5 ratio, and incubate for 5 min at 37 °C.
4. Count viable cells using a hemocytometer.

3.1.2 Tissue Homogenization

1. Thaw tissue samples to room temperature and rinse thoroughly with deionized water.
2. Remove fat tissue and cut the remaining tissue into small pieces and transfer them to tubes.
3. Weigh approximately 50 mg of tissue and suspend it in tissue homogenization buffer (*see* **Note 4**).
4. Homogenize the tissue suspension on ice using a Bio-Gen PRO200 homogenizer.

3.2 Protein Extraction

3.2.1 Cytosolic Proteins

1. Add cold RIPA buffer to sample. Keep the sample on ice for 45 min and vortex every 15 min.
2. Centrifuge the sample at 14,000 *g* for 15 min.
3. Transfer the supernatant to a new tube and determine protein concentration using a BCA protein assay kit.

3.2.2 Membrane Proteins

1. Centrifuge the sample at 10,000 *g* for 10 min, and resuspend the pellet in 500 μL of membrane protein extraction buffer.
2. Incubate the sample on ice for 30 min, and vortex every 10 min, followed by incubation at 37 °C for 10 min (*see* **Note 5**).
3. Centrifuge the mixture at 10,000 *g* for 3 min to separate detergent and aqueous phases.
4. Add 500 μL of 1% extraction buffer and 500 μL of wash buffer to the aqueous and detergent phases, respectively. Then repeat the above incubation and centrifugation steps.
5. Combine the detergent phases and precipitate proteins using cold acetone (pre-frozen at −20 °C for 1 h before use).
6. Allow acetone to evaporate at room temperature.
7. Dissolve the protein pellet in 1% SDS solution.
8. Determine protein concentration of the obtained sample using a BCA protein assay kit.

3.3 In-Solution Tryptic Digestion

1. Mix 100 μL of the sample with 50 μL of 50 mM NH_4HCO_3.
2. Denature proteins at 95 °C for 8 min (*see* **Note 6**).
3. Add 50 μL of 50 mM DTT to the sample, followed by incubation at 60 °C for 30 min.
4. Add 30 μL of 400 mM IAA and incubate at room temperature for 30 min in the dark.
5. Add 50 μL of sequencing grade trypsin solution (1:20 enzyme: protein) and incubate at 37 °C for 24 h.
6. Add 10 μL of 0.1% TFA to stop the reaction.
7. Dry the sample in a vacuum centrifuge.
8. Resuspend the sample in 100 μL of ACN:water (50:50, *v/v*) containing 0.1% FA.

3.4 Desalting

1. Add 100 μL of internal standard solution to the tryptic mixture.
2. Precondition the microspin C18 column (The Nest Group, Inc., MA, USA) with 100 μL of ACN and 100 μL of water in advance.
3. Transfer 50 μL of the sample into the column and centrifuge at 1000 *g* for 1 min.
4. Wash the column with 50 μL of ACN:water (5:95, *v/v*) containing 0.1% TFA and elute it with 50 μL of ACN:water (80:20, *v/v*) containing 0.1% FA.
5. Repeat the above procedure 3–4 times, and finally combine the collections.

3.5 Surrogate Peptide Selection

1. The most critical step in the establishment of a targeted proteomics assay is the selection of proteolytic peptides that (1) are unique to a candidate protein, (2) can provide an adequate response, (3) are completely digested, and (4) can generate high-quality SRM [16, 17] (*see* **Note 7**).
2. The uniqueness of selected surrogate peptides is normally checked using a BLAST search. For example, the peptides of 434STTVQLMQR442, 674GSQAQDR680, and 368IIDNKPSIDSYSK380 were found to be unique to P-gp (accession no. P08183 (MDR1 HUMAN), gi: 2506118) [8].
3. A LC-MS/MS analysis with a list of SRM transition pairs based on either in silico prediction or spectral evidence from public repositories is usually performed to identify the peptide with the greatest abundance [18] (*see* **Note 8**). Synthetic references peptides are usually employed for confirmation (Fig. 3).
4. The digestion efficiency is evaluated using the substrate peptide containing the same peptide sequence (to mimic a piece of the target protein). The digestion efficiency was calculated by comparing the response ratios of the tryptic peptide after digestion and the equimolar synthetic peptide standard in the digestion (Fig. 4) [19].
5. Optimize SRM transitions (*see* **Notes 9–11**].

3.6 Internal Standard

Synthetic stable isotope-labeled peptide is prepared according to the selected surrogate peptide. C_{13} and D stable isotope-labeled amino acids are usually employed. For instance, a stable isotope-labeled valine with an added mass of 8 Da from deuterium was coupled to the peptide sequence STTV*QLMQR at position 4 to yield a molecular mass shift of 8 Da from the non-labeled peptide STTVQLMQR [8].

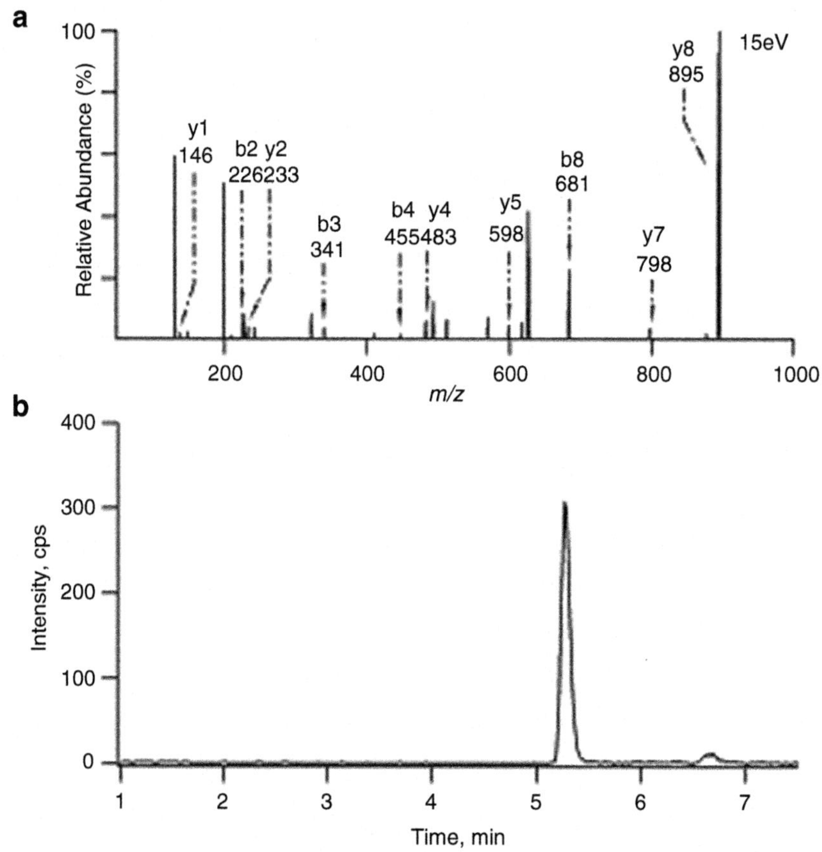

Fig. 3 The product ion spectrum and LC-MS/MS chromatogram of 368IIDNKPSIDSYSK380, a surrogate peptide of P-gp. The characteristic sequence-specific *b* ions and *y* ions, and retention time are indicative of this peptide (reproduced from ref. 9 with permission from Elsevier)

3.7 Immuno-Depleted Matrix Preparation

1. Add BioMagPlus IgG beads that are pre-incubated with anti-target-protein antibody (*see* **Note 12**).
2. Incubate the mixture at 25 °C for 2 h with shaking.
3. Magnetically separate the beads and collect the supernatant.
4. Rinse the beads with 1% SDS solution and combine the eluate with the above supernatant.
5. Examine this synthetic matrix using Western blotting and LC-MS/MS-based targeted proteomics assay (Fig. 5).

3.8 Assay Development and Validation

1. Prepare 1 mg/mL stock solution by weighing the peptide (including internal standard peptide) and dissolving it deionized water. The solution was stored at −20 °C in a brown glass tube to protect it from light.
2. Prepare calibration standards and QC standards by serial dilution of the stock solution using immuno-depleted matrix. For P-gp, the concentrations of the calibration standards are

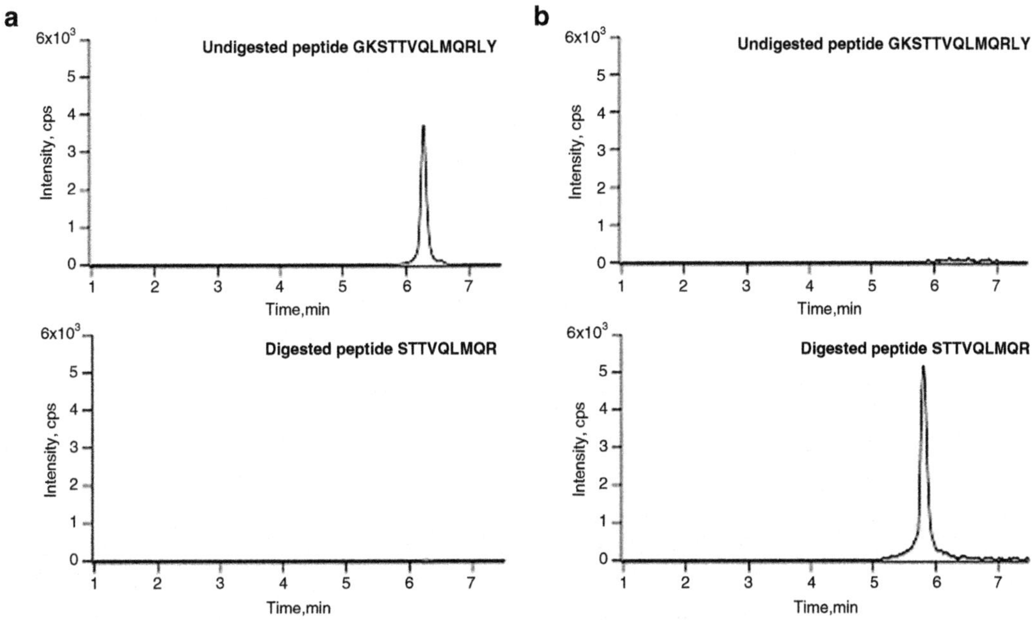

Fig. 4 LC-MS/MS chromatograms of STTVQLMQ and its substrate peptide GKSTTVQLMQRLY (**a**) before and (**b**) after tryptic digestion (reproduced from ref. 8 with permission from Elsevier)

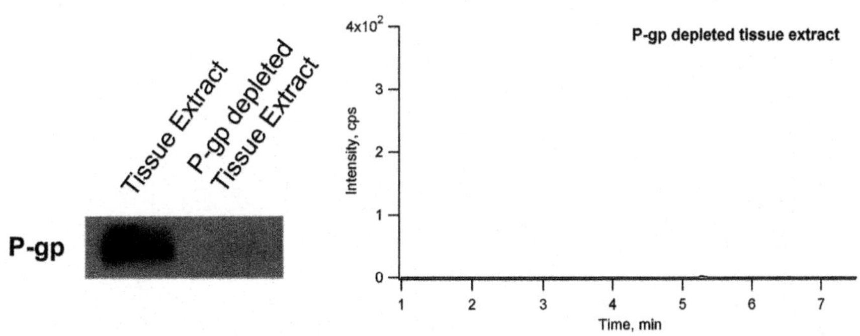

Fig. 5 The Western blotting image and LC-MS/MS chromatogram for P-gp depleted tissue extract (reproduced from ref. 9 with permission from Elsevier)

10, 25, 50, 100, 250, 400, 700, and 1000 ng/mL. The QC standards for the lower limit of quantification (LLOQ), low QC, mid QC, and high QC were prepared at 10, 30, 200, and 800 ng/mL, respectively, and frozen prior to use.

3. Construct the calibration curve using a weighted linear regression model with a weighting factor of $1/x^2$. The relative peak area ratio of the analyte and the stable isotope-labeled internal standard was plotted as a function of concentration (Fig. 6).
4. The precision and accuracy of the assay were assessed by observing the response of the QC samples with four different

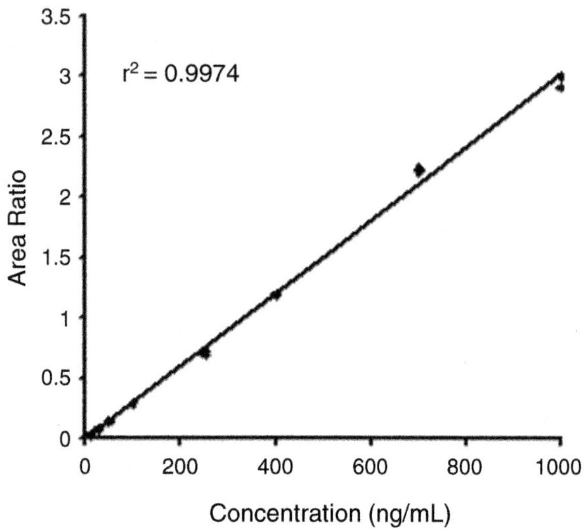

Fig. 6 Representative calibration curves (10–1000 ng/mL) for the standards using P-gp depleted matrices (reproduced from ref. 9 with permission from Elsevier)

Table 1
Accuracy and precision for the QC samples using P-gp depleted matrix (reproduced from ref. 8 with permission from Elsevier)

Nominal concentration	10 ng/mL	30 ng/mL	200 ng/mL	800 ng/mL
Mean	9.33	29.0	200	770
%Bias	−6.6	−3.3	0.0	−3.8
Intra-day Precision (%CV)	4.4	4.6	2.8	2.7
Inter-day precision (%CV)	9.4	4.1	3.1	1.0
n	18	18	18	18
Number of runs	3	3	3	3

concentrations of P-gp in three validation runs. The intra- and inter-day precisions were expressed as the percent coefficient of variation (%CV). The accuracy was obtained by comparing the average calculated concentrations to their nominal values (%bias) (Table 1).

3.9 Sample Analysis

1. Apply the LC-MS/MS-based targeted proteomics assay to analyze the samples.
2. Calculate the amount of target proteins in cell and tissue samples using the calibration curves built above (Figs. 7 and 8) (*see* **Note 13**). The protein amounts can also be monitored in a time manner (Fig. 9).

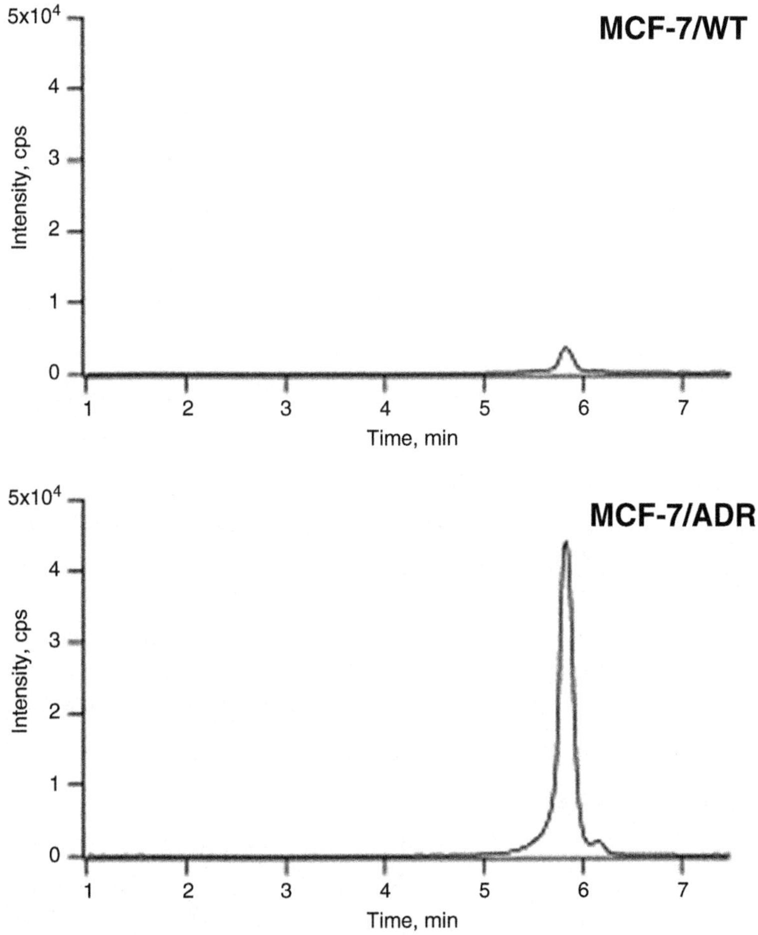

Fig. 7 LC-MS/MS chromatograms of P-gp in cells. The expression levels of P-gp were accurately quantified to be 3.53 fg/cell in MCF-7/WT and 34.5 fg/cell in MCF-7/ADR cells (reproduced from ref. 8 with permission from Elsevier)

4 Notes

1. DTT may not be stable and should not be stored for long periods of time, so make up fresh before use [20].
2. If necessary, add protease and phosphatase inhibitors to RIPA buffer immediately before use.
3. It should be noted that other instrument platforms (e.g., ion trap, Q-TOF) are also capable of performing SRM like experiments; however, quadrupole mass spectrometer is the preferred technology for quantification and the most accessible instrument in routine research and clinical lab [21].
4. Tissue collection must be approved by the Medical Ethics Review.

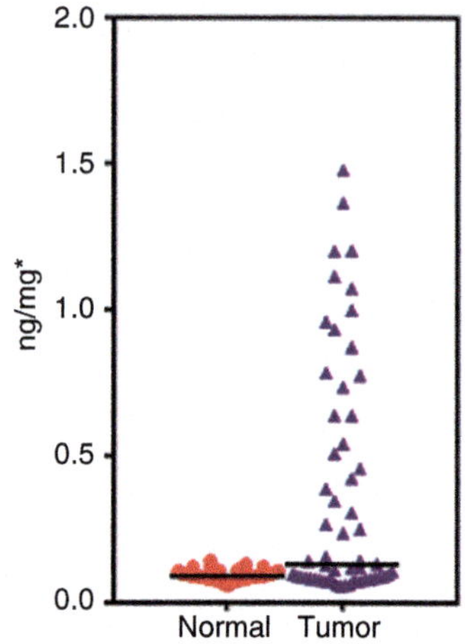

Fig. 8 P-gp amounts in 36 matched pairs of breast tissue samples (reproduced from ref. 9 with permission from Elsevier)

5. Increase incubation time if necessary.
6. Currently, many denaturing conditions are available such as a strong acid or base (e.g., acetic acid), a concentrated inorganic salt (e.g., urea), an organic solvent (e.g., alcohol), or heat [22].
7. There are primarily two opposite approaches, prospective and retrospective, either using in silico prediction by various algorithms or based on spectral evidence using data from either public repositories or in-house experiments (e.g., spectra recorded during global discovery experiments). As a prospective SRM design, it would be possible to predict which peptides and product ions are most appropriate for SRM in protein quantification by several computational tools, for example, ESP predictor, STEPP, Peptide sieve (PAGE-ESI), Peptide detectability. However, it should be noted that the mechanisms of proteolysis, ionization, and fragmentation are not yet sufficiently well understood to produce accurate models from which to make such predictions. The current models can only assist to select high-responding peptides, particularly in the absence of experimental data. The retrospective approach uses experimentally obtained peptide spectra as evidence, and several software tools have been developed. Publicly available spectral repositories include PRIDE, GPMDB, PeptideAtlas, NIST, and MacCoss. The software tools are Targeted Identification for Quantitative Analysis by Multiple reaction monitoring (TIQAM), MRMer, SRMCollider, MaRiMba, MRMaid, Skyline and ATAQS, or commercial with the software platforms provided by mass spectrometer vendors (e.g., SRM Workflow software (based on SIEVE), Pinpoint and P3 predictor (Thermo Scientific), mTRAQ-reagent-based MRMPilot

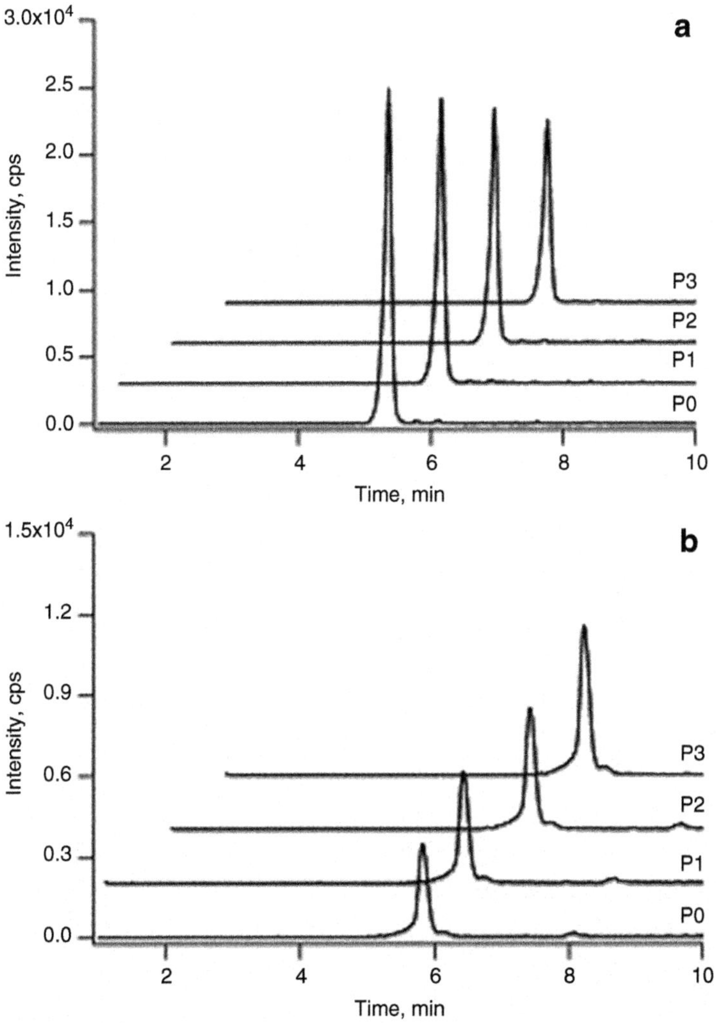

Passages after the treatment of DOX	0	1	2	3
HSP27 (pg/cell)	11.4 ± 0.3	9.73 ± 0.55	8.03 ± 0.25	6.21 ± 0.36
P-gp (fg/cell)	3.62 ± 0.20	4.41 ± 0.20	4.74 ± 0.14	5.83 ± 0.28

Fig. 9 The LC/MS-MS chromatograms of (**a**) HSP27 and (**b**) P-gp in freshly prepared (P0) and passage 1, 2, 3 (P1, P2, P3) MCF-7/WT cells after the treatment of DOX. The amounts are also listed in the table (reproduced from ref. 15 with permission from Elsevier)

software and multiple reaction monitoring initiated detection and sequencing (MIDAS) Workflow Designer (Applied Biosystems), VerifyE and TargetLynx™ Application Manager (Waters), MassHunter Optimizer (Agilent Technologies)).

8. There are several empirical criteria for selection of peptides with high abundance, (1) length between 6 and 16 amino acids, (2) no

methionine or cysteine residues, (3) no post-translational modifications (e.g., proteolysis, phosphorylation, or glycosylation) or single nucleotide polymorphism, (4) no transmembrane region for membrane proteins, (5) no proline residue at the C-terminal side of an arginine or lysine residue, and (6) no continuous sequence of arginine or lysine residues (RR, KK, RK, KR).

9. To achieve a better peak separation and signal sensitivity, it is important to choose appropriate column type and optimize organic composition, flow rate and column temperature.
10. Usually, optimal collision condition for doubly and triply charged peptides ranges between 20 and 40 eV, or even lower (10 eV). Longer peptides and peptides with less charges may need higher collision energy (CE) and *y* ions could require higher CE than *b* ions [19].
11. Singly charged *y* ions are normally the predominant fragments generated by collision-induced dissociation, as compared to *b* ions that are low abundant or even absent in product ion spectrum due to their lower stability and easier decomposition. Product ion with *m/z* value close to the precursor should be avoided because such transitions are usually noisy [19].
12. Antibodies from multiple species and the corresponding IgG beads are eligible for protein depletion [23].
13. SRM transitions can be summed to quantify the peptide, or it is possible to use one or several transitions for quantification and the others for a confirmation of the peptide identity [19].

References

1. Yocum AK, Chinnaiyan AM (2009) Current affairs in quantitative targeted proteomics: multiple reaction monitoring–mass spectrometry. Brief Funct Genomics Proteomics 8 (2):145–157
2. Pan S, Aebersold R, Chen R, Rush J, Goodlett DR, McIntosh MW, Zhang J, Brentnall TA (2008) Mass spectrometry based targeted protein quantification: methods and applications. J Proteome Res 8(2):787–797
3. Parker CE, Pearson TW, Anderson NL, Borchers CH (2010) Mass-spectrometry-based clinical proteomics–a review and prospective. Analyst 135(8):1830–1838
4. Domon B, Aebersold R (2006) Mass spectrometry and protein analysis. Science 312 (5771):212–217
5. Marx V (2013) Targeted proteomics. Nat Methods 10(1):19–22
6. Doerr A (2010) Targeted proteomics. Nat Methods 7(1):34–34
7. Doerr A (2013) Mass spectrometry-based targeted proteomics. Nat Methods 10(1):23–23
8. Yang T, Xu F, Xu J, Fang D, Yu Y, Chen Y (2013) Comparison of liquid chromatography–tandem mass spectrometry-based targeted proteomics and conventional analytical methods for the determination of P-glycoprotein in human breast cancer cells. J Chromatogr B 936:18–24
9. Yang T, Chen F, Xu F, Wang F, Xu Q, Chen Y (2014) A liquid chromatography–tandem mass spectrometry-based targeted proteomics assay for monitoring P-glycoprotein levels in human breast tissue. Clin Chim Acta 436:283–289
10. Yu Y, Xu J, Liu Y, Chen Y (2012) Quantification of human serum transferrin using liquid chromatography–tandem mass spectrometry based targeted proteomics. J Chromatogr B 902:10–15
11. Yang T, Xu F, Zhao Y, Wang S, Yang M, Chen Y (2014) A liquid chromatography-tandem

mass spectrometry-based targeted proteomics approach for the assessment of transferrin receptor levels in breast cancer. Proteom Clin Appl 8(9–10):773–782
12. Yang T, Xu F, Fang D, Chen Y (2015) Targeted proteomics enables simultaneous quantification of folate receptor isoforms and potential isoform-based diagnosis in breast cancer. Sci Rep 5:16733
13. Yang T, Xu F, Sheng Y, Zhang W, Chen Y (2016) A targeted proteomics approach to the quantitative analysis of ERK/Bcl-2-mediated anti-apoptosis and multi-drug resistance in breast cancer. Anal Bioanal Chem 408 (26):7491–7503
14. Zhang W, Zhong T, Chen Y (2017) LC-MS/MS-based targeted proteomics quantitatively detects the interaction between p53 and MDM2 in breast cancer. J Proteome 152:172–180
15. Xu F, Yang T, Fang D, Xu Q, Chen Y (2014) An investigation of heat shock protein 27 and P-glycoprotein mediated multi-drug resistance in breast cancer using liquid chromatography-tandem mass spectrometry-based targeted proteomics. J Proteome 108:188–197
16. Anderson L, Hunter CL (2006) Quantitative mass spectrometric multiple reaction monitoring assays for major plasma proteins. Mol Cell Proteomics 5(4):573–588
17. Prakash A, Tomazela DM, Frewen B, MacLean B, Merrihew G, Peterman S, MacCoss MJ (2009) Expediting the development of targeted SRM assays: using data from shotgun proteomics to automate method development. J Proteome Res 8 (6):2733–2739
18. Picotti P, Aebersold R (2012) Selected reaction monitoring-based proteomics: workflows, potential, pitfalls and future directions. Nat Methods 9(6):555–566
19. Gianazza E, Tremoli E, Banfi C (2014) The selected reaction monitoring/multiple reaction monitoring-based mass spectrometry approach for the accurate quantitation of proteins: clinical applications in the cardiovascular diseases. Expert Rev Proteomics 11 (6):771–788
20. Schmidt C, Urlaub H (2012) Absolute quantification of proteins using standard peptides and multiple reaction monitoring. Methods Mol Biol 893:249–265
21. Dillen L, Cools W, Vereyken L, Lorreyne W, Huybrechts T, de Vries R, Ghobarah H, Cuyckens F (2012) Comparison of triple quadrupole and high-resolution TOF-MS for quantification of peptides. Bioanalysis 4 (5):565–579
22. Mosby I (2006) Mosby's medical dictionary. Mosby
23. Zolotarjova N, Martosella J, Nicol G, Bailey J, Boyes BE, Barrett WC (2005) Differences among techniques for high-abundant protein depletion. Proteomics 5(13):3304–3313

Chapter 18

Metabolomic Investigation of *Staphylococcus aureus* Antibiotic Susceptibility by Liquid Chromatography Coupled to High-Resolution Mass Spectrometry

Sandrine Aros-Calt, Florence A. Castelli, Patricia Lamourette, Gaspard Gervasi, Christophe Junot, Bruno H. Muller, and François Fenaille

Abstract

Staphylococcus aureus is a major human pathogen that can readily acquire antibiotic resistance. For instance, methicillin-resistant *S. aureus* represents a major cause of hospital- and community-acquired bacterial infections. In this chapter, we first provide a detailed protocol for obtaining unbiased and reproducible *S. aureus* metabolic profiles. The resulting intracellular metabolome is then analyzed in an untargeted manner by using both hydrophilic interaction liquid chromatography and pentafluorophenyl-propyl columns coupled to high-resolution mass spectrometry. Such analyses are done in conjunction with our in-house spectral database to identify with high confidence as many meaningful *S. aureus* metabolites as possible. Under these conditions, we can routinely monitor more than 200 annotated *S. aureus* metabolites. We also indicate how this protocol can be used to investigate the metabolic differences between methicillin-resistant and susceptible strains.

Key words *Staphylococcus aureus*, Methicillin resistance, Metabolomics, Liquid chromatography, High-resolution mass spectrometry

1 Introduction

Staphylococcus aureus (*S. aureus*) is a notorious and opportunistic pathogen causing a wide range of diseases and syndromes, including bacteremia, pneumonia, cellulitis, osteomyelitis, and infections affecting skin and soft tissues [1]. In addition, *S. aureus* is well known for its ability to acquire resistance to various kind of antibiotics. Among antibiotic-resistant *S. aureus* strains, Methicillin-Resistant *S. aureus* (MRSA) is one of the most serious threat-level pathogens that had also developed multidrug resistance [2], with increasing resistance to vancomycin [3]. MRSA is involved in most of the global *S. aureus* bacteremia cases, and is often associated with

Xing Wang and Matthew Kuruc (eds.), *Functional Proteomics: Methods and Protocols*, Methods in Molecular Biology, vol. 1871, https://doi.org/10.1007/978-1-4939-8814-3_18,

poor clinical outcomes (~30% mortality rate) [4, 5]. Although the level of MRSA isolates has decreased over time in Europe, some European countries still report 25% or more of invasive *S. aureus* isolates as MRSA [5]. Acquisition of antibiotic resistance can partially be explained by the misuse and mishandling of drugs to treat infections. Unfortunately, first antibiotic resistances often appear very soon after the introduction of an antibiotic. When combined with drug companies' reluctance to develop novel antibiotics, this has led to serious health concerns [6]. Therefore, numerous efforts have been made to better understand antibiotic-resistance mechanisms in order to potentially find new therapeutic strategies [7–10].

In that objective, metabolomics can greatly help digging deeper into the pathogenicity and biochemical mechanisms behind the antibiotic resistance of MRSA strains. Indeed, metabolomics provides the most direct assessment of cellular phenotype and is well suited to the quantitative and dynamic monitoring of bacterial metabolism variations in response to particular environmental conditions. Recently published papers clearly support this statement regarding the study MRSA strains [11–15]. Just to cite a few, the pioneering work of Liebeke et al. regarding *S. aureus* metabolomics demonstrated the possibility to simultaneously analyze about 80 *S. aureus* metabolites by using liquid chromatography coupled to mass spectrometry (LC-MS). They reported that the *S. aureus* central carbon metabolism, which notably includes energy transfer molecules like nucleotides, sugar mono- and bi-phosphates, and cofactors, can be used to monitor metabolic disturbances in response to genetic deletions of serine/threonine kinase and phosphatase [11].

Keaton et al. investigated the changes in metabolic pathways when treating MRSA strains with subinhibitory concentrations of β-lactam antibiotics. They used a combination of LC-MS and gas chromatography coupled to mass spectrometry (GC-MS) techniques to highlight significant increases in tricarboxylic acid (TCA) cycle intermediates under the studied conditions, which tended to demonstrate that the energy production of MRSA strains was redirected to supply the cell wall synthesis/metabolism and so, contributed to their survival in the presence of β-lactam antibiotics [16]. In a recent study, Dorries et al. investigated the impact of five antibiotics with different cellular targets on *S. aureus* metabolism by studying its global intra- and extracellular metabolic profiles thanks to LC-MS, GC-MS, and NMR techniques [13]. Their analytical platforms largely covering primary metabolism allowed them to highlight accumulation as well as depletion of metabolites from various biosynthetic pathways, such as central carbon and amino acid metabolism; peptidoglycan, purine, and pyrimidine synthesis [13]. Also, Ammons et al. successfully implemented an NMR-based strategy to distinguish MRSA from MSSA strains based on the quantitative monitoring of 40 intracellular metabolites (essentially amino acids and TCA-cycle intermediates) [12].

Although successful in distinguishing MRSA from MSSA strains or getting insights into *S. aureus* antibiotic resistance, most of these studies investigated a limited number of metabolites or described metabolites that were only putatively annotated according to the Metabolomic Standard Initiative criteria [17]. Indeed, to be considered as formally identified, metabolites need to have a minimum of two orthogonal physicochemical parameters matching those of an authentic standard analyzed under identical experimental conditions (i.e., retention time and mass spectrum, or accurate mass and tandem mass spectra, for example). In the absence of an available corresponding authentic chemical standard, metabolites of interest can only be regarded as putatively annotated, for example, based on their accurate measured mass and interpretation of the MS/MS spectra when available [17]. Considerable efforts are still required to annotate the metabolome of *S. aureus*, which is predicted to be a very complex microorganism with ~500 to ~1400 expected metabolites, as deduced from genome scale models [18]. About 100–150 *S. aureus* metabolites have been reported as confidently identified in the most exhaustive studies published, this suggests that the availability a fully annotated and comprehensive metabolite library is still many years away.

The prerequisites for obtaining unbiased and most comprehensive intracellular metabolite profiles of *S. aureus* strains are appropriate sample preparation as well as an efficient detection method to cope with the natural chemical diversity of the metabolome. In that context, we have designed and carefully optimized a specific and robust sample preparation protocol to obtain a reliable snapshot of *S. aureus* bacterial metabolism under various experimental conditions. We have also reported the development of two complementary liquid chromatography coupled to high-resolution mass spectrometry (LC-HRMS) platforms to identify as many *S. aureus* metabolites as possible, at the highest confidence level (Fig. 1) [14]. Our metabolite identification workflow followed the criteria reported by the Metabolomics Standard Initiative [17], and has led to the successful characterization of 210 *S. aureus* metabolites with up to 173 formal identifications. We thus demonstrated the ability of the implemented protocol to reproducibly detect differences in metabolic profiles between MRSA and MSSA strains [14].

In this chapter, the step-by-step protocols for the extraction, preparation, and analysis of *S. aureus* intracellular metabolites are described so as to enable the reader to replicate them.

2 Materials

2.1 Bacterial Strains

Methicillin-resistant and susceptible *Staphylococcus aureus* strains were obtained from the bioMérieux collection of microorganisms. Presence of the *mecA/mecC* gene in the studied strains was

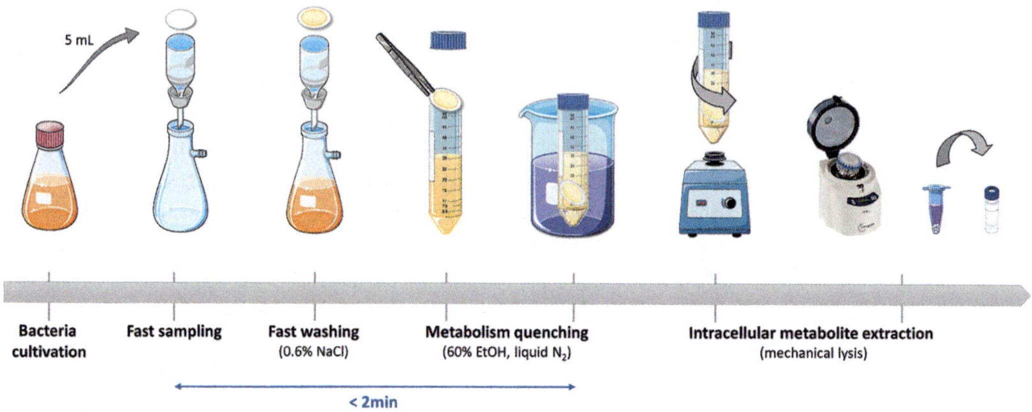

Fig. 1 Experimental workflow for *S. aureus* metabolome analysis by LC-HRMS

evaluated by multiplex PCR assay, while minimal inhibitory concentrations (MIC) were determined using oxacillin Etest® strips (bioMérieux, Marcy l'Etoile, France).

2.2 Bacterial Culture

1. Mueller Hinton II broth (MHII, cation-adjusted) from Becton Dickinson (product reference 212322, Franklin Lakes, NJ). Dissolve 22 g of MHII medium in 1 L of ultrapure water, autoclave the mixture at 120 °C for 20 min, and store at 4 °C until use.
2. Colombia (COS) agar containing 5% sheep blood plates (bioMérieux).
3. Cefoxitin (Sigma-Aldrich, Saint Quentin Fallavier, France).
4. Petri dishes (Sigma-Aldrich).
5. Erlenmeyer baffled cell culture flasks (Sigma-Aldrich).
6. Sterile inoculating loops and needles (VWR, Fontenay-sous-Bois, France).
7. Serological pipettes (5 mL, Sigma-Aldrich).
8. Minitron II rotary shaker (Infors HT, Bottmingen-Basel, Switzerland).
9. Eppendorf BioPhotometer (Eppendorf, Montesson, France) or equivalent spectrophotometer.

2.3 Metabolite Extraction

2.3.1 Bacteria Collection and Metabolism Quenching

1. Polyethersulfone (PES) sterile membrane disc filters (47 mm diameter, 0.45 μm pore size, PALL, Washington, NY).
2. Filtration system (Millipore, Darmstadt, Germany), clamps (Dutscher, Brumath, France), and vacuum pump (KNF, Village-Neuf, France).
3. Washing buffer: 0.6% NaCl. Dissolve 3 g of NaCl in 500 mL of deionized water, filter the resulting solution at 0.22 μm, and store at 4 °C until use.

4. Ice-cold 60% ethanol (Sigma-Aldrich) in deionized water, stored at −20 °C.
5. Liquid nitrogen.

2.3.2 Extraction of Intracellular Metabolites

1. Tissue homogenizer Precellys 24 (Bertin Technologies, Montigny-le-Bretonneux, France).
2. 0.1 mm glass beads and tubes (Bertin Technologies).
3. Turbovap evaporator (Caliper Life Science Inc., Roissy, France).

2.4 LC–MS Analysis of Intracellular Metabolites

2.4.1 Chromatographic Columns

1. Sequant ZIC-pHILIC column, 2.1×150 mm, 5 μm, HPLC PEEK (Merck, Darmstadt, Germany).
2. Discovery HSF5 Pentafluorophenylpropyl (PFPP) column, 2.1 × 150 mm, 5 μm (Sigma-Aldrich).

2.4.2 Mobile Phases

1. Mobile phase A (ZIC-pHILIC): 10 mM ammonium carbonate pH 10.5. Dissolve 960 mg of ammonium carbonate (product reference 68392, Sigma-Aldrich) in 1 L of ultrapure water. Adjust pH to 10.5 using a 28% NH_4OH solution.
2. Mobile phase B (ZIC-pHILIC): Pure acetonitrile (ACN).
3. Mobile phase A (PFPP): Water containing 0.1% formic acid.
4. Mobile phase B (PFPP): ACN containing 0.1% formic acid.

2.4.3 LC-MS(/MS) Systems

1. LC-MS experiments were performed using an Ultimate 3000 chromatographic system (Thermo Fisher Scientific, Courtaboeuf, France) coupled to an Exactive Orbitrap mass spectrometer (Thermo Fisher Scientific) fitted with an electrospray (ESI) source.
2. LC-MS/MS experiments were realized using an Ultimate 3000 chromatographic system coupled to a Q-Orbitrap mass spectrometer (Q-Exactive Plus, Thermo Fisher Scientific) fitted with an electrospray source.
3. The mass spectrometer is calibrated externally once per week in both ESI polarities using the manufacturer's predefined methods and the recommended calibration mixture provided by the manufacturer. Under such routine conditions, absolute mass accuracies are on average below 3 ppm for both negative and positive ionization modes.

2.4.4 Internal Standards (IS)

Solutions of internal standards are prepared in pure water (Table 1).

Table 1
List of internal standards used for LC-MS analysis

Compound name	Concentration of individual stock solutions (mg/mL water)	Compound concentration in the 20× standard mixture (μg/mL)
$^{13}C_1$-alanine	4	200
Ethylmalonic acid	4	30
$^{15}N_1$-aspartic acid	4	200
$^{13}C_1$-glucose	4	200
Ampicillin	2	50
Prednisone	2	10
Dihydrostreptomycin	4	200
Roxithromycin	2	200
$^{15}N_5$-AMP	1	100
$^{15}N_5$-ADP	1	50
$^{15}N_5$-ATP	1	50

All compounds except ^{15}N-labeled AMP and ATP (Euriso-Top, Saint-Aubin, France) were from Sigma-Aldrich

3 Methods

3.1 Bacterial Pre-culture and Culture

1. Preheat liquid and solid cultivation media a few hours at 37 °C to exclude any bacterial contamination.
2. *S. aureus* strains are first isolated from an overnight culture at 37 °C on COS plates.
3. Bacterial pre-culture is performed under aerobic conditions by inoculating few bacterial colonies in 12.5 mL of preheated MHII medium within a 125 mL Erlenmeyer flask (10% of total volume to ensure sufficient aeration). Incubate the resulting culture medium at 37 °C for 12–18 h with vigorous shaking (200 rpm).
4. Withdraw an aliquot and dilute to an optical density at 600 nm (OD_{600}) of ~0.1 in a fresh MHII medium (in a 2 L Erlenmeyer flask). Allow bacteria growth at 37 °C with vigorous shaking (200 rpm). For all the strains studied, the early-exponential phase corresponded to an OD_{600} of 1, which was equivalent to 5×10^8 CFU/mL (*see* **Notes 1** and **2**).

3.2 Bacteria Sampling and Metabolism Quenching

1. Prepare the filtration device, run the vacuum pump, and condition the filter by passing through 5 mL of MHII medium.
2. Withdraw a 5 mL aliquot from the main culture broth (at the targeted OD_{600}) and fill it into the filter system to rapidly separate bacteria from culture medium.
3. Wash the bacteria retained on the filter with 5 mL of 0.6% NaCl to remove culture medium (*see* **Note 3**).
4. The filter is then rapidly transferred to a 50 mL Falcon tube containing 5 mL of ice-cold 60% ethanol (*see* **Note 4**). The tube is subsequently quickly immersed in liquid nitrogen to quench bacterial metabolism (*see* **Note 5**) before mechanical cell disruption. At this step, samples can be conserved several months at −80 °C.

3.3 Mechanical Cell Disruption and Metabolite Extraction

1. Following quenching, tubes containing bacteria on filter in the extraction solution are vortexed 10 times (10 s at 4 °C) to remove cells from the filter.
2. Transfer 1 mL of the bacterial suspension into the Precellys tubes, while the remaining 4 mL are kept at −80 °C.
3. Bacterial lysis is accomplished by performing three cycles in a Precellys 24 homogenizer for 30 s at 3800 rpm and at ~4 °C.
4. Centrifuge the tubes during 5 min at 4 °C and 10,000 *g* to remove glass beads and cell debris.
5. Withdraw 400 μL of the supernatant and transfer into a 1.5 mL Eppendorf tube.
6. Evaporate under a stream of nitrogen.
7. 200 μL of each sample can also be withdrawn and pooled to obtain a Quality Control (QC) sample. 400-μL aliquots of the resulting mixture can then be evaporated under a stream of nitrogen.
8. Store samples and QCs at −80 °C until LC-MS analyses.

3.4 LC-MS Analysis of Intracellular Metabolites

A complete detailed protocol for running LC-MS experiments will not be provided hereafter. Below, we only list the main highlights of our approach involving a combination of two complementary ZIC-pHILIC and PFPP columns coupled to an Orbitrap instrument to obtain an optimal coverage of the *S. aureus* metabolome. For additional details, the reader can refer to Boudah et al. and Aros-Calt et al. [14, 19]. In principle, most of routine methods used in metabolomics laboratories are expected to be used for *S. aureus* metabolite analysis with limited or no modifications, even if they make use of another type of high-resolution mass spectrometer (e.g., Q-TOF).

3.4.1 LC-MS Analysis Using a ZIC-pHILIC Column

1. Dilute eight times the 20× internal standard mixture (Table 1) in 10 mM ammonium carbonate pH 10.5.
2. Estimate the number of CFU in each dried sample tube and adjust the resuspension volume to obtain a concentration of 1.25×10^7 CFU per 10 μL. The protocol provided hereafter is provided for a final volume of 80 μL.
3. Solubilize the dried bacterial extract and QC samples in 32 μL of the diluted internal standard mixture (**step 1**). After vigorous mixing, incubate the resulting mixture in an ultrasonic bath for 5 min.
4. Centrifuge at 10,000 × *g* for 5 min at 4 °C.
5. Transfer the resulting supernatant in an injection vial and add 48 μL of acetonitrile.
6. Inject 10 μL into the LC-MS system.
7. Metabolites are eluted from the column (maintained at 15 °C) at a flow rate of 200 μL/min using the gradient reported in Table 2.
8. The Exactive Orbitrap mass spectrometer is operated in the negative ion mode at a resolution of 50,000 at *m/z* 200 (full width at half-maximum), using the following source parameters: Capillary voltage, −3 kV; capillary temperature, 280 °C; sheath gas pressure, 60 arbitrary units; auxiliary gas pressure, 10 arbitrary units. The detection is performed from *m/z* 75 to 1000, using an injection time set at 100 ms and an AGC target value of 3×10^6.

3.4.2 LC–MS Analysis Using a PFPP Column

1. Dilute 20 times the 20× internal standard mixture (Table 1) in water containing 0.1% formic acid.
2. Estimate the number of CFU in each dried sample tube and adjust the resuspension volume to obtain a concentration of

Table 2
Gradient conditions used for metabolite LC-MS analysis using the ZIC-pHILIC column

Time (min)	% Mobile phase B (ACN)
0	80
2	80
12	40
12.01	0
17	0
17.01	80
42	80

Table 3
Gradient conditions used for metabolite LC-MS analysis using the PFPP column

Time (min)	% Mobile phase B (ACN containing 0.1% formic acid)
0	5
2	5
20	100
24	100
24.01	5
30	5

1.25×10^7 CFU per 10 μL. The protocol provided hereafter is provided for a final volume of 80 μL.

3. Solubilize the dried bacterial extract and QC samples in 80 μL of the diluted internal standard mixture (**step 1**). After vigorous mixing, incubate the resulting mixture in an ultrasonic bath for 5 min.
4. Centrifuge at 10,000 *g* for 5 min at 4 °C.
5. Transfer the resulting supernatant into the injection vial.
6. Inject 10 μL into the LC-MS system.
7. Metabolites are eluted from the column (maintained at 30 °C) at a flow rate of 250 μL/min using the gradient reported in Table 3.
8. The Exactive Orbitrap mass spectrometer is operated in the positive ion mode using the parameters described above, except the source voltage that is set at 5 kV.

3.5 Analysis of LC–MS Data

3.5.1 Determination of the Adenylate Energy Charge (AEC)

1. AMP, ADP, and ATP were quantified by the isotope dilution method using their ^{15}N-labeled homologues, with a procedure similar to that described by Martano et al. [20]. Under our conditions, a mass accuracy better than 3 ppm for the endogenous nucleotides and a perfect coelution with their labeled homologues ensure compound identification and accurate quantification (*see* **Note 6**).
2. Calculate the AEC by using molar concentrations with the following formula (*see* **Note 7**):

$$\text{AEC} = \frac{[\text{ATP}] + 0.5 \times [\text{ADP}]}{[\text{ATP}] + [\text{ADP}] + [\text{AMP}]}$$

3.5.2 Well-Designed Sample Dataset and Quality Control Samples (QCs) for Reliable Untargeted Metabolomics

Obtaining reliable metabolic profiling data is a complex task and particular attention should be paid to the experimental design in order to avoid any instrumental or analytical bias. Well-designed, robust, and reproducible metabolomics workflows commonly involve column conditioning, sample randomization, and use of QC samples. QCs are made from a pooled sample obtained by mixing an equal volume (10–100 μL) of each sample to be studied, and therefore constitute a representative bulk control sample so as the signal variations of any metabolite could be reflected in the QCs [21]. QC samples proved really useful to correct for drifts in MS response, mass measurement accuracy, as well as chromatographic retention time between analytical runs or batches [22, 23]. Another particularly relevant feature of QCs implies dilution of QCs to evaluate the corresponding MS response linearity. Under these conditions, the metabolites showing a linear trend can be considered as analytically relevant. Figure 2 depicts a typical sample run order used for LC-HRMS-based metabolomics.

Injection Order	Sample
1	Blank
2	Blank
3	QC
4	QC
5	QC
6	QC
7	QC
8	Blank
9	8x dil. QC
10	4x dil. QC
11	2x dil. QC
12	QC
13	Blank
14	QC
15	Sample 1
16	Sample 2
17	Sample 3
...	...
24	Sample 10
25	Blank
26	QC
27	Sample 11
...	...
36	Sample 20
37	Blank
38	QC
...	...

Fig. 2 Typical sample run order for LC-HRMS. If needed, additional inter-batches QCs can be added

3.5.3 Data Processing

Data processing as well as statistical analysis can be performed automatically and reproducibly online using the Workflow4metabolomics (W4M) platform for computational metabolomics [24, 25]. W4M is built on the Galaxy environment providing intuitive and powerful features that allow the analyst to set up and run complex workflows (Fig. 3). Four main processing steps can be considered.

1. Preprocessing: Automatic peak detection, alignment, and extraction using the XCMS software.
2. Normalization thanks to QCs: Correction of signal variation, selection of analytically pertinent metabolites.
3. Statistical analyses (univariate and multivariate statistical tools). Other statistical software such as SIMCA-P (Umetrics, Umea, Sweden) can be used to perform multivariate data analyses such as principal component analysis (PCA) and partial least squares discriminant analysis (PLS-DA) or Prism (GraphPad, La Jolla, USA) to perform univariate data analyzes (e.g., *t*-test).
4. Compound annotation is commonly performed considering a ±10 ppm mass tolerance using either our in-house database (including ~1000 metabolites) [14, 19, 26] or publically available databases such as KEGG [27], HMDB [28], or METLIN [29].
5. Confirmation of metabolite annotation and identification of other statistically relevant unknown metabolites is accomplished by MS/MS experiments. Resulting MS/MS spectra are compared to those included in our in-house database or in public databases such as METLIN. Figure 4 summarizes our workflow for formal metabolite identification (*see* **Notes 8** and **9**).

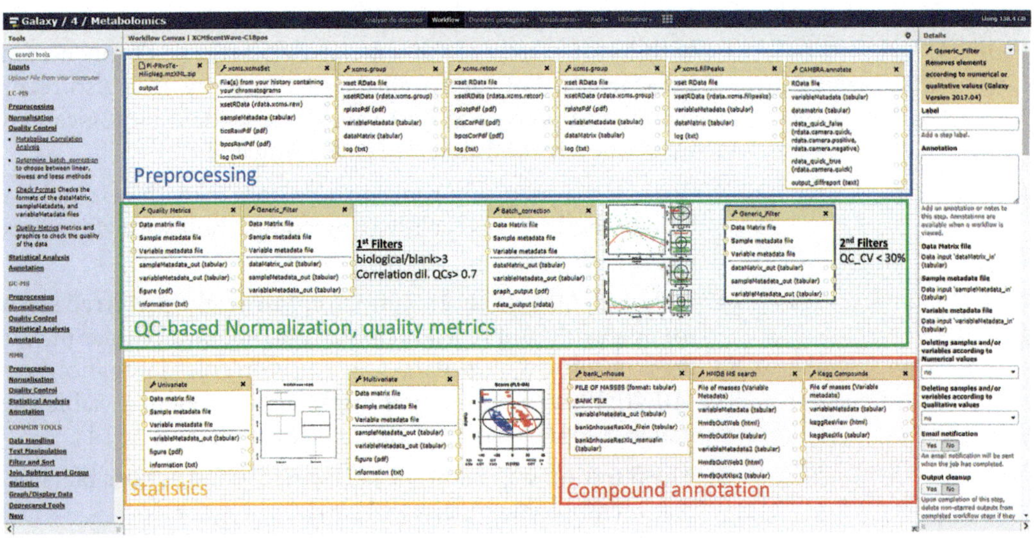

Fig. 3 Data treatment tools available in the W4M infrastructure. The left and right panels give the tools and parameters that can be used to build the workflow in the central panel

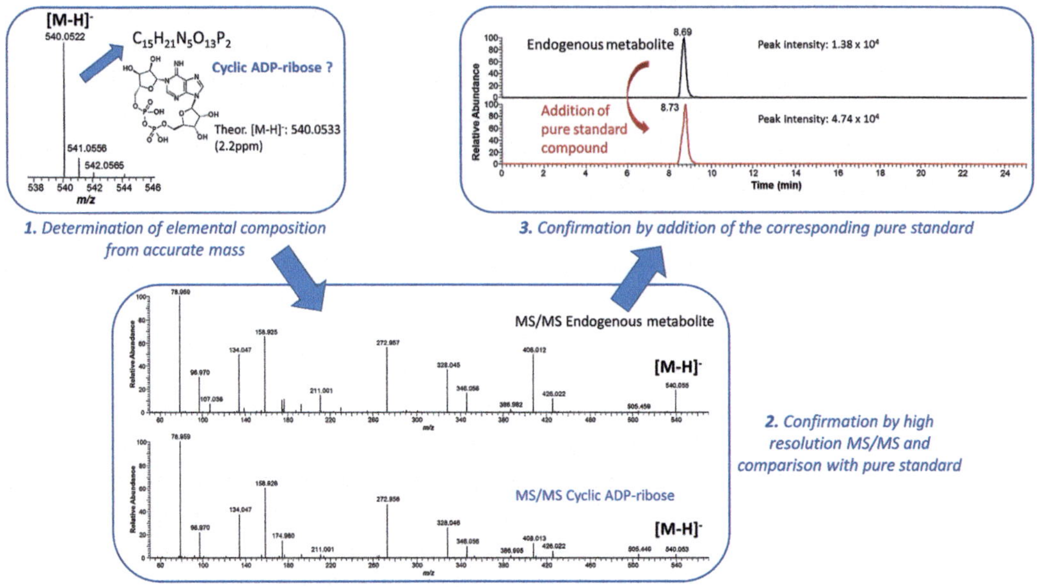

Fig. 4 Analytical workflow for HRMS-based metabolite identification

4 Notes

1. Growth curves need to be carefully determined for each strain studied. They are used to determine the OD_{600} at which the bacteria should be harvested to correspond to early-, mid-, and late-exponential or stationary phases.
2. Bacterial metabolism is a very fast changing and versatile process (especially during the exponential phase). Therefore, the intracellular metabolome of each studied strain needs to be withdrawn exactly at the same growth stage to ensure accurate comparison. For improved data consistency, it is also recommended to prepare at least three independent biological replicates of each condition/strain investigated.
3. A single washing step proved to be optimal with no significant metabolite leakages/cell lysis by comparison with three washing rounds [14].
4. Although 60% ethanol solution represents the preferred solvent for efficiently and reproducibly extracting intracellular metabolites from *S. aureus*, an acetonitrile/methanol/water (40:40:20, v/v/v) mixture might also represent a viable alternative if some particular metabolites are targeted [14].
5. The filtration step has to be performed as fast as possible, not to bias metabolome quality [14, 30].

6. By spiking ^{15}N-labeled nucleotides in bacterial extracts, the limits of detection (LODs) were estimated at 55, 45, and 230 nM for AMP, ADP, and ATP, respectively.
7. Exponentially growing bacterial cells should have a stable AEC above 0.8, while stressed cells would have lower values [31]. Using our optimized fast-filtration protocol, we can routinely and reproducibly obtain an AEC of 0.76 ± 0.02. This value is slightly lower than the theoretically expected value of 0.8 and might seem insufficient at first sight. However, the high reproducibility of the measured AEC values indicated that these values were probably not related with an incorrect handling of the cells during sampling but rather to a particular physiological characteristic of these cells under our growth conditions. Several authors have already reported AEC values significantly and negatively impacted (down to ~0.1) by parameters such as the growth phase and the culture medium itself for cells with maintained metabolic capability [32, 33]. Therefore, the AEC value might not be considered as a strict indicator of metabolic integrity, but rather of particular physiological characteristics observed under specific growth conditions.
8. Under optimal conditions and according to the Metabolomic Standards Initiative criteria [17], we were able to characterize up to 210 metabolites in *S. aureus*, of which 173 and 9 were identified and putatively annotated, respectively, while the remaining 28 were only characterized by their accurate masses [14].
9. As a representative example, Fig. 5 represents the PLS-DA score plot of features from ZIC-pHILIC/MS fingerprints of MRSA and MSSA strains harvested at the mid-exponential growth phase. Each bacterial strain was then treated with cefoxitin

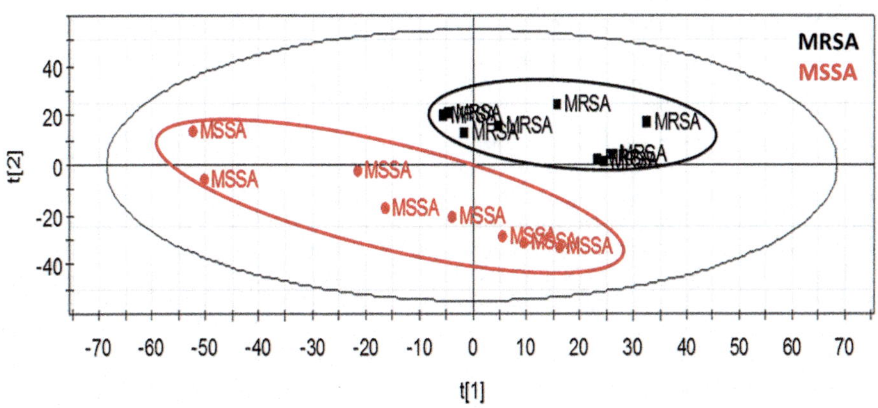

Fig. 5 PLS-DA score plot of features from ZIC-pHILIC/MS fingerprints of MRSA and MSSA harvested at the mid-exponential growth phase and then exposed to cefoxitin. The PLS-DA model was validated by a permutation test (100 times)

(β-lactam) at its minimal inhibitory concentration (MIC × 1) during both pre-culture and culture steps. The discriminant metabolites were selected by combining multivariate variable importance in the projection (VIP) obtained from the PLS-DA model and univariate *p*-values (nonparametric Mann–Whitney statistical test). The metabolites were considered as discriminant when VIP > 1.5 and *p*-value <0.05. Under these conditions, up to 26 metabolites discriminate MSSA and MRSA strains. Thus, intracellular levels of some nucleotides, amino acids, or metabolites belonging to glycolysis of Krebs cycle were also modified. Other metabolites such as N-acetyl-muramoate-6-phosphate, UDP-N-MurNAc-Ala-Glu-Lys-Ala-Ala, N-acetyl-glucosamine/galactosamine, and glycerol-3-phosphate involved in cell-wall biosynthesis were also discriminant. These data tended to reinforce what has been observed without any antibiotic exposure [14].

Acknowledgements

This work was supported by bioMérieux S.A. and the Association Nationale de la Recherche et de la Technologie (ANRT). S.A.-C. is the recipient of a CIFRE fellowship (grant number 2011/1474). This work was also supported by the Commissariat à l'Energie Atomique et aux Energies Alternatives and the MetaboHUB infrastructure (ANR-11-INBS-0010 grant).

References

1. Lowy FD (1998) *Staphylococcus aureus* infections. N Engl J Med 339:520–532
2. Chambers HF, Deleo FR (2009) Waves of resistance: *Staphylococcus aureus* in the antibiotic era. Nat Rev Microbiol 7:629–641
3. Gardete S, Tomasz A (2014) Mechanisms of vancomycin resistance in *Staphylococcus aureus*. J Clin Invest 124:2836–2840
4. Chastre J, Blasi F, Masterton RG et al (2014) European perspective and update on the management of nosocomial pneumonia due to methicillin-resistant *Staphylococcus aureus* after more than 10 years of experience with linezolid. Clin Microbiol Infect Off Publ Eur Soc Clin Microbiol Infect Dis 20(Suppl 4):19–36
5. Hassoun A, Linden PK, Friedman B (2017) Incidence, prevalence, and management of MRSA bacteremia across patient populations-a review of recent developments in MRSA management and treatment. Crit Care Lond Engl 21:211
6. Boucher HW, Corey GR (2008) Epidemiology of methicillin-resistant *Staphylococcus aureus*. Clin Infect Dis Off Publ Infect Dis Soc Am 46(Suppl 5):S344–S349
7. Sommer MOA, Dantas G (2011) Antibiotics and the resistant microbiome. Curr Opin Microbiol 14:556–563
8. Munck C, Gumpert HK, Wallin AIN et al (2014) Prediction of resistance development against drug combinations by collateral responses to component drugs. Sci Transl Med 6:262ra156
9. Belenky P, Ye JD, Porter CBM et al (2015) Bactericidal antibiotics induce toxic metabolic perturbations that lead to cellular damage. Cell Rep 13:968–980
10. Ling LL, Schneider T, Peoples AJ et al (2015) A new antibiotic kills pathogens without detectable resistance. Nature 517:455–459
11. Liebeke M, Meyer H, Donat S et al (2010) A metabolomic view of *Staphylococcus aureus* and its ser/thr kinase and phosphatase deletion

mutants: involvement in cell wall biosynthesis. Chem Biol 17:820–830

12. Ammons MCB, Tripet BP, Carlson RP et al (2014) Quantitative NMR metabolite profiling of methicillin-resistant and methicillin-susceptible *Staphylococcus aureus* discriminates between biofilm and planktonic phenotypes. J Proteome Res 13:2973–2985
13. Dörries K, Schlueter R, Lalk M (2014) Impact of antibiotics with various target sites on the metabolome of *Staphylococcus aureus*. Antimicrob Agents Chemother 58:7151–7163
14. Aros-Calt S, Muller BH, Boudah S et al (2015) Annotation of the *Staphylococcus aureus* Metabolome using liquid chromatography coupled to high-resolution mass spectrometry and application to the study of methicillin resistance. J Proteome Res 14:4863–4875
15. Schelli K, Zhong F, Zhu J (2017) Comparative metabolomics revealing *Staphylococcus aureus* metabolic response to different antibiotics. Microb Biotechnol 10:1764–1774
16. Keaton MA, Rosato RR, Plata KB et al (2013) Exposure of clinical MRSA heterogeneous strains to β-lactams redirects metabolism to optimize energy production through the TCA cycle. PLoS One 8:e71025
17. Sumner LW, Amberg A, Barrett D et al (2007) Proposed minimum reporting standards for chemical analysis chemical analysis working group (CAWG) metabolomics standards initiative (MSI). Metabolomics 3:211–221
18. Liebeke M, Lalk M (2014) *Staphylococcus aureus* metabolic response to changing environmental conditions—a metabolomics perspective. Int J Med Microbiol 304:222–229
19. Boudah S, Olivier M-F, Aros-Calt S et al (2014) Annotation of the human serum metabolome by coupling three liquid chromatography methods to high-resolution mass spectrometry. J Chromatogr B Analyt Technol Biomed Life Sci 966:34–47
20. Martano G, Delmotte N, Kiefer P et al (2015) Fast sampling method for mammalian cell metabolic analyses using liquid chromatography-mass spectrometry. Nat Protoc 10:1–11
21. Naz S, Vallejo M, García A et al (2014) Method validation strategies involved in non-targeted metabolomics. J Chromatogr A 1353:99–105
22. Dunn WB, Broadhurst D, Begley P et al (2011) Procedures for large-scale metabolic profiling of serum and plasma using gas chromatography and liquid chromatography coupled to mass spectrometry. Nat Protoc 6:1060–1083
23. Dunn WB, Wilson ID, Nicholls AW et al (2012) The importance of experimental design and QC samples in large-scale and MS-driven untargeted metabolomic studies of humans. Bioanalysis 4:2249–2264
24. Giacomoni F, Le Corguillé G, Monsoor M et al (2015) Workflow4Metabolomics: a collaborative research infrastructure for computational metabolomics. Bioinformatics 31:1493–1495
25. Guitton Y, Tremblay-Franco M, Le Corguillé G et al (2017) Create, run, share, publish, and reference your LC-MS, FIA-MS, GC-MS, and NMR data analysis workflows with the Workflow4Metabolomics 3.0 Galaxy online infrastructure for metabolomics. Int J Biochem Cell Biol 93:89–101
26. Roux A, Xu Y, Heilier J-F et al (2012) Annotation of the human adult urinary metabolome and metabolite identification using ultra high performance liquid chromatography coupled to a linear quadrupole ion trap-Orbitrap mass spectrometer. Anal Chem 84:6429–6437
27. Kanehisa M, Goto S (2000) KEGG: Kyoto encyclopedia of genes and genomes. Nucleic Acids Res 28:27–30
28. Wishart DS, Tzur D, Knox C et al (2007) HMDB: the human Metabolome database. Nucleic Acids Res 35:D521–D526
29. Smith CA, O'Maille G, Want EJ et al (2005) METLIN: a metabolite mass spectral database. Ther Drug Monit 27:747–751
30. Meyer H, Liebeke M, Lalk M (2010) A protocol for the investigation of the intracellular *Staphylococcus aureus* metabolome. Anal Biochem 401:250–259
31. Chapman AG, Fall L, DE A (1971) Adenylate energy charge in *Escherichia coli* during growth and starvation. J Bacteriol 108:1072–1086
32. van der Werf MJ, Overkamp KM, Muilwijk B et al (2008) Comprehensive analysis of the metabolome of *Pseudomonas putida* S12 grown on different carbon sources. Mol BioSyst 4:315–327
33. Stuani L, Lechaplais C, Salminen AV et al (2014) Novel metabolic features in *Acinetobacter baylyi* ADP1 revealed by a multiomics approach. Metabolomics 10:1223–1238

Chapter 19

Nuts and Bolts of Protein Quantification by Online Trypsin Digestion Coupled LC-MS/MS Analysis

Christopher A. Toth, Zsuzsanna Kuklenyik, and John R. Barr

Abstract

Protein digestion coupled to liquid chromatography and tandem mass spectrometry (LC-MS/MS) detection enables multiplexed quantification of proteins in complex biological matrices. However, the reproducibility of enzymatic digestion of proteins to produce proteotypic target peptides is a major limiting factor of assay precision. Online digestion using immobilized trypsin addresses this problem through precise control of digestion conditions and time. Because online digestion is typically for a short time, the potential for peptide degradation, a major source of measurement bias, is significantly reduced. Online proteolysis requires minimal sample preparation and is easily coupled to LC-MS/MS systems, further reducing potential method variability. We describe herein a method optimized for the multiplexed quantification of several apolipoproteins in human serum using on-column digestion. We highlight key features of the method that enhance assay accuracy and precision. These include the use of value-assigned serum as calibrators and stable isotope-labeled (SIL) peptide analogs as internal standards. We also comment on practical aspects of column switching valve design, instrument maintenance, tandem mass spectrometry data acquisition, and data processing.

Key words Online digestion, Immobilized enzyme reactor, IMER, Apolipoproteins, Quantitative, IDMS, IMER-LC-MS/MS

1 Introduction

Mass spectrometry (MS) approaches have become effective means for selective quantitative measurement of proteins in complex biological matrices [1]. Targeted multiplexed assays for quantification of up to 30–50 proteins involve proteolytic digestion, most often with trypsin, and analysis of resulting unique peptides by liquid chromatography with tandem mass spectrometry (LC-MS/MS) detection. The use of stable isotope-labeled (SIL) peptide or protein analogs as internal standards further enhances precision by counteracting (normalizing) for matrix effects. This approach is commonly referred to as isotope dilution MS (IDMS). The IDMS approach is based on the assumption that there is a proportionate relationship between the measured MS/MS response ratio and

Xing Wang and Matthew Kuruc (eds.), *Functional Proteomics: Methods and Protocols*, Methods in Molecular Biology, vol. 1871, https://doi.org/10.1007/978-1-4939-8814-3_19,

analyte molar ratio both before and after digestion. Furthermore, this relationship needs to hold true both in the calibrators and in the unknown samples regardless of heterogeneous proteoforms [2].

Protein digestion is commonly performed in-solution, on a batch of samples, off-line from IDMS analysis. A typical batch digestion protocol involves denaturation, reduction and alkylation, enzymatic digestion, and some form of sample cleanup, i.e., solid phase extraction. A main challenge of batch digestion is in the selection of optimal digestion conditions for multiple proteins, each with a potentially wide range of concentrations in vivo. Long digestion times and high temperatures typically result in more complete proteolytic cleavage of the target peptides. In practice however, incubation times ≥3 h and temperatures >37 °C give rise to several non-idealities, particularly the aggregation, adsorption, and chemical degradation of cleavage products and their SIL analogs. Peptide decay during in-solution digestion is a main source of calibration bias and method variability [3].

An emerging alternative to in-solution digestion is online or on-column digestion [4], where trypsin is covalently bound to porous particles of packing material in an LC column. This immobilized enzyme reactor (IMER) can be directly connected to the LC-MS/MS system. Continuous plug-flow IMERs digest proteins on-column as they pass through the enzyme laden stationary phase. Samples are easily combined with SIL internal standards immediately prior to digestion.

Theoretically, the recovery of the peptide cleavage products collected at the outlet of the IMER is determined by the flow rate, temperature, void volume, injection volume, and the concentration of the proteins in the injected sample [5]. If all conditions, barring the protein concentration, are held constant and in an optimal range, the measured [peptide cleavage product peak area]/[SIL peptide peak area] ratio has a linear relationship with the [native protein]/[SIL peptide] molar ratios in the injected sample. With constant moles of SIL peptide, the peak area ratio also has a linear correlation with the native protein concentration in the injected sample.

IMER-LC-MS/MS is emerging as a valuable tool to facilitate precise relative quantitation of proteins. The main advantage of online digestion is the accelerated cleavage of the target peptides due to high enzyme–protein ratio inside the IMER. This allows for short reaction times (2–8 min), which reduce the opportunity for chemical degradation of the proteins, the peptide cleavage products and the SIL peptides, enhancing reproducibility. Immobilization greatly reduces the propensity of trypsin for autolysis and denaturation, allowing digestion at temperatures > 37 °C (further enhancing cleavage rates), and the repeated use of the IMER for the consecutive digestion of numerous samples. In addition, minimal sample preparation is required, lending to simple dilute-and-shoot methods. Sample cleanup is also performed online, as target peptides elute from the IMER they are trapped on a short reverse phase LC column where the peptides can be desalted and washed prior to LC-MS/MS analysis. All valve switching intervals, reagent additions, digestion times, and flow rates are controlled in a precise and automated fashion. Overall, relative to off-line batch digestion, the IMER digestion is more simple, reproducible, and economical in terms of labor and reagent cost.

The main limitation of online IMER digestion is that in many cases complete proteolysis cannot be achieved. Therefore, to obtain quantitative accuracy requires the use of matrix and concentration range representative protein calibrators. Additionally, the IMER-LC-MS/MS technique typically works best if the target proteins are present at >100 nmol/L concentration in serum [5].

While the use of online trypsin digestion with LC-MS/MS for qualitative analysis of proteins is relatively well demonstrated in the literature, its application for quantitative protein analysis is rarely discussed [6, 7]. In this chapter, we use a panel of apolipoproteins in serum as an example to address the practical aspects of online trypsin digestion for multiplexed protein quantification.

2 Materials

2.1 Reagents and Consumables

1. LC-MS grade acetonitrile and 2-propanol.
2. Ultrapure deionized water.
3. Diluting buffer: 10 mM $NaHCO_3$, 150 mM NaCl, pH 7.4 (*see* **Note 1**).
4. Detergent solution: 0.45% w/v Zwittergent 3–12 in sample buffer (*see* **Note 2**).
5. 0.5 mL 96-well round bottom microplates (Agilent).
6. Slit Seal 96-well microplate covers (BioChromato).
7. 1.5 mL standard polypropylene autosampler vials and caps (Thermo Fisher).
8. SIL (^{13}C, ^{15}N) analogs of the proteotypic peptides of interest (*see* **Note 3**).

2.2 HPLC Columns

1. IMER: Trypsin column, 33 × 2.1 mm id (Perfinity Biosciences) (*see* **Note 4**).
2. Trapping/desalting column: Halo® ES-C18 peptide, 4.6 × 5 mm id, 2.7 μm particle diameter (*see* **Note 5**).
3. Analytical column: Halo® C18 core shell, 2.1 mm × 100 mm, 2.7 μm particle size or similar.

2.3 HPLC Solvents

1. Digestion buffers (quaternary pump).
   (a) Digestion buffer: 50 mM Tris–HCl, 2 mM $CaCl_2$, pH 8.4, in deionized water.
   (b) Wash buffer: 50 mM Tris–HCl, 2 mM $CaCl_2$, 25% 2-propanol, pH 8.4, in deionized water.
2. Gradient Eluents (binary pump).
   (a) Eluent A: 0.1% formic acid in water.
   (b) Eluent B: 0.1% formic acid in acetonitrile.

2.4 LC-MS/MS Instrumentation

1. Shimadzu 20/30 series HPLC system (Perfinity Biosciences) (*see* **Note 6**).
2. Mass spectrometer: QTrap® 6500+ (SCIEX) or any suitable triple quadrupole mass analyzer.
3. Ionization source: Electrospray ionization.
4. LC Control Software: Labsolutions (Shimadzu).
5. MS Control Software: Analyst (SCIEX).

2.5 Data Processing Software

1. Skyline (MacCoss Lab) (*see* **Note 7**).
2. MultiQuant (SCIEX) (*see* **Note 8**).

2.6 Human Serum

Frozen human serum from individual donors was purchased from a specimen depository (Bioreclamation IVT, Westbury, NY).

3 Methods

3.1 Preparation of SIL Peptide Stock Solutions

1. Dissolve 2–5 mg of lyophilized powder in 100 mL of 5% acetonitrile/0.1% formic acid in water (*see* **Note 9**).
2. Distribute each peptide into 200–500 μL aliquots and store at −80 °C (*see* **Note 10**).

3.2 Preparation of Working SIL Peptide Mix

1. Thaw one aliquot of each SIL stock solution on the bench, vortex-mix, and spin down.
2. To a 15 mL Falcon™ conical centrifuge tube, add 20–450 μL of each individual SIL peptide stock solution (*see* **Notes 11** and **22**).
3. Add 100 μL of a 0.1% Zwittergent 3–12/0.1% formic acid in water solution, cap, and vortex-mix (*see* **Note 12**).
4. Bring to 10 mL volume with 0.1% formic acid in water.
5. Store the SIL peptide mix for maximum 4 weeks at 4 °C (*see* **Note 13**).
6. Before analysis dispense a 1.5 mL aliquot into a standard polypropylene autosampler vial and place into the reagent tray of the autosampler.

3.3 Preparation of Calibrator Pool and Storage

1. Allow frozen serum from individual donors to thaw on ice.
2. Merge equal volumes from each donor (*see* **Note 14**).
3. Vacuum filter the pool using a Nalgene™ Rapid-Flow™ filter (0.45 μm pore size) or equivalent.
4. Continuously stir the pool at 250 rpm in an ice bath on a magnetic stir plate while distributing a series of 100, 50, 20, 10, and 5 μL aliquots using a positive displacement repeater pipette into 2 mL storage vials.
5. Verify the volume of each aliquot by mass using an analytical balance and record.
6. Store the aliquots at −80 °C until use.
7. For the value assignment of the calibrator pool by standard addition methodology, *see* **Note 15**.

3.4 Preparation of Standard Series

1. Pull a set of stored Calibrator Pool Aliquots (100, 50, 20, 10 and 5 μL) from the freezer and allow to thaw on ice.
2. Bring to volume with diluting buffer (Subheading 2.1, **step 3**) following an appropriate dilution level (*see* **Note 16**).
3. Gently invert each tube 10 times to mix and spin down.
4. Use immediately or store at 4 °C for no more than 4 days (*see* **Note 17**).

3.5 Sample Preparation for IMER-LC-MS/MS Analysis

1. Pipette 10 μL of serum sample to a microcentrifuge tube.
2. Add 990 μL of sample diluent (Subheading 2.1, **step 3**).
3. Invert 10 times to mix and briefly spin down.
4. Transfer 100 μL of diluted sample to a standard PP autosampler vial or microplate.
5. Add 50 μL 0.45% w/v Zwittergent 3–12 solution (Subheading 2.1, **step 4**) to achieve a final concentration of 0.15% w/v.
6. Place on shaker plate for 5 min at 500 rpm.
7. Store at 4 °C prior to IMER-LC-MS/MS analysis or place immediately into chilled autosampler compartment.

3.6 IMER-LC-MS/MS Operation

1. Set the temperature of the column compartment to 50 °C and autosampler compartment to 4 °C.
2. Program the autosampler via the Labsolutions software to draw 5 μL from the reagent vial containing the SIL mix, followed by of 50 μL sample (*see* **Note 18**).
3. Digest the sample for 7 min using a flow rate of 20 μL/min (*see* **Note 19**).
4. Peptide separation: The binary gradient flow rate should be set to 0.5 mL/min. The starting composition should be set to ≤2% Solvent B. The low initial organic composition improves

focusing and can be quickly ramped to achieve fast and efficient peptide separation. The start of gradient elution must be synchronized with the trap column switching valve. To enhance sample throughput, the plumbing of the column switching system can be modified for dual operation mode, with simultaneous digestion/trapping and gradient elution of consecutive samples (*see* **Notes 20** and **21**).

5. The optimal operation of the mass spectrometer; 6500+ QTRAP® (Sciex, Foster City, CA) with a heated electrospray ionization probe in positive ion mode requires the setting of the following parameters: Ion spray voltage 5500 V, ion source heater temperature 450 °C, source gas 50 psi, curtain gas 35 psi. The native and the isotopically labeled IS peptide chromatograms are acquired by multiple reaction monitoring (MRM) with unit mass resolution, in scheduled 60 s acquisition windows with a 0.65 s target scan time (*see* **Note 22**).
6. For the highest accuracy and precision, each sample unknown should be analyzed in triplicate, and calibrators in duplicate. After the highest calibrator, duplicate blank injections are recommended to avoid carry-over from trypsin column (*see* **Note 23**).

3.7 Data Processing

1. Pull the raw mass spectrometry data into MultiQuant (SCIEX) for the calculation of [native cleavage peptide peak area]/[SIL peptide peak area] ratios.
2. Examine the automated peak integrations.
3. Before reporting concentrations review the data based on acceptable range of SIL peak intensities, product ion ratios, and deviations between technical repeats.

4 Notes

1. Phosphate containing buffer should not be used to dilute samples as it quickly forms insoluble precipitates with $CaCl_2$ in the digestion buffer inside the flow lines.
2. The addition of a mild detergent at near critical micelle concentration was found to be critical to enhance method sensitivity. As shown in Fig. 1 using bovine serum albumin as an example, addition of detergent gave a marked increase (20-fold in some cases) in peptide peak areas.
3. Recommendations for the selection, preparation, and storage of labeled synthetic peptides have been previously described [8]. In addition, a list of target peptides specific for apolipoprotein quantification and corresponding SIL peptides are described in a previous publication [7].

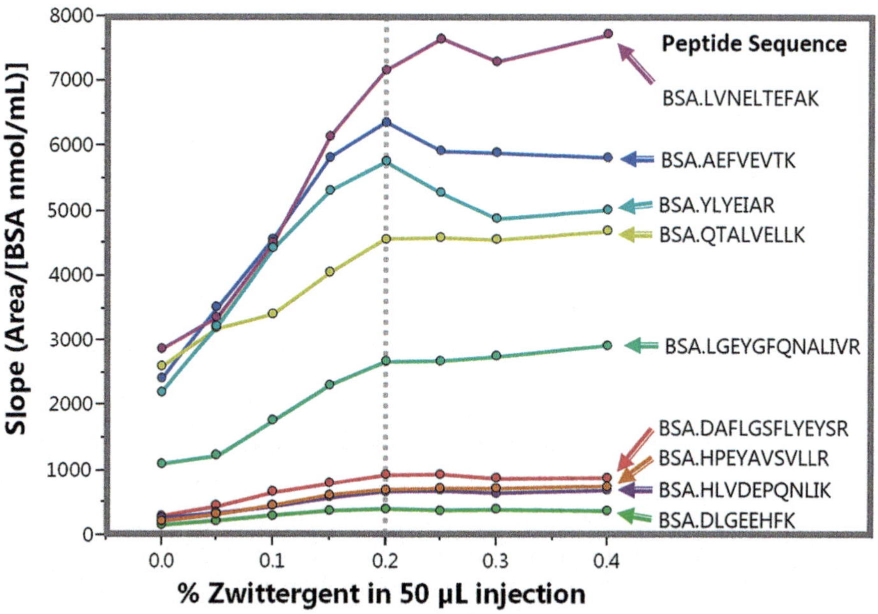

Fig. 1 The effect of Zwittergent 3-12 concentration on the digestion efficacy is demonstrated with bovine serum albumin as a model protein. As the final detergent concentration in the sample well increases, the peptide peak area/protein concentration increases for all tryptic peptides. This effect levels off just beyond the critical micelle concentration (CMC) of Zwittergent 3-12 (4 mM, 0.15% w/v)

4. Trypsin columns available from Perfinity Biosciences could be reliably used for 2000 injections of dilute human serum and can be kept in a 50 °C column compartment for 60 days without significant loss of efficiency. A decrease in digestion efficacy, less than 30%, can be corrected for by the measurement of the calibrator pool with every batch.
5. A 4.6 mm id trapping column had twice the life span compared to 2.1 mm id trap columns. Each 4.6 mm id trapping column reliably functioned for 200 injections of 1:100 diluted human serum before chromatographic artifacts developed and warranted replacement.
6. The fully integrated digestion and LC platform (Perfinity Biosciences/Shimadzu Scientific Instruments) consisted of an autosampler (SIL-20 ACHT with pretreatment option), column oven (CT0-20AC), a low pressure ($\leq$ 3000 psi) quaternary pump with solvent selection valve to alternate between digestion and wash buffers (LC-20 AD), two high-pressure HPLC pumps (LC-20ADXR) and control module (CBM-20A). The system must be operated in "XL mode" in order to program the required pretreatment operations with the SIL-20 ACHT autosampler. The LC and MS systems were controlled from separate computers with synchronization

through a trigger cable between the LC control module and the MS instrument.

7. Skyline software was used for the generation of the scheduled MRM acquisition method including declustering potentials and collision energies, and were used without further optimization.
8. Multiquant software was used for MS/MS peak integration and assessment of accuracy and precision of concentration measurements.
9. Peptide stock solutions should be relatively concentrated (0.5–2 nmol/μL) for accurate determination of purity by amino acid analysis and to reduce the significance of peptide adsorption to the container wall during storage. Some hydrophobic peptides may require between 5 and 30% acetonitrile and 0.1–1% formic acid to fully dissolve [8].
10. For long term storage (>1 year at −20 to −80 °C) lyophilization of precise aliquot volumes is recommended [8]. The concentration of SIL peptide is not considered when native protein concentration is calculated. However, initial and periodical amino acid analysis is good practice to confirm SIL peptide purity and storage stability.
11. As a general rule, the concentration of the SIL peptides should be optimized to match the SIL peak areas with the median native peptide peak areas from the digestion of 1:100 diluted samples and the calibrator pool.
12. The addition of 0.001% w/v Zwittergent 3–12 reduces peptide adsorption to container walls during storage.
13. Because the absolute concentration of the SIL peptides are not factored into the calculation of the protein concentration minor peptide degradation of SIL peptides in the working mix over the course of 30 days storage at 4 °C did not negatively affect assay performance. For longer storage (>4 weeks), the working SIL peptide mix should be aliquotted and stored frozen (−20 to −80 °C) until use. Avoid multiple freeze/thaw cycles if possible.
14. Recommendations for the preparation of calibrator pools have been proposed [9].
15. The absolute concentration of several apolipoproteins in the calibrator pool were determined using standard addition methodology. Commercially available purified proteins were purchased from Academy Biomedical, Novoprotein Scientific, and Sigma Aldrich. Lyophilized powder was preferred where possible and purities ranged from 95 to 98% as stated by the manufacturer. The protein stock solutions were prepared in

sodium bicarbonate buffered saline (10 mM $NaHCO_3$, 150 mM NaCl, pH 7.4) to approximate concentrations of 600 nmol/L apoA-II, 55 nmol/L apoA-IV, 150 nmol/L apoC-I, 80 nmol/L apoC-II, 450 nmol/L apoC-III, and 90 nmol/L apoE3. The concentrations were chosen to approximate 2–5 times the expected concentration in 1:100 diluted matrix. Concentrations of the individual protein stock solutions were measured by amino acid analysis by Midwest Bio-Tech (Fishers, IN, USA). For each protein of interest, a value-assigned protein stock solution was diluted in triplicate to a 7-point series with 1, 1.25, 1.67, 2.5, 5, 10, and 20-fold dilution factors, in addition to a buffer blank. A 50 μL aliquot at each dilution level was added to the 50 μL aliquots of 1:100 diluted serum and 50 μL of 0.45% Zwittergent 3–12. The plate was covered with a slit-seal cover, mixed on a shaker plate for 5 min at 500 rpm, then stored in an autosampler at 4 °C prior to analysis. Samples were analyzed by IMER-LC-MS/MS. Light/heavy area ratio vs. protein dilution ratio linear regression curves were calculated for each peptide transition, of which one peptide transition for each protein was selected for quantification (Fig. 2). The concentration of each protein of

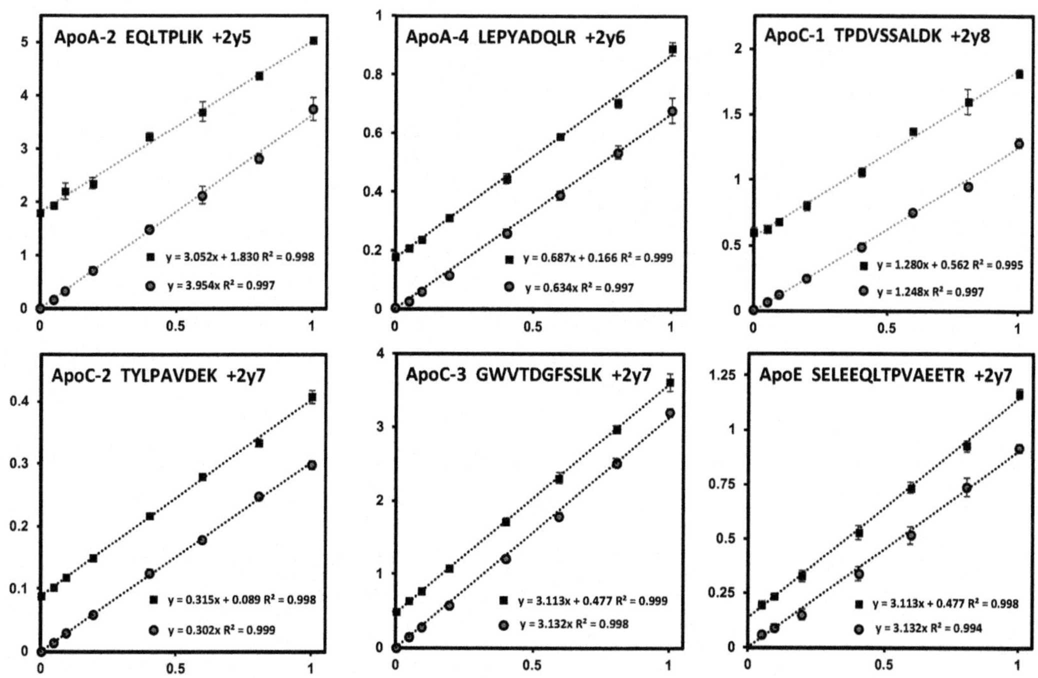

Fig. 2 Area ratio (*Y*-axis) versus concentration ratio (*X*-axis) for standard addition of purified recombinant proteins to a spiked matrix (square) and solvent blank (circle) for slope comparison. Error bars indicate standard deviation ($N = 3$). The comparable slope indicates that spiked recombinant proteins digest in an analogous way to endogenous protein in diluted matrix, a critical assumption for this method

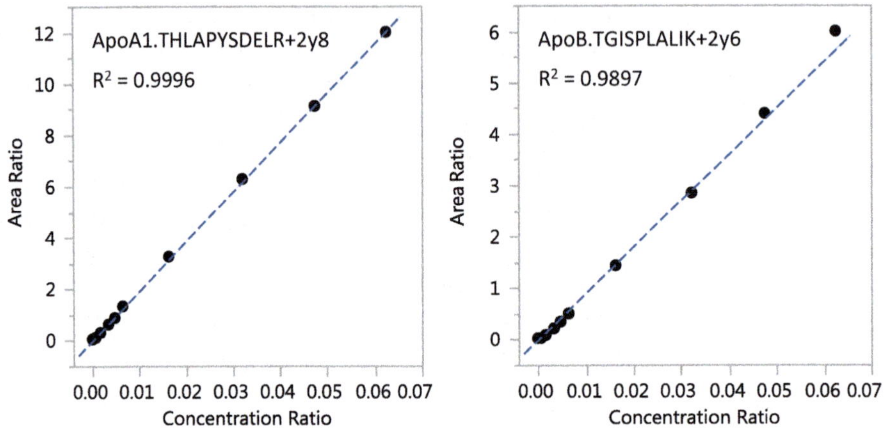

Fig. 3 Demonstration of the linear dynamic range for two selected peptide transitions, apoA1: THLAPYSDELR 2y8 and apoB-100: TGISPLALIK 2y6 in diluted serum calibrator. The nine point calibration curve represents 1:15 and 1:1500 fold dilutions at the ULOQ and LLOQ, respectively

interest in the calibrator serum pool was calculated by dividing the area ratio y-intercept with the regression slope, multiplied by the concentration of the protein stock solution (from AAA) and the dilution factor. To compare the digestion efficiency between endogenous proteins and their spiked recombinant analogs, standard addition was also performed in triplicate on a buffer blank. In the case of apoA-I and apoB-100, standard addition was performed using certified reference materials (CRM) from the World Health Organization (WHO), SP1-01 and SP3-08, respectively, to assign apoA-I and apoB-100 concentrations to the calibrator serum pool.

16. *See* Fig. 3.
17. When the calibrator pool was diluted to prepare the standard series, these materials were stable for only 4 days at 4 °C. The frequent re-preparation of the standard series was one of the main sources of method variability. Storing the calibrator pool in a range of volumes verified by mass, and dilution only after thawing provided better calibration reproducibility.
18. For addition of internal standard, the injection sequence should be programmed to draw the SIL peptide mix and the protein sample sequentially into the sample loop, then transfer both simultaneously into the IMER by the 20 μL/min digestion buffer flow. The SIL peptides and the peptide cleavage product are retained together on the trapping column and eluted through the analytical column into the LC-MS interface. The authors preferred this method over addition of SIL peptide mix directly into the sample well, as some SIL peptides show signs of degradation in dilute endogenous matrix. However, SIL peptides kept in a concentrated mixture within the

control rack remained stable for several weeks at 4 °C. In addition, the automated addition of internal standard improved method precision and reduced sample preparation time and pipetting steps. The Labsolutions pretreatment program is shown in Table 1.

19. The combined sample & SIL mix injection-plug is carried through the IMER by digestion buffer (Subheading 2.3, **step 2a**) at a rate of 20 μL/min for 7 min (Fig. 4). After exiting the

Table 1
Labsolutions pretreatment program

Line #	Value
1	al=1001
2	vial al
3	n.strk ns
4	aspir 5,5
5	air.a 1,5
6	rinse 200,50
7	vial sn
8	n.strk ns
9	aspir iv,ss
10	air.a 1,5
11	rinse 200,50
12	inj.p
13	s.inj
14	end

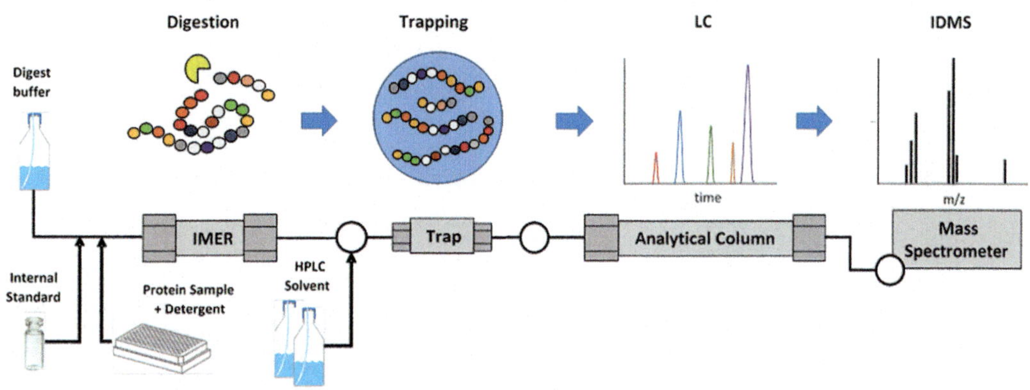

Fig. 4 Simplified diagram of the IMER-LC-MS/MS system showing sample flow path and valve positions

IMER, the native cleavage products, the SIL peptides, and remaining matrix components are retained on the trapping/desalting column, while salts and unretained matrix components pass into waste. From 7 to 7.5 min, the flow rate is increased to 0.5 mL/min to purge out the remaining cleavage products from the IMER and flow lines into the trap. At 7.5 min, the loop encompassing the trapping column is idled (with cleavage products and SIL peptides), while the IMER is washed at 2 mL/min with wash buffer (Subheading 2.3, **step 2b**) for 2 min, then re-equilibrated with digestion buffer for 2 min.

20. The start of gradient elution is synchronized with the turn of the trap column switching valve. This allows the elution flow of the binary gradient pumps to carry the trapped cleavage products and SIL peptides to the analytical column and to the LC-MS interface. To initiate the peptide transfer from the trapping column to the analytical column, Solvent B was ramped in 0.25 min to 8% after injection. Solvent B is then increased 8–16% during 0.25–7 min, 16–25% during 7–8 min, and 25–95% during 8–9 min, held at 95% during 9–10.5 min before re-equilibrating to initial conditions during 10.5–11.5. For more polar peptides than those described in this method, initial conditions may require longer loading times at lower initial solvent B content to avoid significant peak broadening and/or inconsistent retention times. Example LC parameters and a corresponding solvent composition curve are shown in Fig. 5 (Table 2).

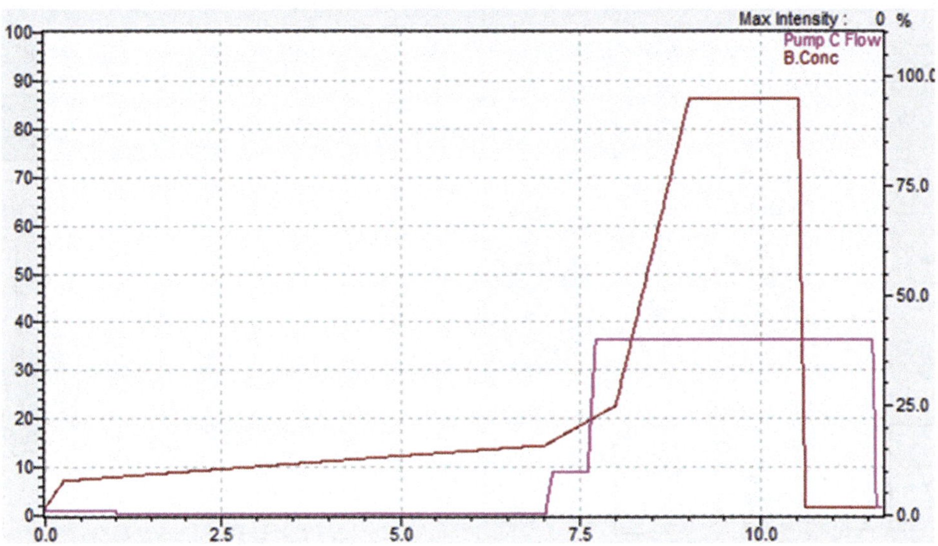

Fig. 5 LC solvent composition curve

Table 2
LC instrument parameters

Line	Time (min)	Hardware	Parameter	Value
1	0.01	Controller	Event	13
2	0.01	Pumps	Pump C Flow	0.05
3	0.01	Controller	Start	
4	0.25	Pumps	Pump B Conc.	8
5	1	Pumps	Pump C Flow	0.05
6	1.01	Pumps	Pump C Flow	0.02
7	7	Pumps	Pump B Conc.	16
8	7.01	Pumps	Pump C Flow	0.02
9	7.11	Pumps	Pump C Flow	0.5
10	7.61	Controller	Event	123
11	7.61	Pumps	Pump C Flow	0.5
12	7.61	Pumps	SV(Pump C)	D
13	7.71	Pumps	Pump C Flow	2
14	8	Pumps	Pump B Conc.	25
15	9.01	Pumps	Pump B Conc.	95
16	9.61	Pumps	SV(Pump C)	B
17	10.51	Pumps	Pump B Conc.	95
18	10.61	Pumps	Pump B Conc.	2
19	11.55	Pumps	Pump C Flow	2
20	11.61	Pumps	Pump C Flow	0.1
21	11.7	Controller	Stop	

21. The modified valve design shown in Fig. 6a, b allows simultaneous digestion and LC separation. This is accomplished by employing a 10-port switching valve and two identical trapping columns. While the cleavage products and SIL peptides are collected on one column, the cleavage products and SIL peptide from the previous sample eluted to the analytical column and to the LC-MS interface. This alternating setup effectively reduces total analysis time to the length of the LC gradient. Additionally, the valve design reverses the direction of the flow on the trapping column during desalting and elution, refocusing analytes onto the analytical column and reducing peak broadening.
22. *See* Fig. 7 and Table 3.

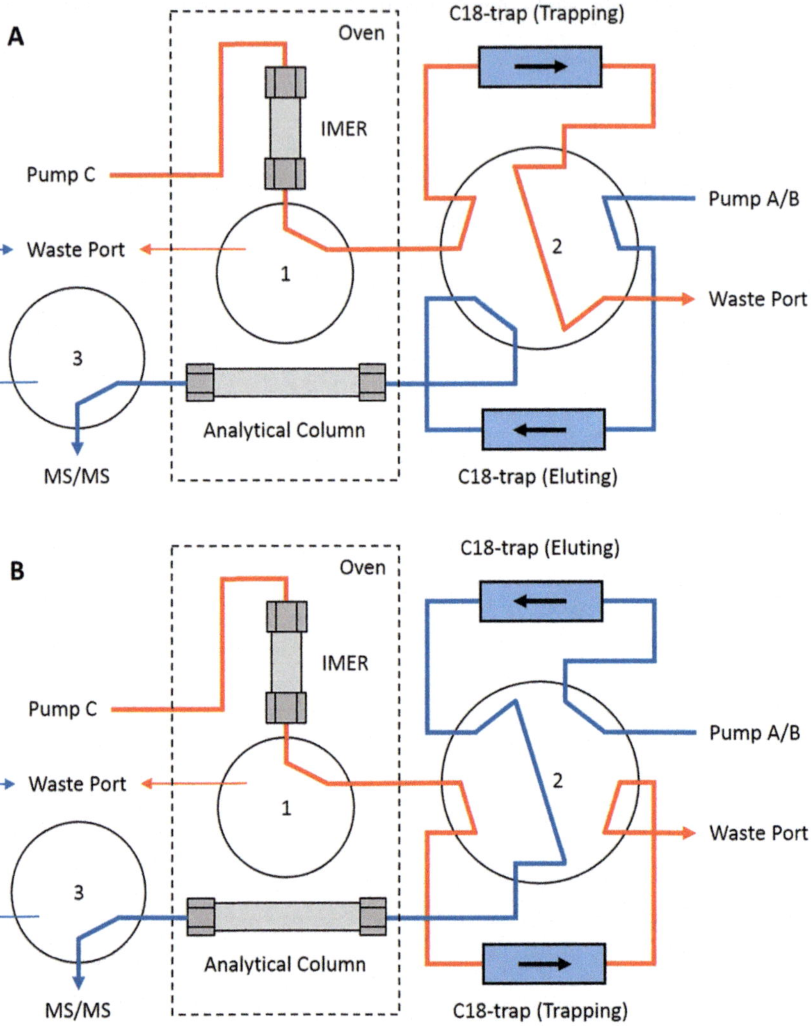

Fig. 6 Dual trapping valve design and valve positions corresponding to LC methods A and B

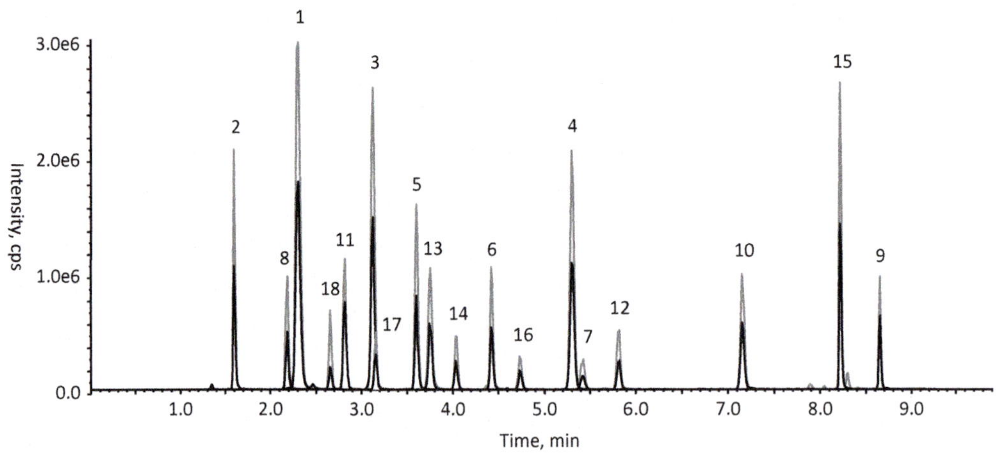

Fig. 7 Representative chromatogram: 50 μL of 1:100 diluted serum calibrator (gray) and 5 μL SIL mix (black) injected into system. Corresponding peak IDs are listed in Table 3

Table 3
MRM instrument parameters for native and labeled peptides

		Native						Label			
				Fragmention						Fragmention	
ID	Protein	Peptide sequence	Precursor ion *m/z*	1	2	DP	CE	Peptide sequence	Precursor ion *m/z*	1	2
1	apoA-I	AKPALEDLR	506.8	716.4	813.4*	68.1	27.1	AKPALED(L+7)R	510.3	723.4	820.5
2		ATEHLSTLSEK	608.3	664.4*	777.4	75.5	30.8	ATEHLST(L+7)SEK	611.8	671.4	784.5
3		THLAPYSDELR	651.3	950.5*	1063.5	78.6	32.3	THLAPYSDE(L+7)R	654.8	957.5	1070.6
4	apoA-II	EQLTPLIK	471.3	470.3	571.4*	65.5	25.8	EQLTPL(I+7)K	474.8	477.4	578.4
5	apoA-IV	LEPYADQLR	552.8	765.4*	862.4	71.4	28.8	LEPYADQ(L+7)R	556.3	772.4	869.5
6		LTPYADEFK	542.3	772.4*	869.4	70.6	28.4	LTPYADEF(K+8)	546.3	780.4	877.4
7	apoB	ATGVLYDYVNK	621.8	801.4*	914.5	76.4	31.2	ATGVLYDY(V+6)NK	624.8	807.4	920.5
8		AAIQALR	371.7	487.3	600.4*	58.2	22.2	AAIQA(L+7)R	375.2	494.3	607.4
9		TGISPLALIK	506.8	654.5*	741.5	68.1	27.1	TGISPLAL(I+7)K	510.3	661.5	748.5
10	apoC-I	EWFSETFQK	601.3	739.4*	886.4	75	30.5	EWFSETFQ(K+8)	605.3	747.4	894.4
11		TPDVSSALDK	516.8	620.3	834.4*	68.8	27.5	TPDVSSA(L+7)DK	520.3	627.3	841.4
12	apoC-II	ESLSSYWESAK	643.8	870.4*	957.4	78	32	ESLSSYWESA(K+8)	647.8	878.4	965.4
13		TYLPAVDEK	518.3	658.3	771.4*	68.9	27.5	TYLPA(V+6)DEK	521.3	664.4	777.4
14	apoC-III	DALSSVQESQVAQQAR	858.9	887.5*	1016.5	93.7	39.8	DALSSVQESQVAQ(Q+7)AR	862.4	894.5	1023.5
15		GWVTDGFSSLK	598.8	753.4*	854.4	74.8	30.4	GWVTDGFSS(L+7)K	602.3	760.4	861.4
16	apoE	AATVGSLAGQPLQER	749.4	770.4	827.4*	85.7	35.8	AATVGSLAGQPL(Q+7)ER	752.9	834.4	905.5
17		LGPLVEQGR	484.8	588.3*	489.2	66.5	26.3	LGPLVE(Q+7)GR	488.3	496.3	595.3
18		LQAEAFQAR	517.3	721.4	792.4*	68.8	27.5	LQAEAF(Q+7)AR	520.8	728.4	799.4

Product ions marked with (*) were chosen for quantitation

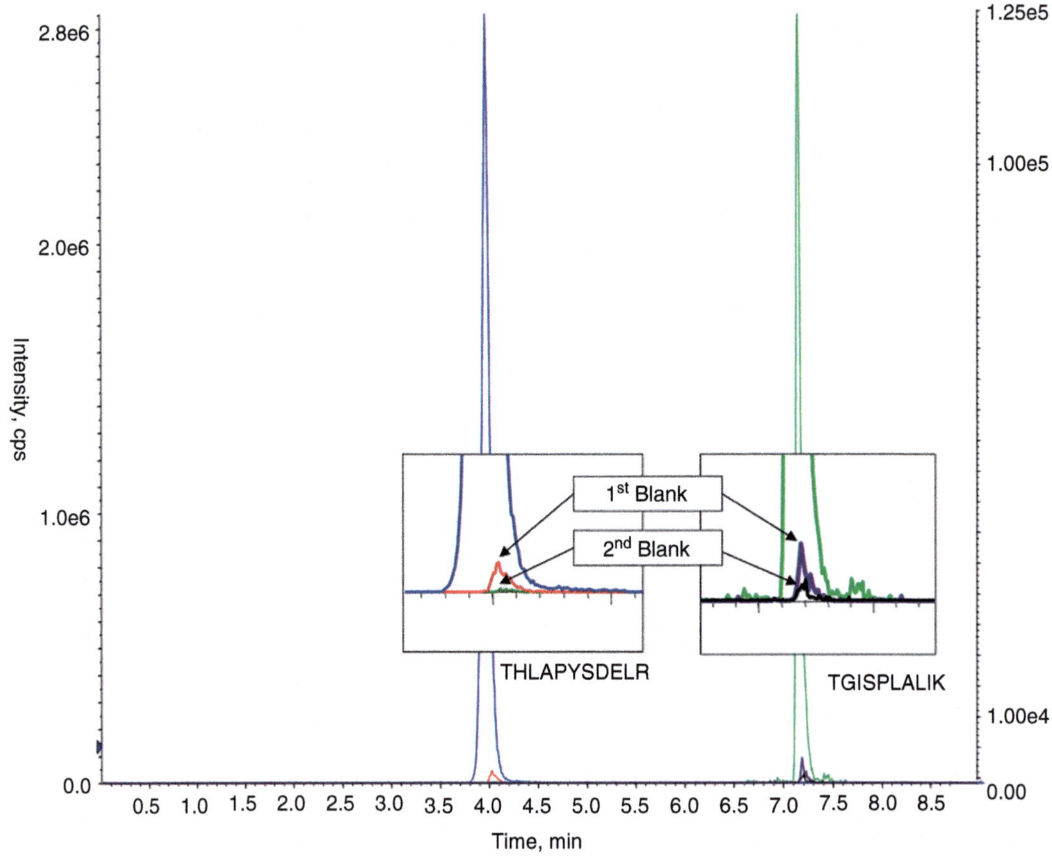

Fig. 8 Example of carry-over contribution to the native signal in one and two blank injections (50 μL of digest buffer) immediately following a highly concentrated sample (50 μL of 1:15 diluted human serum on-column). The intensities are plotted for two peptide transitions with contrasting native protein concentrations, THLAPYSDELR (apoA-I; left axis) and TGISPLALIK (apoB; right axis)

23. Carry-over contribution for the two peptides was calculated at 0.2 and 1.3% after one blank injection, respectively, and 0.05 and 1.0% after two blank injections, respectively (Fig. 8).

Disclaimer

References in this article to any specific commercial products, process, service, manufacturer, or company do not constitute an endorsement or a recommendation by the U.S. Government or the Centers for Disease Control and Prevention. The findings and conclusions in this report are those of the authors and do not necessarily represent the views of CDC.

References

1. Picotti P, Aebersold R (2012) Selected reaction monitoring-based proteomics: workflows, potential, pitfalls and future directions. Nat Methods 9(6):555–566. https://doi.org/10.1038/nmeth.2015
2. Shuford CM, Walters JJ, Holland PM, Sreenivasan U, Askari N, Ray K, Grant RP (2017) Absolute protein quantification by mass spectrometry: not as simple as advertised. Anal Chem 89(14):7406–7415. https://doi.org/10.1021/acs.analchem.7b00858
3. Shuford CM, Sederoff RR, Chiang VL, Muddiman DC (2012) Peptide production and decay rates affect the quantitative accuracy of protein cleavage isotope dilution mass spectrometry (PC-IDMS). Mol Cell Proteomics 11 (9):814–823. https://doi.org/10.1074/mcp.O112.017145
4. Regnier FE, Kim JH (2014) Accelerating trypsin digestion: the immobilized enzyme reactor. Bioanalysis 6(19):2685–2698. https://doi.org/10.4155/bio.14.216
5. Kuklenyik Z, Jones JI, Toth CA, Gardner MS, Pirkle JL, Barr JR (2017) Optimization of the linear quantification range of an online trypsin digestion coupled liquid chromatography–tandem mass spectrometry (LC–MS/MS) platform. Instrum Sci Technol 46:1–13. https://doi.org/10.1080/10739149.2017.1311912
6. Bonichon M, Combès A, Desoubries C, Bossée A, Pichon V (2016) Development of immobilized-pepsin microreactors coupled to nano liquid chromatography and tandem mass spectrometry for the quantitative analysis of human butyrylcholinesterase. J Chromatogr A 1461:84–91. https://doi.org/10.1016/j.chroma.2016.07.058
7. Toth CA, Kuklenyik Z, Jones JI, Parks BA, Gardner MS, Schieltz DM, Rees JC, Andrews ML, McWilliams LG, Pirkle JL, Barr JR (2017) On-column trypsin digestion coupled with LC-MS/MS for quantification of apolipoproteins. J Proteome 150:258–267. https://doi.org/10.1016/j.jprot.2016.09.011
8. Hoofnagle AN, Whiteaker JR, Carr SA, Kuhn E, Liu T, Massoni SA, Thomas SN, Reid Townsend R, Zimmerman LJ, Boja E, Chen J, Crimmins DL, Davies SR, Gao Y, Hiltke TR, Ketchum KA, Kinsinger CR, Mesri M, Meyer MR, Qian WJ, Schoenherr RM, Scott MG, Shi T, Whiteley GR, Wrobel JA, Wu C, Ackermann BL, Aebersold R, Barnidge DR, Bunk DM, Clarke N, Fishman JB, Grant RP, Kusebauch U, Kushnir MM, Lowenthal MS, Moritz RL, Neubert H, Patterson SD, Rockwood AL, Rogers J, Singh RJ, Van Eyk JE, Wong SH, Zhang S, Chan DW, Chen X, Ellis MJ, Liebler DC, Rodland KD, Rodriguez H, Smith RD, Zhang Z, Zhang H, Paulovich AG (2016) Recommendations for the generation, quantification, storage, and handling of peptides used for mass spectrometry-based assays. Clin Chem 62(1):48–69. https://doi.org/10.1373/clinchem.2015.250563
9. Grant RP, Hoofnagle AN (2014) From lost in translation to paradise found: enabling protein biomarker method transfer by mass spectrometry. Clin Chem 60(7):941–944. https://doi.org/10.1373/clinchem.2014.224840

Chapter 20

Proteases: Pivot Points in Functional Proteomics

Ingrid M. Verhamme, Sarah E. Leonard, and Ray C. Perkins

Abstract

Proteases drive the life cycle of all proteins, ensuring the transportation and activation of newly minted, would-be proteins into their functional form while recycling spent or unneeded proteins. Far from their image as engines of protein digestion, proteases play fundamental roles in basic physiology and regulation at multiple levels of systems biology. Proteases are intimately associated with disease and modulation of proteolytic activity is the presumed target for successful therapeutics. "Proteases: Pivot Points in Functional Proteomics" examines the crucial roles of proteolysis across a wide range of physiological processes and diseases. The existing and potential impacts of proteolysis-related activity on drug and biomarker development are presented in detail. All told the decisive roles of proteases in four major categories comprising 23 separate subcategories are addressed. Within this construct, 15 sets of subject-specific, tabulated data are presented that include identification of proteases, protease inhibitors, substrates, and their actions. Said data are derived from and confirmed by over 300 references. Cross comparison of datasets indicates that proteases, their inhibitors/promoters and substrates intersect over a range of physiological processes and diseases, both chronic and pathogenic. Indeed, "Proteases: Pivot Points . . ." closes by dramatizing this very point through association of (pro)Thrombin and Fibrin(ogen) with: hemostasis, innate immunity, cardiovascular and metabolic disease, cancer, neurodegeneration, and bacterial self-defense.

Key words Protease, Peptidase, Proteolysis, Protease inhibitor, Protease promoter, Digestion, Hemostasis, Complement system, Immune regulation, Signaling, Cell migration, Cell proliferation, Programmed cell death, Protein secretion, DNA replication, DNA repair, DNA processing, Intranuclear proteolysis, Transmembrane proteolysis, Intramembrane proteolysis, Cytosolic proteolysis, Epigenetics, Inflammation, Cardiovascular disease, Metabolic disease, Stroke, Cancer, Neurodegenerative disease, Autoimmune disease, Infectious organisms, Drug target, Drug development, Biomarker development, Precision medicine

1 Introduction

1.1 Proteases: More Than Just Protein Digestion

One or more proteolytic events initiate the active life for many proteins and proteolytic events terminate the relatively short life of all proteins. Roughly a third of proteins, as translated, include peptides that influence proteins' life span (initiator methionine), govern their transport (signal peptide), direct them to the appropriate organelle (transit peptide), and produce their active form (propeptide). These processes are merely those proteolytic events

Xing Wang and Matthew Kuruc (eds.), *Functional Proteomics: Methods and Protocols*, Methods in Molecular Biology, vol. 1871, https://doi.org/10.1007/978-1-4939-8814-3_20,

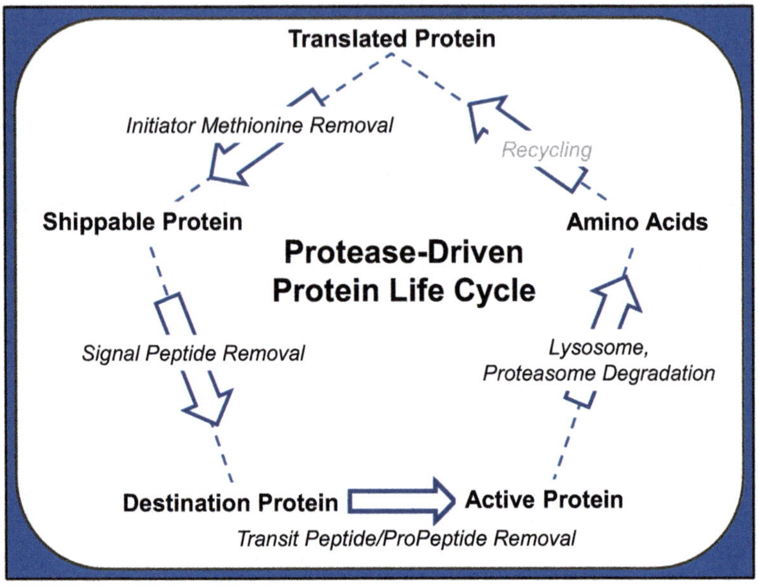

Fig. 1 Stages of the Protease-Driven Protein Life Cycle

that begin the active life of proteins. Over the course of lifetimes that may be as brief as 11 min or as long as 4 months, additional proteolytic events remake individual proteins—one example being the much studied A4-Human amyloid precursor transmembrane protein whose 13 proteolytic products include a family of pathology-related, β amyloid peptides. Having served their temporary purposes, all proteins are once again substrates for proteases with digestion proceeding to amino acids recycled as raw materials for ongoing expression of new proteins. In the light of these varied and continuous processes, proteolytic activity literally reshapes the proteome. Of course, this brief discourse is but the tip of the iceberg. Proteolytic enzymes—and their promoters and inhibitors—are essential actors in nearly all biological processes. This chapter addresses many of those processes especially in relation to disease and development of new therapeutics. Truly, proteases give definition to the Functional Proteome—one protein at a time (Fig. 1).

In the 1830s, parallel studies in Europe and the USA demonstrated that digestion of proteins in food required a compound in the gastric mucosa, in addition to gastric juice hydrochloric acid [1]. The discovery of pepsin triggered the formulation of popular therapeutic preparations for dyspepsia, and intensified pepsin purification efforts. However, it would take nearly a century before crystallization of pure pepsin was achieved. Northrop, Kunitz, and Herriott published the earliest characterization of classical digestive proteases such as pepsin, chymotrypsin, trypsin, pancreatic carboxypeptidases, and their zymogens as pure compounds

[2]. The protease field has come a long way since the discovery of pepsin. Rapidly developing methods for protein purification, sequencing, structure-function analysis, X-ray crystallography, and synthetic substrate development have since then accelerated the identification of countless mammalian, plant, fungal, and bacterial proteolytic enzymes [3], and the number of identified sequences is growing exponentially. As of September 2017, the MEROPS database of proteolytic enzymes (https://www.ebi.ac.uk/merops/) listed more than one million sequences [4].

The complete ensemble of human proteases, known as the protease degradome, currently consists of 588 proteases, organized into five classes: aspartic, cysteine, metallo-, serine, and threonine proteases [5]. They represent >2% of the human genome, and a database is maintained by the lab of López-Otín (http://degradome.uniovi.es) [6]. Protease systems have gained considerable recognition since the realization that they are important regulators of countless biological mechanisms, and not simply part of a machinery for nonspecific protein digestion. Blood coagulation, fibrinolysis, complement activation, peptide hormone processing, protein secretion and degradation, DNA replication and repair, cell signaling and proliferation, and programmed cell death are just a few of these processes in which physiological proteases target specific substrates. The peptidase field is rapidly evolving, and extensive volumes have already been dedicated to cataloguing and documenting its recently discovered and previously known members [7]. Here we focus on novel aspects of the intimate connection between disease and proteolysis, and the potential of mechanism-based drug targeting.

1.2 Physiological and Regulatory Roles

Physiological protease activity is strictly controlled to avoid indiscriminate and unwanted protein degradation [8]. Proteases may be present as inactive zymogens requiring a proteolytic activation step followed by a conformational rearrangement to form the active site (hemostatic protease zymogens); complexed with an inhibitory propeptide or N-terminal domain that blocks the active site (matrix metalloproteinases or MMPs); or in a low-reactive state requiring allosteric activation (factor VII and tissue factor), proteolysis (single chain tPA), or di/multimerization (caspase-8 and -9, proteasome proteases). Active proteases may be pH-controlled, and examples of this are lysosomal cathepsins and gastric pepsin. Many proteases are transcriptionally regulated and only expressed in specific cells and tissues, sometimes temporally restricted, whereas housekeeping proteases are expressed constitutively. Physiological triggers such as inflammation may temporarily upregulate zymogen expression, as seen for the precursors of cathepsins and MMPs.

Because many proteases are associated with other proteins and act within networks of effectors, ligands, and receptors, studying their biological functions has become increasingly complex. In vitro delineation of structural and kinetic properties of a protease is required for defining its mechanisms of substrate and inhibitor specificity, but its catalytic efficiency is often profoundly affected in vivo by complex macromolecular interactions. Protease activity may be up- or downregulated by physiological feedback mechanisms, processes at the genomic level, and stressors in the molecular environment. Depending on the molecular environment and the binding partners, proteases may catalyze reactions that result in opposite physiological processes. Thrombin, the central protease in hemostasis, is a good example of this. Uncomplexed thrombin is procoagulant, and cleaves fibrinogen to fibrin to form a thrombus. However, when bound to thrombomodulin, thrombin activates the anticoagulant protein C. The proteases of the contact activation system in the intrinsic coagulation pathway aid in sustaining clotting once the extrinsic pathway is activated, but they also activate plasminogen in the fibrinolytic pathway.

Protease networks are regulated by intrinsic inhibitors of a protein or polypeptide nature. Serine proteases feature predominantly in coagulation, fibrinolysis, and the complement system, and also play roles in digestion, late stage apoptosis, development, fertilization, and membrane-associated signaling. They are typically inhibited by serpins (serine proteinase inhibitors) in irreversible covalent complexes, or by polypeptide inhibitors of the Kunitz, Kazal, or elafin types, with a protease-binding loop in a conserved canonical backbone conformation, complementary to the protease active site. Inhibitory serpins belong to a superfamily of ubiquitous proteins with a typical fold that includes a metastable reactive center loop (RCL) for baiting their target proteases, and a core consisting of 3 β-sheets and 8 or more α-helices [9]. Upon cleavage of the RCL, the protease stays attached, and the resulting acyl-enzyme undergoes a dramatic conformational rearrangement during which the cleaved RCL inserts as an additional strand in β-sheet A, and the attached protease translocates to the distal end of the serpin. Deformation of the active site prohibits completion of hydrolysis, with a stable covalent complex as the end product. Kunitz, Kazal, and elafin-like inhibitors act by a different mechanism of tight binding in a substrate-like mode, followed by slow and reversible cleavage of the reactive bond [10]. The cleaved inhibitor stays bound to the protease but typically loses inhibitory potency by several orders of magnitude.

Analysis of known cleavage events has shown that regulation of the proteome in vivo occurs through an interconnected protease-inhibitor web [11], with protein protease inhibitors being protease substrates themselves. Many inhibitors target groups of related proteases rather than single enzymes, and proteolytic inactivation of one inhibitor may represent a key on/off switch for an entire protease subnetwork. Multiple regulatory mechanisms of proteolysis have been identified, such as the presence of PEST (proline, glutamic acid, serine, threonine) sequences in intracellular proteins targeted for rapid degradation; KFERQ motifs (lysine, phenylalanine, glutamic acid, arginine, glutamine) guiding cytosolic proteins to the endosome or lysosome for degradation; and RxxLxxIxN destruction box motifs that tag proteins for degradation by the ubiquitin-proteasome system [12]. Whereas the protease/substrate/inhibitor interconnectedness adds many levels of complexity, it also provides more opportunities for design of new therapeutic approaches.

1.3 Proteases and Disease

Protease functionality often depends on the concerted action of a catalytic domain and specific nonenzymatic domains and modules, either incorporated in the protein or as separate entities associated with the catalytic domain. Anion-binding sites, kringle and apple domains, epidermal growth factor (EGF) and fibronectin domains, thrombospondin repeats and transmembrane domains have diverse functions such as localization; recognition of substrates, inhibitors, and effectors; and interactions with various ligands, cofactors, and other proteins. Many of these domains are evolutionary conserved, and are present in diverse proteases either as single units (e.g., EGF domains) or as repeats (kringle and apple domains, thrombospondin domains). Not surprisingly, defects in the catalytic domain or any of the regulatory domains result in a physiologically dysfunctional protease. Intrinsic protease dysregulation is a hallmark of many pathologies such as inflammation, cancer, hemostatic and autoimmune diseases, and neurodegeneration [13, 14]. A selection of recent findings with regard to abnormal protease activity in these pathologies is discussed below. The López-Otín degradome database currently lists 124 hereditary diseases due to protease mutations, and many other pathologies are associated with posttranslational and epigenetic changes in protease activity. Although much progress has recently been made in inflammation-related protease research, many processes are still unclear with respect to upregulation of protease activity as a cause or consequence. However there is ample consensus about the importance of proteases as attractive drug targets in many disease states. Effector ligands and substrates may provide additional targets. Protease up- or downregulation may be used as a diagnostic, and proteolytic generation of signal peptides may provide a previously underappreciated arsenal of biomarkers in various disease states.

2 Regulation of Physiological Processes

2.1 Hemostasis

If you are familiar with only one proteolytically regulated process in the human body, it is likely hemostasis. Featured prominently in documentaries about European royalty and in nightly medication commercials, the formation and degradation of blood clots has been intensely studied. It is a model system for illustrating the complexity of proteolysis: multiple pathways acting both in concert and in opposition to each other, dozens of actors, extensive regulation at each step, and even individual actors performing opposing roles under different conditions. Each step is an opportunity for both dysfunction and intervention, and both affects and is affected by processes in cells throughout the body (Table 1).

In primary hemostasis, vascular injury exposes the highly thrombogenic subendothelium, and platelets are recruited to the site of injury, where they are activated and form a platelet plug. In the "secondary hemostasis" cascade model of coagulation, proposed more than 50 years ago, vascular injury triggers a stepwise amplification of clotting factor activity. This culminates in the formation of thrombin, the central coagulation protease, that cleaves fibrinogen to fibrin [15]. In the extrinsic pathway, plasma factor VIIa forms a highly reactive complex with tissue factor exposed

Table 1
Hemostasis clot formation: activities and Uniprot ID codes (where applicable) of proteases, protease inhibitors, and cofactors associated with hemostasis, specific to clot formation

Protease/inhibitor	Uniprot	Action
Thrombin (Prothrombin)	P00734	Converts fibrinogen to fibrin Activates factors V, VII, VIII, XI, XIII Complexes with thrombomodulin Thrombin/thrombomodulin activates protein C
Plasma factor VIIa (Coagulation factor VII)	P08709	Complexes with Tissue Factor VIIa/TF converts/activates X to Xa VIIa/TF converts/activates IX to IXa
Tissue factor	P13726	Propagates coagulation protease cascade Complexes with phospholipids Complexes with circulating factor VII or VIIa VIIa/TF converts/activates X to Xa VIIa/TF converts/activates IX to IXa
Coagulation factor X	P00742	Converts/activates prothrombin to thrombin Complexes with phospholipids & calcium Activates factor VII to form factor VIIa
Prekallikrein (Plasma kallikrein) Coagulation factor XII	P03952 P00748	Reciprocal activation of factor XII Reciprocal activation of Prekallikrein
Coagulation factor IX	P00740	Converts/activates factor X Activates factor VII to form factor VIIa Activates factor X to form factor Xa
Antithrombin (Antithrombin-III)	P01008	Inhibits thrombin and factors IXa, Xa and XIa Activity enhanced by heparin
Heparin cofactor II	P05546	Inhibits thrombin and factors IXa, Xa and XIa Inhibits chymotrypsin
TFPI (Tissue factor pathway inhibitor)	P10646	Inhibits factor X (Xa) TFPI + Xa inhibits VIIa/tissue factor

during injury, and sequential activation of factor X and prothrombin leads to clot formation. The intrinsic or contact activation pathway consists of prekallikrein, and factors XII and IX. Stepwise activation of factors XII and XI generates factor IXa that sustains formation of factor Xa. Except for factor XIIIa, a transaminase that crosslinks fibrin, all the hemostatic enzymes are serine proteases. Intrinsic serine protease inhibitors (antithrombin, heparin cofactor II) and the Kunitz inhibitor TFPI provide regulatory control. Phospholipid surfaces and the nonenzymatic cofactors VIII and V are required for activation of factor X and prothrombin, respectively. This "waterfall" mechanism did not explain why factor XII-deficient patients do not have a bleeding tendency, and a new "hemostatic network" mechanism was discovered in which thrombin activates factor XI to sustain hemostasis [16, 17]. Whereas the contact activation proteases are not critical in normal hemostasis, animal studies have shown that they are important contributors to formation of pathologic intravascular thrombi [18], and may be suitable targets for therapeutic inhibitors.

When the blood clot has served its purpose, endothelial cells release tPA that converts plasminogen to plasmin on the fibrin surface. Both tPA and plasminogen bind to fibrin through their kringle structures (binding domains specific to blood clotting factors), resulting in significant enhancement of the rate of plasmin formation. Plasmin degrades fibrin which exposes additional carboxy-terminal lysines for interaction with the kringles on tPA, plasminogen and plasmin, ultimately resulting in accelerated fibrin degradation. In a regulatory process, carboxypeptidase U, also known as thrombin-activatable fibrinolysis inhibitor (TAFIa), removes carboxy-terminal lysine residues and stabilizes fibrin thrombi. TAFI proved to be a poor substrate for thrombin in the absence of thrombomodulin, and a riveting chronicle of the lengthy process of its discovery, purification and characterization again demonstrates the gratifying aspect of understanding the molecular basis of a biochemical process [19]. The serpins plasminogen activator inhibitor-1 (PAI-1) and α2-antiplasmin (α2-AP), respectively targeting tPA and plasmin, provide additional regulation of fibrin degradation. Plasmin bound to fibrin is protected from α2-AP, whereas TAFI decreases this protection by eliminating plasmin-binding lysine residues on fibrin (Table 2).

Platelets can both promote and counteract fibrinolysis. Activated platelets localize plasminogen and its physiological activators tPA and urokinase via the GPIIb/IIIa (integrin αIIbβ3) complex. Thrombospondin, released from platelet granules and exposed on the platelet surface, also binds plasminogen and enhances its activation. Hence activated platelets provide an alternative surface for promoting fibrinolysis. As part of a regulatory mechanism, platelets secrete two antifibrinolytic serpins, PAI-1 and α2-AP, and as a result platelet-rich thrombi resist fibrinolysis.

Table 2
Hemostasis clot degradation: activities and Uniprot ID codes (where applicable) of proteases, protease inhibitors, and cofactors associated with hemostasis, specific to clot degradation

Protease/inhibitor	Uniprot	Action
tPA (Tissue-type plasminogen activator)	P00750	Converts plasminogen to plasmin on the fibrin surface Displaces plasmin from fibrin, promoting inhibition by alpha-2-antiplasmin
Plasmin (Plasminogen) carboxypeptidase U (carboxypeptidase B2, thrombin-activatable fibrinolysis inhibitor3)	P00747 Q96IY4	Dissolves the fibrin of blood clots Removes C-terminal lysine residues from fibrin Down-regulates fibrinolysis Cleaves complement proteins C3a and C5a Activated by thrombin/thrombomodulin complex
Plasminogen activator inhibitor 1	P05121	'Bait' for tissue plasminogen activator, urokinase, protein C and matriptase-3/TMPRSS7
α2-antiplasmin (Alpha-2-antiplasmin)	P08697	Inhibits plasmin and trypsin Inactivates matriptase-3/TMPRSS7 and chymotrypsin
Protein C (Vitamin K-dependent protein C)	P04070	Inactivates factors Va and VIIIa in the presence of calcium ions and phospholipids Activated by Thrombin/thrombomodulin complex

Thrombin may act both as a procoagulant and as an anticoagulant enzyme, and has been called a "Janus-headed" protease [20]. Crystal structures of α-thrombin bound to numerous effectors have aided in identifying extended recognition sites on its surface. By combining these discrete functional surface regions thrombin interacts with various substrates and ligands in a selective and specific manner. Binding of Na^+ causes thrombin to adopt a "fast" conformation which rapidly cleaves procoagulant substrates. In the Na^+-free state, thrombin bound to thrombomodulin preferentially initiates the protein C anticoagulant pathway in which factors V and VIII are proteolytically inactivated [21, 22].

2.2 The Complement System and Immune Regulation

The plasma complement system regulates the innate immune defense by opsonization and elimination of pathogens, cell debris, and host cells that have undergone alterations [23–25]. Activation of the complement system occurs via three pathways: in the classical pathway, the recognition protein C1q binds to antigen-antibody complexes, the C1 complex is activated, and a series of serine protease activation reactions leads to formation of C3- and ultimately C5-convertase; in the lectin pathway, mannose-binding lectin binds to mannose on the surface of pathogens as a trigger to formation of C3/C5-convertase; in the alternative pathway, continuous low level activation of C3 and binding to the pathogen leads to C3/C5-convertase. C3-convertases cleave C3 into the anaphylatoxin C3a and the opsonin C3b which is deposited on the pathogen surface and facilitates targeting by macrophages. C5-convertases generate the pro-inflammatory anaphylatoxins C5a and C5b, and a membrane attack complex (MAC) is formed by C5b–C9 assembly. This complex forms a pore in the membrane that kills the pathogen or the targeted cell. Anaphylatoxins C3a and C5a promote chemotaxis of immune cells. The complement serine proteases involved in these complicated interrelated processes include C1r, C1s, MASPs 1-3 [26], C2, and Factors B, D, and I, all with restricted specificity. The plasma serpin, C1-inhibitor, covalently inactivates C1r, C1s, and MASPs 1 and 2. C2 and Factor B activity is controlled by the Regulation of Complement Activation (or RCA) proteins, and no endogenous inhibitors are known for Factors I and D. Complement deficiency causes increased susceptibility to infection, and clearance impairment of immune complexes and apoptotic cells results in the development of systemic lupus erythematosus (SLE). However, excessive complement activation is also associated with autoimmune diseases such as SLE, rheumatoid arthritis, and certain cancers. The monoclonal antibody Eculizumab inhibits C5, and was recently approved for the treatment of complement hyperactivation in paroxysmal nocturnal hemoglobinuria and atypical hemolytic uremic syndrome. It may ultimately prove useful in the treatment of SLE as well [27] (Table 3).

Human cytotoxic T lymphocytes and natural killer cells secrete five types of granzymes (A, B, H, K, M), serine proteases that aid in the neutralization of virus-infected and tumor cells. Only granzyme B and M have known intracellular inhibitors, serpinB9 (PI-9) and serpinB4 (SCCA2), respectively [28]. Increased PI-9 expression may be an immune evasion mechanism used by lung cancer cells for protection from granzyme B-mediated cytotoxicity [29].

Immune regulation is tightly associated with proteolytic processes in the gut. In immune diseases of the gut, cytokines upregulate protease activity, resulting in inflammation and exacerbated immune response [30]. In inflammatory bowel disease, MMPs, neutrophil elastase, and cathepsins are typically overexpressed in the gut epithelium and basement membrane. The nature of the gut microbiome is equally important for maintaining immune homeostasis, and commensal and pathogenic bacteria produce a wide range of proteases that differentially affect the integrity of the intestinal mucosa.

The serpin α1-antitrypsin (now renamed α1-proteinase inhibitor or α1-PI), produced in the liver, protects the lungs from inflammatory neutrophil elastase damage, and it is also an acute-phase protein that reduces pro-inflammatory cytokine production, inhibits apoptosis, blocks leukocyte degranulation and migration, and

Table 3
Complement system and immune regulation: activities and Uniprot ID codes (where applicable) of proteases, protease inhibitors, and cofactors associated with the complement system and immune regulation

Protease/inhibitor	Uniprot	Action
C1R (Complement C1r subcomponent, Classical)	P00736	Cleaves/Activates C1s
C1S (Complement C1s subcomponent, Classical)	P09871	Cleaves/Activates C2 and C4
C2a (Complement C2, C1s cleavage product)	P06681	Combines with C4b to form C3 convertase (classical, lectin)
Factor D (Complement factor D, Bb fragment)	P00746	Cleaves/Activates Complement Factor B
Factor B (Complement factor B, Bb fragment)	P00751	Cleavage Product Bb combines with C3b to form C3 Convertase (Alternative)
C3-convertase (Classical, Lectin: C4bC2a)		Cleaves C3 into the anaphylatoxin C3a and the opsonin C3b
C3-convertase (Alternative: C3bBb)		Cleaves C3 into the anaphylatoxin C3a and the opsonin C3b
C3-convertase (Aqueous: C3: H2O)		Cleaves C3 into the anaphylatoxin C3a and the opsonin C3b
C5-convertase (Classical: Cell Membrane, C4b2b3b)		Cleaves C5 into the anaphylatoxin C5a and the MAC component C5b
C5-convertase (Alternative: Cell Membrane, C3bBbC3b)		Cleaves C5 into the anaphylatoxin C5a and the MAC component C5b
C5-convertase (Classical: Fluid, C4b2boxy3b)		Cleaves C5 into the anaphylatoxin C5a and the MAC component C5b
MASP 1 (Mannan-binding lectin serine protease 1)	P48740	Activates MASP2 or C2 or C3
MASP 2 (Mannan-binding lectin serine protease 2)	O00187	Cleaves/Activates C2 and C4
MASP 3 (Mannan-binding lectin serine protease 3)	P48740[a]	Cleaves/Activates complement pro-factor D[a] Alternative Splicing Product
C1-inhibitor (Plasma protease C1 inhibitor)[a]	P05155	Complexes with/Inactivates C1r, C1s, MASP 1, MASP 2, chymotrypsin, kallikrein, fXIa, FXIIa
Factor I (Complement factor I)	P05156	Cleaves/Inactivates C3b, iC3b, and C4b

[a]Complement activation, blood coagulation, fibrinolysis, and the generation of kinins

modulates local and systemic inflammatory responses [31]. In monocytes, α1-antitrypsin increases intracellular cAMP, regulates CD14 expression, and suppresses NF-κB nuclear translocation. These functions may be related to the inhibitory activity of antitrypsin, protein-protein interactions, or both. Preclinical use of antitrypsin in autoimmunity and transplantation models showed that it is capable of preventing or reversing autoimmune disease and graft loss.

2.3 Proteolytic Processing

The lysosome and the ubiquitin-proteasome are the two major intracellular proteolytic systems keeping the protein pool in balance. Originally considered strictly degradative, these systems have revealed regulatory functions beyond catabolism, and their molecular defects are associated with various disease states. Lysosomes contain cathepsins B, D, and L in addition to lipases, nucleases, glycosidases, phospholipases, phosphatases, and sulfatases that are active in an acidic milieu [32]. Lysosomes regulate autophagy during nutrient starvation, and participate in development and differentiation, induction of cathepsin-dependent cell death, and degradation of apoptotic cells. Cancer cell lysosomes have a higher membrane permeability ("leaky") and express more cathepsin than those of normal cells, and this property may be exploited in cancer treatment. Agents such as tetrahydrocannabinol and chloroquine may disrupt the lysosome and trigger killing of the cancer cells (Table 4).

The ubiquitin-proteasome, an intracellular high molecular weight protease complex predominantly located in the cytosol, selectively degrades proteins tagged with ubiquitin at lysine residues [33]. Its "central pore" contains several inward facing protease active sites, with caspase-, trypsin-, and chymotrypsin-like specificity. This multi-protein construct is capped by one or two activator complexes that conformationally regulate access of protein substrates to the pore. Whereas the physiological function of the proteasome was originally thought to be restricted to intracellular protein catabolism, new functions have been discovered with respect to regulation of the cell cycle progression, gene expression, and responses to cellular stress [34]. Protein ubiquitination is reversible, and more than 100 potentially regulatory deubiquitinase (DUB) genes have been identified, mainly cysteine and metalloproteinases (MMPs). DUBs rescue proteins from degradation and reverse ubiquitination-induced signaling. The immunoproteasome, containing specific subunits with increased chymotrypsin- and trypsin-like activities, and decreased caspase-like activity, participates in production of peptide epitopes for cytotoxic T lymphocytes. In the thymoproteasome, chymotryptic activity is attenuated but the caspase- and trypsin-like activities are conserved. Its peptide products are MHC class I ligands with moderate avidity, which supports positive selection of CD8+ T cells [35]. Foreign peptides, generated during the breakdown of virus and cancer cell proteins, bind MHC class I molecules on the cell surface, and the cells are recognized by cytotoxic T cells as potentially dangerous and are destroyed. The proteasome discards misfolded proteins, and

Table 4
Proteolytic processing: activities and Uniprot ID codes (where applicable) of proteases, protease inhibitors, and cofactors associated with proteolytic processing

Protease/inhibitor	Uniprot	Action
Cathepsin B	P07858	Intracellular degradation and protein turnover Upregulation of Cathepsin D, matrix metalloproteinase, and urokinase Implicated in metastasis and immune resistance
Cathepsin D	P07339	Intracellular degradation and protein turnover Used by macrophages to degrade bacterial proteins Activates ADAM30, implicated in Alzheimer's progression Implicated in metastasis in breast cancer
Cathepsin L1	P07711	Intracellular degradation and protein turnover Degrades collagen and elastin Degrades alpha-1 protease inhibitor
Deubiquitinase		Cleaves ubiquitin from proteins and other molecules Group of approx. 102 cysteine proteases and metalloproteases
Immunoproteasome		Degrades proteins into peptide ligands for Major Histocompatibility Complex (MHC) Proteasome with β1i, β2i, and β5i subunits
Thymoproteasome		Degrades proteins into peptide ligands for MHC 1, selective for CD8+ T cells Unique to thymic cortex
Trypsin-1	P07477	Degradation of food proteins in small intestine
Trypsin-2	P07478	Degradation of food proteins in small intestine
Mesotrypsin	P35030	Degradation of antitrypsin inhibitors
Chymotrypsinogen B1	P17538	Degradation of food proteins in small intestine
Chymotrypsinogen B2	Q6GPI1	Degradation of food proteins in small intestine
Serine Protease Inhibitor Kazal-type 1 (SPINK1)	P00995	Trypsin inhibitor, in pancreas protects against self-activated trypsin Inhibits calcium binding and NO production in sperm

proteasome defects may contribute to the pathogenesis of neurodegenerative diseases such as Parkinson's, Huntington's, Alzheimer's, and ALS. A decline in proteasome activity is also a hallmark of aging cells.

The gastrointestinal tract contains the highest concentrations of endogenous and exogenous proteases. The intestinal mucosa is constantly exposed to low level protease activity, from bacteria in the lumen, immune and mesenchymal cells in the basement membrane, and epithelial cells at the brush border membrane. Protease activity is tightly controlled, as the mucosal barrier is thin and susceptible to proteolysis. The intestinal epithelium is at the interface of digestive, absorptive, and secretory functions, and signaling processes to the mucosal immune, vascular, and nervous systems. Endogenous growth factors, cytokines and extracellular matrix (ECM) proteins that modulate these functions, are digestive protease substrates [36]. The biochemistry of digestive protein processing has been studied for more than 80 years since the first crystallographic studies were published, and the functions of pancreatic trypsin, chymotrypsin and elastase and their inhibitors are well known. Pancreatic trypsinogen is activated to trypsin by membrane-bound enterokinase in the small intestine. Trypsin activates pancreatic chymotrypsinogen, procarboxypeptidases, proelastases, and prolipases. Turnover of intestinal epithelium is rapid, requiring tight control of gut protease activity under normal physiological conditions. Pancreatic PRSS3/mesotrypsin, discovered in the late 1970s, is an atypical trypsin with an evolutionary mutation that renders the protease resistant to inactivation by the physiological Kazal inhibitor, pancreatic secretory trypsin inhibitor (SPINK1), and endows it with specific digestive trypsin inhibitor-degrading properties [37]. In pancreatitis, trypsin is activated in the pancreas, causing tissue destruction and inflammation. Mesotrypsin is upregulated in certain cancers, and SPINK1 deficiency is associated with hereditary pancreatitis.

2.4 Tissue Remodeling, Signaling, Cell Migration, and Proliferation

Zinc proteases feature prominently in these biological processes. They can be subdivided according to the structure of their catalytic sites and their domain organization [38]. The human ADAM family (a disintegrin and metalloprotease) currently counts 13 proteolytically active transmembrane and secreted members. ADAMs are largely tissue-specific and play roles in fertilization, proliferation, migration, and cell adhesion. Transmembrane ADAMs act as sheddases, *i.e.*, proteases that cause extracellular shedding of adjacent transmembrane proteins by proteolytic cleavage at the membrane. Examples of such activated proteins are TNF-α and the ErbB family of receptor tyrosine kinases; and EGF receptor ligands such as TGF-α, heparin-binding EGF-like growth factor, betacellulin, epiregulin, and amphiregulin. ADAM-mediated shedding is often followed by RIPping, or **R**egulated **I**ntramembrane **P**roteolysis, in which the intracellular portion of these transmembrane proteins is cleaved off by aspartyl proteases, S2P-metalloproteases, and rhomboid serine proteases. The released intracellular domain participates in signaling to the nucleus to modify gene expression. Processing of amyloid precursor protein (APP) and Notch signaling are typical examples of RIPping. The 19 known human ADAM-TS proteases have a similar architecture to the ADAMs, except for the presence of thrombospondin repeats instead of a transmembrane domain, which makes them extracellular. They process procollagen and von Willebrand factor, and cleave extracellular matrix aggrecan, versican, brevican, and neurocan. Matrix metalloproteinases (MMPs) typically have three common domains: the N-terminal propeptide that keeps the protease in an inactive form, the catalytic domain containing the Zn^{2+} ion, and a C-terminal hemopexin-like β-propeller domain for protein-protein interactions. MMPs are not only instrumental in matrix remodeling and tissue maintenance, but also as regulators of signaling pathways [39] (Table 5).

MMPs, originally thought to degrade extracellular matrix proteins in a rather indiscriminate fashion, were later shown to have specific physiological roles in shedding, activation, and inactivation of proteins such as growth factors and cytokines. They cleave their substrates using a HExxHxxGxxH motif which contains three zinc-binding histidines and a glutamate that acts as a general base/acid during catalysis. To date, there are 23 known human MMPs, organized in four classes according to their substrate specificity: collagenases (MMP-1, -8, and -13), gelatinases (MMP-2 and -9), stromelysins (MMP-3, -10, and -11), and a group containing matrilysin (MMP-7), metallo-elastase (MMP-12), enamelysin (MMP-20), matrilysin-2 (MMP-26), and epilysin (MMP-28). Pro-MMPs are activated by proteolytic removal of the N-terminal prodomain that keeps the zymogen inactive by using a cysteine switch to bind the catalytic zinc ion. MMPs participate in multiple processes that involve tissue remodeling, *e.g.*, embryo implantation, wound healing, cell proliferation, bone ossification, and blood vessel remodeling; signaling by all 54 human chemokines; and innate immune defense [39]. MMP activity is regulated by endogenous, tight-binding tissue inhibitors of metalloproteinases (TIMPs). Abnormal MMP expression and activity have been observed in cancer, cardiac remodeling and aneurysm formation, impaired wound healing, neurodegeneration, and after UV radiation exposure of the aging skin and the cornea [40]. The family of endogenous tissue-inhibitors of metalloproteinases (TIMPs) consists of 4 proteins that target protease activity of MMPs, ADAMs, and ADAM-TSs. They also affect cell growth and differentiation, cell migration, anti-angiogenesis, anti- and pro-apoptosis, and synaptic plasticity in biological processes different from protease inhibition [41].

Protease signaling is a relatively new concept, and in contrast with other types like receptor or kinase signaling the process is irreversible [8]. The major immediate results of protease signaling are target protein activation or inactivation, exposure of cryptic sites, shedding of transmembrane proteins, and receptor agonist/antagonist interconversion. These processes may initiate downstream signaling, resulting in a wide variety of physiological or pathological responses. Selection of a physiological substrate is facilitated by protease–substrate colocalization; substrate specificity, as determined by the complementarity of the protease active site to the reaction transition state; interactions immediately distal from

Table 5
Tissue remodeling, signaling, cell migration, and proliferation: activities and Uniprot ID codes (where applicable) of proteases, protease inhibitors, and cofactors associated with tissue remodeling, signaling, cell migration, and proliferation

Protease/inhibitor	Uniprot	Action
ADAM family (a disintegrin and metalloprotease)		Play roles in fertilization, proliferation, migration and cell adhesion Not all are proteases, those which are act as sheddases
ADAM-TS (a disintegrin and metalloproteinase with thrombospondin motifs		Process procollagen and von Willebrand factor, and cleave extracellular matrix aggrecan, versican, brevican, and neurocan
MMPs (Matrix metalloproteinases)		All implicated in metastasis except 12, 20, and 28
TIMPs (tissue-inhibitors of metalloproteinases)		Endogenous inhibitors of MMPs
MMP-1	P03956	Cleaves collagens I, II, III, VII, and X Mediates neurotoxicity of HIV viral Tat protein
MMP-8	P22894	Degrades fibrillar collagens I, II, and III
MMP-13	P45452	Cleaves collagens I, II, III, IV, XIV, and X Degrades fibrillar collagen, fibronectin, tenascin C, and aggrecan
MMP-2	P08253	Degrades extracellular matrix proteins, including collagen I and IV
MMP-9	P14780	Cleaves collagen IV and V and fibronectin Implicated in neovascularization in malignant gliomas
MMP-3	P08254	Degrades fibronectin, laminin, gelatins of type I, III, IV, and V; collagens III, IV, X, and IX, and cartilage proteoglycans Activates MMPs 1, 7, and 9
MMP-10	P09238	Degrades fibronectin, gelatins of type I, III, IV, and V, collagens III, IV, and V Activates procollagenase
MMP-11	P24347	Cleaves alpha 1-proteinase inhibitor, activated intracellularly by furin
MMP-7	P09237	Degrades casein, gelatins of types I, III, IV, and V, and fibronectin Activates procollagenase Activates MMP-2 and MMP-9
MMP-12	P39900	Cleaves elastin, implicated in aneurysm formation
MMP-20	O60882	Degrades amelogenin, aggrecan, and cartilage oligomeric matrix protein (COMP)
MMP-26	Q9NRE1	Degrades collagen type IV, fibronectin, fibrinogen, beta-casein, type I gelatin and alpha-1 proteinase inhibitor Activates progelatinase B
MMP-28	Q9H239	Degrades casein

the substrate-binding pocket; and interactions with protease exosites remote from the active site.

Protease-activated receptors (PARs) are prototypical examples of protease signaling. These four G protein-coupled receptors are activated irreversibly by extracellular proteases, by cleavage of the N-terminal ectodomain and exposure of a tethered peptide ligand. Transmembrane signaling is initiated by binding of this tethered peptide to the body of the receptor [42]. PAR1, PAR3, and PAR4 are activated by thrombin, and signaling occurs during tissue injury, hemostasis, and inflammation. Signaling is regulated by rapid internalization of spent receptors. PAR1 and PAR4 cleavage on platelets causes robust platelet activation. Thrombin has a higher affinity for PAR1, and the PAR1 antagonist vorapaxar was approved in 2014 as antiplatelet drug. However, major bleeding side effects prompted the development of PAR4 antagonists which are currently in clinical testing. PAR4 signaling promotes vascular disease and cardiac post-infarction remodeling, and these antagonists are promising candidates for safer antithrombotic and anti-inflammatory therapy [43]. PAR1 on endothelial cells is productively cleaved by activated protein C (APC) in the presence of the endothelial protein C receptor. This triggers expression of monocyte chemoattractant protein-1, acting as a protective component during sepsis. PAR2 is activated by trypsin, tryptase, the coagulation factors VIIa and Xa, and matriptase. PAR2 signaling is thought to regulate epithelial growth and function. Thrombin-mediated PAR activation has been implicated in vascular smooth muscle cell migration and proliferation as causative processes in restenosis after stent placement [44], and in tumor metastasis, where a simultaneous requirement for PARs and fibrinogen was found [45]. Thrombosis and cancer have since long been recognized as interconnected pathologies, and in this light argatroban, the tight-binding competitive inhibitor of active thrombin, has been re-evaluated as a clinically useful antiproliferative and antimetastatic agent [46].

2.5 Programmed Cell Death

For obvious reasons, the process of apoptosis must be both highly regulated and a model of organized efficiency once initiated. It is governed by a cascade of caspases which, though they sometimes have other functions, thoroughly dismantle the innards of a cell so that it may be phagocytosed by immune cells without releasing cytosolic components into the extracellular space. Caspases are cysteine aspartate proteases involved in cell death, cellular remodeling, stem cell fate determination, spermatogenesis, and red blood cell differentiation. Their sets of substrates with regard to apoptosis are well defined, and cooperative cleavage of these substrate sets triggers apoptosis. With regard to apoptosis, their functions fall into one of two categories: Initiator (caspases 2, 8, 9, and 10) or Executioner (caspases 3, 6, and 7). Caspases 1, 4, 5, and 12L are considered inflammatory [47]. Activation of the Initiator caspases can be induced intrinsically, by release of cytochrome c into the cytosol by mitochondria, or extrinsically by ligand binding to Death Receptors. When cytochrome c is released, it binds to adaptor protein APAF-1, inducing it to form the apoptosome oligomer, which then binds to the caspase activation domain (CARD) of procaspase 9, inducing oligomerization of the procaspase. This induces autoproteolysis of caspase 9 to activate it [48, 49]. This process can be suppressed by the presence of the β transcription variant of caspase 9 which lacks a catalytic domain [50, 51]. Caspase 9 then activates procaspases 3 and 7 by cleaving them at a L-G-H-D-(cut)-X sequence [52, 53]. Caspase 9 can be downregulated by phosphorylation, or inhibited by proteins in the Inhibitor of Apoptosis (IAP) family [54]. Activated Caspase 3 is known to inhibit the function of IAPs [55], ensuring that once the cascade is initiated it progresses rapidly, and to activate caspases 6, 7, and 9, which further accelerates the process. Extrinsic activation is initiated by binding of Death Factors, such as FasL, to Death Receptors, such as FasR. The conformation change in the receptor's cytoplasmic Death Domain (DD) induces a change in the bound adaptor protein FADD which recruits procaspase 8 to bind at its Death Effector Domain (DED, proving gallows humor is irresistible even at the cellular level), activating caspase 8 [56]. Caspase 8 activates caspases 3, 4, 6, 7, 9, and 10. Caspase 10 then activates Caspases 3, 4, 6, 7, 8, and 9. Caspase 6 has a limited capacity to autoproteolyze and activate itself [57], and is known to target both Huntingtin and Amyloid Precursor Protein (APP), linking it to neurodegenerative diseases [58]. In apoptosis it is responsible for disinhibition of the immune system by cleaving interleukin-10 and interleukin-1 receptor-associated kinase 3 (IRAK3) [59]. The Executioner Caspases, as a group, are responsible for cleavage of over 600 other proteins [60]. Caspase 3 can also be activated by Granzyme B, allowing T lymphocytes and natural killer cells to initiate apoptosis in target cells. The substrate landscape in non-apoptotic events may be much broader [61], as suggested by recent global proteomics studies. The "forward" approach involves triggering endogenous caspases to identify native substrates in intact cells, whereas in the "reverse" approach exogenous caspases are added to cell lysates. Isolated cleavage products are digested and identified by tandem mass spectrometry. The forward method allows identification of substrates in intact cells, rather than which caspase performs the cleavage. In the reverse method, specific caspases are tested for their ability of cleaving substrates, but in a cell lysate with destroyed organelles, endogenous proteases may contribute to substrate cleavage, requiring the need for strict controls. Moreover, proteolysis in organelle membranes may be missed due to removal of insoluble material before analysis. A combination of current methods has yielded several hundred potential native substrates for caspases, and measuring rates and extent of substrate cleavage allows distinguishing functional from bystander targets [62]. Eight human endogenous inhibitors of apoptosis (IAPs) have been identified, and their inhibitory activity is neutralized by the mitochondrial protein Smac/DIABLO. Development of Smac/DIABLO-like peptidomimetics has been proposed as a potential therapeutic approach in cancer treatment. Because of the roles of caspases in inflammation, caspase inhibitors may also prove beneficial in treating sepsis [63]. In addition to caspases, other proteases are also associated with apoptosis, *e.g.*, calpains, cathepsins, granzymes, and the proteasome. These are regulated by their respective endogenous inhibitors: calpastatins, cystatins, the serpin PI-9, and various macromolecular proteins (Table 6).

2.6 Protein Secretion

Many current cardiovascular biomarkers are secreted proteins, generated by cleavage of their pro-proteins at the endoplasmic reticulum (ER). Upon release of the mature protein, the signal peptide is proteolytically separated from the ER by signal peptidase, an intramembrane aspartic protease. The long-held belief that signal peptides are invariably recycled or degraded by ubiquitin-proteasome related factors has been challenged during recent years, and several were shown to remain intact after cleavage in the ER [64]. These signal peptides have biological functions of their own, and play roles in regulation of immunity, trafficking, and other processes. In type I diabetes, a signal peptide fragment from pre-proinsulin, presented at the surface of the pancreatic β-cells, acts as antigen and flags the cell for destruction by cytotoxic T cells. Identifying agents to control β-cell destruction may be a new therapeutic strategy. Similarly, a signal peptide fragment from pre-procalcitonin, highly abundant in several lung cancers and medullary thyroid cancers, is an epitope for T-killer cells. This knowledge may aid in the development of treatments of these cancers. Rapidly increasing plasma levels of N-terminal adducted signal peptide fragments from A-, B-, and C-type natriuretic peptides are characteristic of ST-elevated myocardial infarction. The nature of the N-terminal adducts may be useful in assay design and disease assessment, and development of fast biomarker assays for these signal peptide fragments may ultimately be beneficial in clinical decision-making (Table 7).

2.7 DNA Replication, Repair, and Processing

DNA damage hinders replication, and may lead to strand breaks, genomic instability, aging, and cancer [65]. DNA-topoisomerase 1 crosslinks (DPC) are bulky lesions that trap otherwise transient covalent DNA-protein intermediates, and inhibit movement of polymerases and helicase, causing stalling of the replication fork. In yeast models, the protease Wss1 was identified as effector of DPC repair. BLAST searches revealed a conserved family of DPC proteases, with Spartan being the human member of this class [66]. Spartan was recently characterized as a DNA replication-

Table 6
Programmed cell death: activities and Uniprot ID codes (where applicable) of proteases, protease inhibitors, and cofactors associated with programmed cell death

Protease/inhibitor	Uniprot	Action
Caspase-2	P42575	Function uncertain, sequence homology with initiator caspases
Caspase-8	Q14790	Activates caspases 3, 4, 6, 7, 9, and 10 Activated by death receptors via FADD
Apoptotic protease activating factor 1 (APAF-1)	O14727	Forms apoptosome complex Activated by binding of cytochrome c and ATP
Caspase-9	P55211	Activates caspase 3 Activated by the apoptosome complex Cleaves poly(ADP-ribose) polymerase (PARP) Implicated in activation of Abelson murine leukemia viral oncogene homolog 1 (ABL1)
Caspase-10	Q92851	Activates caspases 3, 4, 6, 7, 8, and 9 Activated by caspase 8
Caspase-3	P42574	Activates caspases 6, 7, and 9 Activated by caspases 8 and 9 Cleaves poly(ADP-ribose) polymerase (PARP) Cleaves and activates sterol regulatory element-binding proteins (SREBPs) Implicated in Huntington's disease
Caspase-6	P55212	Dis-inhibits immune system, cleaves interleukin-10 and interleukin-1 receptor-associated kinase 3 Cleaves poly(ADP-ribose) polymerase (PARP) and lamins Implicated in Huntington's and Alzheimer's
Caspase-7	P55210	Degradation of cellular proteins in apoptosis Cleaves poly(ADP-ribose) polymerase (PARP) Cleaves and activates sterol regulatory element-binding proteins (SREBPs)
Apoptosome		Heptameric complex of APAF-1, activates caspase 9
FasL	P48023	Tumor necrosis factor ligand, activates death receptors to initiate apoptosis
FasR	P25445	Death receptor, tumor necrosis factor receptor, activates caspase 8

coupled metalloprotease for DPC repair and restoration of genomic stability [67, 68]. Mutations have been associated with premature aging and early onset hepatocellular carcinoma, suggesting Spartan as a tumor suppressor, and DPC repair as a protective antitumor mechanism. Double-strand DNA breaks are repaired by the DNA damage response, in pathways that are tightly controlled by ubiquitinylation and deubiquitinylation events. The

Table 7
Protein secretion: activities and Uniprot ID codes (where applicable) of proteases, protease inhibitors, and cofactors associated with protein secretion

Protease/inhibitor	Uniprot	Action
Signal Peptidase		Removes amino terminal signal sequences from secretory pro-proteins Group of aspartic proteases
Insulin	P01308	increases cell permeability to monosaccharides, amino acids, and fatty acids accelerates glycolysis, the pentose phosphate cycle, and glycogen synthesis in liver
Calcitonin	P01258	Promotes rapid incorporation of calcium and phosphate into bone
Atrial natriuretic peptide (ANP)	P01160	Peptide hormone regulates of natriuresis, diuresis, and vasodilation promotes trophoblast invasion and spiral artery remodeling in uterus during pregnancy Binds and stimulates the cGMP production of the NPR1 receptor Binds the clearance receptor NPR3
Brain natriuretic peptide (BNP)	P16860	Peptide hormone Regulates natriuresis, diuresis, vasorelaxation, and inhibition of renin and aldosterone secretion Binds and stimulates the cGMP production of the NPR1 receptor Binds the clearance receptor NPR3
Natriuretic peptide precursor C (NPPC)	P23582	Peptide hormone Regulates of cartilaginous growth plate chondrocytes proliferation and differentiation Binds and stimulates the cGMP production of the NPR2 receptor

latter are catalyzed by deubiquitinases (DUBs) that can positively or negatively affect the damage response. The proteasomal deubiquitinating enzyme POH1 promotes double-strand DNA break repair [69]. Various DUBs are also associated with transcriptional and epigenetic control of gene expression, DNA damage repair pathways, and cell cycle checkpoint control, often deregulated in tumor cells [70]. These DUBs may be potential targets for therapeutic inhibition, and they are currently the subject of small molecule screening. DNA processing is also under indirect control of proteases. DNA fragmentation and chromatin condensation are final processes in apoptosis. The DNase CAD, catalyzing these reactions, is normally under tight control of its ligand ICAD that acts as a chaperone and inhibitor when bound to CAD. This prevents spontaneous activation of CAD in non-apoptotic cells. ICAD cleavage by caspase-3 during the apoptotic execution phase liberates active CAD that enters the nucleus to fragment DNA and catalyze chromatin condensation [71]. Human Lon protease binds to mitochondrial single-stranded DNA sequences with a propensity for forming G-quadruplexes [72]. The precise role of human Lon protease is not yet clear, but it has been suggested that DNA-bound Lon may process proteins involved in mitochondrial DNA and RNA metabolism.

2.8 Intranuclear, Cytosolic, Transmembrane, and Intramembrane Proteolysis

In addition to Spartan and the proteasome involved in DNA processing, other proteases are known to have intranuclear functions. Interleukin-1β-converting enzyme (ICE), also known as caspase-1, plays a role in the inflammatory immune response. The N-terminal prodomain of its precursor, procaspase-1, possesses a nuclear localization signal. Tumor necrosis factor induces translocation of procaspase-1 from the cytosol to the nucleus where it is activated by proteolytic removal of the intact prodomain [73]. Cell transfection studies showed that the prodomain alone is capable of triggering apoptosis, which suggests separate nuclear roles for the prodomain and active caspase-1. Caspase-3 also effects nuclear changes in apoptotic cells. Its inactive precursor is cleaved in the cytoplasm by initiator caspases in response to death signals, and also by cytosolic MMP-2 and -9. The active caspase-3 dimerizes and translocates to the nucleus by an active transport system. Caspase-7 is only found in the cytoplasm, suggesting that translocation is caspase-3 specific and not the result of simple diffusion after the nuclear-cytoplasmic barrier is disrupted [74]. The extracellular properties of MMPs as tissue remodeling proteases are well documented but less is known about their intracellular functions. Various MMPs have been detected in the intranuclear space, and they are mostly associated with pathological processes [75, 76]. Oxygen-glucose deprivation in ischemic stroke induces an intrinsic, caspase-independent apoptotic pathway in neurons, characterized by elevated intranuclear MMP-2 and -9 activity that targets nuclear DNA repair proteins. Similarly, MMP-2 in the nuclei of stressed cardiac myocytes can induce apoptosis. Intranuclear MMP-3 in osteoarthritis and certain cancers upregulates connective tissue growth factor, a mediator of cell migration, proliferation, and pathological fibrosis. During viral infections, macrophage-secreted MMP-12 translocates to the nucleus of infected cells, and enhances transcription of IκBα, a protein that promotes antiviral interferon-α (IFN-α) secretion. Extracellular MMP-12 degrades excess secreted IFN-α, thus limiting its systemic toxicity. It is becoming increasingly clear that the same protease may exert significantly divergent functions, depending on its microenvironment, substrates, and effectors. Understanding these different mechanisms of interaction on a molecular level is ultimately the key to successful design of therapeutics (Table 8).

Promotion of angiogenesis is regulated by binding of single chain urokinase-type plasminogen activator (scuPA) to its receptor on the endothelial cell surface, and subsequent transport of the

Table 8
Intranuclear, transmembrane, intramembrane, and cytosolic proteolysis: activities and Uniprot ID codes (where applicable) of proteases, protease inhibitors, and cofactors associated with intranuclear, transmembrane, intramembrane, and cytosolic proteolysis

Protease/inhibitor	Uniprot	Action
Matriptase (Suppressor of tumorigenicity 14 protein)	Q9Y5Y6	Degrades extracellular matrix, trypsin-like activity Promotes epithelial differentiation and possibly growth Implicated in metastasis
Matriptase 2	Q8IU80	Cleaves collagen I, fibronectin, and fibrinogen Involved in matrix remodeling processes in liver Regulates the expression of the iron absorption-regulating hormone hepcidin/HAMP
Prostasin	Q16651	Stimulates epithelial sodium channel (ENaC) activity through activating cleavage of the gamma subunits (SCNN1G) Also found in seminal fluid
hepatocyte growth factor (HGF) activator inhibitor type2 (SPINT2)	O43291	Inhibits HGF, possibly inhibits serine proteases generally Implicated in suppression of liver cancer
hepatocyte growth factor activator inhibitor type1 (SPINT1)	O43278	Inhibits HGF and matriptase

protease to the nucleus [77]. scuPA de-represses transcription of the VEGF receptor 1 (VEGFR1) and 2 (VEGFR2) genes by interfering with the proline-rich homeodomain protein that represses the activity of vegfr1 and vegfr2 gene promoters. The VEGF growth factors are known targets for control of pathologic angiogenesis in macular degeneration, and the discovery of the scuPA-mediated pathway may offer additional avenues for therapeutic intervention.

Transmembrane proteases may be anchored to the membrane by a C-terminal domain (Type I), an N-terminal domain with cytoplasmic extension (Type II), or by glycosylphosphatidylinositol (GPI) [78]. Their catalytic domains are extracellular. Among the zinc-dependent proteases are MMP-14, 15, -16, -24, ADAM-10 and -17, meprins α and β (Type I), MMP-23 (Type II) and MMP-17 and -25 (GPI). The Type I zinc proteases act as sheddases upon proteolytic removal of their N-terminal propeptide. The majority of transmembrane serine proteases are classified as Type II subfamilies: hepsin/TMPRSS (transmembrane protease/serine), matriptase, corin, and HAT/DESC (human airway trypsin-like protease/differentially expressed in squamous cell carcinoma). Corin in cardiomyocytes activates the atrial natriuretic factor (ANF), a cardiac hormone that regulates blood pressure and cardiac function by promoting natriuresis, diuresis, and vasodilation.

Tryptase γ1 is the only known Type I serine protease, and prostasin and testisin are GPI-anchored. Prostasin plays a role in epithelial sodium channel regulation, and testisin regulates germ cell maturation. All these proteases are involved in physiological development, but also in pathological processes of inflammation and cancer. They activate peptide hormones, growth and differentiation factors, receptors, enzymes, adhesion molecules, and viral coat proteins. Matriptases 1 and 2, and prostasin are expressed in human epithelium, and inhibited by their cognate Kunitz-type inhibitors, membrane-anchored hepatocyte growth factor activator inhibitors (HAI) 1 and 2 [79]. Matriptase overexpression elicits signaling via the PAR-2, and promotes fibroblast activation, proliferation, and migration in idiopathic pulmonary fibrosis [80]. Matriptase proteolytically activates hepatocyte growth factor (HGF) that binds to its receptor c-Met, a receptor tyrosine kinase. This activates critical signaling pathways in organ development. Abnormal c-Met signaling is associated with cell proliferation, migration and invasion, and progression of lung, breast, ovary, kidney, colon, thyroid, liver, and gastric carcinomas. There is also some matriptase crosstalk with the hemostatic system with regard to epithelial defense and repair after injury and infection. Exposure of membrane-anchored tissue factor (TF) in damaged vascular endothelium to factor VIIa triggers the extrinsic coagulation pathway and formation of factor Xa. The TF: fVIIa complex and factor Xa activate epithelial pro-matriptase to matriptase which cleaves the PAR2 receptor. This induces enhancement of the epithelial barrier function [81].

Intramembrane proteases (IMPs), a fairly recently discovered class, are embedded in lipid bilayers and their catalytic site is formed by residues in different transmembrane helices [82]. The four IMP families are metallo-, serine, aspartate, and glutamate proteases, and they are found in the Golgi apparatus, endosomes and lysosomes, the plasma membrane, endoplasmic reticulum, and the inner mitochondrial membrane. IMPs cleave their substrates with a fairly high specificity, given the fact that of the ~2500 identified single-pass transmembrane proteins only a limited number are identified as IMP substrates. The function of IMPs is diverse, ranging from transcription factor signaling, mitochondrial remodeling, and protein maturation to regulation of immunity, and quorum sensing and parasite-host interactions in pathogens. Many IMP defects are associated with pathogenesis.

Site-2 protease (S2P) is the only member of the metalloproteinase IMP group, and mutations cause ichthyosis follicularis and osteogenesis imperfecta. The HIV inhibitor Nelfinavir was found to inhibit S2P in castration-resistant prostate cancer cell lines; however, due to the multitude of other Nelfinavir targets and the less than impressive effects on PC-3 cancer cells this drug may not be specific enough for use in prostate cancer. Five human rhomboid serine IMPs are known, but no function or substrates have been identified for rhomboid 1 and 3. Epidermal growth factor and thrombomodulin are rhomboid 2 substrates, and the protease is thought to control cell migration and proliferation [83]. Lowered expression impairs wound healing, and overexpression may be linked to tumor metastasis. Rhomboid 4 upregulation is associated with poor outcome in colorectal cancer, but there is no clear consensus on the molecular mechanism of this process. Rhomboid 4 cleaves amyloid precursor protein (APP) within its ectodomain and reduces formation of Aβ38, 40, and 42 peptides. This pathway may be an alternative to the pathological processing of APP by γ-secretase in Alzheimer's disease. The role of the mitochondrial rhomboid protease PARL in Parkinson's disease is controversial: some studies attribute a protective function to PARL in inducing removal of defective mitochondria through autophagy, whereas others suggest that PARL knockdown is responsible for this process. A potential link of low level PARL activity with type 2 diabetes was first seen in obese sand rats with diet-induced diabetes. Normal PARL levels and insulin sensitivity were restored when the rats were put on an exercise regimen. Similarly, in humans with type 2 diabetes, PARL mRNA and mitochondrial DNA are reduced in skeletal muscle. PARL was recently identified as a pro-apoptotic protease because it cleaves mitochondrial Smac/DIABLO [84]. The processed protein is released into the cytosol and binds an apoptosis inhibitor, thereby triggering the caspase cascade. The most studied member of the aspartate IMPs is the γ-secretase complex, with presenilin as the intramembrane catalytic subunit. Presenilin and APP mutations are linked to familial Alzheimer's disease (AD), and γ-secretase was considered as an anti-AD drug target. However, severe side effects limit the use of γ-secretase inhibitors. Processing of Notch by γ-secretase releases the intracellular Notch domain for intranuclear modulation of gene expression. Alterations of this pathway are linked to several types of cancer [85], and γ-secretase inhibitors are now in clinical trials as potential anticancer drugs. The multifunctionality of these proteases illustrates the caveats in developing drugs that indiscriminately target protease activity without taking specific molecular mechanisms and the protease microenvironment into account.

3 Proteases and Disease

3.1 Epigenetics and Disease

Many inheritable diseases are directly related to DNA modification; however, epigenetic processes are equally prominent in the disease state, and they are essential contributors to normal physiological development. Environmental factors, diet, aging, and diseases such as cancer may contribute to positive or negative changes in gene expression that are passed onto daughter cells: DNA (hydroxyl) methylation, covalent histone modification and chromatin remodeling, arrangement of the chromatin-histone nucleosomes along the DNA sequence, gene-activating transcription factor activity of new gene products, and downregulation of messenger RNA by noncoding microRNA. Formation of prion structures in transmissible spongiform encephalopathies is also considered an epigenetic phenomenon.

Proteases play regulatory roles in epigenetic mechanisms of altered gene expression. Various nonspecific and N-tail-specific histone proteases are thought to assist in fertilization, histone turnover, gene de-repression, histone removal during spermatogenesis, and reversal of N-tail methylation [86]. In N-tail clipped histone H3, lysines for acetylation have been removed, which may result in transcriptionally inactive chromatin. This process may be in part responsible for age-related declining gene expression. Cathepsin L, susceptible to stefin B inhibition, was identified as the histone H3-cleaving protease during stem cell differentiation in mice; however, the corresponding H3-clipping protease in human embryonic stem cells is refractory to specific cathepsin L inhibitors and remains to be identified [87].

3.2 Inflammation as an Over-Arching Symptom of Disease

A comprehensive, recent overview of protease activity in inflammation is given by Deraison et al. [88]. The host inflammatory response is accompanied by release of proteases from neutrophil granulocytes, macrophages, and mast cells. These proteases form a first line of defense in bacterial infections but, if left uncontrolled, are also damaging to host tissues. Human neutrophil elastase, present in six isoforms, and the related cathepsin G and proteinase-3 are serine proteases localizing to neutrophil extracellular traps (NETs) in a defense mechanism against bacterial invasion. They cleave collagen-IV and elastin, and excessive secretion may cause idiopathic pulmonary fibrosis, rheumatoid arthritis, and adult respiratory distress syndrome. Alpha1-proteinase inhibitor (α1-PI) and leukocyte elastase inhibitor are their major endogenous irreversible serine proteinase inhibitors (serpins). Historically α1-PI was identified as a trypsin inhibitor and later found to be more specific for elastase inactivation. Smoking-induced oxidative inactivation or mutation-induced misfolding of α1-PI causes emphysema and cystic fibrosis, and may be treated with α1-PI augmentation therapy [89], as gene therapy for α1-PI deficiency is still in the Clinical Phase II stage. Increased risk for lung cancer has also been observed in α1-PI-deficient patients who never smoked [90]. Proteinase-3 generates antimicrobial peptides by cleaving cathelicidin in neutrophils, but it is also predominant in anti-neutrophil cytoplasmic antibody (ANCA)-associated vasculitis, a severe multisystem autoimmune disease with poor prognosis [91] (Table 9).

Tryptase (6 isoforms), chymase, granzyme B, and carboxypeptidase A, released by mast cells, degrade extracellular

Table 9
Inflammation as an over-arching symptom of disease: activities and Uniprot ID codes (where applicable) of proteases, protease inhibitors, and cofactors associated with inflammation, specifically as related to symptoms of disease

Protease/inhibitor	Uniprot	Action
Neutrophil Elastase	P08246	Broad substrate specificity, chymotrypsin family Inflammatory response triggers bacterial and host tissue destruction Inhibits C5a-dependent neutrophil enzyme release and chemotaxis Proteolysis of collagen-IV and elastin of the extracellular matrix Degrades pathogenic outer membrane proteins and virulence factors
Cathepsin G	P08311	Degrades ingested host pathogens Breaks down ECM components at inflammatory sites Cleaves complement component C3 Inhibited by Rv3364c (*M. tuberculosis* protein) Indirect suppression of macrophage apoptosis Converts angiotensin I to angiotensin II
Proteinase-3 (Myeloblastin)	P24158	Degrades elastin, fibronectin, and collagen (in vitro) Target antigen for anti-neutrophil cytoplasmic antibodies (ANCA)
Alpha1-proteinase inhibitor (a1-PI, Alpha-1-antitrypsin)	P01009	Inhibits elastase, plasmin and thrombin Irreversibly inhibits trypsin, chymotrypsin, and plasminogen activator Possible non-protease inhibitor activities: Anti- and pro-inflammation Anti-apoptosis
Leukocyte Elastase Inhibitor	P30740	Inhibits neutrophil elastase, cathepsin G, proteinase-3, chymase, chymotrypsin, and kallikrein-3 Potent intracellular inhibitor of granzyme H
Tryptase:		Major neutral protease present in mast cells Resistant to endogenous proteinase inhibitors Active only as heparin-stabilized tetramers Six Isoforms: Tryptase alpha/beta-1(Q15661) Tryptase beta-2(P20231), Tryptase delta(Q9BZJ3) Tryptase gamma(Q9NRR2), Brain-specific serine protease 4(Q9GZN4)
Chymase (CMA1)	P23946	Major secreted protease of mast cells Release may promote inflammatory response Converts angiotensin I to angiotensin II
Granzyme B (GZMB)	P10144	Unique to cytolytic T-lymphocytes and natural killer cells Activates caspases 3, 7, 9, and 10 Cleaves/activates BH3 interacting-domain death agonist (BID) Cleaves/activates Inhibitor of caspase-activated DNase (ICAD) Generate a cytotoxic level of mitochondrial reactive oxygen species

(continued)

Table 9
(continued)

Protease/inhibitor	Uniprot	Action
Carboxypeptidase A3 (CPA3)	P15088	Cleaves C-terminal aromatic or aliphatic residue Unique to mast cells Upregulated in sepsis and anaphylaxis Implicated in autoimmune diseases
Alpha 1-antichymotrypsin (SERPINA3)	P01011	Inhibits neutrophil cathepsin G and mast cell chymase
Cathepsin L1	P07711	Intracellular degradation and protein turnover Degrades collagen and elastin Degrades alpha-1 proteinase inhibitor
Cathepsin B	P07858	Intracellular degradation and protein turnover Upregulation of cathepsin D, matrix metalloproteinase, and urokinase Implicated in metastasis and immune resistance
Cathepsin D	P07339	Intracellular degradation and protein turnover Used by macrophages to degrade bacterial proteins Activates ADAM30, implicated in Alzheimer's progression Implicated in metastasis in breast cancer
Trypsin-3 (PRSS3) (meso-trypsinogen)	P35030	Degrades trypsin inhibitors
Serine Protease Inhibitor Kazal-type 1 (SPINK1)	P00995	Trypsin inhibitor, specifically inhibits autoactivated trypsin in the pancreas Inhibits calcium binding and NO production in sperm
Caspase-3	P42574	Activates caspases 6, 7, and 9 Activated by caspases 8 and 9 Cleaves poly(ADP-ribose) polymerase (PARP) Cleaves and activates sterol regulatory element-binding proteins (SREBPs) Implicated in Huntington's disease

matrix components. An imbalance between tryptase and its endogenous inhibitors is characteristic in rheumatoid arthritis [92]. Experimental tryptase inhibition in an in vivo model alleviated some, but not all, symptoms, suggesting the necessity for a multidrug approach. Cell surface proteoglycan-bound chymase is partially protected from endogenous inactivation by α1-PI and α1-antichymotrypsin, and inhibitory caging by α2-macroglobulin. Mast cell infiltration of atherosclerotic plaque aggravates the local inflammatory status by causing smooth muscle cell (SMC) apoptosis in a chymase-dependent process: fibronectin cleavage by chymase unmasks pro-apoptotic epitopes and disrupts the p-FAK-dependent cell-survival signaling cascade, leading to SMC cell death [93]. Mast cell tryptase and chymase activity is also associated with several other pathological processes in atherosclerosis, abdominal aortic aneurysm (AAA) formation and metabolic disease, as discussed below [94].

Macrophages release matrix metalloproteinases (MMPs), cysteine proteinases (caspases and cathepsin L), and cathepsin D, an aspartyl protease also present in lysosomes. Their proteolytic power is a part of the diverse arsenal of mechanisms macrophages utilize in vessel wall-localized pro-inflammatory processes associated with the development of vasculitis, a fast-developing pathology, and atherosclerosis, a disease developing over decades [95].

Cathepsin B, a lysosomal cysteine proteinase, preferentially activates mesotrypsinogen in pancreatic acinar cells. Increased mesotrypsin (trypsin-3) activity as a result from overexpressed cathepsin B lowers protective SPINK1 levels and initiates apoptosis by activating caspase-3 [96]. Both processes contribute to the development of human pancreatitis. Crystallization studies identified diminazene analogues as small molecule inhibitors of mesotrypsin, and these structures may form the basis for developing selective, tight-binding drugs [97]. In irritable bowel syndrome (IBS) the intestinal epithelium overproduces mesotrypsin, which increases intestinal epithelium permeability, signals to human submucosal enteric neurons and induces visceral hypersensitivity by a protease-activated receptor-2-dependent mechanism [98]. Mesotrypsin may be a suitable biomarker for IBS as well as a target for novel, specific drugs.

Inflammation and complement activation are interconnected, and the inflammatory environment in degenerative diseases, cancer, transplant rejection, and exposure to chronic external stimuli is characterized by excessive activation or insufficient control of complement activation [99]. The physiological response of complement involves self-recognition of normal cells; immune recognition and clearance of diseased cells, apoptotic cell debris, and immune complexes; elimination and danger signaling of pathogens; and tolerance to transplants and biomaterials. Excessive complement activation triggers inflammatory reactions seen in danger signaling and attack of "self" cells resulting in autoimmune disease, pathogen infection, and tissue/biomaterial rejection. The proteases and other components of the complement system may be attractive candidates for novel anti-inflammatory drug intervention; however, the complexity and extensive cross talk of this network poses significant challenges in the development of specific inhibitors devoid of off-target side effects.

3.3 Cardiovascular and Metabolic Diseases, and Stroke

Due to the interconnectedness of cardiovascular and metabolic disease, a new term, cardiometabolic disease has been coined, and recent studies underscore the critical roles of MMPs, calpains, cathepsins, and caspases in disease progression [100]. Physiological hemostasis is a delicate balance between cascading serine protease zymogen activation, positive and negative feedback mechanisms, protease inhibition, and thrombus dissolution. Many clotting and bleeding disorders are the result of gene defects affecting the expression or function of protease zymogens, or the substrates, cofactors, and inhibitors of the active proteases. In the extrinsic pathway, deficiency of factors VII, IX (hemophilia B), and X causes bleeding. The G20210A mutation in the 3' untranslated region of the prothrombin gene stabilizes the precursor mRNA which results in increased prothrombin levels and venous thrombosis risk. Up to 8% of the Caucasian population is heterozygous for this mutation. In the contact activation pathway, deficiency of factor XI causes mild to moderate bleeding, whereas factor XII and prekallikrein deficiencies are generally asymptomatic. This is consistent with a fairly recently discovered role for thrombin, generated through the extrinsic pathway, as a factor XI activator in the absence of factor XII [16]. Pathogen infections may also trigger hypercoagulability, as seen in coagulase-positive *Staphylococcus aureus* infections. Staphylocoagulase is not a protease but activates host prothrombin conformationally, resulting in deposition of fibrin/bacterial vegetations that can embolize to the lungs, brain, and other parts of the body [101] (Table 10).

The proteases of the contact pathway play a role in inflammatory and immune processes as well as in maintaining hemostasis [102]. Kallikrein cleaves high molecular weight kininogen to the pro-inflammatory bradykinin. This nonapeptide is a vasodilator, increases vascular permeability and contributes to inflammatory pain by binding to bradykinin receptors. Elevated bradykinin levels are seen in rheumatoid arthritis and IBS. The serpin, C1-inhibitor is the main physiological inhibitor of factor XIIa and kallikrein. An overactive contact pathway can be the result of decreased or dysfunctional C1-inhibitor, or a factor XII mutation leading to a more active fXIIa form. The pathological manifestation is hereditary angioedema, with sometimes life-threatening swelling in the upper respiratory tract or the intestinal mucosa. Abnormal activation of factor XII by β-amyloid triggers inflammation in Alzheimer's patients [103]. These findings suggest that drug targeting of the contact pathway proteases and regulators may prove beneficial in a variety of pathologies.

In anticoagulant protein C deficiency, insufficient proteolytic inactivation of factors V and VIII weakens this negative regulatory feedback, causing thrombophilia. Protein S is an obligatory cofactor in this reaction, and its deficiency also results in thrombophilia even at normal levels of functional protein C [104]. Patients with the Arg506Gln factor V Leiden mutation at one of the cleavage sites for activated protein C have a higher risk for developing venous thromboembolism due to decreased proteolytic factor Va processing [105]. About 40–50% of inherited thrombophilia cases are due to factor V Leiden, and 4–10% of the Caucasian population

Table 10
Cardiovascular and metabolic diseases, and stroke: activities and Uniprot ID codes (where applicable) of proteases, protease inhibitors, and cofactors associated with cardiovascular disease, metabolic diseases, and stroke

Protease/inhibitor	Uniprot	Action
Factor VII (F7)	P08709	Complexes with Tissue Factor VIIa/TF converts/activates X to Xa VIIa/TF converts/activates IX to IXa
Factor IX (or Christmas factor) (F9)	P00740	Converts/activates factor X Activates factor VII to form factor VIIa Activates factor X to form factor Xa Deficiency results in hemophilia B
Factor X (Stuart–Prower factor) (F10)	P00742	Converts/activates prothrombin to thrombin Complexes with phospholipids and calcium Activates factor VII to form factor VIIa
Thrombin (Prothrombin)	P00734	Converts fibrinogen to fibrin Activates factors V, VII, VIII, XI, XIII Complexes with thrombomodulin Thrombin/thrombomodulin activates protein C
Factor XI (Plasma thromboplastin antecedent)	P03951	Activates factor IX Inhibited by protein Z-dependent protease inhibitor (ZPI)
Coagulation factor XII (Hageman factor)	P00748	Reciprocal activation of Prekallikrein
Prekallikrein (Plasma kallikrein)	P03952	Reciprocal activation of factor XII
C1-inhibitor (Plasma protease C1 inhibitor)[a]	P05155	Complexes with/Inactivates C1r, C1s, MASP 1, MASP 2, chymotrypsin, kallikrein, fXIa, FXIIa
Protein C (Vitamin K-dependent protein C)	P04070	Inactivates factors Va and VIIIa in the presence of calcium ions and phospholipids Activated by Thrombin/thrombomodulin complex
Factor V	P12259	Cofactor required by factor Xa Activated by Thrombin Degraded by protein C
Factor VIII (FVIII)	P00451	Cofactor required by factor IXa Deficiency results in hemophilia A High levels implicated in deep vein thrombosis and pulmonary embolism
Antithrombin (Antithrombin-III)	P01008	Inhibits thrombin and factors IXa, Xa, and XIa Activity enhanced by heparin
Heparin cofactor II	P05546	Inhibits thrombin and factors IXa, Xa, and XIa Inhibits chymotrypsin
Alpha 2-antiplasmin (or α2-antiplasmin or plasmin inhibitor)	P08697	Inhibits plasmin and trypsin Inactivates matriptase-3/TMPRSS7 and chymotrypsin

(continued)

Table 10
(continued)

Protease/inhibitor	Uniprot	Action
Plasmin	P00747	Dissolves the fibrin of blood clots
tPA (Tissue plasminogen activator)	P00750	Converts plasminogen to plasmin on the fibrin surface Displaces plasmin from fibrin, promoting inhibition by alpha-2-antiplasmin
Mast cell chymase	P23946	Major secreted protease of mast cells Release may promote inflammatory response Converts angiotensin I to angiotensin II
Tryptase {6 isoforms, see previous list table}		Major neutral protease present in mast cells Resistant to endogenous proteinase inhibitors Active only as heparin-stabilized tetramers Six Isoforms: Tryptase alpha/beta-1(Q15661) Tryptase beta-2(P20231), Tryptase delta(Q9BZJ3) Tryptase gamma(Q9NRR2), Brain-specific serine protease 4(Q9GZN4)
pro-MMP-9 (MMP-9)	P14780	Cleaves collagen IV and V and fibronectin Implicated in neovascularization in malignant gliomas
pro-MMP-1 (MMP-1)	P03956	Cleaves collagens I, II, III, VII, and X Mediates neurotoxicity of HIV viral Tat protein
pro-MMP-2 (MMP-2)	P08253	Degrades extracellular matrix proteins, including collagen I and IV
pro-MMP-3 (MMP-3)	P08254	Degrades fibronectin, laminin, gelatins of type I, III, IV, and V; collagens III, IV, X, and IX, and cartilage proteoglycans Activates MMPs 1, 7, and 9
Proteinase-3 (PRTN3)	P24158	Degrades elastin, fibronectin, and collagen (in vitro) Target antigen for anti-neutrophil cytoplasmic antibodies (ANCA)
Kallikrein 13	Q9UKR3	Cleaves kininogen to the pro-inflammatory bradykinin
Cathepsin A	P10619	Protects beta-galactosidase and neuraminidase
Cathepsin C (CTSC)	P53634	Activates elastase, cathepsin G, granzymes A and B, neuraminidase, factor XIII, chymase, and tryptase
Cathepsin D	P07339	Intracellular degradation and protein turnover Used by macrophages to degrade bacterial proteins Activates ADAM30, implicated in Alzheimer's progression Implicated in metastasis in breast cancer
Cathepsin L1	P07711	Intracellular degradation and protein turnover Degrades collagen and elastin Degrades alpha-1 protease inhibitor
Cathepsin X/Z/P	Q9UBR2	Lysosomal protease, cleaves C-terminal residue

(continued)

Table 10 (continued)

Protease/inhibitor	Uniprot	Action
Calpain-10	Q9HC96	Limited proteolysis of substrates involved in cytoskeletal remodeling and signal transduction
Cathepsin K	P43235	Cleaves elastin, collagen, and gelatin Involved in breakdown of bone for remodeling Implicated in emphysema Activated by inflammatory cytokines Degraded by cathepsin S
Caspase-3 (CASP3)	P42574	Activates caspases 6, 7, and 9 Activated by caspases 8 and 9 Cleaves poly(ADP-ribose) polymerase (PARP) Cleaves and activates sterol regulatory element-binding proteins (SREBPs) Implicated in Huntington's disease
Caspase-6	P55212	Dis-inhibits immune system, cleaves interleukin-10 and interleukin-1 receptor-associated kinase 3 Cleaves poly(ADP-ribose) polymerase (PARP) and lamins Implicated in Huntington's and Alzheimer's
Caspase-8	Q14790	Activates caspases 3, 4, 6, 7, 9, and 10 Activated by death receptors via FADD
Factor VII activating protease (FSAP)	Q14520	Activates factor VII and pro-urokinase May act as tumor suppressor
Tissue inhibitor of metalloproteinase-1 (TIMP-1)	P01033	Irreversibly inhibits MMP 1, 2, 3, 7, 8, 9, 10, 11, 12, 13, and 16 Activates integrin signaling via CD63 and ITGB1

[a]Complement activation, blood coagulation, fibrinolysis and the generation of kinins

is heterozygous. The heterozygosity has been suggested as a potentially protective, evolutionarily conserved factor against excessive blood loss during child birth, and was shown to protect from post-cardiac surgery hemorrhage [106].

Venous thrombosis can also originate from functional or expression defects of antithrombin, the endogenous serpin that irreversibly inactivates thrombin, and factors Xa, IXa, and XIa in reactions that are dramatically accelerated by heparin. The antithrombin mutation database currently lists 127 different mutations [107], and major functional impairments are due to missense mutations in the reactive site, the heparin-binding site and serpin-protease contact regions important for serpin folding and stable covalent complex formation. Heparin cofactor II (HCII) is an equally potent but highly specific thrombin inhibitor, in the presence of cell surface dermatan and heparan sulfate, and small over-sulfated molecules that do not affect the antithrombin–thrombin interaction [108–111]. HCII deficiency is associated with arterial thrombosis, development of atherosclerosis and in-stent restenosis [44, 112]. As 60% of HCII is extravascular, it may control thrombin's signaling properties, and other extravascular serine proteases may be yet unidentified HCII targets. A few serpin-related bleeding disorders are known: α1-PI Pittsburgh has a Met358Arg mutation in its reactive site, which shifts its specificity from elastase to thrombin, thereby impairing normal blood clotting; and congenital α2-antiplasmin deficiency results in premature lysis of hemostatic plugs by excess plasmin. Selective inhibition of the anticoagulant activated protein C by mutating the reactive site-flanking residues of α1-PI Pittsburgh to lysines has been shown successful in normalizing bleeding in a hemophilia B mouse model, and may show promise as a novel hemophilia drug [113].

Hyperfibrinolysis by uncontrolled plasminogen activation is characterized by excessive bleeding mimicking hemophilia. In the congenital disease α2-AP or PAI-1 are deficient, and acquired hyperfibrinolysis may occur in liver disease, trauma, or during surgery. Treatment with tranexamic acid, ε-aminocaproic acid, or other lysine analogs inhibits plasminogen activation by tPA on the surface of the fibrin clot by occupying the lysine-binding sites on plasminogen. This displaces plasminogen from the fibrin surface and inhibits plasmin formation. Tranexamic acid may be anti-inflammatory by inhibiting plasmin-dependent activation of complement, monocytes and neutrophils. The lysine analogs also block conformational plasminogen activation by the bacterial nonenzymatic cofactor streptokinase (SK), a fibrinolytic that has been displaced by tPA in the USA but is still used in many European and non-Western countries. SK has a C-terminal lysine residue that binds to plasmin(ogen) kringles, thereby increasing the affinity of the plasmin(ogen) complexes with SK and the rate of plasminogen activation [114–116].

Many cardiovascular and metabolic pathologies have an inflammatory component throughout the development of the disease, and extravascular infiltration of circulatory hemostatic proteases upon tissue damage contributes to inflammation. Cellular proteases produced by white blood cells also feature prominently in inflammation. Macrophages transform into cholesterol- and lipid-laden foam cells in the atherosclerotic vascular wall. Monocytes, neutrophils, lymphocytes, and mast cells in particular play a role in foam cell formation in the arterial intima [94]. Mast cell chymase converts angiotensin-I in vascular cells to the potent pro-inflammatory angiotensin II that upregulates expression of redox-sensitive cytokines, chemokines, and growth factors implicated in the formation of atherosclerotic lesions [117]. Elevated angiotensin II causes arterial hypertension, and has been implicated in vascular proliferation, aortic valve disease, myocardial infarction, heart failure, and abdominal aortic aneurysm (AAA). Mice with angiotensin

II-induced hypertension develop arterial vascular inflammation, dependent on thrombin-triggered activation of factor XI bound to platelets via its receptor glycoprotein Iba [118]. Patients with uncontrolled arterial hypertension also exhibit factor XI-dependent, amplified platelet-localized thrombin generation which may serve as an inflammatory marker of high blood pressure. Blocking factor XIa activity in combination with inhibition of the renin-angiotensin system may show promise in treating hypertension and associated vascular inflammation. Inhibition of the renin-angiotensin system in animal models and humans also diminishes plaque formation, and may provide an avenue for treatment and prevention of atherosclerosis.

Chymase and tryptase degrade ApoE and HDL3, thereby decreasing cholesterol efflux from foam cells and impairing cholesterol reverse transport. Chymase induces SMC apoptosis, inhibits SMC growth and collagen synthesis, and degrades endothelin-1, leading to impaired vasodilation. Chymase activates pro-MMP-9 whereas tryptase activates pro-MMP-1, -2, and -3, all involved in the development of atherosclerosis and abdominal aortic aneurysm. Extracellular matrix degradation by elevated MMPs facilitates chemokine- and angiogenic factor-triggered migration of leukocytes and endothelial cells, which accompanies neovascularization and growth of the atherosclerotic lesion, and eventually facilitates plaque rupture [119]. Chymase-activated TGF-β1 disrupts endothelial function and also contributes to intima thickening. Elevated plasma chymase and tryptase levels were detected in patients with acute myocardial infarction (MI) or unstable angina pectoris but not in stable angina, indicating a correlation with plaque instability. The mast cell inflammatory cytokines IL-6, TNF-α, and IFN-γ induce smooth muscle cell and endothelial cell expression of chymase and tryptase, and their plasma levels correlate directly with the AAA expansion rate. Elevated levels of matrix MMPs are associated with the development of AAA, and high plasma MMP-1 and -9 concentrations are indicative of poor outcomes after aneurysm rupture [120]. Anti-angiogenic drugs showed adverse effects in clinical trials with cancer patients suffering from atherosclerosis, and drugs targeting proteases may be an alternative to help combat atherosclerosis. Caspase-mediated apoptosis occurs in atherosclerosis, and both beneficial and harmful caspase effects have been reported. In a population study, apoptotic markers of 4284 subjects were measured, and at the mean 19-year follow-up, 381 patients presented with adverse cardiovascular events. Elevated caspase-8 at baseline was strongly correlated with their incidence [121]. Macrophage apoptosis in atherosclerosis may have both pro- and anti-atherogenic effects, and more studies are needed to elucidate these complex mechanisms.

Plasma chymase is elevated in type 2 diabetes and prediabetes, and clinical trials currently evaluate chymase and tryptase as drug targets for small molecule inhibitors. Chymase-generated angiotensin II contributes to islet disorganization and high risk for cardiovascular events in diabetic patients. Urinary extracellular vesicles of patients with diabetic nephropathy contain elevated levels of MMP-9, proteinase-3, kallikrein 13, and cathepsins A, C, D, L and X/Z/P compared to controls [122], and these proteases derive from neutrophils and monocytes, recruited to the glomerular endothelial cells. The profiles may have prognostic and diagnostic value in the assessment of kidney damage in type 1 and 2 diabetes. Proteinase-3 cleaves insulin-like growth factor 1 and promotes glomerular inflammation. These combined findings illustrate how endothelial dysfunction and inflammation may be predictors of diabetic nephropathy.

A strong association exists between calpain-10 and type 2 diabetes, and blocking calpain activation prevents diabetes-associated cardiac injury. Mast cells and macrophages produce cathepsins, among which L and K are associated with obesity. In mouse models of obesity, L and K knockout mice or wild-type mice given L- and K-selective small molecule inhibitors were significantly leaner than control mice and had improved glucose sensitivity [94]. Cathepsin K is a marker of adiposity, and recent findings report cathepsin S and D association with human obesity. Maternal diabetes may cause embryonic neural tube defects, characterized by elevated levels of caspase-3, -6, and -8. The mechanism involves proteolytic activation of the effector caspases-3 and -6 by initiator caspase-8 [123]. Caspase-3 is also a major effector of insulin-producing pancreatic β-cell apoptosis in type 1 diabetes [124].

In ischemic stroke, arterial blockage can be caused by a thrombus formed within the brain, or an embolus formed elsewhere in the body. Hemorrhagic stroke, less frequent but often more severe, is the result of a ruptured blood vessel. In patients on anticoagulants or fibrinolytics, ischemic stroke may develop a hemorrhagic component, known as hemorrhagic transformation. Factor VII activating protease (FSAP) is a plasma serine protease that activates pro-urokinase (pro-uPA) rather than factor VII. The FSAP-Marburg I polymorphism (1704G > A), which reduces FSAP activity, increases stroke risk and mortality but seems to lower the risk of developing carotid restenosis in atherosclerotic patients [125]. Ischemic stroke triggers uncontrolled MMP-2 and MMP-9 activity, associated with disruption of the blood-brain barrier and onset of edema, and MMP-9 is also elevated in hemorrhagic transformation [126]. Expression of the endogenous tissue inhibitor of metalloproteinase-1 (TIMP-1) is observed in conjunction with elevated MMP-9, as a protective response to tissue injury. Neutrophils, rather than resident brain cells are the main source of pro-MMP-9 following stroke, and upon degranulation, the proenzyme is proteolytically activated in the extracellular space. MMP-2 and TIMPs are ubiquitously expressed in tissues of the central nervous system [127]. These findings suggest that focusing on endothelial cells, pericytes, astrocytes, and infiltrating leukocytes, rather than neurons, may prove to be more successful in identifying new therapeutic targets. Understanding the relationship between MMP-9 and neutrophils may help elucidate mechanisms involved in disruption of the blood-brain barrier, and lead to more successful therapeutic approaches.

3.4 Cancer

The long-held concept that somatic mutations are the causal event in the majority of cancers has recently come under scrutiny [128, 129]. Priming the cellular microenvironment for development of cancer is characterized by a sequence of events that precede the transformation of a normal cell into a cancer cell, and somatic mutations are actually later events in the development of many cancers. Chronic inflammation and fibrosis have been identified as two of these events. Hemostatic proteases have recently been recognized to contribute to inflammatory processes in cancer. Disruption of the endothelial barrier during tissue damage allows hemostatic zymogens to be activated. Not only do these proteases contribute to extravascular coagulation and fibrinolysis, they also trigger signaling through cell surface activation of PAR receptors, binding to uPAR and LRP-1, and activation of MMPs [130]. Inflammation also triggers the release of TGF-β that potently induces MMP-2 and MMP-9 expression. In turn, MMP-2, -9, and -14 proteolytically activate latent TGF-β in the ECM. Transmembrane MMP-14 and several members of the ADAM family are localized on invadopodia of migrating cells. The involvement of MMPs in extracellular matrix remodeling facilitates tumor invasion, and MMPs also figure prominently in cancer-related signaling. Whereas many MMPs are recognized as pro-tumorigenic, some may negatively affect cancer progression, depending on the microenvironment of the cell [131]. The MMP and ADAM inhibitor Marimastat showed no broad therapeutic anticancer potential due to lack of specificity; however, it inhibited ADAM-17, highly expressed in renal cell carcinoma [132]. This inhibition downregulated Notch pathway-mediated cell proliferation and invasion more effectively than γ-secretase inhibition. Hence, Marimastat may have therapeutic potential in renal cell cancer. Tissue inhibitors of metalloproteinases (TIMPs) are differentially expressed in cancer: high TIMP1 expression is associated with fibrotic processes and poor outcome, and TIMP3 silencing indicates advanced disease [133]. TIMPs figure prominently in other pathologies such as cardiovascular disease and sepsis, and fibrosis as measured by TIMP1 levels was recently shown to predict all-cause mortality in the AGES-Reykjavik Study [134] (Table 11).

Epigenetic processes are increasingly recognized as essential in carcinogenesis. During epithelial-mesenchymal transition (EMT) in cancer initiation, progression, and metastasis, epigenetic

Table 11
Cancer: activities and Uniprot ID codes (where applicable) of proteases, protease inhibitors, and cofactors associated with cancer

Protease/inhibitor	Uniprot	Action
MMP-2	P08253	Degrades extracellular matrix proteins, including collagen I and IV
MMP-9	P14780	Cleaves collagen IV and V and fibronectin Implicated in neovascularization in malignant gliomas
MMP-14	P50281	Degrades extracellular matrix proteins Activates progelatinase A and MMP-15 Inhibits angiogenesis via cleavage of ADGRB1
ADAM-17 (ADAM metallopeptidase domain 17)	P78536	Activates tumor necrosis factor alpha Activates Notch Pathway Sheddase, activates multiple growth factors Implicated in tumor resistance to radiotherapy
Tissue inhibitor of metalloproteinase-1 (TIMP-1)	P01033	Irreversibly inhibits MMP 1, 2, 3, 7, 8, 9, 10, 11, 12, 13, and 16 Activates integrin signaling via CD63 and ITGB1
TIMP-3 (Metalloproteinase inhibitor 3)	P35625	Irreversibly inhibits MMP 1, 2, 3, 7, 9, 13, 14, and 15
Thrombin (Prothrombin)	P00734	Converts fibrinogen to fibrin Activates factors V, VII, VIII, XI, XIII Complexes with thrombomodulin Thrombin/thrombomodulin activates protein C
PAR-1 (Proteinase-activated receptor 1 or coagulation factor II (thrombin) receptor)	P25116	Stimulates phosphoinositide hydrolysis Activated by thrombin May play a role in vascular development
SUMO (Sentrin-specific protease 7)	Q9BQF6	Removes SUMO (Small Ubiquitin-like Modifier protein) 2 and 3
Serine Protease HTRA1	Q92743	Degrades extracellular matrix Degrades insulin-like growth factor receptors and tuberin
Cathepsin L1	P07711	Intracellular degradation and protein turnover Degrades collagen and elastin Degrades alpha-1 protease inhibitor
Cathepsin B	P07858	Intracellular degradation and protein turnover Upregulation of Cathepsin D, matrix metalloproteinase, and urokinase Implicated in metastasis and immune resistance

(continued)

Table 11 (continued)

Protease/inhibitor	Uniprot	Action
Cathepsin D	P07339	Intracellular degradation and protein turnover Used by macrophages to degrade bacterial proteins Activates ADAM30, implicated in Alzheimer's progression Implicated in metastasis in breast cancer
Mesotrypsin (Trypsin-3)	P35030	Degradation of natural trypsin inhibitors
Matriptase (Suppressor of tumorigenicity 14 protein)	Q9Y5Y6	Degrades extracellular matrix, trypsin-like activity Promotes epithelial differentiation and possibly growth Implicated in metastasis
HAI-1 (hepatocyte growth factor activator inhibitor type1 (SPINT1))	O43278	Inhibits HGF and matriptase
HAI-2 (hepatocyte growth factor activator inhibitor type2 (SPINT2))	O43291	Inhibits HGF, possibly inhibits serine proteases generally Implicated in suppression of liver cancer
Proteasome		Group of massive protease complexes, stacked ring structure Degrade proteins "tagged" with multiple ubiquitins Critical in protein turnover, apoptosis, and adaptive immune response
Kallikrein-3 (hK3)	P07288	Liquifies seminal fluid, degrades cervical mucus Elevated levels associated with prostate cancer
Kallikrein-5 (hK5)	Q9Y337	Degrades extracellular matrix proteins in epithelium, leads to cell shedding
Gingipains (RgpA, RgpB, and Kgp)		
rgpA (Gingipain R1) <human>	P28784	Bacterial thiol proteases Degrade host tissue proteins and cytokines
SPINK6 (Serine protease inhibitor Kazal-type 6)	Q6UWN8	Inhibits KLK4, KLK5, KLK6, KLK7, KLK12, KLK13, and KLK14

mechanisms such as DNA methylation and histone modifications regulate EMT-related genes [135]. The transformation of epithelial cells into migratory fibroblasts and mesenchymal cells is a hallmark of metastasis, and various protease activities are associated with this process. In human gastric cancer cells, thrombin-catalyzed activation of PAR-1 is thought to trigger EMT [136], and in breast cancer, induction of the SUMO-specific protease 7 long variant promotes gene expression favoring cell proliferation and EMT [137]. Mammalian intracellular high-temperature requirement A (HtrA) serine proteases contain a chymotrypsin-like domain and play a role in protein quality control. Epigenetic silencing of the *HTRA1* gene in cancer cells may be caused by histone deacetylase targeting of the promoter, or transcription repression of the methylated promoter by binding of methyl-CpG-binding domain protein 2 (MBD2) [138]. Silencing of tumor suppressor microRNAs via protease-activated PAR and Nf-κB signaling, and of caspase-8 expression by DNA methylation are yet a few other epigenetic mechanisms associated with the development of certain cancers [139, 131].

Altered expression of secreted lysosomal cysteine cathepsins has been associated with a variety of cancers, and several studies correlate either overexpression or gene knockout with progression of malignancy, depending on the type of cathepsin, and the nature and localization of the cancer. Overexpression of tumor tissue cathepsins B and L is detected in ovarian cancer but not in benign tumors and control tissue, and plasma cathepsin L is elevated in patients with malignant tumors. These proteases may be useful biomarkers [140]. Native and mutated forms of the aspartic protease cathepsin D precursor feature prominently in metastatic breast cancer. This tumor marker is present at higher levels in invasive ductal carcinomas, lymph node metastases, and hormone-receptor negative cancers than in lobular cancers and nodal positive carcinomas [141].

About 90% of the cancers originate in the epithelium. Epithelial cells express mesotrypsin and matriptase, and upregulation of these proteases is observed in epithelial cancers. Increased mesotrypsin activity is an indicator of poor prognosis in breast, prostate, pancreatic, and many other cancers [142]. Mesotrypsin is unusual in that it is not inhibited by Kunitz and Kazal-like trypsin inhibitors, but rather recognizes these proteins as substrates. It exhibits specificity for Arg/Lys-Ser/Met bonds, targets thrombin substrates such as the PAR1, 3, and 4 receptors, and is not inhibited by α1-antitrypsin (Met-Ser reactive bond) but by α1-antitrypsin Pittsburgh (Arg-Ser). Engineering a triple mutant M17G/I18F/F34V of the human amyloid precursor protein Kunitz protease inhibitor domain (APPI) created a selective, tight-binding inhibitor with an inhibition constant (K_i) of 89 pM, active in cell-based models of mesotrypsin-dependent prostate cancer cell invasiveness [143]. The crystal structure of the APPI M17G/I18F/F34V/mesotrypsin complex shows unique active site features that may be critical in driving metastasis, given the observation that other trypsins do not contribute to the invasive prostate cancer phenotype. This structural information may facilitate development of therapeutic peptide inhibitors, complementary to the mesotrypsin active site. Matriptase is the best-known member of the serine proteases with a type II N-terminal transmembrane domain, and its endogenous inhibitors are the Kunitz-type hepatocyte growth factor activator inhibitor types 1 and 2 (HAI-1 and 2). In normal tissue, matriptase proteolytic activity is tightly regulated by excess HAI-1 and 2, whereas in cancer tissue this balance is tilted heavily toward excess matriptase. Transgenic expression of epidermis matriptase in a squamous cell carcinoma mouse model causes tumor formation that is inhibited by HAI-1 or HAI-1 co-expression [144], strongly suggesting matriptase proteolytic activity as an essential trigger for malignancy. In highly aggressive inflammatory breast cancer (IBC), matriptase proteolytically activates pro-hepatocyte growth factor (pro-HGF). Binding of HGF to the receptor tyrosine kinase, c-Met, activates signaling pathways leading to cell proliferation, migration, morphogenesis, and invasion [145]. Both matriptase and c-Met are membrane-bound in IBC cells, and upregulated in cancer cells of IBC patients. Proliferation and invasion of IBC cells is halted by silencing with RNAi or treatment with synthetic matriptase inhibitors, illustrating their potential merit in IBC therapy.

Excessive proteasome activity occurs in certain blood cancers. Degradation of pro-apoptotic factors such as p53 impairs programmed cell death in cancer cells, and proteasome inhibition was proposed as a potential cancer treatment. The proteasome inhibitors bortezomib, carfilzomib, and ixazomib were approved by the FDA for the treatment of multiple myeloma, and are currently in clinical trials of blood, lung, and breast cancers. These potent inhibitors of the β5 peptidase activity of the 26S proteasome have only modest activity against β1 and β2 peptidases, which appears to limit their usefulness to multiple myeloma [146]. The β5 peptidase inhibitors have not been successful in treatment of solid tumors, and in recent studies triple negative breast cancer cell lines only responded to bortezomib or carfilzomib after CRISPR gene editing to inactivate β2 [147]. Development of dual β2/β5 inhibitors, while conceptually attractive, may prove daunting, and combination therapy with the β5 and known β2 inhibitors is a more realistic approach.

Balanced complement-associated inflammation can be of advantage in potentiation of immunotherapy, whereas an imbalance may sustain tumor cell proliferation, migration, invasiveness, and metastasis [148]. Genetic and epigenetic changes mark tumor cells as nonself, and the innate immune cells assist in clearing opsonized tumor cells through concerted actions of antitumor monoclonal antibodies (mAb) and complement cytotoxicity. Rituximab and ofatumumab, FDA-approved chimeric anti-CD20 mAbs were developed for the treatment of B cell lymphomas and chronic lymphocytic leukemia. Their targeting of tumor antigens elicits complement-dependent phagocytosis. Binding of C1q to the Fc portion of the mAbs results in formation of the proteolytically active C1 complex that initiates the cascade. Other tumor-specific mAbs recognize CD38 and CD52 as epitopes highly expressed on B cell- or T cell-derived tumors, and they are well studied in terms of their ability to elicit complement-dependent cytotoxicity. However, some solid tumors downregulate complement cytotoxicity and opsonization by overexpressing or sequestering surface proteins, thus limiting the efficacy of therapeutic antibodies.

The connection between complement activation, chronic inflammation, and cancer is becoming increasingly evident [149]. Complement factors and their active cleavage products themselves contribute to mitogenic signaling cascades and growth factor production (C3a, C5a, and the membrane attack complex MAC), angiogenesis (C3, C3a, C5a, MAC), protection from anti-growth signals and apoptosis (C3a, C4, C5a, MAC), cellular invasion and migration through the extracellular matrix (C1q, C1s, factor B, C3, C3a, C3d, C5, C5a, C9), proliferation (C3, C3a, C4, C5a, MAC), and suppression of antitumor immunity (C5a). From these observations it is clear that complement itself can promote cancer, under the "right" circumstances. The physiological role of the MAC complex is to disrupt the cell membrane and cause lysis; however, subthreshold MAC activity does not kill the cell but activates the cell cycle and triggers proliferation. Hence inhibition of complement may become an emerging strategy for the fight against cancer.

Excessive complement activation in colorectal, breast, pancreatic, lung, prostate, esophageal cancer, lymphoma, and leukemia has suggested the use of C3 activation fragments as diagnostic or prognostic biomarkers. However, C3 is abundant in plasma, and mass spectrometry quantitation may not be straightforward. Intratumoral C3 expression in ovarian cancer has been linked with disease prognosis, and a C3 fragment is found in prostatic fluid from cancer patients. Prostate-specific antigen (PSA) cleaves C3 and C5, and may act pro-tumorigenic by proteolysis of complement proteins.

Human tissue kallikreins (hK) are secreted serine proteases that are differentially expressed in many endocrine cancers. The *KLK3, 8, 10, 13,* and *14* genes are thought to encode tumor suppressor proteins, illustrating the recently recognized concept that protease upregulation does not always reflect tumor progression [131]. With this in mind, defining target protease specificity is crucial when developing protease inhibitors as potential anticancer drugs. hK3 or prostate-specific antigen (PSA) is the best-known biomarker for screening, diagnosis, and monitoring of prostate cancer. PSA is also elevated in benign prostatic hyperplasia, and its extent of complex formation with the serpin, $\alpha 1$-antichymotrypsin, differentiates between both pathologies. Tissues and plasma of prostate cancer patients contain higher levels of the complex than those of patients without cancer, and a level of 25% or more of free PSA activity is generally a good indicator of benign hyperplasia [150]. Many other kallikreins may also be suitable cancer biomarkers. hK5 proteolytically activates PAR-2, leading to Nf-κB activation and downregulation of tumor suppressor microRNAs in oral squamous cell carcinoma [139]. Plasma kallikrein is capable of activating the complement system, linking inflammatory responses to many cancer-related processes [151].

A very strong correlation has also been observed between cancer and hemostasis, and cancer patients invariably exhibit hypercoagulability, contributing to mortality and morbidity [152]. This prothrombotic state is attributed to the ability of tumor cells to activate coagulation, by producing procoagulant factors and inflammatory cytokines; interacting with monocytes, platelets, neutrophils, and vascular cells; and triggering acute-phase reactants and necrosis. Various anticoagulant therapies with heparins, vitamin K antagonists, or direct oral anticoagulants have proven to be beneficial in the treatment of cancer patients [153].

3.5 Neurodegenerative Diseases

Patients with Alzheimer's disease (AD), representing up to 70% of the dementia cases, have brain tissue containing amyloid plaque composed of toxic Aβ peptides, and neurofibrillary tangles composed of tau protein. Tau is localized in neuronal axons, promotes tubulin polymerization, and stabilizes microtubules. Amyloid β A4 precursor protein (APP) is a highly conserved synaptic integral membrane protein, thought to regulate synapse formation, neural plasticity, and maintenance of homeostasis in the central nervous system. Normal proteolytic processing of APP occurs via cleavage by α-secretase to release extracellular APPsα, followed by cleavage by intramembrane γ-secretase, with release of an extracellular p3 fragment, and an intracellular AICD fragment [154]. APP processing is different in AD: sequential cleavage by β- and γ-secretase releases the extracellular APPsβ, $A\beta_{1\text{-}40}$ and $A\beta_{1\text{-}42}$ peptides, and intracellular AICD. Heterogeneous proteolytic degradation yields several other extracellular species, ranging from 37 to 49 residues. $A\beta_{1\text{-}40}$ and $A\beta_{1\text{-}42}$ are considered neurotoxic, and form plaques in the brain. The APPsβ fragment oligomerizes and mediates death receptor signaling. The $A\beta_{1\text{-}42}$ fragment in cerebrospinal fluid is routinely used as a biomarker, combined with the measurement of total and hyperphosphorylated tau protein. This combination assay can diagnose AD in an early stage, and provide a prognosis of disease progression [155]. N-truncated Aβ peptides with cyclized terminal glutamate residues figure prominently in amyloid deposits, and are particularly useful additional biomarkers. However, no clear correlation could be made between plasma Aβ fragment concentrations and AD, dementia and various stages of cognitive decline, perhaps in part due to the limited sensitivity of the current analytical methods [156]. Due to observed discrepancies between AD dementia and amyloid deposition, some groups have suggested that either soluble oligomeric Aβ peptides may be more toxic, or that tau neurofibrillary tangles may be the pathogenic species [157] (Table 12).

Table 12
Neurodegenerative diseases: activities and Uniprot ID codes (where applicable) of proteases, protease inhibitors, a critical disease-related protease target, and cofactors associated with neurodegenerative diseases

Protease/inhibitor	Uniprot	Action
Amyloid β A4 precursor protein (APP)	P05067	Cell surface receptor involved in neuronal growth, adhesion, and motility Upregulated in neuronal repair Proteolysis generates amyloid β peptides (Abeta) from 37 to 49 residues in length, Abeta 40 and Abeta 42 implicated in Alzheimer's disease
α-Secretase		Group of ADAM family sheddases which cleave the amyloid precursor protein (APP)
γ-Secretase		Cleaves single-pass transmembrane proteins, including APP Transmembrane complex of presenilin-1 (PSEN1) (P49768), nicastrin (Q92542), anterior pharynx-defective 1 (APH-1) (Q5TB21), and presenilin enhancer 2 (PNE-2) (Q9NZ42)
β-Secretase 1 (BACE1)	P56817	Transmembrane aspartic-acid protease important for myelin sheath formation Cleaves APP to form Abeta 40 and Abeta 42 Several BACE1 inhibitors currently being tested as Alzheimer's treatments
ADAM10	O14672	Primary alpha secretase in platelets
MMP-9	P14780	Cleaves collagen IV and V and fibronectin Implicated in neovascularization in malignant gliomas
TIMPs (tissue-inhibitors of metalloproteinases)		Endogenous inhibitors of MMPs
Caspase-6	P55212	Dis-inhibits immune system, cleaves interleukin-10 and interleukin-1 receptor-associated kinase 3 Cleaves poly(ADP-ribose) polymerase (PARP) and lamins Implicated in Huntington's and Alzheimer's
Calpain (family)		Family of calcium-dependent, non-lysosomal cysteine proteases Excessive activity implicated in cytoskeletal degradation and altered calcium homeostasis in Alzheimer's disease
Calpastatin	P20810	Endogenous calpain inhibitor
m-AAA (Mitochondrial AAA proteases)		Group of ATP-dependent mitochondrial proteases
HtrA2 (high temperature requirement)	O43464	Mitochondrial serine protease, initiates cell death by binding IAPs (Inhibitors of Apoptosis Proteins) Implicated in Parkinson's disease
Presenilins-associated rhomboid-like protein (PARL)	Q9H300	Antiapoptotic, activates optic atrophy 1 (OPA1) which prevents release of cytochrome C into cytosol Mutation controversially implicated in Parkinson's disease
SUMO-specific protease 2 (SENP2)	Q9HC62	Processes SUMO1, SUMO2, and SUMO3 into mature proteins Deconjugates SUMO1, SUMO2, and SUMO3 from target proteins

APP is also produced in platelets, and cleaved by platelet α-secretase to release soluble APPsα in the circulation upon platelet activation [158]. APP isoforms of 130 kDa (intact) and 106–110 kDa (cleaved) can be detected by immunoblotting of platelet lysates. Patients with AD and mild cognitive impairment (MCI) have a significantly lower ratio of intact vs. cleaved APP than healthy controls. The decrease in ratio parallels cognitive decline and may predict conversion from MCI to AD. The metalloprotease ADAM10 in the brain processes APP through an alternative, non-amyloidogenic pathway [159], and platelets have the same proteolytic machinery as neurons for processing APP. The increase in the amyloidogenic pathway in AD patients is reflected by a decrease of platelet α-secretase and ADAM10 activity, and an increase of platelet β-secretase activity. These observations suggest that platelet biomarker assays for AD may be feasible, and raised the question if Aβ peptides generated by platelets can enter the brain and contribute to neuronal deficit. A recent elegant in vivo study showed indeed that Aβ peptides, originating from transgenic AD mice in prolonged parabiosis with healthy wild-type mice, accumulated in the brain of the healthy mice [160]. This finding strongly suggests a biological connection between altered platelet and neuronal protease expression in AD patients.

Clinical trials have focused on the development of β- and γ-secretase inhibitors, and antibodies targeting Aβ peptides with the goal of peptide removal as a means of reducing plaque formation. To date the results have been disappointing, except for clinical trials of the monoclonal Aducanumab, directed against aggregated and soluble Aβ peptides [161]. In contrast, the monoclonal Solanezumab, targeting soluble monomeric Aβ, failed in three consecutive trials, the latest one Expedition3, which was halted in January 2018 [162]. Small molecule inhibitors of β-secretase were also put to the test: negative results halted the Verubecestat Epoch trial in February 2017, but two trials studying the compound JNJ-54861911 are set to run until 2023. The 2010 phase III failure of Semagacestat, a γ-secretase inhibitor, was due to side effects of blocking Notch signaling; however, some questions remain about the design of the clinical trial and potential optimization of the drug dosage [163]. Although chronic but partial lowering of γ-secretase activity in heterozygous knockout mice does not cause a diseased phenotype and might have been tolerated in humans, the trial design opted for short peaks of complete γ-secretase inhibition in the brain alternated with periods of normal activity. This proved to wreak havoc on the ultradian oscillation of Notch signaling, as corroborated by severe Notch phenotypes in complete γ-secretase KO mice. Plasma concentrations of the drug were ~360-fold higher than the IC_{50} for γ-secretase inhibition in cell culture, and the side effects on the skin, gastrointestinal system, and weight loss due to very high intermittent drug dosage might also have contributed to poor performance in cognitive tests. With these caveats, there are still unexplored options for the development of γ-secretase inhibitors as drugs for cognitive decline.

Other proteases have been implicated in neurogenerative diseases. Several MMPs cleave APP in vitro, raising the question whether they also do so in vivo, and if there is a correlation with circulating MMPs and their inhibitors in AD. Expression of MMP-9 and the tissue inhibitors (TIMPs) was found to be elevated in postmortem AD brain tissue [164]. Significantly higher levels of MMP-9, but not of MMP-2 or the TIMPs were found in the plasma of AD patients, suggesting that MMP-9 may contribute to AD. Caspase-6 is found in non-apoptotic brain tissue of Huntington's disease and AD patients, indicating a function other than its executioner role [58]. Caspase-6 is implicated in axonal degeneration and neuronal loss in both diseases, and it cleaves tau, CREB-binding protein (CBP) which regulates transcription in cortical neurons, and NF-κB. Hence selective caspase-6 inhibitors may have therapeutic potential. CREB is indispensable for synaptic plasticity, and its impaired activation contributes to AD [165]. CREB is a substrate of the neutral, cytosolic cysteine protease calpain, and inhibition of this protease restores synaptic plasticity in a mouse model of familial AD. Calpain also cleaves tau protein, and upregulation or decreased degradation of the endogenous calpain inhibitor calpastatin has been a therapeutic goal in AD. Defective mitochondrial proteases can cause neuronal cell death and axonal dysfunction [166], and the human proteases m-AAA, the serine protease HTRA2 (high temperature requirement) and the rhomboid protease PARL have been identified in neurodegenerative processes. Two human m-AAA isoenzymes are differentially involved in neurodevelopment and protection against neurodegeneration, by preventing accumulation of misfolded polypeptides, and regulating mitochondrial protein synthesis, transport and proteolytic control of gatekeeping functions to prevent Ca^{2+} overload in the neuron [167]. Mutations in m-AAA cause hereditary spastic paraplegia and spinocerebellar ataxia. HTRA2 and PARL increase the susceptibility of neurons to apoptotic cell death. HTRA2 is involved in caspase-dependent apoptosis and in Parkinson's disease [168], but the role of PARL is still controversial. Posttranslational modification of proteins by small ubiquitin-related modifier (SUMO) can be reversed by SUMO-specific protease 2 (SENP2), and accumulation of SUMO-conjugated proteins is observed in patients with neurodegeneration. A knockout mouse model confirmed that disruption of this mitochondrial protease causes neuronal cell death [169]. Finally, the link between neurodegeneration and dysregulated complement activity has been firmly established. Acute brain injury triggers uncontrolled complement activation, flooding of the injury site with inflammatory anaphylatoxins and phagocytes, and blood brain barrier (BBB) damage

[170]. However, normal complement function plays a role in brain development (wiring), and brain homeostasis and repair during adulthood. Therapeutic approaches of complement modulation will therefore depend on its acute, subacute, and chronic nature of activation, and will require selective targeting of complement components.

3.6 Autoimmune Diseases

Upregulation and activation of pro-inflammatory cytokines and chemokines, uncontrolled endogenous protease activity, inflammation and antibodies/T lymphocytes against "self" antigens are hallmarks of autoimmune diseases [171]. Chemokines recruit leukocytes to release MMP-9 that generates peptides with immunodominant epitopes. These epitopes are presented to autoreactive T lymphocytes and stimulate B cells to produce autoantibodies. The "Remnant Epitopes Generate Autoimmunity" (REGA) model, based on cytokine, chemokine, and protease action, has been validated for multiple sclerosis, rheumatoid arthritis, and diabetes. According to this model, potential strategies for disease treatment may involve the use of anti-inflammatory cytokines, and the inhibition of pro-inflammatory and protease-inducing cytokines and chemokines. MMP-9 cleavage of collagen in rheumatoid arthritis and of insulin in autoimmune pancreatitis was found to generate remnant epitopes. Inflammasomes are large macromolecular complexes involved in activation of procaspase-1. Caspase-1 proteolytically activates the precursors of the pro-inflammatory cytokines IL-1β, IL-18, and IL33, and has been implicated in various autoimmune diseases. IL-1β blockade in autoimmune diseases can be accomplished with IL-1 receptor antagonists, neutralizing monoclonal antibodies, and the injectable IL-1β inhibitor Rilonacept [172]. Clinical trials of Pralnacasan, an oral caspase-1 inhibitor for treatment of rheumatoid arthritis, were halted in 2003 after liver toxicity was observed in animal studies (Table 13).

B cells contribute to autoimmune diseases by secretion of autoantibodies, presentation of autoantigen, and inflammatory cytokine secretion. Antibody therapy with Rituximab targets CD20 on the B cell surface, triggering cell death, and is used for B cell depletion to treat rheumatoid arthritis, idiopathic thrombocytopenic purpura, pemphigus vulgaris, and myasthenia gravis. The recent discovery that the intramembrane signal peptide peptidase-like protease SPPL2A promotes B cell differentiation by cleavage of CD74 suggested that SPPL2A may be a suitable target for inhibition in the treatment of autoimmune diseases [82]. Major histocompatibility complex (MHC) class II-mediated priming of T and B lymphocytes occurs in systemic lupus erythematosus (SLE) and lupus nephritis. The cysteine protease cathepsin S degrades CD74 during MHC II assembly with antigenic peptide in antigen-presenting cells, and cathepsin S inhibition might be therapeutic in SLE [173]. In some cases, deficiency of protease activity is

Table 13
Autoimmune diseases: activities and Uniprot ID codes (where applicable) of proteases, protease inhibitors, and cofactors associated with autoimmune diseases

Protease/inhibitor	Uniprot	Action
MMP-9	P14780	Cleaves collagen IV and V and fibronectin Implicated in neovascularization in malignant gliomas
Caspase-1 (Interleukin-1 converting enzyme) CASP1	P29466	Activates interleukin 1β and interleukin 18, initiating inflammation Activates gasdermin D, initiating lytic cell death Activated by incorporation into inflammasome complex, initiated by NOD-like receptors or AIM-1 like receptors Inhibited by CARD only proteins (COPs), COPs prevent formation of inflammasome
SPPL2A (Signal peptide peptidase-like 2A)	Q8TCT8	Cleaves type II membrane signal peptides, such as tumor necrosis factor alpha (TNF), the Fas antigen ligand (FASLG), and Cluster of Differentiation 74 (CD74) Initiation of innate immune response through CD74 activation implicated in autoimmune diseases
Cathepsin S	P25774	Cleaves proteins into peptides for presentation as antigens in macrophages, B-lymphocytes, microglia, and dendritic cells
ADAMTS13 (a disintegrin and metalloproteinase with a thrombospondin type 1 motif, member 13)	Q76LX8	Degrades von Willebrand factor, adversely affects clot formation

associated with autoimmune disease. The plasma of thrombotic thrombocytopenic purpura (TTP) patients contains unusually large forms of von Willebrand factor multimers. Most TTP cases arise from autoantibody-mediated inhibition or accelerated clearance of ADAMTS13 [174]. Highly similar anti-ADAMTS13 autoantibodies were found in unrelated TTP patients, suggesting that this autoimmune response is antigen-driven.

3.7 Proteases, Inhibitors, and Cofactors in Infectious Organisms

Infectious organisms employ their own arsenal of proteases for propagation and virulence, such as HIV protease [175], *Trypanosoma cruzi* cruzipain [176], *Porphyromonas gingivalis* gingipains [177], and *Bacillus anthracis* lethal factor [178]. Several bacterial, viral, protozoan, and fungal proteases trigger inflammation by activating the intrinsic coagulation pathway [102], or act as procoagulants by non-canonical, direct activation of prothrombin [179]. Large panels of small molecule inhibitors of the proteasome in pathogenic organisms are currently being screened for potential therapeutic benefit and minimal toxicity toward the cellular machinery of the host [180]. Bacterial infections are associated with increased thrombotic risk; however, this correlation is not restricted to pathogenic bacteria. Subtilisin, produced by the non-virulent bacterium *Bacillus subtilis*, can cleave prothrombin to an active thrombin-like species that converts fibrinogen to fibrin [181]. Deregulation of the intestinal microbiota is typical in IBS, and excess protease production by commensal enteric bacteria has been proposed to promote adherence to and invasion of intestinal epithelial cells, activate protease-activated receptors (PARs), disrupt the intestinal barrier, and facilitate bacterial interaction with immune cells, leading to inflammation [182] (Table 14).

Pathogens may use mechanisms other than direct proteolytic activity to enhance their virulence or facilitate dissemination and propagation. The streptococcal and staphylococcal cofactors streptokinase (SK) and staphylocoagulase (SC) are not enzymes themselves, and respectively bind and activate host plasminogen and prothrombin in a non-proteolytic fashion by inserting their N-termini into the zymogen activation pocket. This triggers a conformational change that forms the active site in the zymogen [101, 183]. The cofactor complexes with the zymogens as well as the active proteases are very tight, and refractive to endogenous serpins that inactivate plasmin and thrombin, thus adding to the bacterial virulence. The prothrombin•SC and thrombin•SC complexes cleave host fibrinogen to form fibrin barriers, protecting the pathogen from the host immune system. Upon activation of free host plasminogen to plasmin by the plasminogen•SK complex [183], the tighter binding plasmin•SK complex is formed, and degrades the host extracellular matrix to facilitate pathogen invasion and dissemination. Numerous streptococcal strains also increase their invasiveness by recruiting host plasminogen and plasmin to bacterial cell wall M-proteins [184]. Von Willebrand factor-binding protein (VWbp) is another conformational prothrombin activator secreted by *S. aureus*, and belongs to the family of staphylococcal and streptococcal homologs named zymogen activator and adhesion proteins (ZAAPs), based on the SC structure [185]. Staphylokinase (SAK) bears no sequence similarity to SK, but shares a similar domain fold. It does not activate plasminogen conformationally, but forms a tight plasmin•SAK complex that cleaves plasminogen as a substrate [186]. The skizzle (SkzL) protein, secreted by *Streptococcus agalactiae*, has moderate sequence identity to SK and SAK [187]. SkzL binds host plasminogen, and enhances its activation by the plasminogen activators uPA and single chain tPA, and plasma clot lysis by these plasminogen activators. *S. agalactiae* pathogenesis likely includes SkzL to enhance bacterial spreading through fibrinolytic enhancement. These are prime examples of

Table 14
Proteases, inhibitors and cofactors in infectious organisms: activities and Uniprot ID codes (where applicable) of proteases, protease inhibitors, and cofactors associated with infectious organisms

Protease/inhibitor	Uniprot	Action
HIV protease (HIV-1)	P04585	Cleaves viral polyprotein into individual proteins, including itself Critical for viral replication, prominent drug target
Trypanosoma cruzi cruzipain	P25779	cysteine protease expressed by *Trypanosoma cruzi*, vital to the parasitic protozoa's life cycle
Porphyromonas gingivalis gingipains	P28784	Bacterial thiol proteases Degrade host tissue proteins and cytokines
Bacillus anthracis lethal factor		Anthrax protein which degrades mitogen-activated protein kinase kinase, disrupting function of mitogen-activated protein kinases (MAPKs)
Subtilisin		Nonspecific bacterial protease known to activate thrombin
Staphylocoagulase (1 and 2)	P07767 and P17855	Activates prothrombin through binding, not a protease
Streptokinase (SK)		Activates plasmin
Von Willebrand factor-binding protein	A0A1D4Z3F9	Staphylococcus protein promotes clot formation
Staphylokinase (SAK)	P68802	Staphylococcus protein, plasmin-SAK complex activates plasminogen
skizzle (SkzL)	Q8DZH4	Streptococcus protein, enhances activation of plasminogen by uPA and sc-tPA
ADAMTS7	Q9UKP4	Degrades cartilage oligomeric matrix protein (COMP) Implicated in cancer, arthritis, and coronary artery disease Required for influenza A virus replication
Carboxypeptidase E	P16870	Cleaves C-terminal arginine of lysine residue Processes most neuropeptides and peptide hormones Required for influenza A virus replication
Dipeptidyl peptidase 3	Q9NY33	Degrades angiotensin, Leu-enkephalin, and Met-enkephalin Implicated in ovarian cancer Required for influenza A virus replication
Macrophage stimulating 1 protease (macrophage stimulating protein)	P26927	Unknown, sequence homology with hepatocyte growth factor Required for influenza A virus replication
Neurotrypsin	PRSS12	Cleaves agrin Multidomain serine protease expressed in the nervous system Required for influenza A virus replication
Ubiquitin-specific protease 14	P54578	Proteasome-associated deubiquitinase, prevents ubiquitin digestion Prevents degradation of prion protein

pathogens exerting virulence by hijacking the host coagulation and fibrinolytic systems [188].

The human host proteases ADAM-TS7, carboxypeptidase E, dipeptidyl peptidase 3, macrophage stimulating 1 protease, and neurotrypsin are required for influenza A virus replication, and are under control of eight host miRNAs regulating gene expression during virus replication [189]. These host genes and microRNAs may provide new therapeutic targets. The ubiquitin-specific protease 14 (USP14), a deubiquitinating enzyme, prevents degradation of prion protein by rescuing it from the proteasome, and may be a suitable target in the development of therapeutic strategies for prion diseases [190].

Porphyromonas gingivalis is prevalent in periodontitis, a risk factor for oral and gastric tract tumors, and also lung cancer, as recently identified in a follow-up of the Atherosclerosis Risk in Communities (ARIC) study [191]. *P. gingivalis* gingipains are cysteine proteases associated with this type of chronic inflammation, and they are the only bacterial proteases that degrade SPINK6, a Kazal-type inhibitor of various human kallikreins in skin and oral epithelium. Loss of this proteolytic control has been suggested as a link between periodontal disease and tumor development [177].

4 Proteolysis-Related Processes as Drug Targets

4.1 Overexpressed or Impaired Endogenous Proteolytic Activity

As of 2010, an estimated 5–10% of all drugs under development were targeted toward proteases [192], many of them small molecules designed to block the protease active site. Among past and present commercially successful protease inhibitors are blood pressure regulators (e.g., captopril and aliskiren) which respectively inhibit the metalloprotease angiotensin-converting enzyme (ACE), and the aspartic protease renin, by competitive binding to the protease active site; dipeptidyl peptidase-4 inhibitors (e.g., sitagliptin) to combat type 2 diabetes; the threonine protease inhibitor bortezomib as a cancer drug directed against the proteasome; the direct oral anticoagulants (DOACs), thrombin and factor Xa inhibitors (argatroban, dabigatran, apixaban, rivaroxaban, edoxaban) that bind tightly and reversibly to the protease active site; tight-binding hirudin-based thrombin inhibitors (lepirudin, desirudin, bivalirudin) for patients with heparin sensitivity.

Several endogenous protease, cofactor, and inhibitor deficiencies are treated by augmentation therapy. Hemophilia A is the deficiency of factor VIII, the essential cofactor of factor IXa to activate factor X; and hemophilia B patients lack functional factor IX. Both deficiencies prevent the formation of the intrinsic Xase complex that is responsible for generation of the majority of active factor Xa, with as end-result the impairment of clot formation.

Intravenous replacement with plasma-derived or recombinant fVIII and factor IX requires frequent injections, although preparations with longer half-life are being developed. Gene therapy for hemophilia B, based on in vivo gene transfer with adeno-associated viral (AAV) vectors to the liver has been in clinical trials for 16 years, with partial success due to cellular immune responses [193]. However, as of December 2017 the results of two small cohort studies are promising: 52 weeks after infusion of a single intravenous dose of an AAV5 vector encoding factor VIII, no cellular immune response, liver toxicity, or inhibitory antibodies were observed in a clinical trial for treatment of severe hemophilia A [194]; and in a small-scale hemophilia B patient study, a high level of expression of functional factor IX was seen after a single injection of an AAV vector containing the hyperfunctional factor IX Padua gene, targeting the liver [195]. Deficiency of factor XI, also known as hemophilia C, is a rare bleeding disorder, often seen in Ashkenazi Jewish populations, and does not cause bleeding in the joints. Tranexamic acid is administered to control traumatic bleeding incidents and during dental procedures, whereas fresh frozen plasma or recombinant factor XI may be used during surgery.

In sepsis, the systemic response by the host to pathogenic invasion triggers activation of inflammatory and coagulation pathways and inhibition of fibrinolysis. In this regard, administration of recombinant human-activated protein C (drotrecogin alpha activated, DAA) as an anticoagulant was deemed a useful strategy, and in 2001 it became the first biologic approved for treatment of severe sepsis [196]. Although a first trial indicated reduction in mortality, later trials failed to confirm these findings, and DAA was withdrawn in 2011. Observational trials consistently showed a benefit while randomized trials did not. The difficulties associated with obtaining reproducibility in these trials may be attributed to a variety of reasons: differences in acute illness of patient subgroups, perhaps as a result of conscious or subconscious patient selection; midway amendment through the first trial, changing inclusion/exclusion criteria, the type of placebo and the drug formulation, the combination of which favored the use of DAA and led to early termination; and differences in the timing and appropriateness of antibiotic administration and fluid resuscitation. New drug development for targeting severe sepsis will undoubtedly benefit from targeting pathophysiologic pathways characterized by specific biomarkers rather than heterogeneous patient populations grouped by clinical phenotypes, and DAA may yet be found beneficial for well-defined target groups.

Dysfunctional or poorly expressed serpins, inhibitors of serine proteases, cause a variety of severe diseases. COPD, emphysema, cystic fibrosis, liver disease and panniculitis due to functional α1-PI deficiency and accumulation of inhibitor polymers are alleviated by intravenous administration of plasma-derived α1-PI. Experimental approaches include aerosolized formulations of plasma or recombinant inhibitor, and direct delivery to the lung is expected to circumvent short half-life issues plaguing the intravenous formulation. However, no clinical trial reports are available to date. Intravenous recombinant α1-PI formulations are in the experimental stage, and conjugation with polyethylene glycol may delay rapid renal clearance [89]. It has been recognized that α1-PI inhibits proteases other than elastase and trypsin, namely proteinase-3, kallikreins 7 and 14, matriptase, caspase-3, and the metallopeptidase ADAM17 [197]. This opens up new avenues for modulating the activities of these proteases in disease states. Recombinant human antithrombin (Atryn) is purified from the milk of transgenic goats, and used to avoid perioperative and peripartum clotting complications in patients with hereditary antithrombin deficiency. It is not indicated for treatment of thromboembolic events in these patients. Its glycosylation profile differs from that of plasma-derived antithrombin, with increased heparin affinity as a result. The modification ensures efficient inhibition of elevated thrombin and factor Xa.

C1-inhibitor is a serpin targeting C1 esterase of the complement system, and it is also the physiological inhibitor of kallikrein, and factors XIIa and XIa of the contact activation pathway of coagulation. Both inherited and acquired C1-inhibitor deficiency can lead to life-threatening angioedema [198]. Inherited, heterozygous deficiency results in lack of transcription, translation or secretion, or in expression of a mutated, dysfunctional inhibitor. Acquired deficiency is the result of inhibitor depletion due to autoantibody formation or accelerated consumption in lymphoproliferative diseases. Elevated kallikrein activity causes unregulated cleavage of high molecular weight kininogen and release of bradykinin, the mediator of angioedema. Acute attacks of angioedema are treated with C1-inhibitor concentrate from plasma, recombinant inhibitor, and the kallikrein inhibitor ecallantide. Prophylactic treatment with the antifibrinolytic agents ε-aminocaproic acid and tranexamic acid regulate the fibrinolytic system which is continuously activated in autoimmune angioedema. Aprotinin (Trasylol), or bovine pancreatic trypsin inhibitor, is a Kunitz-type inhibitor of kallikrein and plasmin. It was used to treat laryngeal edema until its temporary withdrawal from the market in 2007 due to reports of increased death risk as a bleeding preventative during cardiac surgery. As aprotinin was derived from bovine lung tissue, concerns for allergic reactions and bovine spongiform encephalopathy prompted its discontinuation in Italy. In 2012 the European Medicines Agency proposed to lift the ban, and aprotinin is currently marketed by Nordic.

The proteases of the complement system have increasingly been recognized as potentially attractive points of interference for mitigation of inflammatory diseases. Recent developments in complement therapeutics focus on the proteases of the initiation pathways, with C1-inhibitor targeting C1r/s and mannan-binding lectin serine protease (MASP) (*e.g.*, Cinryze, Berinert, Cetor, Ruconest), and antibodies targeting C1q, C1s, C2, MASP-2 and -3 (ANX-005, TNT009, OMS721, CLG561, NM9401) [199].

4.2 Protease Inhibitors as Drugs: Some Caveats

Undesired properties or side effects of many of these therapies underscore the need for continued mechanism-based drug design. In 2014 the renin inhibitor aliskiren was placed on the list of drugs to avoid, due to severe side effects in patients with diabetes and kidney impairment [200]. Hirudin derivatives are an attractive alternative for treatment of patients with heparin hypersensitivity or thrombocytopenia; however, they have a short half-life. Hirudins are cleared through the kidneys, and dose adjustment is required in patients with renal impairment. Some small molecule drugs have limited bioavailability and solubility, and their efficiency may be mitigated by resistance mutations in the target proteases. Fast-acting DOACs are at least as effective as warfarin, with reduced risk for intracranial bleeding, and are prescribed for stroke prevention in atrial fibrillation, thromboprophylaxis in hip or knee replacement surgery, and for treatment and secondary prevention of venous thromboembolic disease [201]. Unlike with vitamin K antagonists, no routine monitoring of coagulation is needed. A recent disadvantage was the lack of antidotes for direct factor Xa inhibitors in case of traumatic bleeding. Activated prothrombin complex concentrates and recombinant activated factor VIIa have been proposed to reverse DOAC action; the monoclonal antibody idarucizumab was approved by the FDA in 2015 as a dabigatran antidote; and andexanet alfa, a factor Xa decoy, in May 2018 to counteract apixaban and rivaroxaban.

MMP-dependent degradation of extracellular matrix proteins is associated with angiogenesis and metastasis in cancer, and MMP inhibitors were proposed as suitable anticancer drugs. The zinc ion in MMPs was the first target, but small molecule peptidomimetic inhibitors based on zinc-targeting warheads (e.g., batimastat) had limited selectivity, failed to distinguish between different MMP classes also involved in the Notch-, Wnt-, and NFκB-signaling pathway, and were fraught with off-target side effects [202, 203]. A novel class of small molecules blocking the hydrophobic S1′ specificity pocket, exosites, and other MMP domains yielded reasonably specific inhibitors for several MMPs; however, their efficacy as second generation drugs has not yet been demonstrated. None of the 50 completed clinical trials with MMP inhibitors were successful, due to off-target toxicity or absence of efficacy. A recent study reports effective allosteric prevention of pro-MMP-9 activation in a mouse neuroinflammatory model by a small, highly selective heterocyclic chemical inhibitor [204]. This orally administered compound does not prevent activation of the structurally related pro-MMP-2, and does not inhibit catalytically active MMP-1, -2, -3, -9, or -14. Although these findings are encouraging for future drug development, the efficacy of this compound needs to be tested in other models, relevant to cancer, fibrosis, and neurodegeneration.

Indiscriminate targeting of overexpressed MMPs in certain cancers may not always yield desired results [205, 206]. Pancreatic ductal adenocarcinoma cells (PDAC) overexpress MMP-9, thought to play a role in invasion and metastasis. However, systemic knockout of MMP-9 in a PDAC mouse model caused increased interleukin-6 (IL-6) expression, and induced invasive growth and STAT3 activation in PDAC cells via IL-6 receptor signaling. The model system, animal genetic background and other experimental conditions may influence the effect of MMP activity, resulting in tumor-promoting, -inhibitory or null effect, as seen in various mouse models of breast cancer. In the light of many controversial experimental results, systemic MMP inhibitors should be used with caution, and information obtained from clinical studies associating specific MMPs with diseases profiles will be critical for identifying the proper MMPs as therapeutic targets. The metalloprotease ADAM10 has been associated with various disease states [159]. In the brain, it cleaves APP via a non-amyloidogenic pathway, with formation of the neuroprotective soluble ectodomain, and decrease of the toxic Aβ fragment. It may also slow down progression of fibrosis in chronic liver inflammation. However, it acts as a sheddase for the cellular prion protein, potentially promoting spreading, and increased activity is seen in synaptic dysfunction linked to Huntington's disease. Respectively, ADAM10 inhibition would be therapeutic or detrimental in these pathologies. ADAM10 is upregulated in various cancers, atherosclerosis, and various autoimmune diseases, suggesting a potential benefit of inhibition. The broad substrate specificity of ADAM10 and its similarity with ADAM17, its systemic presence, and its divergent effects in various diseases pose significant problems in targeting ADAM10, either for upregulation or inhibition. Ideally, drugs would be needed that regulate ADAM10 activity in a tissue and substrate specific manner. Focusing on the interaction of ADAM10 with specific substrates or regulatory partner proteins may provide some promise.

Efforts to develop γ-secretase inhibitors for treating Alzheimer's disease turned out to miss the mark, mainly due to the fact that the mechanism of APP processing by γ-secretase was not fully understood. Mutations in its presenilin domain were originally interpreted to enhance γ-secretase activity; however, the γ-secretase inhibitor Semagacestat showed worsening of familial Alzheimer's patients in a phase III clinical trial that was halted in 2010. In addition to issues with the design and dosage, this result was in part also explained by a recent in vivo study showing that the presenilin-1 mutations inactivate rather than enhance γ-secretase activity, impair hippocampal memory and synaptic function, and cause neurodegeneration [207]. γ-Secretase inhibitors have since then been repurposed as potential cancer therapeutics due to their inhibition of the Notch signaling pathway, upregulated in many cancers. However the panel of known inhibitors shows a wide range of activities toward cleavage of various other γ-secretase substrates, and off-target interference is likely to cause major side effects, thus limiting the long-term clinical usefulness of these inhibitors [208].

Osteoclasts express the cysteine protease cathepsin K, which degrades type I collagen in bone. Selective inhibition of cathepsin K increases bone mass, improves bone strength, decreases bone resorption, and contributes to bone formation [209]. The small molecule inhibitors relacatib, balicatib, and odanacatib were initially tested clinically as potential drugs for treatment of postmenopausal osteoporosis, with odanacatib ultimately making it through Phase II and III trials. The phase III trial was halted early after reports of positive efficacy and safety; however, a more thorough analysis discovered an increased risk of atrial fibrillation and stroke [210]. Odanacatib development was discontinued in 2016, after more than 12 years of research.

In the treatment of ischemic stroke, an active protease, rather than an inhibitor, is used as a therapeutic. Tissue-type plasminogen activator (tPA) was approved in 1996 for clot dissolution in ischemic stroke; however, its use is restricted to the first 3 h poststroke, and carries an established risk for bleeding. One study attributed the increased blood-brain barrier permeability to tPA-catalyzed activation of platelet-derived growth factor-CC, with Mac-1 integrin and LRP1 acting as cofactors in this reaction [211]. Another study suggested increased MMP-9 activity as a potential cause for increased bleeding risk [212], while yet another report pointed toward a connection of tPA-induced bleeding with hyperglycemia [213].

These are but a few examples of disappointing outcomes, or interference of significant off-target and other side effects in protease-related drug development. A recurring theme is that of insufficient knowledge of the underlying biochemical mechanisms and of the interconnectedness of protease and inhibitor activity in the proteome network. With the increasing availability of large online platforms and databases such as the National Center for Biotechnology Information, RCSB Protein Databank, UniProt, and MEROPS, discovery of such interconnectedness should prove increasingly less challenging.

4.3 Targeting Exogenous Proteolytic Activity

As antibiotic-resistant infections are on the rise, targeting proteases and protease-related processes in pathogenic bacteria may offer novel avenues for drug development. Proteases and protease cofactors that function as virulence factors are obviously first choice targets. Progress has been made in developing small molecule compounds that inhibit the expression of SK in Group A streptococci [214]. In addition, the proteolytic complexes Lon, ClpXP, HtrA, the proteasome, and signal peptidases are good candidates for disruption of bacterial mechanisms necessary for survival and pathogenicity [178]. Caseinolytic proteases (ClpPs) are conserved multimeric complexes that are conformationally activated by binding of Hsp100 ATPases. This binding aligns the catalytic triad of the proteases, and energy provided by ATP is used to unfold protein substrates for entry into the pore and subsequent degradation in an energy-dependent manner. Several mechanisms for ClpP deregulation have been proposed: (a) inhibitors such as phenyl esters and β-lactones can directly interact with the catalytic residues and halt protein degradation; (b) blocking of ATPase binding and uncoupling of ClpP and Hsp100 ATPase activity by acyldepsipeptides (ADEPs) and macrocyclic peptides may result in continuous ClpP activation and promiscuous protein degradation. However, many of these compounds are not 100% efficient due to limitations of stability, solubility, resorption, and half-life. Small molecule inhibitors have limited specificity and may not distinguish between bacterial protease complexes and their human orthologs. Recently developed ADEP derivatives are active against methicillin-, vancomycin-, and penicillin-resistant pathogens, and work well in combination therapy. The natural oligopeptide compounds cyclomarin A, ecumicin, and lassomycin target *Mycobacterium tuberculosis* but do not kill commensal members of the human microbiota. Encouraging in vitro and in vivo results have also been reported for small molecule inhibitors of *B. anthracis* lethal factor, and the cysteine protease domains of *Clostridium difficile* toxins. Specific interference with proteases required for bacterial viability offers an attractive possibility for developing a novel class of therapeutics, less prone to trigger resistance.

Parasitic cysteine proteases may be attractive targets for treatment of Chagas disease, African sleeping sickness, and leishmaniasis. Recent progress was made in the design of more efficient peptide inhibitors of cruzipain, by modifying the electrophilic warhead group that forms a covalent bond with the catalytic cysteine [215]. Classical nitrile group warheads are metabolically stable, polar, and small, but proved to be less potent than oxime and aldehyde warheads. Substitutions at the P_1 and P_3 residues alter the inhibitory potency and provide a means of modulating specificity. Ideally, a successful drug specifically recognizes the parasitic proteases over the host caspases, calpains, and cathepsins by forming specific non-covalent interactions adjacent to the active site, but obtaining this level of selectivity is challenging. Covalent binding is often irreversible, and one drawback of permanent attachment to peptide fragments after protein degradation is the immunogenicity of these fragments.

Currently ten FDA-approved HIV protease inhibitors are approved, and nine available, with structures that were thought to mimic the substrate transition state [216]. HIV protease inhibitors show off-target interference with proteases required for maturation of SREBP-1, a transcription factor that regulates gene expression in lipogenesis, with as result lipodystrophy syndrome; and blocking of glucose transporter-4 with as result insulin resistance; inhibition of the proteasome, resulting in metabolic complications, increased ER stress and autophagy; and caspase-dependent apoptosis, the discovery of which triggered interest in HIV protease inhibitors as potential anticancer drugs. The emergence of HIV-1 strains that are resistant to the current protease inhibitor drugs prompted the design of novel compounds with broad-spectrum activity against these variants [217, 218]. Small non-peptide molecules with substituted pyrrolidines, piperidines, and thiazolidines as P2-P3 ligands for binding to the S2-S3 specificity site, and flexible macrocyclic P1'-P2' tethers were good candidates, with inhibition constants (K_i) and IC_{50} values in the nanomolar range. Incorporation of heteroatoms in the macrocyclic skeleton yielded inhibitors with picomolar K_i and nanomolar IC50 antiviral activity. Biological evaluation, structure-activity relationships, and X-ray studies of the protease-inhibitor complexes validated the design approach, illustrating the power of structure-based molecular design. Peptide ketoamide-based NS3/4A serine protease inhibitors (boceprevir and telaprevir) are used for treating genotype 1 hepatitis C (HCV) [219]. In August 2017 Abbvie released Mavyret, a combination drug of glecaprevir/pibrentasvir targeting NS3/4A serine protease activity and the NS5A replication complex, and suitable for treatment of all HCV genotypes. However, complications may arise in HCV/hepatitis B (HBV) co-infected patients who completed treatment with HCV direct-acting antivirals and were not receiving HBV antiviral therapy. Fulminant hepatitis, hepatic failure, hepatitis flare, HBV reactivation, and death have been reported.

5 The Future of Proteolysis-Related Drug and Diagnostic Development

5.1 Active Site Targeting, Exosite, and Effector-Binding Sites

Active site targeting is a main component in many established approaches of drug development for controlling protease activity. However, small molecule inhibitors that are limited to interference with the conserved catalytic machinery of an entire class of proteases may have severe limitations due to their broad-spectrum activity, resulting in off-target protease inhibition. This was illustrated by the failure of many small molecule and zinc-targeting MMP inhibitors in clinical trials. Early irreversible inhibitors used active site targeting of nucleophilic proteases with electrophilic alkylating agents such as diazo- or halo-ketone warheads; however, the necessity of attaching a sizeable peptide to the warhead for specific protease binding proved impractical as large libraries of peptide warheads were required to identify effective inhibitors. The development of sitagliptin, an inhibitor of dipeptidyl peptidase 4 (DPP4) in type 2 diabetes, is a success story after a few consecutive setbacks [220]. DPP-4 inhibitors threo- and allo-isoleucyl thiazolidide initially showed significant animal toxicity due to off-target reaction with DPP8 and DPP9. Alpha amino compounds related to isoleucyl thiazolidide proved nonselective, and structure-activity screening identified a highly selective and rapidly metabolized beta-amino acid piperazine series. Bicyclic derivatization resulted in triazolopiperazine compounds that had suitable preclinical pharmacokinetic properties. Optimization led to the discovery of the highly selective sitagliptin.

Pitfalls in inhibitor design based on substrate interaction with the active site are illustrated by the development of transition state analogs against HIV protease. Enzyme transition states are very short-lived, on the femto- to picosecond timescale, but binding of transition state analogs converts these to a stable thermodynamic state. Kinetic isotope effects and computational chemistry identify which chemical steps are involved in transition state binding. Typically, these analogs can bind up to millions of times tighter than substrates, making them attractive compounds in drug development [221]. The HIV-1 protease–substrate complex has three transition states with partial bonds in the reaction coordinate, and two intermediates with equilibrated bonds. The high-energy intermediates bind tightly to the protease, as do inhibitors designed as mimics of these intermediates. Ten FDA-approved competitive HIV-1 protease inhibitors, with Saquinavir as first and prototypical drug, were originally considered transition state analogs because they have an sp3 center to mimic the geometry of the transition state, but they were later actually found to be intermediate mimics. Ile84Val and Leu90Met and several other mutations in HIV protease were symptoms of emerging drug resistance toward these inhibitors [222], prompting investigators to take a closer look at the protease transition states. The crystal structures of the transition states of both native and protease-inhibitor-resistant HIV-1 proteases showed that they are chemically and structurally identical, which means that resistance is due to changes outside the true transition state [223]. Mimicking specific chemical features of the true transition state may solve this resistance problem.

The active site of proteases forms a groove that accommodates several substrate residues adjacent to the scissile P_1-$P_{1'}$ bond. Substrates exhibit complementarity with the protease residues S_4-S_3-S_2-S_1-$S_{1'}$-$S_{2'}$-$S_{3'}$-$S_{4'}$ in the binding site, causing a favorable binding interaction. The architecture of the S_1 or specificity pocket often defines the nature of substrate cleavage, *e.g.,* hydrophobic and aromatic P_1 substrate residues for chymotrypsin-like proteases, basic P_1 residues for trypsin-like proteases, small aliphatic P_1 residues for elastase-like proteases, and hydrophobic bonds for aspartate proteases. Cysteine proteases prefer bulky nonpolar residues at the P_2 position. The MMP substrate specificity is more involved, with the $S_{1'}$ pocket selectively accommodating the substrate residue immediately after the scissile bond [224]. MMP-1 and -7 have small $S_{1'}$ pockets preferring small hydrophobic residues, whereas MMP-2, -3, -8, -9, and -13 have large pockets and bind a diverse array of amino acids. Additional allosteric, exosite, and effector-binding site interactions are expected to contribute significantly to exclusive selection of the target protease. Structural conservation of the specificity site throughout a protease family with diverse catalytic properties and biological targets may pose a problem in designing specific drugs, and a promising alternative was developed by targeting zymogen activation rather than the active protease [204]. A highly selective compound that allosterically inhibits MMP-9 activation by binding to a pocket near the zymogen cleavage site may be a first viable drug candidate.

Identification of protease subsite preferences can be achieved by positional scanning to identify the best fit. Peptides spanning the active site cleft carry a fluorophore and an internal quencher, and preferential cleavage of peptide libraries is determined from fluorescence yields upon proteolytic removal of the quencher [192]. High-throughput screening (HTS) and fragment-based screening do not require previous knowledge of substrate specificity and may yield rapid results, but need appropriate filtering by functional activity to eliminate nonselective reactions. Fragment-based screening using NMR, mass spectrometry, or differential scanning fluorimetry identifies moderate to weak binders that can be optimized into more potent inhibitors. The use of X-ray crystallographic structural information coupled with in silico drug design is getting a boost from rapidly developing high-throughput X-ray crystallization and structure determination, using TRAP screens with the most successful crystallization conditions [225]. High-throughput crystallographic screening of brominated fragment libraries, based on anomalous scattering to localize bromine, successfully identified targets for HIV protease, and detected novel binding sites in the surface-exposed active site glycine-rich β-hairpin flap region and the exosite region [226].

The importance of exosite interactions in protease inhibition is illustrated by the thrombin inhibitor, hirudin, a polypeptide produced by the salivary glands of medicinal leeches such as *Hirudo medicinalis.* Hematophagous animals have a need for a natural anticoagulant to prevent their food from clotting. The crystal structure of the hirudin–thrombin complex shows a globular N-terminal domain making contact with the active site, and a 17-residue extended C-terminal chain wrapping around thrombin exosite I, the anion-binding site (ABE I) that binds the thrombin substrate, fibrinogen [227]. For this reason hirudins are sometimes referred to as bivalent direct thrombin inhibitors [228]. This dual interaction confers tight binding, and the classification as a slow tight-binding inhibitor indicates that the off-rate for inhibitor dissociation is extremely slow. Several recombinant hirudins are on the market. They have a short half-life, but may be preferred to heparin as an anticoagulant in patients with heparin-induced thrombocytopenia. Thrombin exosites also figure prominently in its irreversible inactivation by the serpins antithrombin and heparin cofactor II (HCII). Thrombin ABE I binds fibrinogen, and exosite II (ABE II) binds heparin. Because heparin accelerates the irreversible inactivation of thrombin by endogenous serpins, it is sometimes called an indirect anticoagulant. A long heparin template binds both thrombin via ABE II and antithrombin or HCII via their heparin-binding site. This approximation causes a dramatic increase in the thrombin inactivation rate. HCII provides an extra interaction by binding of its N-terminus to thrombin ABE I. Fragment screening against caspase-7 identified two small molecule noncompetitive inhibitors with potential for drug development [229]. X-ray crystallography showed allosteric binding at the caspase dimer interface, more than 17 Å removed from the active site. This recent finding illustrates that allosteric control is yet another approach toward drug development.

Lysosomal cysteine cathepsins are required for normal lipid metabolism and cholesterol homeostasis, proper function of mitochondria, and clearance of apoptotic cells [230]. A deregulated lipid metabolism, vascular inflammation, arterial remodeling, neovascularization, autophagy, and necrosis are all hallmarks of atherosclerosis, and these processes are linked to upregulated cysteine cathepsin activity. Because vascular lesions in atherosclerotic patients may remain silent until the disease is well advanced, suitable biomarkers for these dysregulated pathways would be useful. Detection of cysteine cathepsins in macrophages allows distinguishing between stable and unstable lesions in excised carotid plaques, and this information from the plaque microenvironment may be harnessed in the development of molecular imaging. With this noninvasive technique, localized cleavage or covalent retention of specific protease substrates and inhibitors carrying fluorophores, reporter groups, or contrast agents can be detected. Substrate-based probes change their spectral properties upon cleavage, and the commercial self-quenched poly-lysine cathepsin probe Prosense has shown usefulness in preclinical cardiovascular imaging studies for detection of vascular inflammation, macrophage concentration, and cathepsin activity. Newer, lipidated cathepsin substrates show promise due to their improved homing properties. Quenched activity-based probes (qABPs) for cysteine cathepsin activity have the quencher QSY21 attached to a near-infrared fluorophore Cy5-labeled acyloxymethyl-ketone analog. The cell-permeable probe covalently modifies the target cathepsin, resulting in loss of the quenching group, and formation of a fluorescently labeled target protease [231]. This type of probe has a low nonspecific fluorescent background and has been successfully applied for detecting cysteine cathepsin activity in tumor cells, and in preclinical models of atherosclerosis. Reduced light penetrance is a major issue, necessitating further development of multimodal ABPs for PET-CT application [232]. This imaging approach has already been used successfully to localize idiopathic pulmonary fibrosis in patients, and may be adapted to detect rupture-prone atherosclerotic lesions.

5.2 Indirect, Mechanism-Based Targeting

Cytokines and chemokines are activated by uncontrolled protease action in many immune pathologies, and these proteins and their receptors may be potential therapeutic targets themselves. Cytokine storms typically occur as a reaction of a healthy immune system against new, highly pathogenic invaders, and are thought to be responsible for many deaths in influenza pandemics. Cleavage of the influenza hemagglutinin by host trypsin-like proteases is required for infectivity, and IL-1β was identified as the major cytokine that upregulates host trypsin expression and triggers formation of more IL-1β [233]. An anti- IL-1β antibody successfully suppressed upregulation of pro-inflammatory cytokines and trypsin in a mouse model, and antibodies against IL-1β and its receptor have been proposed as potential therapeutics. Drug targeting of non-protease components of the complement network involve the activation and amplification pathways, with peptides (AMY-101, APL-1 and -2), proteases (CB2782) and protein inhibitors (AMY-201, Mirococept) targeting C3 and C3b; and inhibitors of the terminal pathways, with antibodies (eculizumab), a tick saliva protein (Coversin) and aptamers (Zimura), binding to C5 and preventing C5 convertase activity [199]. The complement component C5a, generated in the terminal pathway of complement activation, is a chemokine, and attracts leukocytes to the inflammatory focus. Patients with paroxysmal nocturnal hemoglobinuria and atypical hemolytic uremic syndrome exhibit uncontrolled complement activation, and the recombinant tick protein prevents release of C5a and formation of C5b–9, the membrane attack complex (MAC). The natural function of the saliva protein in *Ornithodoros moubata* is suppression of the host immune response when the tick is feeding. The therapeutic protein is also effective in patients with C5 polymorphisms who are resistant to eculizumab.

Protease-activated receptors have long been considered potential drug targets, and the therapeutic applications of PAR-1 and PAR-4 antagonists are discussed in a previous section. In preclinical models, a PAR-2 antagonist inhibited tumor growth and the formation of new blood vessels in cancer, and inflammation in rheumatoid arthritis and acute inflammation models, perhaps indicating a potential link between inflammation and cancer [234, 235]. Strategies for developing suitable PAR antagonists include modified peptidomimetics such as trans-cinnamoyl-YPGKF-NH_2 that bind but do not activate the receptor; low molecular weight heterocyclic structures, *e.g.*, 1-benzyl-3-(ethoxycarbonylphenyl)-indazole; N-terminal palmitate-modified oligopeptides (pepducins) that anchor the peptide to the cell membrane; and specific function-blocking monoclonal antibodies [236]. The majority of these efforts is ongoing, and focused on PAR-4 antagonists as novel antiplatelet agents.

5.3 Drugging the Undruggable: A Familiar Theme, with Some Fancy Targeting

Designing substrates, inhibitors, and activity-based probes with specificity toward a single protease is a major challenge when the protease shares a similar catalytic mechanism and substrate specificity with other proteases in the same family, but is functionally completely different [237]. The main approaches utilized to date for optimizing complementarity with the protease specificity sites include positional scanning synthetic combinatorial libraries with coumarin-derived reporters (PS-SCL) [238], phage display, hybrid combinatorial substrate libraries using unnatural amino acids that allow more thorough scanning of the active site (HyCoSuL), counter selection substrate libraries (CoSeSuL), internally quenched fluorescent substrate or fluorescence resonance energy transfer libraries (IQF or FRET), proteomics and exopeptidase fingerprinting. HyCoSuL screening of the P_3 and P_4 positions, respectively, with methionine sulfone and 2-amino-6-benzyloxy-hexanoic acid allowed distinguishing between neutrophil elastase and proteinase-3 for the first time, and tracking neutrophil elastase activity in neutrophil traps [239]. Whereas PS-SCL, HyCoSuL, and CoSeSuL allow determination of protease preferences only in prime active site pockets, IQF can be used to refine complementarity to both prime and non-prime pockets. The advantage of phage display is its ability to generate large and diverse substrate arrays, up to 10^{10} peptides, and enrich for specificity after each cycle, which is not feasible with chemical synthesis. The label-free nature of this method requires individual kinetic analysis and reporter group labeling of positive hits. Multiplex substrate profiling with liquid chromatography–tandem mass spectrometry sequencing proved successful for distinguishing neutral serine protease activity in human neutrophil extracellular traps toward a large and diverse tetradecapeptide panel, and ranking granzyme B substrate efficiency, using label-free quantitation of precursor-ion abundance [240]. The above techniques have also allowed differentiation between various diverse members of the granzyme, kallikrein, caspase, metalloproteinase, exopeptidase, deubiquitinating and desumoylating protease families. A novel approach of profiling protease specificity is currently being developed, combining yeast endoplasmic reticulum (ER) sequestration with next-generation sequencing

(YESS-NGS) [241]. A substrate library, targeted to the ER, is exposed to ER proteases as it transports through the secretory pathway, and cleaved/uncleaved substrates localize to the cell surface. FACS analysis of cells labeled with fluorophore-conjugated antibodies specifically detects substrate cleavage. Characterization of proteolytic processing in secretory pathways may be useful to detect changes in the secretome in various disease states.

Recent, targeted "intractable" protein degradation methodology uses heterobifunctional chemistry for simultaneous binding of proteins inside cells, which tags them for degradation by the cell's own ubiquitin-proteasome system. The small molecule drug candidate uses a protein-binding domain linked to a ubiquitin ligase-binding domain, with the goal of eliminating pathological or defective intracellular proteins. This approach is currently developed by Kymera. A variant of this method is targeted proteolysis of endogenous proteins via the affinity-directed protein missile (AdPROM) system [242], that harbors the von Hippel-Lindau (VHL) protein, the substrate receptor of the Cullin2 (CUL2) E3 ligase complex, tethered to polypeptide binders that selectively bind and recruit endogenous target proteins to the CUL2-E3 ligase complex for ubiquitination and proteasomal degradation. Synthetic monobodies and a camelid-derived VHH nanobody were used in a feasibility model to target the tyrosine phosphatase SHP2 and the inflammasome protein ASC for degradation. This method has advantages over CRISPR/Cas9-mediated gene knockouts that are irreversible and may not always be feasible, and over RNA interference that requires prolonged treatment and may be incomplete. Both methods may have off-target effects. A possible breakthrough in nanomedicine may be the construction of a "DNA vault", a DNA origami nanodevice that locks up a single enzyme molecule, and that can be opened and closed by DNA locks to regulate access to substrate [243]. In a proof-of-principle model, chymotrypsin was covalently anchored to an open DNA vault, and after closure a FITC-casein substrate was added, together with the opening key or a control key. Enzymatic activity was detected predominantly in the reaction containing the opening key. This technology can be refined to program natural enzymes to operate as signal amplifiers for diagnostics applications and as delivery vehicles for therapeutic applications.

5.4 Discovery of Protease and Inhibitor Biomarkers for Disease

Conventional shotgun proteomics for biomarker discovery lack sensitivity and selectivity because they focus mainly on quantitative rather than qualitative differences between the normal and the diseased state. These approaches are also ill-suited for detecting pathologic posttranslational modifications. Novel MS-based proteomics techniques have the potential of identifying newly generated N- and C-termini that are characteristic in disease-related proteolysis. The TAILS and C-TAILS techniques enrich for protein N- and C-terminal peptides by polymer-based removal of internal peptides generated by tryptic digestion, and improve detection limits by several orders of magnitude [154]. The differential processing of APP protein in Alzheimer's disease, and of the chemokines CCL7 and SDF1 in arthritis and HIV-associated dementia, respectively, are prime examples of neo-terminus formation. N- and C-terminal peptide removal may activate or inactivate chemokines, turn them into receptor antagonists or change their receptor specificity. As a result, simple quantitation of chemokine expression is not sufficient to define the extent of the pathological inflammatory response, but quantitation of neo-terminus formation is. A fraction of Aβ peptides in Alzheimer's patients is N-truncated, with a cyclized terminal glutamate posttranslational modification ($A\beta_{3(Pe)}$), and this species may be a promising biomarker for early detection of Alzheimer's disease. A search of the TopFIND public database [244] shows that many FDA-approved biomarker proteins have multiple different N- and C-termini, potentially affecting their biological activity and "visibility" in current assays. Specific disease-related protease assays may be confounded by endogenous and exogenous inhibitors, and varying concentrations of cofactors and competing proteases in biological samples. A combination of targeted MS and neo-terminal-directed antibody assays may vastly improve the development of reliable biomarkers.

Novel functional proteomics techniques such as bead-based proteome enrichment combined with the 2D-electrophoresis-based Protein Elution Plate (PEP) also prove extremely useful for rapid, convenient profiling of functional protease activity in physiological and pathological patient samples [245]. This versatile plate reader technology uses specific protease substrates with various degrees of sensitivity and allows functional analysis of the protease landscape beyond the information available from protein abundance measurements.

The pursuit of biomarker development within a Functional Proteomics context is addressed elsewhere in this book ("Making the Case for Functional Proteomics" and "Methods to Monitor the Functional Subproteomes of SERPIN Protease Inhibitors"). Some of the observations of those sections are re-viewed here through the lens of proteolytic activity. Historically, biomarkers that drive drug and/or diagnostic discovery comprise a single entity, be it a metabolite, waste product or protein. In no small part that pattern reflects technology limitations of the time, not a preferred indicator of biological functionality. Further, of the approximately 2000 in vitro protein assays approved by the US Food and Drug Administration, only seven assays reference protease activity (and some of those are redundant for a given assay). This apparent lack of emphasis also points to limitations in proteolytic assay procedures, especially in media such as body fluids. Finally, as is abundantly apparent from the information presented in this chapter, the activity of a single protease is an unreliable indicator of physiological activity. A wider net must be cast.

For any functionally defined biological process, e.g., apoptosis or pathogen response, multiple proteolytic events occur in (relatively) linear sequences or in (parallel) cascades. Thus, any proposed therapeutic, regardless of selectivity at the level of a single entity, modulates multiple outcomes. And, any diagnostic must reflect that multi-entity process. The odds that a single proteolytic event or single intermediate or end product will define that process are slim indeed. The logic seems inescapable that biomarker development of the future must focus on biological networks, not single entities. Pioneering efforts are occurring as this is written.

An example: one diagnostic in development [246] for cancer "Stroma Liquid Biopsy™" demonstrates a network (or pattern [247])-based analysis of LC/MS data, connecting multiple networks or pathways that rely profoundly on proteolytic activity (coagulation, complement, and inflammation). The discovery and development sequence for such biomarker development is addressed elsewhere in this book ("Making the Case . . ."). Application of the diagnostic in the clinic relies on selection of a test "subproteome" as described in "Methods to Monitor the Functional Subproteomes of SERPIN Protease Inhibitors." It's noted that not only does the "Stroma Liquid Biopsy™" apparently differentiate serum from cancer patients, the pattern on which it is based informs cancer biology. Further, the identified subproteome (s) become a relevant test bed for screening new therapeutics.

The future is integrated discovery of network-based biomarkers that support therapeutic and diagnostic discovery while informing both biology and disease. Detection of protease activity per se in complex media will likely require development of new methods.

5.5 A Two-Protein Wrap-Up

This chapter presents, in text and tabulated formats, the diversity of proteases and protease-related activity. Perhaps the most compelling overall observation is the degree of overlap among physiological processes and disease states at the level of individual proteases. This is most readily seen in the tabulated summary of proteases, inhibitors and activity. A thumbnail analysis shows that half of all listed proteases are active in multiple physiological processes and/or associated with multiple disease states. Chief among the multi-process activities are the cathepsins, followed by the caspases and MMPs. The chapter closes with an examination of the multiple associations with physiological processes and disease states of two related proteins: the protease thrombin and the proteolysis product fibrin (gene expression products, prothrombin and fibrinogen).

Textbooks cite the cleavage of fibrinogen by thrombin to form fibrin as a central event in formation of blood clots formed in response to injury. This common response is certainly beneficial. However, other actions and associations of thrombin and fibrin are of a darker or unexpected nature (Fig. 2). In addition to its participation in blood clot formation, thrombin also triggers proteases that attenuate clot formation [20]. Thus thrombin (and

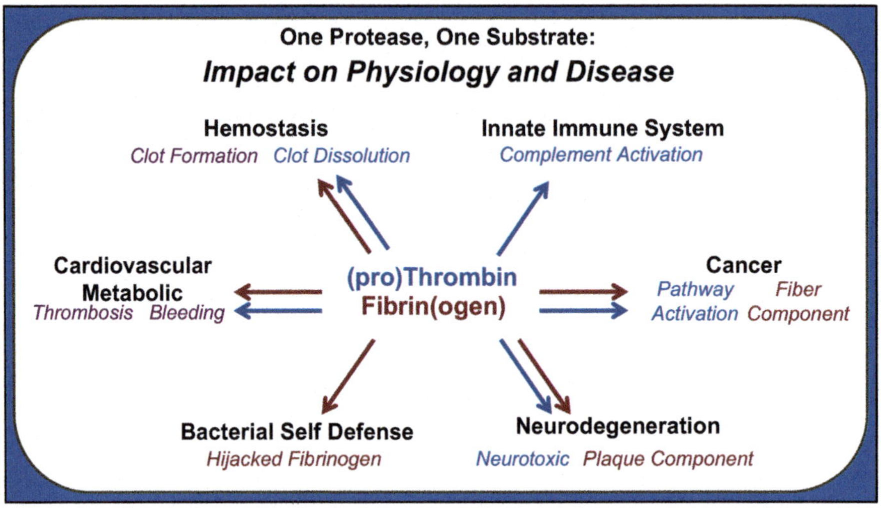

Fig. 2 Multiple Physiological and Pathological Roles of the (pro)Thrombin - Fibrin(ogen) Interaction

prothrombin) is associated with both thrombosis and clot dissolution. Elevated fibrin(ogen) levels track CV disease severity [248], and elevated levels of fibrin are found in diabetic patients [249]. In cancer, thrombin activates key pathways and fibrin is found in tumors. Thrombin is often cited as a neurotoxin [250] and is implicated in both Alzheimer's disease and peripheral neuropathy. Fibrin is found in AD plaques and may be oligomerized by amyloid beta [251]. Fibrinogen is hijacked by bacteria and through nonenzymatic prothrombin activation forms fibrin barriers to evade cellular defenses [101, 185]. Finally, thrombin cleaves complement C3 [252], thus igniting the innate immune system independent of any triggers associated with infection. This latter event may explain, in part, thrombin's apparent and destructive ability to trigger inflammatory and immune responses. Proteolytic events govern the life cycle of all proteins, as was noted in the Introduction to this chapter.

The clear lesson of the contents of this chapter is that proteolysis is central to the functioning of the organism, indeed can reshape the organism. Proteolytic activity is essential to multiple physiological processes and, specific proteases are central to multiple processes. It is not surprising, therefore, that disease is also strongly connected to proteolysis. Indeed, even at the resolution of a single protease, activity is not always to the benefit of the organism. As was demonstrated in the paragraph immediately above, what might be called the side effects of proteolysis can prove quite harmful. Of particular interest is the tantalizing correlation of proteolytic activity and chronic disease.

Within the context of Functional Proteomics, whether the scale is that of reshaping individual proteins or determining the state of the organism, proteolytic events are foundational.

Acknowledgements

I.M.V. is supported by NIH/NHLBI grants R01 HL071544 and R01 HL130018.

References

1. Fruton JS (2002) A history of pepsin and related enzymes. Q Rev Biol 77(2):127–147
2. Northrop JH, Kunitz M, Herriott RM (1938) Crystalline enzymes. Columbia Univ. Press, New York
3. Neurath H (1999) Proteolytic enzymes, past and future. Proc Natl Acad Sci U S A 96 (20):10962–10963
4. Rawlings ND, Barrett AJ, Finn R (2016) Twenty years of the MEROPS database of proteolytic enzymes, their substrates and inhibitors. Nucleic Acids Res 44(D1): D343–D350. https://doi.org/10.1093/nar/gkv1118
5. Kappelhoff R, Puente XS, Wilson CH, Seth A, Lopez-Otin C, Overall CM (2017) Overview of transcriptomic analysis of all human proteases, non-proteolytic homologs and inhibitors: organ, tissue and ovarian cancer cell line expression profiling of the human protease degradome by the CLIP-CHIP DNA microarray. Biochim Biophys Acta 1864(11 Pt B):2210–2219. https://doi.org/10.1016/j.bbamcr.2017.08.004
6. Perez-Silva JG, Espanol Y, Velasco G, Quesada V (2016) The Degradome database: expanding roles of mammalian proteases in life and disease. Nucleic Acids Res 44(D1): D351–D355. https://doi.org/10.1093/nar/gkv1201
7. Rawlings ND, Salvesen G (2013) Handbook of proteolytic enzymes, 3rd edn. Elsevier/AP, Amsterdam
8. Turk B, Turk D, Turk V (2012) Protease signalling: the cutting edge. EMBO J 31 (7):1630–1643. https://doi.org/10.1038/emboj.2012.42
9. Gettins PG, Olson ST (2016) Inhibitory serpins. New insights into their folding, polymerization, regulation and clearance. Biochem J 473(15):2273–2293. https://doi.org/10.1042/BCJ20160014
10. Laskowski M Jr, Kato I (1980) Protein inhibitors of proteinases. Annu Rev Biochem 49:593–626. https://doi.org/10.1146/annurev.bi.49.070180.003113
11. Fortelny N, Cox JH, Kappelhoff R, Starr AE, Lange PF, Pavlidis P, Overall CM (2014) Network analyses reveal pervasive functional regulation between proteases in the human protease web. PLoS Biol 12(5):e1001869. https://doi.org/10.1371/journal.pbio.1001869
12. Rechsteiner M, Rogers SW (1996) PEST sequences and regulation by proteolysis. Trends Biochem Sci 21(7):267–271
13. Lopez-Otin C, Bond JS (2008) Proteases: multifunctional enzymes in life and disease. J Biol Chem 283(45):30433–30437. https://doi.org/10.1074/jbc.R800035200
14. Chakraborti S, Chakraborti T, Dhalla NS (eds) (2017) Proteases in human diseases. Springer Singapore, New York
15. Macfarlane RG (1964) An enzyme cascade in the blood clotting mechanism, and its function as a biochemical amplifier. Nature 202:498–499
16. Gailani D, Broze GJ Jr (1993) Factor XI activation by thrombin and factor XIa. Semin Thromb Hemost 19(4):396–404. https://doi.org/10.1055/s-2007-993291
17. Gailani D, Renne T (2007) Intrinsic pathway of coagulation and arterial thrombosis. Arterioscler Thromb Vasc Biol 27(12):2507–2513. https://doi.org/10.1161/ATVBAHA.107.155952
18. Gailani D, Renne T (2007) The intrinsic pathway of coagulation: a target for treating thromboembolic disease? J Thromb Haemost 5(6):1106–1112. https://doi.org/10.1111/j.1538-7836.2007.02446.x
19. Nesheim M, Bajzar L (2005) The discovery of TAFI. J Thromb Haemost 3(10):2139–2146. https://doi.org/10.1111/j.1538-7836.2005.01280.x
20. Bode W (2006) The structure of thrombin: a janus-headed proteinase. Semin Thromb

Hemost 32(Suppl 1):16–31. https://doi.org/10.1055/s-2006-939551
21. Huntington JA (2008) How Na+ activates thrombin—a review of the functional and structural data. Biol Chem 389 (8):1025–1035. https://doi.org/10.1515/BC.2008.113
22. Di Cera E (2007) Thrombin as procoagulant and anticoagulant. J Thromb Haemost 5 (Suppl 1):196–202. https://doi.org/10.1111/j.1538-7836.2007.02485.x
23. Trouw LA, Pickering MC, Blom AM (2017) The complement system as a potential therapeutic target in rheumatic disease. Nat Rev Rheumatol 13(9):538–547. https://doi.org/10.1038/nrrheum.2017.125
24. Sim RB, Laich A (2000) Serine proteases of the complement system. Biochem Soc Trans 28(5):545–550
25. Cooper NR, Muller-Eberhard HJ (1970) The reaction mechanism of human C5 in immune hemolysis. J Exp Med 132(4):775–793
26. Dobo J, Szakacs D, Oroszlan G, Kortvely E, Kiss B, Boros E, Szasz R, Zavodszky P, Gal P, Pal G (2016) MASP-3 is the exclusive pro-factor D activator in resting blood: the lectin and the alternative complement pathways are fundamentally linked. Sci Rep-UK 6. https://doi.org/10.1038/srep31877
27. Sciascia S, Radin M, Yazdany J, Tektonidou M, Cecchi I, Roccatello D, Dall'Era M (2017) Expanding the therapeutic options for renal involvement in lupus: eculizumab, available evidence. Rheumatol Int 37 (8):1249–1255. https://doi.org/10.1007/s00296-017-3686-5
28. de Koning PJ, Kummer JA, de Poot SA, Quadir R, Broekhuizen R, McGettrick AF, Higgins WJ, Devreese B, Worrall DM, Bovenschen N (2011) Intracellular serine protease inhibitor SERPINB4 inhibits granzyme M-induced cell death. PLoS One 6(8): e22645. https://doi.org/10.1371/journal.pone.0022645
29. Soriano C, Mukaro V, Hodge G, Ahern J, Holmes M, Jersmann H, Moffat D, Meredith D, Jurisevic C, Reynolds PN, Hodge S (2012) Increased proteinase inhibitor-9 (PI-9) and reduced granzyme B in lung cancer: mechanism for immune evasion? Lung Cancer 77(1):38–45. https://doi.org/10.1016/j.lungcan.2012.01.017
30. Biancheri P, Di Sabatino A, Corazza GR, MacDonald TT (2013) Proteases and the gut barrier. Cell Tissue Res 351(2):269–280. https://doi.org/10.1007/s00441-012-1390-z
31. Ehlers MR (2014) Immune-modulating effects of alpha-1 antitrypsin. Biol Chem 395 (10):1187–1193. https://doi.org/10.1515/hsz-2014-0161
32. Boya P (2012) Lysosomal function and dysfunction: mechanism and disease. Antioxid Redox Signal 17(5):766–774. https://doi.org/10.1089/ars.2011.4405
33. Schmidt M, Finley D (2014) Regulation of proteasome activity in health and disease. Biochim Biophys Acta 1843(1):13–25. https://doi.org/10.1016/j.bbamcr.2013.08.012
34. Goldberg AL (2005) Nobel committee tags ubiquitin for distinction. Neuron 45 (3):339–344. https://doi.org/10.1016/j.neuron.2005.01.019
35. Tanaka K (2009) The proteasome: overview of structure and functions. Proc Jpn Acad Ser B Phys Biol Sci 85(1):12–36
36. Antalis TM, Shea-Donohue T, Vogel SN, Sears C, Fasano A (2007) Mechanisms of disease: protease functions in intestinal mucosal pathobiology. Nat Clin Pract Gastroenterol Hepatol 4(7):393–402. https://doi.org/10.1038/ncpgasthep0846
37. Alloy AP, Kayode O, Wang RY, Hockla A, Soares AS, Radisky ES (2015) Mesotrypsin has evolved four unique residues to cleave trypsin inhibitors as substrates. J Biol Chem 290(35):21523–21535. https://doi.org/10.1074/jbc.M115.662429
38. Giebeler N, Zigrino P (2016) A disintegrin and metalloprotease (ADAM): historical overview of their functions. Toxins (Basel) 8 (4):122. https://doi.org/10.3390/toxins8040122
39. Rodriguez D, Morrison CJ, Overall CM (2010) Matrix metalloproteinases: what do they not do? New substrates and biological roles identified by murine models and proteomics. Biochim Biophys Acta 1803(1):39–54. https://doi.org/10.1016/j.bbamcr.2009.09.015
40. Freitas-Rodriguez S, Folgueras AR, Lopez-Otin C (2017) The role of matrix metalloproteinases in aging: Tissue remodeling and beyond. Biochim Biophys Acta 1864(11 Pt A):2015–2025. https://doi.org/10.1016/j.bbamcr.2017.05.007
41. Brew K, Nagase H (2010) The tissue inhibitors of metalloproteinases (TIMPs): an ancient family with structural and functional diversity. Biochim Biophys Acta 1803 (1):55–71. https://doi.org/10.1016/j.bbamcr.2010.01.003
42. Coughlin SR (2005) Protease-activated receptors in hemostasis, thrombosis and

vascular biology. J Thromb Haemost 3 (8):1800–1814. https://doi.org/10.1111/j.1538-7836.2005.01377.x
43. Fender AC, Rauch BH, Geisler T, Schror K (2017) Protease-activated receptor par-4: an inducible switch between thrombosis and vascular inflammation? Thromb Haemost 117 (11):2013–2025. https://doi.org/10.1160/TH17-03-0219
44. Takamori N, Azuma H, Kato M, Hashizume S, Aihara K, Akaike M, Tamura K, Matsumoto T (2004) High plasma heparin cofactor II activity is associated with reduced incidence of in-stent restenosis after percutaneous coronary intervention. Circulation 109(4):481–486. https://doi.org/10.1161/01.CIR.0000109695.39671.37
45. Nierodzik ML, Karpatkin S (2006) Thrombin induces tumor growth, metastasis, and angiogenesis: Evidence for a thrombin-regulated dormant tumor phenotype. Cancer Cell 10 (5):355–362. https://doi.org/10.1016/j.ccr.2006.10.002
46. Asanuma K, Wakabayashi H, Okamoto T, Asanuma Y, Akita N, Yoshikawa T, Hayashi T, Matsumine A, Uchida A, Sudo A (2013) The thrombin inhibitor, argatroban, inhibits breast cancer metastasis to bone. Breast Cancer 20(3):241–246. https://doi.org/10.1007/s12282-012-0334-5
47. McIlwain DR, Berger T, Mak TW (2013) Caspase functions in cell death and disease. Cold Spring Harb Perspect Biol 5(4): a008656. https://doi.org/10.1101/cshperspect.a008656
48. Li P, Nijhawan D, Budihardjo I, Srinivasula SM, Ahmad M, Alnemri ES, Wang X (1997) Cytochrome c and dATP-dependent formation of Apaf-1/caspase-9 complex initiates an apoptotic protease cascade. Cell 91 (4):479–489
49. Zhivotovsky B, Samali A, Gahm A, Orrenius S (1999) Caspases: their intracellular localization and translocation during apoptosis. Cell Death Differ 6(7):644–651. https://doi.org/10.1038/sj.cdd.4400536
50. Li P, Zhou L, Zhao T, Liu X, Zhang P, Liu Y, Zheng X, Li Q (2017) Caspase-9: structure, mechanisms and clinical application. Oncotarget 8(14):23996–24008. https://doi.org/10.18632/oncotarget.15098
51. Vu NT, Park MA, Shultz JC, Goehe RW, Hoeferlin LA, Shultz MD, Smith SA, Lynch KW, Chalfant CE (2013) hnRNP U enhances caspase-9 splicing and is modulated by AKT-dependent phosphorylation of hnRNP L. J Biol Chem 288(12):8575–8584. https://doi.org/10.1074/jbc.M112.443333
52. Kuida K (2000) Caspase-9. Int J Biochem Cell Biol 32(2):121–124
53. Blasche S, Mortl M, Steuber H, Siszler G, Nisa S, Schwarz F, Lavrik I, Gronewold TM, Maskos K, Donnenberg MS, Ullmann D, Uetz P, Kogl M (2013) The E. coli effector protein NleF is a caspase inhibitor. PLoS One 8(3):e58937. https://doi.org/10.1371/journal.pone.0058937
54. Li P, Nijhawan D, Wang X (2004) Mitochondrial activation of apoptosis. Cell 116 (2 Suppl):S57–59, 52 p following S59
55. Denault JB, Eckelman BP, Shin H, Pop C, Salvesen GS (2007) Caspase 3 attenuates XIAP (X-linked inhibitor of apoptosis protein)-mediated inhibition of caspase 9. Biochem J 405(1):11–19. https://doi.org/10.1042/BJ20070288
56. Creagh EM (2014) Caspase crosstalk: integration of apoptotic and innate immune signalling pathways. Trends Immunol 35 (12):631–640. https://doi.org/10.1016/j.it.2014.10.004
57. Wang XJ, Cao Q, Liu X, Wang KT, Mi W, Zhang Y, Li LF, LeBlanc AC, Su XD (2010) Crystal structures of human caspase 6 reveal a new mechanism for intramolecular cleavage self-activation. EMBO Rep 11(11):841–847. https://doi.org/10.1038/embor.2010.141
58. Graham RK, Ehrnhoefer DE, Hayden MR (2011) Caspase-6 and neurodegeneration. Trends Neurosci 34(12):646–656. https://doi.org/10.1016/j.tins.2011.09.001
59. Bartel A, Gohler A, Hopf V, Breitbach K (2017) Caspase-6 mediates resistance against Burkholderia pseudomallei infection and influences the expression of detrimental cytokines. PLoS One 12(7):e0180203. https://doi.org/10.1371/journal.pone.0180203
60. Sollberger G, Strittmatter GE, Garstkiewicz M, Sand J, Beer HD (2014) Caspase-1: the inflammasome and beyond. Innate Immun 20(2):115–125. https://doi.org/10.1177/1753425913484374
61. Duclos C, Lavoie C, Denault JB (2017) Caspases rule the intracellular trafficking cartel. FEBS J 284(10):1394–1420. https://doi.org/10.1111/febs.14071
62. Julien O, Wells JA (2017) Caspases and their substrates. Cell Death Differ 24 (8):1380–1389. https://doi.org/10.1038/cdd.2017.44
63. Aziz M, Jacob A, Wang P (2014) Revisiting caspases in sepsis. Cell Death Dis 5:e1526. https://doi.org/10.1038/cddis.2014.488
64. Pemberton CJ (2014) Signal peptides: new markers in cardiovascular disease? Biomark

Med 8(8):1013–1019. https://doi.org/10.2217/bmm.14.64

65. Morocz M, Zsigmond E, Toth R, Enyedi MZ, Pinter L, Haracska L (2017) DNA-dependent protease activity of human Spartan facilitates replication of DNA-protein crosslink-containing DNA. Nucleic Acids Res 45(6):3172–3188. https://doi.org/10.1093/nar/gkw1315
66. Stingele J, Habermann B, Jentsch S (2015) DNA-protein crosslink repair: proteases as DNA repair enzymes. Trends Biochem Sci 40(2):67–71. https://doi.org/10.1016/j.tibs.2014.10.012
67. Vaz B, Popovic M, Newman JA, Fielden J, Aitkenhead H, Halder S, Singh AN, Vendrell I, Fischer R, Torrecilla I, Drobnitzky N, Freire R, Amor DJ, Lockhart PJ, Kessler BM, McKenna GW, Gileadi O, Ramadan K (2016) Metalloprotease SPRTN/DVC1 orchestrates replication-coupled DNA-protein crosslink repair. Mol Cell 64(4):704–719. https://doi.org/10.1016/j.molcel.2016.09.032
68. Maskey RS, Flatten KS, Sieben CJ, Peterson KL, Baker DJ, Nam HJ, Kim MS, Smyrk TC, Kojima Y, Machida Y, Santiago A, van Deursen JM, Kaufmann SH, Machida YJ (2017) Spartan deficiency causes accumulation of Topoisomerase 1 cleavage complexes and tumorigenesis. Nucleic Acids Res 45(8):4564–4576. https://doi.org/10.1093/nar/gkx107
69. Butler LR, Densham RM, Jia J, Garvin AJ, Stone HR, Shah V, Weekes D, Festy F, Beesley J, Morris JR (2012) The proteasomal de-ubiquitinating enzyme POH1 promotes the double-strand DNA break response. EMBO J 31(19):3918–3934. https://doi.org/10.1038/emboj.2012.232
70. Pinto-Fernandez A, Kessler BM (2016) DUBbing cancer: deubiquitylating enzymes involved in epigenetics, DNA damage and the cell cycle as therapeutic targets. Front Genet 7:133. https://doi.org/10.3389/fgene.2016.00133
71. Enari M, Sakahira H, Yokoyama H, Okawa K, Iwamatsu A, Nagata S (1998) A caspase-activated DNase that degrades DNA during apoptosis, and its inhibitor ICAD. Nature 391(6662):43–50
72. Venkatesh S, Lee J, Singh K, Lee I, Suzuki CK (2012) Multitasking in the mitochondrion by the ATP-dependent Lon protease. Biochim Biophys Acta 1823(1):56–66. https://doi.org/10.1016/j.bbamcr.2011.11.003
73. Mao PL, Jiang Y, Wee BY, Porter AG (1998) Activation of caspase-1 in the nucleus requires nuclear translocation of pro-caspase-1 mediated by its prodomain. J Biol Chem 273(37):23621–23624
74. Kamada S, Kikkawa U, Tsujimoto Y, Hunter T (2005) Nuclear translocation of caspase-3 is dependent on its proteolytic activation and recognition of a substrate-like protein(s). J Biol Chem 280(2):857–860. https://doi.org/10.1074/jbc.C400538200
75. Hill JW, Poddar R, Thompson JF, Rosenberg GA, Yang Y (2012) Intranuclear matrix metalloproteinases promote DNA damage and apoptosis induced by oxygen-glucose deprivation in neurons. Neuroscience 220:277–290. https://doi.org/10.1016/j.neuroscience.2012.06.019
76. Eguchi T, Calderwood SK, Takigawa M, Kubota S, Kozaki KI (2017) Intracellular MMP3 promotes HSP gene expression in collaboration with chromobox proteins. J Cell Biochem 118(1):43–51. https://doi.org/10.1002/jcb.25607
77. Stepanova V, Jayaraman PS, Zaitsev SV, Lebedeva T, Bdeir K, Kershaw R, Holman KR, Parfyonova YV, Semina EV, Beloglazova IB, Tkachuk VA, Cines DB (2016) Urokinase-type plasminogen activator (uPA) promotes angiogenesis by attenuating proline-rich homeodomain protein (PRH) transcription factor activity and de-repressing vascular endothelial growth factor (VEGF) receptor expression. J Biol Chem 291(29):15029–15045. https://doi.org/10.1074/jbc.M115.678490
78. Antalis TM, Bugge TH, Wu Q (2011) Membrane-anchored serine proteases in health and disease. Prog Mol Biol Transl Sci 99:1–50. https://doi.org/10.1016/B978-0-12-385504-6.00001-4
79. Friis S, Sales KU, Schafer JM, Vogel LK, Kataoka H, Bugge TH (2014) The protease inhibitor HAI-2, but not HAI-1, regulates matriptase activation and shedding through prostasin. J Biol Chem 289(32):22319–22332. https://doi.org/10.1074/jbc.M114.574400
80. Bardou O, Menou A, Francois C, Duitman JW, von der Thusen JH, Borie R, Sales KU, Mutze K, Castier Y, Sage E, Liu L, Bugge TH, Fairlie DP, Konigshoff M, Crestani B, Borensztajn KS (2016) Membrane-anchored serine protease matriptase is a trigger of pulmonary fibrogenesis. Am J Respir Crit Care Med 193(8):847–860. https://doi.org/10.1164/rccm.201502-0299OC
81. Le Gall SM, Szabo R, Lee M, Kirchhofer D, Craik CS, Bugge TH, Camerer E (2016) Matriptase activation connects tissue factor-

dependent coagulation initiation to epithelial proteolysis and signaling. Blood 127 (25):3260–3269. https://doi.org/10.1182/blood-2015-11-683110

82. Verhelst SHL (2017) Intramembrane proteases as drug targets. FEBS J 284 (10):1489–1502. https://doi.org/10.1111/febs.13979

83. Dusterhoft S, Kunzel U, Freeman M (2017) Rhomboid proteases in human disease: Mechanisms and future prospects. Biochim Biophys Acta 1864(11 Pt B):2200–2209. https://doi.org/10.1016/j.bbamcr.2017.04.016

84. Saita S, Nolte H, Fiedler KU, Kashkar H, Venne AS, Zahedi RP, Kruger M, Langer T (2017) PARL mediates Smac proteolytic maturation in mitochondria to promote apoptosis. Nat Cell Biol 19(4):318–328. https://doi.org/10.1038/ncb3488

85. Ranganathan P, Weaver KL, Capobianco AJ (2011) Notch signalling in solid tumours: a little bit of everything but not all the time. Nat Rev Cancer 11(5):338–351. https://doi.org/10.1038/nrc3035

86. Chauhan S, Mandal P, Tomar RS (2016) Biochemical analysis reveals the multifactorial mechanism of histone H3 clipping by chicken liver histone H3 protease. Biochemistry 55 (38):5464–5482. https://doi.org/10.1021/acs.biochem.6b00625

87. Vossaert L, Meert P, Scheerlinck E, Glibert P, Van Roy N, Heindryckx B, De Sutter P, Dhaenens M, Deforce D (2014) Identification of histone H3 clipping activity in human embryonic stem cells. Stem Cell Res 13 (1):123–134. https://doi.org/10.1016/j.scr.2014.05.002

88. Deraison C, Bonnart C, Vergnolle N (2018) Proteases. In: Cavaillon J-M, Singer M (eds) Inflammation: from molecular and cellular mechanisms to the clinic. Wiley-VCH, Weinheim, Germany, pp 727–766

89. Chotirmall SH, Al-Alawi M, McEnery T, McElvaney NG (2015) Alpha-1 proteinase inhibitors for the treatment of alpha-1 antitrypsin deficiency: safety, tolerability, and patient outcomes. Ther Clin Risk Manag 11:143–151. https://doi.org/10.2147/TCRM.S51474

90. Torres-Duran M, Ruano-Ravina A, Parente-Lamelas I, Abal-Arca J, Leiro-Fernandez V, Montero-Martinez C, Pena C, Castro-Anon O, Golpe-Gomez A, Gonzalez-Barcala FJ, Martinez C, Guzman-Taveras R, Provencio M, Mejuto-Marti MJ, Fernandez-Villar A, Barros-Dios JM (2015) Alpha-1 antitrypsin deficiency and lung cancer risk: a case-control study in never-smokers. J Thorac Oncol 10(9):1279–1284. https://doi.org/10.1097/JTO.0000000000000609

91. Soderberg D, Segelmark M (2016) Neutrophil extracellular traps in ANCA-associated vasculitis. Front Immunol 7:256. https://doi.org/10.3389/fimmu.2016.00256

92. Denadai-Souza A, Ribeiro CM, Rolland C, Thouard A, Deraison C, Scavone C, Gonzalez-Dunia D, Vergnolle N, Avellar MCW (2017) Effect of tryptase inhibition on joint inflammation: a pharmacological and lentivirus-mediated gene transfer study. Arthritis Res Ther 19. https://doi.org/10.1186/s13075-017-1326-9

93. Leskinen MJ, Lindstedt KA, Wang Y, Kovanen PT (2003) Mast cell chymase induces smooth muscle cell apoptosis by a mechanism involving fibronectin degradation and disruption of focal adhesions. Arterioscler Thromb Vasc Biol 23(2):238–243

94. He A, Shi GP (2013) Mast cell chymase and tryptase as targets for cardiovascular and metabolic diseases. Curr Pharm Des 19 (6):1114–1125

95. Shirai T, Hilhorst M, Harrison DG, Goronzy JJ, Weyand CM (2015) Macrophages in vascular inflammation—from atherosclerosis to vasculitis. Autoimmunity 48(3):139–151. https://doi.org/10.3109/08916934.2015.1027815

96. Sendler M, Maertin S, John D, Persike M, Weiss FU, Kruger B, Wartmann T, Wagh P, Halangk W, Schaschke N, Mayerle J, Lerch MM (2016) Cathepsin B activity initiates apoptosis via digestive protease activation in pancreatic acinar cells and experimental pancreatitis. J Biol Chem 291 (28):14717–14731. https://doi.org/10.1074/jbc.M116.718999

97. Kayode O, Huang Z, Soares AS, Caulfield TR, Dong Z, Bode AM, Radisky ES (2017) Small molecule inhibitors of mesotrypsin from a structure-based docking screen. PLoS One 12(5):e0176694. https://doi.org/10.1371/journal.pone.0176694

98. Rolland-Fourcade C, Denadai-Souza A, Cirillo C, Lopez C, Jaramillo JO, Desormeaux C, Cenac N, Motta JP, Larauche M, Tache Y, Berghe PV, Neunlist M, Coron E, Kirzin S, Portier G, Bonnet D, Alric L, Vanner S, Deraison C, Vergnolle N (2017) Epithelial expression and function of trypsin-3 in irritable bowel syndrome. Gut 66(10):1767–1778. https://doi.org/10.1136/gutjnl-2016-312094

99. Ricklin D, Lambris JD (2013) Complement in immune and inflammatory disorders: pathophysiological mechanisms. J Immunol 190(8):3831–3838. https://doi.org/10.4049/jimmunol.1203487
100. Hua Y, Nair S (2015) Proteases in cardiometabolic diseases: pathophysiology, molecular mechanisms and clinical applications. Biochim Biophys Acta 1852(2):195–208. https://doi.org/10.1016/j.bbadis.2014.04.032
101. Friedrich R, Panizzi P, Fuentes-Prior P, Richter K, Verhamme I, Anderson PJ, Kawabata S, Huber R, Bode W, Bock PE (2003) Staphylocoagulase is a prototype for the mechanism of cofactor-induced zymogen activation. Nature 425(6957):535–539. https://doi.org/10.1038/nature01962
102. Weidmann H, Heikaus L, Long AT, Naudin C, Schluter H, Renne T (2017) The plasma contact system, a protease cascade at the nexus of inflammation, coagulation and immunity. Biochim Biophys Acta 1864(11 Pt B):2118–2127. https://doi.org/10.1016/j.bbamcr.2017.07.009
103. Zamolodchikov D, Renne T, Strickland S (2016) The Alzheimer's disease peptide beta-amyloid promotes thrombin generation through activation of coagulation factor XII. J Thromb Haemost 14(5):995–1007. https://doi.org/10.1111/jth.13209
104. Esmon CT, Vigano-D'Angelo S, D'Angelo A, Comp PC (1987) Anticoagulation proteins C and S. Adv Exp Med Biol 214:47–54
105. Bertina RM, Koeleman BP, Koster T, Rosendaal FR, Dirven RJ, de Ronde H, van der Velden PA, Reitsma PH (1994) Mutation in blood coagulation factor V associated with resistance to activated protein C. Nature 369 (6475):64–67. https://doi.org/10.1038/369064a0
106. Kujovich JL (1993) Factor V Leiden thrombophilia. In: Adam MP, Ardinger HH, Pagon RA et al (eds) GeneReviews(R). University of Washington, Seattle, WA
107. Lane D Antithrombin mutation database
108. Verhamme IM, Olson ST, Tollefsen DM, Bock PE (2002) Binding of exosite ligands to human thrombin. Re-evaluation of allosteric linkage between thrombin exosites I and II. J Biol Chem 277(9):6788–6798. https://doi.org/10.1074/jbc.M110257200
109. Sarilla S, Habib SY, Kravtsov DV, Matafonov A, Gailani D, Verhamme IM (2010) Sucrose octasulfate selectively accelerates thrombin inactivation by heparin cofactor II. J Biol Chem 285(11):8278–8289. https://doi.org/10.1074/jbc.M109.005967
110. Raghuraman A, Mosier PD, Desai UR (2010) Understanding dermatan sulfate-heparin cofactor II interaction through virtual library screening. ACS Med Chem Lett 1 (6):281–285. https://doi.org/10.1021/ml100048y
111. Tollefsen DM, Maimone MM, McGuire EA, Peacock ME (1989) Heparin cofactor II activation by dermatan sulfate. Ann N Y Acad Sci 556:116–122
112. Aihara K, Azuma H, Takamori N, Kanagawa Y, Akaike M, Fujimura M, Yoshida T, Hashizume S, Kato M, Yamaguchi H, Kato S, Ikeda Y, Arase T, Kondo A, Matsumoto T (2004) Heparin cofactor II is a novel protective factor against carotid atherosclerosis in elderly individuals. Circulation 109(22):2761–2765. https://doi.org/10.1161/01.CIR.0000129968.46095.F3
113. Polderdijk SG, Adams TE, Ivanciu L, Camire RM, Baglin TP, Huntington JA (2017) Design and characterization of an APC-specific serpin for the treatment of hemophilia. Blood 129(1):105–113. https://doi.org/10.1182/blood-2016-05-718635
114. Panizzi P, Boxrud PD, Verhamme IM, Bock PE (2006) Binding of the COOH-terminal lysine residue of streptokinase to plasmin(ogen) kringles enhances formation of the streptokinase.plasmin(ogen) catalytic complexes. J Biol Chem 281(37):26774–26778. https://doi.org/10.1074/jbc.C600171200
115. Verhamme IM, Bock PE (2008) Rapid-reaction kinetic characterization of the pathway of streptokinase-plasmin catalytic complex formation. J Biol Chem 283 (38):26137–26147. https://doi.org/10.1074/jbc.M804038200
116. Verhamme IM, Bock PE (2014) Rapid binding of plasminogen to streptokinase in a catalytic complex reveals a three-step mechanism. J Biol Chem 289(40):28006–28018. https://doi.org/10.1074/jbc.M114.589077
117. Weiss D, Sorescu D, Taylor WR (2001) Angiotensin II and atherosclerosis. Am J Cardiol 87(8A):25C–32C
118. Kossmann S, Lagrange J, Jackel S, Jurk K, Ehlken M, Schonfelder T, Weihert Y, Knorr M, Brandt M, Xia N, Li H, Daiber A, Oelze M, Reinhardt C, Lackner K, Gruber A, Monia B, Karbach SH, Walter U, Ruggeri

ZM, Renne T, Ruf W, Munzel T, Wenzel P (2017) Platelet-localized FXI promotes a vascular coagulation-inflammatory circuit in arterial hypertension. Sci Transl Med 9 (375). https://doi.org/10.1126/scitranslmed.aah4923
119. Camare C, Pucelle M, Negre-Salvayre A, Salvayre R (2017) Angiogenesis in the atherosclerotic plaque. Redox Biol 12:18–34. https://doi.org/10.1016/j.redox.2017.01.007
120. Wilson WRW, Anderton M, Choke EC, Dawson J, Loftus IM, Thompson MM (2008) Elevated plasma MMP1 and MMP9 are associated with abdominal aortic aneurysm rupture. Eur J Vasc Endovasc 35 (5):580–584. https://doi.org/10.1016/j.ejvs.2007.12.004
121. Xue L, Borne Y, Mattisson IY, Wigren M, Melander O, Ohro-Melander M, Bengtsson E, Fredrikson GN, Nilsson J, Engstrom G (2017) FADD, caspase-3, and caspase-8 and incidence of coronary events. Arterioscler Thromb Vasc Biol 37 (5):983–989. https://doi.org/10.1161/ATVBAHA.117.308995
122. Musante L, Tataruch D, Gu D, Liu X, Forsblom C, Groop PH, Holthofer H (2015) Proteases and protease inhibitors of urinary extracellular vesicles in diabetic nephropathy. J Diabetes Res 2015:289734. https://doi.org/10.1155/2015/289734
123. Zhao Z, Yang P, Eckert RL, Reece EA (2009) Caspase-8: a key role in the pathogenesis of diabetic embryopathy. Birth Defects Res B Dev Reprod Toxicol 86(1):72–77. https://doi.org/10.1002/bdrb.20185
124. Augstein P, Bahr J, Wachlin G, Heinke P, Berg S, Salzsieder E, Harrison LC (2004) Cytokines activate caspase-3 in insulinoma cells of diabetes-prone NOD mice directly and via upregulation of Fas. J Autoimmun 23(4):301–309. https://doi.org/10.1016/j.jaut.2004.09.006
125. Trompet S, Pons D, Kanse SM, de Craen AJ, Ikram MA, Verschuren JJ, Zwinderman AH, Doevendans PA, Tio RA, de Winter RJ, Slagboom PE, Westendorp RG, Jukema JW (2011) Factor VII activating protease polymorphism (G534E) is associated with increased risk for stroke and mortality. Stroke Res Treat 2011:424759. https://doi.org/10.4061/2011/424759
126. Turner RJ, Sharp FR (2016) Implications of MMP9 for blood brain barrier disruption and hemorrhagic transformation following ischemic stroke. Front Cell Neurosci 10:56. https://doi.org/10.3389/fncel.2016.00056
127. Crocker SJ, Pagenstecher A, Campbell IL (2004) The TIMPs tango with MMPs and more in the central nervous system. J Neurosci Res 75(1):1–11. https://doi.org/10.1002/jnr.10836
128. Brucher BL, Jamall IS (2016) Somatic mutation theory—why it's wrong for most cancers. Cell Physiol Biochem 38(5):1663–1680. https://doi.org/10.1159/000443106
129. Brucher BL, Jamall IS (2014) Epistemology of the origin of cancer: a new paradigm. BMC Cancer 14:331. https://doi.org/10.1186/1471-2407-14-331
130. Schuliga M (2015) The inflammatory actions of coagulant and fibrinolytic proteases in disease. Mediators Inflamm 2015:437695. https://doi.org/10.1155/2015/437695
131. Fan J, Ning B, Lyon CJ, Hu TY (2017) Circulating peptidome and tumor-resident proteolysis. Enzyme 42:1–25. https://doi.org/10.1016/bs.enz.2017.08.001
132. Guo Z, Jin X, Jia H (2013) Inhibition of ADAM-17 more effectively down-regulates the Notch pathway than that of gamma-secretase in renal carcinoma. J Exp Clin Cancer Res 32:26. https://doi.org/10.1186/1756-9966-32-26
133. Jackson HW, Defamie V, Waterhouse P, Khokha R (2017) TIMPs: versatile extracellular regulators in cancer. Nat Rev Cancer 17 (1):38–53. https://doi.org/10.1038/nrc.2016.115
134. LaRocca G, Aspelund T, Greve AM, Eiriksdottir G, Acharya T, Thorgeirsson G, Harris TB, Launer LJ, Gudnason V, Arai AE (2017) Fibrosis as measured by the biomarker, tissue inhibitor metalloproteinase-1, predicts mortality in Age Gene Environment Susceptibility-Reykjavik (AGES-Reykjavik) Study. Eur Heart J 38(46):3423–3430. https://doi.org/10.1093/eurheartj/ehx510
135. Lee JY, Kong G (2016) Roles and epigenetic regulation of epithelial-mesenchymal transition and its transcription factors in cancer initiation and progression. Cell Mol Life Sci 73(24):4643–4660. https://doi.org/10.1007/s00018-016-2313-z
136. Otsuki T, Fujimoto D, Hirono Y, Goi T, Yamaguchi A (2014) Thrombin conducts epithelialmesenchymal transition via proteaseactivated receptorl in human gastric cancer. Int J Oncol 45(6):2287–2294. https://doi.org/10.3892/ijo.2014.2651
137. Bawa-Khalfe T, Lu LS, Zuo Y, Huang C, Dere R, Lin FM, Yeh ET (2012) Differential expression of SUMO-specific protease 7 variants regulates epithelial-mesenchymal

transition. Proc Natl Acad Sci U S A 109 (43):17466–17471. https://doi.org/10.1073/pnas.1209378109

138. Schmidt N, Irle I, Ripkens K, Lux V, Nelles J, Johannes C, Parry L, Greenow K, Amir S, Campioni M, Baldi A, Oka C, Kawaichi M, Clarke AR, Ehrmann M (2016) Epigenetic silencing of serine protease HTRA1 drives polyploidy. BMC Cancer 16:399. https://doi.org/10.1186/s12885-016-2425-8
139. Johnson JJ, Miller DL, Jiang R, Liu Y, Shi Z, Tarwater L, Williams R, Balsara R, Sauter ER, Stack MS (2016) Protease-activated receptor-2 (PAR-2)-mediated Nf-kappaB activation suppresses inflammation-associated tumor suppressor MicroRNAs in oral squamous cell carcinoma. J Biol Chem 291 (13):6936–6945. https://doi.org/10.1074/jbc.M115.692640
140. Zhang W, Wang S, Wang Q, Yang Z, Pan Z, Li L (2014) Overexpression of cysteine cathepsin L is a marker of invasion and metastasis in ovarian cancer. Oncol Rep 31 (3):1334–1342. https://doi.org/10.3892/or.2014.2967
141. Dian D, Heublein S, Wiest I, Barthell L, Friese K, Jeschke U (2014) Significance of the tumor protease cathepsin D for the biology of breast cancer. Histol Histopathol 29 (4):433–438. https://doi.org/10.14670/HH-29.10.433
142. Cohen I, Kayode O, Hockla A, Sankaran B, Radisky DC, Radisky ES, Papo N (2016) Combinatorial protein engineering of proteolytically resistant mesotrypsin inhibitors as candidates for cancer therapy. Biochem J 473(10):1329–1341. https://doi.org/10.1042/BJ20151410
143. Salameh MA, Radisky ES (2013) Biochemical and structural insights into mesotrypsin: an unusual human trypsin. Int J Biochem Mol Biol 4(3):129–139
144. Tanabe LM, List K (2017) The role of type II transmembrane serine protease-mediated signaling in cancer. FEBS J 284 (10):1421–1436. https://doi.org/10.1111/febs.13971
145. Zoratti GL, Tanabe LM, Hyland TE, Duhaime MJ, Colombo E, Leduc R, Marsault E, Johnson MD, Lin CY, Boerner J, Lang JE, List K (2016) Matriptase regulates c-Met mediated proliferation and invasion in inflammatory breast cancer. Oncotarget 7(36):58162–58173. https://doi.org/10.18632/oncotarget.11262
146. Rolfe M (2017) The holy grail: solid tumor efficacy by proteasome inhibition. Cell Chem Biol 24(2):125–126. https://doi.org/10.1016/j.chembiol.2017.01.007
147. Weyburne ES, Wilkins OM, Sha Z, Williams DA, Pletnev AA, de Bruin G, Overkleeft HS, Goldberg AL, Cole MD, Kisselev AF (2017) Inhibition of the proteasome beta2 site sensitizes triple-negative breast cancer cells to beta5 inhibitors and suppresses Nrf1 activation. Cell Chem Biol 24(2):218–230. https://doi.org/10.1016/j.chembiol.2016.12.016
148. Reis ES, Mastellos DC, Ricklin D, Mantovani A, Lambris JD (2018) Complement in cancer: untangling an intricate relationship. Nat Rev Immunol 18(1):5–18. https://doi.org/10.1038/nri.2017.97
149. Rutkowski MJ, Sughrue ME, Kane AJ, Mills SA, Parsa AT (2010) Cancer and the complement cascade. Mol Cancer Res 8 (11):1453–1465. https://doi.org/10.1158/1541-7786.MCR-10-0225
150. Zhu L, Jaamaa S, Af Hallstrom TM, Laiho M, Sankila A, Nordling S, Stenman UH, Koistinen H (2013) PSA forms complexes with alpha1-antichymotrypsin in prostate. Prostate 73(2):219–226. https://doi.org/10.1002/pros.22560
151. DiScipio RG (1982) The activation of the alternative pathway C3 convertase by human plasma kallikrein. Immunology 45 (3):587–595
152. Caine GJ, Stonelake PS, Lip GY, Kehoe ST (2002) The hypercoagulable state of malignancy: pathogenesis and current debate. Neoplasia 4(6):465–473. https://doi.org/10.1038/sj.neo.7900263
153. Amiral J, Seghatchian J (2017) Monitoring of anticoagulant therapy in cancer patients with thrombosis and the usefulness of blood activation markers. Transfus Apher Sci 56 (3):279–286. https://doi.org/10.1016/j.transci.2017.05.010
154. Huesgen PF, Lange PF, Overall CM (2014) Ensembles of protein termini and specific proteolytic signatures as candidate biomarkers of disease. Proteomics Clin Appl 8 (5-6):338–350. https://doi.org/10.1002/prca.201300104
155. Kang JH, Korecka M, Toledo JB, Trojanowski JQ, Shaw LM (2013) Clinical utility and analytical challenges in measurement of cerebrospinal fluid amyloid-beta(1-42) and tau proteins as Alzheimer disease biomarkers. Clin Chem 59(6):903–916. https://doi.org/10.1373/clinchem.2013.202937
156. Janelidze S, Stomrud E, Palmqvist S, Zetterberg H, van Westen D, Jeromin A,

Song L, Hanlon D, Tan Hehir CA, Baker D, Blennow K, Hansson O (2016) Plasma beta-amyloid in Alzheimer's disease and vascular disease. Sci Rep 6:26801. https://doi.org/10.1038/srep26801

157. Roher AE, Kokjohn TA, Clarke SG, Sierks MR, Maarouf CL, Serrano GE, Sabbagh MS, Beach TG (2017) APP/Abeta structural diversity and Alzheimer's disease pathogenesis. Neurochem Int 110:1–13. https://doi.org/10.1016/j.neuint.2017.08.007
158. Evin G, Li QX (2012) Platelets and Alzheimer's disease: potential of APP as a biomarker. World J Psychiatry 2(6):102–113. https://doi.org/10.5498/wjp.v2.i6.102
159. Wetzel S, Seipold L, Saftig P (2017) The metalloproteinase ADAM10: A useful therapeutic target? Biochim Biophys Acta 1864 (11 Pt B):2071–2081. https://doi.org/10.1016/j.bbamcr.2017.06.005
160. Bu XL, Xiang Y, Jin WS, Wang J, Shen LL, Huang ZL, Zhang K, Liu YH, Zeng F, Liu JH, Sun HL, Zhuang ZQ, Chen SH, Yao XQ, Giunta B, Shan YC, Tan J, Chen XW, Dong ZF, Zhou HD, Zhou XF, Song W, Wang YJ (2017) Blood-derived amyloid-beta protein induces Alzheimer's disease pathologies. Mol Psychiatry. https://doi.org/10.1038/mp.2017.204
161. Budd Haeberlein S, O'Gorman J, Chiao P, Bussiere T, von Rosenstiel P, Tian Y, Zhu Y, von Hehn C, Gheuens S, Skordos L, Chen T, Sandrock A (2017) Clinical development of aducanumab, an anti-abeta human monoclonal antibody being investigated for the treatment of early Alzheimer's disease. J Prev Alzheimers Dis 4(4):255–263. https://doi.org/10.14283/jpad.2017.39
162. Honig LS, Vellas B, Woodward M, Boada M, Bullock R, Borrie M, Hager K, Andreasen N, Scarpini E, Liu-Seifert H, Case M, Dean RA, Hake A, Sundell K, Poole Hoffmann V, Carlson C, Khanna R, Mintun M, DeMattos R, Selzler KJ, Siemers E (2018) Trial of solanezumab for mild dementia due to Alzheimer's disease. N Engl J Med 378 (4):321–330. https://doi.org/10.1056/NEJMoa1705971
163. De Strooper B (2014) Lessons from a failed gamma-secretase Alzheimer trial. Cell 159 (4):721–726. https://doi.org/10.1016/j.cell.2014.10.016
164. Lorenzl S, Albers DS, Relkin N, Ngyuen T, Hilgenberg SL, Chirichigno J, Cudkowicz ME, Beal MF (2003) Increased plasma levels of matrix metalloproteinase-9 in patients with Alzheimer's disease. Neurochem Int 43 (3):191–196
165. Siklos M, BenAissa M, Thatcher GR (2015) Cysteine proteases as therapeutic targets: does selectivity matter? A systematic review of calpain and cathepsin inhibitors. Acta Pharm Sin B 5(6):506–519. https://doi.org/10.1016/j.apsb.2015.08.001
166. Martinelli P, Rugarli EI (2010) Emerging roles of mitochondrial proteases in neurodegeneration. Biochim Biophys Acta 1797 (1):1–10. https://doi.org/10.1016/j.bbabio.2009.07.013
167. Konig T, Troder SE, Bakka K, Korwitz A, Richter-Dennerlein R, Lampe PA, Patron M, Muhlmeister M, Guerrero-Castillo S, Brandt U, Decker T, Lauria I, Paggio A, Rizzuto R, Rugarli EI, De Stefani D, Langer T (2016) The m-AAA protease associated with neurodegeneration limits MCU activity in mitochondria. Mol Cell 64(1):148–162. https://doi.org/10.1016/j.molcel.2016.08.020
168. Strauss KM, Martins LM, Plun-Favreau H, Marx FP, Kautzmann S, Berg D, Gasser T, Wszolek Z, Muller T, Bornemann A, Wolburg H, Downward J, Riess O, Schulz JB, Kruger R (2005) Loss of function mutations in the gene encoding Omi/HtrA2 in Parkinson's disease. Hum Mol Genet 14 (15):2099–2111. https://doi.org/10.1093/hmg/ddi215
169. Fu J, Yu HM, Chiu SY, Mirando AJ, Maruyama EO, Cheng JG, Hsu W (2014) Disruption of SUMO-specific protease 2 induces mitochondria mediated neurodegeneration. PLoS Genet 10(10):e1004579. https://doi.org/10.1371/journal.pgen.1004579
170. Orsini F, De Blasio D, Zangari R, Zanier ER, De Simoni MG (2014) Versatility of the complement system in neuroinflammation, neurodegeneration and brain homeostasis. Front Cell Neurosci 8:380. https://doi.org/10.3389/fncel.2014.00380
171. Descamps FJ, Van den Steen PE, Nelissen I, Van Damme J, Opdenakker G (2003) Remnant epitopes generate autoimmunity: from rheumatoid arthritis and multiple sclerosis to diabetes. Adv Exp Med Biol 535:69–77
172. Dinarello CA, Simon A, van der Meer JW (2012) Treating inflammation by blocking interleukin-1 in a broad spectrum of diseases. Nat Rev Drug Discov 11(8):633–652. https://doi.org/10.1038/nrd3800
173. Rupanagudi KV, Kulkarni OP, Lichtnekert J, Darisipudi MN, Mulay SR, Schott B, Gruner S, Haap W, Hartmann G, Anders HJ (2015) Cathepsin S inhibition suppresses systemic lupus erythematosus and lupus nephritis because cathepsin S is essential for MHC

class II-mediated CD4 T cell and B cell priming. Ann Rheum Dis 74(2):452–463. https://doi.org/10.1136/annrheumdis-2013-203717
174. Schaller M, Vogel M, Kentouche K, Lammle B, Kremer Hovinga JA (2014) The splenic autoimmune response to ADAMTS13 in thrombotic thrombocytopenic purpura contains recurrent antigen-binding CDR3 motifs. Blood 124(23):3469–3479. https://doi.org/10.1182/blood-2014-04-561142
175. Sadiq SK, Noe F, De Fabritiis G (2012) Kinetic characterization of the critical step in HIV-1 protease maturation. Proc Natl Acad Sci U S A 109(50):20449–20454. https://doi.org/10.1073/pnas.1210983109
176. Duschak VG, Couto AS (2009) Cruzipain, the major cysteine protease of Trypanosoma cruzi: a sulfated glycoprotein antigen as relevant candidate for vaccine development and drug target. A review. Curr Med Chem 16 (24):3174–3202
177. Plaza K, Kalinska M, Bochenska O, Meyer-Hoffert U, Wu Z, Fischer J, Falkowski K, Sasiadek L, Bielecka E, Potempa B, Kozik A, Potempa J, Kantyka T (2016) Gingipains of porphyromonas gingivalis affect the stability and function of serine protease inhibitor of kazal-type 6 (SPINK6), a tissue inhibitor of human kallikreins. J Biol Chem 291 (36):18753–18764. https://doi.org/10.1074/jbc.M116.722942
178. Culp E, Wright GD (2017) Bacterial proteases, untapped antimicrobial drug targets. J Antibiot (Tokyo) 70(4):366–377. https://doi.org/10.1038/ja.2016.138
179. Chang AK, Kim HY, Park JE, Acharya P, Park IS, Yoon SM, You HJ, Hahm KS, Park JK, Lee JS (2005) Vibrio vulnificus secretes a broad-specificity metalloprotease capable of interfering with blood homeostasis through prothrombin activation and fibrinolysis. J Bacteriol 187(20):6909–6916. https://doi.org/10.1128/JB.187.20.6909-6916.2005
180. Bibo-Verdugo B, Jiang Z, Caffrey CR, O'Donoghue AJ (2017) Targeting proteasomes in infectious organisms to combat disease. FEBS J 284(10):1503–1517. https://doi.org/10.1111/febs.14029
181. Pontarollo G, Acquasaliente L, Peterle D, Frasson R, Artusi I, De Filippis V (2017) Non-canonical proteolytic activation of human prothrombin by subtilisin from Bacillus subtilis may shift the procoagulant-anticoagulant equilibrium toward thrombosis. J Biol Chem 292(37):15161–15179. https://doi.org/10.1074/jbc.M117.795245
182. Carroll IM, Maharshak N (2013) Enteric bacterial proteases in inflammatory bowel disease- pathophysiology and clinical implications. World J Gastroenterol 19 (43):7531–7543. https://doi.org/10.3748/wjg.v19.i43.7531
183. Boxrud PD, Verhamme IM, Bock PE (2004) Resolution of conformational activation in the kinetic mechanism of plasminogen activation by streptokinase. J Biol Chem 279 (35):36633–36641. https://doi.org/10.1074/jbc.M405264200
184. Chandrahas V, Glinton K, Liang Z, Donahue DL, Ploplis VA, Castellino FJ (2015) Direct host plasminogen binding to bacterial surface M-protein in pattern D strains of streptococcus pyogenes is required for activation by its natural coinherited SK2b protein. J Biol Chem 290(30):18833–18842. https://doi.org/10.1074/jbc.M115.655365
185. Panizzi P, Friedrich R, Fuentes-Prior P, Bode W, Bock PE (2004) The staphylocoagulase family of zymogen activator and adhesion proteins. Cell Mol Life Sci 61 (22):2793–2798. https://doi.org/10.1007/s00018-004-4285-7
186. Parry MA, Zhang XC, Bode I (2000) Molecular mechanisms of plasminogen activation: bacterial cofactors provide clues. Trends Biochem Sci 25(2):53–59
187. Wiles KG, Panizzi P, Kroh HK, Bock PE (2010) Skizzle is a novel plasminogen- and plasmin-binding protein from Streptococcus agalactiae that targets proteins of human fibrinolysis to promote plasmin generation. J Biol Chem 285(27):21153–21164. https://doi.org/10.1074/jbc.M110.107730
188. Verhamme IM, Panizzi PR, Bock PE (2015) Pathogen activators of plasminogen. J Thromb Haemost 13(Suppl 1):S106–S114. https://doi.org/10.1111/jth.12939
189. Meliopoulos VA, Andersen LE, Brooks P, Yan X, Bakre A, Coleman JK, Tompkins SM, Tripp RA (2012) MicroRNA regulation of human protease genes essential for influenza virus replication. PLoS One 7(5):e37169. https://doi.org/10.1371/journal.pone.0037169
190. Homma T, Ishibashi D, Nakagaki T, Fuse T, Mori T, Satoh K, Atarashi R, Nishida N (2015) Ubiquitin-specific protease 14 modulates degradation of cellular prion protein. Sci Rep 5:11028. https://doi.org/10.1038/srep11028
191. Michaud DS, Lu J, Peacock-Villada AY, Barber JR, Joshu CE, Prizment AE, Beck JD, Offenbacher S, Platz EA (2018) Periodontal disease assessed using clinical dental

measurements and cancer risk in the ARIC study. J Natl Cancer Inst. https://doi.org/10.1093/jnci/djx278

192. Drag M, Salvesen GS (2010) Emerging principles in protease-based drug discovery. Nat Rev Drug Discov 9(9):690–701. https://doi.org/10.1038/nrd3053
193. Herzog RW (2015) Hemophilia gene therapy: caught between a cure and an immune response. Mol Ther 23(9):1411–1412. https://doi.org/10.1038/mt.2015.135
194. Rangarajan S, Walsh L, Lester W, Perry D, Madan B, Laffan M, Yu H, Vettermann C, Pierce GF, Wong WY, Pasi KJ (2017) AAV5-factor VIII gene transfer in severe hemophilia A. N Engl J Med. https://doi.org/10.1056/NEJMoa1708483
195. George LA, Sullivan SK, Giermasz A, Rasko JEJ, Samelson-Jones BJ, Ducore J, Cuker A, Sullivan LM, Majumdar S, Teitel J, McGuinn CE, Ragni MV, Luk AY, Hui D, Wright JF, Chen Y, Liu Y, Wachtel K, Winters A, Tiefenbacher S, Arruda VR, van der Loo JCM, Zelenaia O, Takefman D, Carr ME, Couto LB, Anguela XM, High KA (2017) Hemophilia B gene therapy with a high-specific-activity factor IX variant. N Engl J Med 377(23):2215–2227. https://doi.org/10.1056/NEJMoa1708538
196. Lai PS, Thompson BT (2013) Why activated protein C was not successful in severe sepsis and septic shock: are we still tilting at windmills? Curr Infect Dis Rep 15(5):407–412. https://doi.org/10.1007/s11908-013-0358-9
197. Janciauskiene SM, Bals R, Koczulla R, Vogelmeier C, Kohnlein T, Welte T (2011) The discovery of alpha1-antitrypsin and its role in health and disease. Respir Med 105(8):1129–1139. https://doi.org/10.1016/j.rmed.2011.02.002
198. Carugati A, Pappalardo E, Zingale LC, Cicardi M (2001) C1-inhibitor deficiency and angioedema. Mol Immunol 38(2-3):161–173
199. Ricklin D, Lambris JD (2016) New milestones ahead in complement-targeted therapy. Semin Immunol 28(3):208–222. https://doi.org/10.1016/j.smim.2016.06.001
200. Towards better patient care: drugs to avoid in 2014 (2014). Prescrire Int 23(150):161–165
201. Adcock DM, Gosselin R (2015) Direct Oral Anticoagulants (DOACs) in the Laboratory: 2015 Review. Thromb Res 136(1):7–12. https://doi.org/10.1016/j.thromres.2015.05.001
202. Tallant C, Marrero A, Gomis-Ruth FX (2010) Matrix metalloproteinases: fold and function of their catalytic domains. Biochim Biophys Acta 1803(1):20–28. https://doi.org/10.1016/j.bbamcr.2009.04.003
203. Gomis-Ruth FX (2017) Third time lucky? Getting a grip on matrix metalloproteinases. J Biol Chem 292(43):17975–17976. https://doi.org/10.1074/jbc.H117.806075
204. Scannevin RH, Alexander R, Haarlander TM, Burke SL, Singer M, Huo C, Zhang YM, Maguire D, Spurlino J, Deckman I, Carroll KI, Lewandowski F, Devine E, Dzordzorme K, Tounge B, Milligan C, Bayoumy S, Williams R, Schalk-Hihi C, Leonard K, Jackson P, Todd M, Kuo LC, Rhodes KJ (2017) Discovery of a highly selective chemical inhibitor of matrix metalloproteinase-9 (MMP-9) that allosterically inhibits zymogen activation. J Biol Chem 292(43):17963–17974. https://doi.org/10.1074/jbc.M117.806075
205. Grunwald B, Vandooren J, Gerg M, Ahomaa K, Hunger A, Berchtold S, Akbareian S, Schaten S, Knolle P, Edwards DR, Opdenakker G, Kruger A (2016) Systemic ablation of MMP-9 triggers invasive growth and metastasis of pancreatic cancer via deregulation of IL6 expression in the bone marrow. Mol Cancer Res 14(11):1147–1158. https://doi.org/10.1158/1541-7786.MCR-16-0180
206. Radisky ES, Raeeszadeh-Sarmazdeh M, Radisky DC (2017) Therapeutic potential of matrix metalloproteinase inhibition in breast cancer. J Cell Biochem 118(11):3531–3548. https://doi.org/10.1002/jcb.26185
207. Xia D, Watanabe H, Wu B, Lee SH, Li Y, Tsvetkov E, Bolshakov VY, Shen J, Kelleher RJ 3rd (2015) Presenilin-1 knockin mice reveal loss-of-function mechanism for familial Alzheimer's disease. Neuron 85(5):967–981. https://doi.org/10.1016/j.neuron.2015.02.010
208. Ran Y, Hossain F, Pannuti A, Lessard CB, Ladd GZ, Jung JI, Minter LM, Osborne BA, Miele L, Golde TE (2017) gamma-Secretase inhibitors in cancer clinical trials are pharmacologically and functionally distinct. EMBO Mol Med 9(7):950–966. https://doi.org/10.15252/emmm.201607265
209. Duong le T, Leung AT, Langdahl B (2016) Cathepsin K inhibition: a new mechanism for the treatment of osteoporosis. Calcif Tissue Int 98(4):381–397. https://doi.org/10.1007/s00223-015-0051-0
210. Drake MT, Clarke BL, Oursler MJ, Khosla S (2017) Cathepsin K inhibitors for

osteoporosis: biology, potential clinical utility, and lessons learned. Endocr Rev 38 (4):325–350. https://doi.org/10.1210/er.2015-1114

211. Su EJ, Cao C, Fredriksson L, Nilsson I, Stefanitsch C, Stevenson TK, Zhao J, Ragsdale M, Sun YY, Yepes M, Kuan CY, Eriksson U, Strickland DK, Lawrence DA, Zhang L (2017) Microglial-mediated PDGF-CC activation increases cerebrovascular permeability during ischemic stroke. Acta Neuropathol 134(4):585–604. https://doi.org/10.1007/s00401-017-1749-z

212. Lakhan SE, Kirchgessner A, Tepper D, Leonard A (2013) Matrix metalloproteinases and blood-brain barrier disruption in acute ischemic stroke. Front Neurol 4:32. https://doi.org/10.3389/fneur.2013.00032

213. Hafez S, Coucha M, Bruno A, Fagan SC, Ergul A (2014) Hyperglycemia, acute ischemic stroke, and thrombolytic therapy. Transl Stroke Res 5(4):442–453. https://doi.org/10.1007/s12975-014-0336-z

214. Sun H, Xu Y, Sitkiewicz I, Ma Y, Wang X, Yestrepsky BD, Huang Y, Lapadatescu MC, Larsen MJ, Larsen SD, Musser JM, Ginsburg D (2012) Inhibitor of streptokinase gene expression improves survival after group A streptococcus infection in mice. Proc Natl Acad Sci U S A 109(9):3469–3474. https://doi.org/10.1073/pnas.1201031109

215. Silva DG, Ribeiro JFR, De Vita D, Cianni L, Franco CH, Freitas-Junior LH, Moraes CB, Rocha JR, Burtoloso ACB, Kenny PW, Leitao A, Montanari CA (2017) A comparative study of warheads for design of cysteine protease inhibitors. Bioorg Med Chem Lett 27(22):5031–5035. https://doi.org/10.1016/j.bmcl.2017.10.002

216. Lv Z, Chu Y, Wang Y (2015) HIV protease inhibitors: a review of molecular selectivity and toxicity. HIV AIDS (Auckl) 7:95–104. https://doi.org/10.2147/HIV.S79956

217. Ghosh AK, Brindisi M, Nyalapatla PR, Takayama J, Ella-Menye JR, Yashchuk S, Agniswamy J, Wang YF, Aoki M, Amano M, Weber IT, Mitsuya H (2017) Design of novel HIV-1 protease inhibitors incorporating isophthalamide-derived P2-P3 ligands: Synthesis, biological evaluation and X-ray structural studies of inhibitor-HIV-1 protease complex. Bioorg Med Chem 25 (19):5114–5127. https://doi.org/10.1016/j.bmc.2017.04.005

218. Ghosh AK, Sean Fyvie W, Brindisi M, Steffey M, Agniswamy J, Wang YF, Aoki M, Amano M, Weber IT, Mitsuya H (2017) Design, synthesis, X-ray studies, and biological evaluation of novel macrocyclic HIV-1 protease inhibitors involving the P1′-P2′ ligands. Bioorg Med Chem Lett 27 (21):4925–4931. https://doi.org/10.1016/j.bmcl.2017.09.003

219. McCauley JA, Rudd MT (2016) Hepatitis C virus NS3/4a protease inhibitors. Curr Opin Pharmacol 30:84–92. https://doi.org/10.1016/j.coph.2016.07.015

220. Thornberry NA, Weber AE (2007) Discovery of JANUVIA (Sitagliptin), a selective dipeptidyl peptidase IV inhibitor for the treatment of type 2 diabetes. Curr Top Med Chem 7 (6):557–568

221. Schramm VL (2013) Transition States, analogues, and drug development. ACS Chem Biol 8(1):71–81. https://doi.org/10.1021/cb300631k

222. Mitsuya H, Maeda K, Das D, Ghosh AK (2008) Development of protease inhibitors and the fight with drug-resistant HIV-1 variants. Adv Pharmacol 56:169–197. https://doi.org/10.1016/S1054-3589(07)56006-0

223. Kipp DR, Hirschi JS, Wakata A, Goldstein H, Schramm VL (2012) Transition states of native and drug-resistant HIV-1 protease are the same. Proc Natl Acad Sci U S A 109 (17):6543–6548. https://doi.org/10.1073/pnas.1202808109

224. Overall CM (2002) Molecular determinants of metalloproteinase substrate specificity: matrix metalloproteinase substrate binding domains, modules, and exosites. Mol Biotechnol 22(1):51–86. https://doi.org/10.1385/MB:22:1:051

225. Skarina T, Xu X, Evdokimova E, Savchenko A (2014) High-throughput crystallization screening. Methods Mol Biol 1140:159–168. https://doi.org/10.1007/978-1-4939-0354-2_12

226. Tiefenbrunn T, Forli S, Happer M, Gonzalez A, Tsai Y, Soltis M, Elder JH, Olson AJ, Stout CD (2014) Crystallographic fragment-based drug discovery: use of a brominated fragment library targeting HIV protease. Chem Biol Drug Des 83(2):141–148. https://doi.org/10.1111/cbdd.12227

227. Rydel TJ, Tulinsky A, Bode W, Huber R (1991) Refined structure of the hirudin-thrombin complex. J Mol Biol 221 (2):583–601

228. Warkentin TE (2004) Bivalent direct thrombin inhibitors: hirudin and bivalirudin. Best Pract Res Clin Haematol 17(1):105–125. https://doi.org/10.1016/j.beha.2004.02.002

229. Vance NR, Gakhar L, Spies MA (2017) Allosteric tuning of caspase-7: a fragment-based drug discovery approach. Angew Chem Int Ed Engl 56(46):14443–14447. https://doi.org/10.1002/anie.201706959
230. Weiss-Sadan T, Gotsman I, Blum G (2017) Cysteine proteases in atherosclerosis. FEBS J 284(10):1455–1472. https://doi.org/10.1111/febs.14043
231. Lee S, Xie J, Chen X (2010) Activatable molecular probes for cancer imaging. Curr Top Med Chem 10(11):1135–1144
232. Ren G, Blum G, Verdoes M, Liu H, Syed S, Edgington LE, Gheysens O, Miao Z, Jiang H, Gambhir SS, Bogyo M, Cheng Z (2011) Non-invasive imaging of cysteine cathepsin activity in solid tumors using a 64Cu-labeled activity-based probe. PLoS One 6(11): e28029. https://doi.org/10.1371/journal.pone.0028029
233. Indalao IL, Sawabuchi T, Takahashi E, Kido H (2017) IL-1beta is a key cytokine that induces trypsin upregulation in the influenza virus-cytokine-trypsin cycle. Arch Virol 162 (1):201–211. https://doi.org/10.1007/s00705-016-3093-3
234. Kelso EB, Lockhart JC, Hembrough T, Dunning L, Plevin R, Hollenberg MD, Sommerhoff CP, McLean JS, Ferrell WR (2006) Therapeutic promise of proteinase-activated receptor-2 antagonism in joint inflammation. J Pharmacol Exp Ther 316(3):1017–1024. https://doi.org/10.1124/jpet.105.093807
235. Vergnolle N (2009) Protease-activated receptors as drug targets in inflammation and pain. Pharmacol Ther 123(3):292–309. https://doi.org/10.1016/j.pharmthera.2009.05.004
236. French SL, Hamilton JR (2016) Protease-activated receptor 4: from structure to function and back again. Br J Pharmacol 173 (20):2952–2965. https://doi.org/10.1111/bph.13455
237. Kasperkiewicz P, Poreba M, Groborz K, Drag M (2017) Emerging challenges in the design of selective substrates, inhibitors and activity-based probes for indistinguishable proteases. FEBS J 284(10):1518–1539. https://doi.org/10.1111/febs.14001
238. Harris JL, Backes BJ, Leonetti F, Mahrus S, Ellman JA, Craik CS (2000) Rapid and general profiling of protease specificity by using combinatorial fluorogenic substrate libraries. Proc Natl Acad Sci U S A 97(14):7754–7759. https://doi.org/10.1073/pnas.140132697
239. Kasperkiewicz P, Poreba M, Snipas SJ, Parker H, Winterbourn CC, Salvesen GS, Drag M (2014) Design of ultrasensitive probes for human neutrophil elastase through hybrid combinatorial substrate library profiling. Proc Natl Acad Sci U S A 111 (7):2518–2523. https://doi.org/10.1073/pnas.1318548111
240. O'Donoghue AJ, Eroy-Reveles AA, Knudsen GM, Ingram J, Zhou M, Statnekov JB, Greninger AL, Hostetter DR, Qu G, Maltby DA, Anderson MO, Derisi JL, McKerrow JH, Burlingame AL, Craik CS (2012) Global identification of peptidase specificity by multiplex substrate profiling. Nat Methods 9 (11):1095–1100. https://doi.org/10.1038/nmeth.2182
241. Li Q, Yi L, Hoi KH, Marek P, Georgiou G, Iverson BL (2017) Profiling protease specificity: combining yeast ER sequestration screening (YESS) with next generation sequencing. ACS Chem Biol 12(2):510–518. https://doi.org/10.1021/acschembio.6b00547
242. Fulcher LJ, Hutchinson LD, Macartney TJ, Turnbull C, Sapkota GP (2017) Targeting endogenous proteins for degradation through the affinity-directed protein missile system. Open Biol 7(5). https://doi.org/10.1098/rsob.170066
243. Grossi G, Dalgaard Ebbesen Jepsen M, Kjems J, Andersen ES (2017) Control of enzyme reactions by a reconfigurable DNA nanovault. Nat Commun 8(1):992. https://doi.org/10.1038/s41467-017-01072-8
244. Lange PF, Huesgen PF, Overall CM (2012) TopFIND 2.0—linking protein termini with proteolytic processing and modifications altering protein function. Nucleic Acids Res 40(Database issue):D351–D361. https://doi.org/10.1093/nar/gkr1025
245. Wang X, Davies M, Roy S, Kuruc M (2015) Bead based proteome enrichment enhances features of the protein elution plate (PEP) for functional proteomic profiling. Proteomes 3(4):454–466. https://doi.org/10.3390/proteomes3040454
246. Zheng H, Roy S, Soherwardy A, Rahman S, Kuruc M (2017) Stroma liquid biopsy—proteomic profiles for cancer biomarkers. Poster reprint first presented at NJ Cancer Retreat, May 25, 2017 New Brunswick, NJ, USA
247. Rifai N, Gillette MA, Carr SA (2006) Protein biomarker discovery and validation: the long and uncertain path to clinical utility. Nat Biotechnol 24(8):971–983. https://doi.org/10.1038/nbt1235
248. Koenig W (2003) Fibrin(ogen) in cardiovascular disease: an update. Thromb Haemost 89 (4):601–609

249. Dunn EJ, Ariens RA, Grant PJ (2005) The influence of type 2 diabetes on fibrin structure and function. Diabetologia 48 (6):1198–1206. https://doi.org/10.1007/s00125-005-1742-2
250. Grammas P, Martinez JM (2014) Targeting thrombin: an inflammatory neurotoxin in Alzheimer's disease. J Alzheimers Dis 42 (Suppl 4):S537–S544. https://doi.org/10.3233/JAD-141557
251. Ahn HJ, Zamolodchikov D, Cortes-Canteli-M, Norris EH, Glickman JF, Strickland S (2010) Alzheimer's disease peptide beta-amyloid interacts with fibrinogen and induces its oligomerization. Proc Natl Acad Sci U S A 107(50):21812–21817. https://doi.org/10.1073/pnas.1010373107
252. Amara U, Rittirsch D, Flierl M, Bruckner U, Klos A, Gebhard F, Lambris JD, Huber-Lang M (2008) Interaction between the coagulation and complement system. Adv Exp Med Biol 632:71–79

Chapter 21

The Use of Combinatorial Hexapeptide Ligand Library (CPLL) in Allergomics

Youcef Shahali, Hélène Sénéchal, and Pascal Poncet

Abstract

The recent progress of proteomic protocols led to more efficient protein extraction and concentration procedures to remove nonprotein interfering compounds present in the starting material and to increase the concentration of underrepresented proteins. Combinatorial hexapeptide ligand libraries (CPLL) were recently applied to both plant- and animal-derived tissues for capturing the low- and very low-abundance allergens. Several IgE-binding proteins which were previously absent or poorly represented by using conventional proteomics tools have been detected and characterized through a CPLL-based approach. In the present chapter, a protocol based on improved protein extraction and enrichment by CPLL, allowing the immunochemical characterization of several "hidden allergens" in cypress pollen, is described in detail.

Key words Hexapeptide ligand libraries, Low-abundance allergens, Pollen allergens, Mass spectrometry, Proteomics

1 Introduction

Over the last decade, the study of allergens, defined as allergomics, allergenomics, or IgE immunoproteomics, increases in relevance and precision with the development of proteomics tools allowing the discovery of trace allergens as well as allergenic proteins present in lower concentrations in various biological samples [1–4]. It is estimated that no more than 30% of expressed proteins are detectable using standard analytical methods [1]. A complete knowledge of the proteome compositions of complex biological samples is thus the starting challenge for deciphering allergenic proteins to which people are exposed and sensitized. Although most known allergens have an average concentration and could be identified without the use of enrichment tools, the presence of high-abundance proteins in extracts from allergenic sources often preclude proper detection of the entire allergen repertoire responsible for allergic sensitization [2, 3]. This is particularly important knowing that low- and very

Xing Wang and Matthew Kuruc (eds.), *Functional Proteomics: Methods and Protocols*, Methods in Molecular Biology, vol. 1871, https://doi.org/10.1007/978-1-4939-8814-3_21,

low-abundance allergens can generate strong immunogenic reactions in spite of their low concentration [3–7].

To pave the way for the identification of underrepresented allergens in allergenic extracts, bead-based combinatorial hexapeptide ligand library (CPLL) has been successfully applied to allergomics studies [3–11]. This method has the advantage of being well compatible with a large panel of profiling techniques, including all types of gel electrophoresis and mass spectrometry-based identifications such as MALDI and SELDI or LC-MSMS [12]. Hexapeptide library coupled to carrier beads provides non-covalent binding sites for a broad range of proteins due to their diverse physicochemical properties resulting from their differential combination of amino acids. As each specific combination is limited to a given amount of bead volume, hexapeptides having an affinity to high-abundant proteins are rapidly saturate, whereas low-abundance proteins are quantitatively trapped and concentrated (*see* Fig. 1). Thus, this method circumvents the drawbacks of the depletion or chromatographic fractionation methods where proteins of interest could be removed in an uncontrolled manner or diluted and denatured prior to analyses. Therefore, CPLL represent an efficient tool to get access to low-abundant IgE-binding proteins and a method of choice for detecting trace allergens in food. Over the last few years, peptide libraries help to the discovery of novel allergens in various extracts. One of the first allergenic sources investigated using CPLL was cow's milk whey [5]. In this pioneer study, a number of novel IgE-binding proteins were evidenced after CPLL treatment, especially a polymorphic immunoglobulin [5]. The same approach was fruitfully applied by the same authors to explore the allergen repertoire of other allergy-causing substances such as latex [6] and maize [7] leading to the identification of new IgE-binding proteins. Other authors adopted the CPLL technology for the detection of peanut [8] and casein [9] traces in baked cookies and wines, respectively. The use of CPLL in allergomics studies also led to the detection of fungal allergens present in blood of patients with invasive aspergillosis [10] and novel allergen candidates in hen's egg [11].

The CPLL technology has been used by our team for the in-depth exploration of the cypress pollen (CP) allergen composition [3]. The analysis of CP proteome was known to be difficult due to the relatively poor protein content of CP extracts and the inherent characteristics of its matrix containing high amounts of interfering compounds such as carbohydrates and pigments [13, 14]. Protein enrichment by CPLL was performed at three different pHs (*see* Fig. 2) for the capture of the largest number of proteins [15]. Sample treatment using CPLL led to an impressive extension of the number of identified proteins (*see* Fig. 3) and the characterization of novel candidate allergens (*see* Fig. 4). This study showed that individual reactions of allergic patients against this

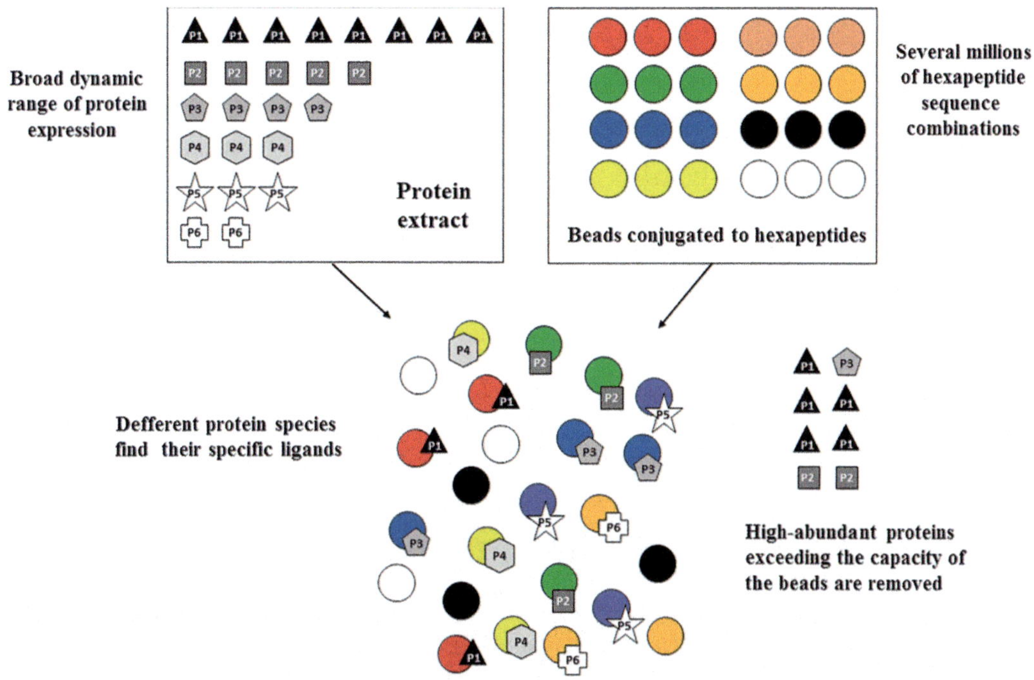

Fig. 1 Schematic illustration of the enrichment process of low-abundant protein species using ProteoMiner beads. A high number of hexapeptide ligands are used in a single analysis which independently bind specific proteins until they reach their maximum capacity. Hexapeptides having an affinity to high-abundant proteins (e.g., P1 and P2) are rapidly saturate, whereas low-abundance proteins (e.g., P5 and P6) are quantitatively trapped and concentrated. High-abundant proteins exceeding the capacity of the beads are removed

pollen are not necessarily against dominant known proteins. Here, the CPLL-based sample preparation protocol used in this study is described in detail.

2 Materials

2.1 Equipment

1. Combinatorial peptide bead libraries are generally provided as an aqueous solution containing preserving agents against potential bacterial contaminations. This material could also be available as powdered dry beads. In both cases, CPLL beads need to be conditioned prior to use. Prepacked spin-columns filled with hexapeptide-coupled beads for different starting protein amounts as well as bulk beads for customized applications are available under the commercial trademark ProteoMiner™ (Bio-Rad Laboratories, Hercules, CA, USA). In this study, we used bulk beads in order to capture CP proteins at three different pHs for the enrichment of the largest number of proteins as previously described [4].
2. Vortex.

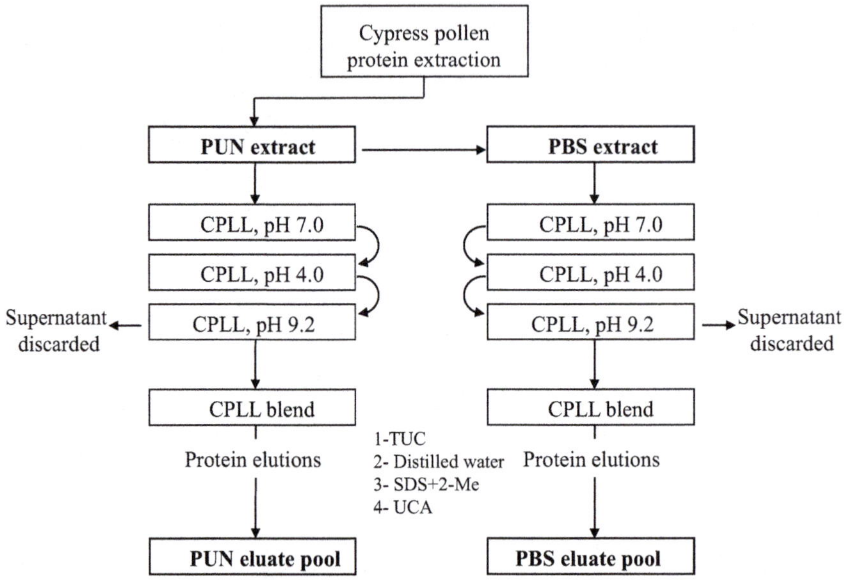

Fig. 2 Schematic experimental procedure of pollen extraction, CPLL treatment, and protein elution from the ProteoMiner beads (adapted from Shahali et al.) [3]. Since the pH influences the protein capture by CPLL, protein enrichment by CPLL was performed at three different pHs. In order to ensure complete protein stripping from the beads, four elution steps were sequentially performed

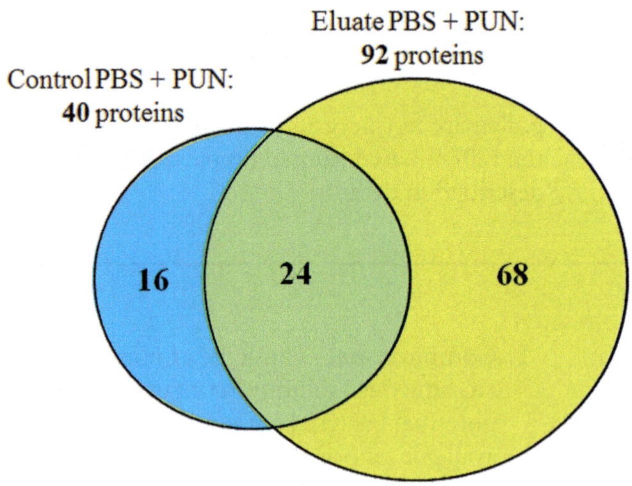

Fig. 3 Overlapping Venn diagram of the total proteins detected by LC-MS/MS analysis. LC-MS analysis was performed in control, untreated PUN and PBS extracts (blue circle) and in pooled eluates (yellow circle) from hexapeptide ligand libraries (from PUN and PBS). The green represented common proteins found in the two types of extraction products. Adapted from Shahali et al. [3]

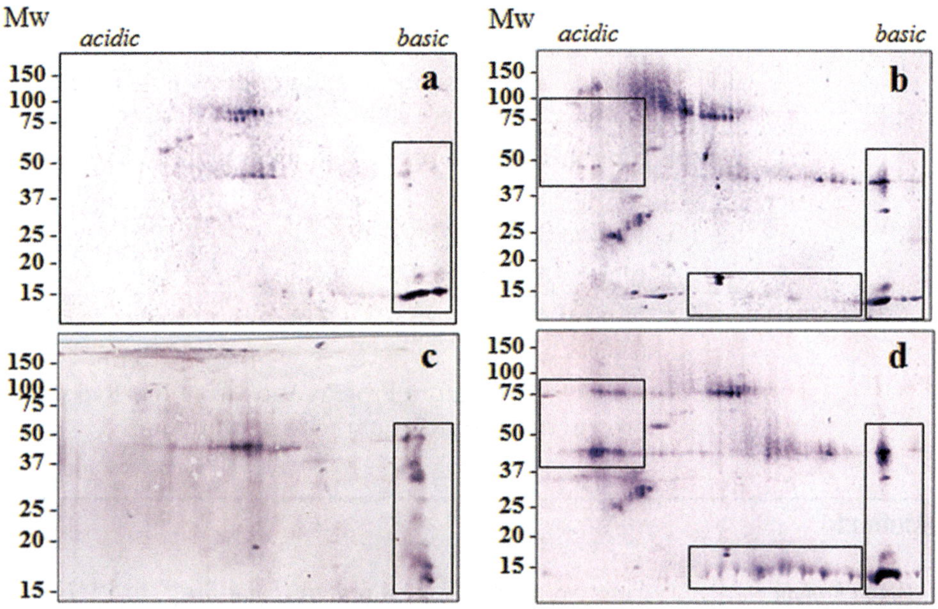

Fig. 4 Two-dimensional IgE immunoblotting probed with the serum of a cypress pollen allergic patient (adapted from Shahali et al.) [3]. PBS control extract: a; PBS eluate: c; PUN control extract: b; PUN eluate: d. Novel IgE-binding were identified in CPLL eluates (outlined in boxes)

3. Refrigerated centrifuge with gravitational force of at least $10,000 \times g$.
4. A set of micropipette.
5. Sonication device.
6. Tube rotator.
7. Magnetic stirrer.

2.2 Pollen Proteins Extraction

1. *Cupressus sempervirens* pollen was supplied by Allergon AB (Ängelholm, Sweden).
2. Chemicals composing the PUN buffer: 10 mM phosphate buffer pH 7.06 containing 3 M urea and 0.2% NP-40 (Nonidet™ 40) were from Sigma-Aldrich (St Louis, MO, USA).
3. Chemicals composing the phosphate buffered saline (PBS): 150 mM NaCl, 7.8 mM Na_2HPO_4 and 0.51 mM KH_2PO4 pH 7.4 were all from Sigma-Aldrich.
4. Both ethylene diamine tetra acetic acid (EDTA) and phenyl methane sulfonyl fluoride (PMSF) as anti-protease cocktail were supplied by Sigma-Aldrich.

2.3 Hexapeptide Ligand Library Sample Preparation

1. Ammonium sulfate for protein precipitation was supplied by Sigma-Aldrich.
2. Bulk ProteoMiner™ beads were from Bio-Rad Laboratories.

3. Thiourea, urea, 3-[3-cholamidopropyl dimethylammonio]-1-propansulfonate (CHAPS), acid acetic, 2-mercaptoethanol (2-Me), ammonia, and sodium dodecyl sulfate (SDS) used for sample elution were all from Sigma-Aldrich (*see* **Note 1**).
4. Elution 1: 2 M thiourea, 7 M urea, and 2% CHAPS in distilled water (TUC).
5. Elution 2: distilled water.
6. Elution 3: 6% SDS, 2% 2-Me in distilled water (SDS + 2-Me).
7. Elution 4: 8 M urea, 2% CHAPS, and 5% acetic acid pH 3.3 (UCA).
8. Quick Start Bradford Protein Assay from Bio-Rad was used for protein determination of their concentration.

3 Method

3.1 Pollen Protein Extraction

1. Slurry ten times 5 g of pollen (total 50 g) in PUN buffer (*see* **Note 2**).
2. Shake tubes containing the pollen suspension for 2–3 min to wet the pollen grains.
3. Sonicate the suspension for 20 s at 4 °C.
4. Place the tube on a rotator and gently shake the suspension overnight at room temperature.
5. Centrifuge the suspension at 10,000 × *g* and 4 °C during 15 min.
6. Pipet the supernatant into a new tube, label PUN extract and date.
7. Add 1 mL of PUN buffer on remaining pollen pellets.
8. Homogenize and centrifuge the suspension at 10,000 × *g* and 4 °C during 15 min.
9. Pool the supernatant with the PUN extract previously obtained in **step 6**. The overall volume of the PUN extract was 100 mL.
10. Add 30 mL of PBS pH 7.5 to each pollen pellet.
11. Place the tube on a rotator and shake the suspension overnight at room temperature.
12. Collect the supernatants by centrifugation at 10,000 × *g* and 4 °C during 15 min.
13. Pool all PBS extracts. The resulting total volume was 125 mL.

3.2 Removal of Interfering Substances Prior to CPLL Analysis

1. Add ammonium sulfate to both PUN and PBS extracts under stirring to reach up to 90% saturation or, respectively, 66 g (for 100 mL) and 82.5 g (for 125 mL).
2. Gently agitate the two mixtures at 4 °C overnight.

3. Centrifuge the two mixtures during 30 min at 18,000 × *g* at 4 °C.
4. Discard the supernatants.
5. Dissolve the pellets of PUN extract in 10 mL of PUN buffer.
6. Dissolve the pellet of PBS extract in 9 mL of PBS.
7. Add EDTA and PMSF up to 1 mM concentration for both.
8. Dialyze (cutoff 3500 Da) PUN solution against 3 M urea overnight.
9. Dialyze PBS solution against PBS (1 L) overnight at 4 °C (*see* **Note 3**).
10. After dialyzing, add again to the same concentration as before dialysis protease inhibitors (PMSF and EDTA).
11. Perform Bradford assay on recovered PUN and PBS extracts and keep them at −20 °C before use for analysis.

3.3 Sample Treatment with CPLL

1. Wash three times one hundred microliters of ProteoMiner beads with 200 μL PUN solution (*see* **Note 4**).
2. Centrifuge the suspension at 1000 × *g* for 30–60 s at room temperature to remove the PUN washing solution.
3. After elimination of washing solution excess, add 30 mL of PUN protein extract to the beads (*see* **Note 5**).
4. Check the pH at 7.0 and shake the mixture under stirring for 2 h at room temperature.
5. Separate beads by centrifugation at 1000 × *g* for 30–60 s at 15 °C.
6. Transfer the supernatants to a new tube, label and store at 4 °C until use for a subsequent treatment in different pH (*see* **Note 6**).
7. Wash the beads with distilled water and then separate by centrifugation at 1000 × *g* for 30–60 s.
8. Discard excess water and store the bead pellet at 4 °C for further use (*see* **Note** 7).
9. Mix the supernatant collected in the **step 6** with a second sample of 100 μL prewashed ProteoMiner beads.
10. Adjust the pH of the suspension to 4.0 by addition of acetic acid 1 M dropwise.
11. Shake the mixture under stirring for 2 h at room temperature.
12. Separate beads by centrifugation at 1000 × *g* for 30–60 s at 15 °C.
13. Transfer the supernatants to a new tube, label, and store at 4 °C until use for a subsequent treatment in different pH.

14. Wash the beads with distilled water and then separate the bead pellet by centrifugation at 1000 × *g* for 30–60 s at 15 °C.
15. Discard excess water and store the bead pellet at 4 °C for further use.
16. Mix the supernatant collected in the **step 13** with a third sample of 100 μL prewashed ProteoMiner beads.
17. Adjust the pH of the suspension to 9.25 by dropwise addition of 4 M ammonia.
18. Gently shake the resulting suspension at room temperature for 2 h.
19. Separate beads by centrifugation at 1000 × *g* for 30–60 s and discard the supernatant.
20. Wash the beads once with distilled water and then separate the bead pellet by centrifugation at 1000 × *g* for 30–60 s at 15 °C.
21. Discard excess water and store the bead pellet at 4 °C for further use.

3.4 Protein elution from CPLL Beads

1. Blend together the three bead aliquots of 100 μL each (collected from **steps 8**, **15**, and **21** of the CPLL treatment).
2. Rapidly wash once the beads with 600 μL of distilled water and discard supernatants.
3. Add 500 μL of TUC elution solution to 300 μL bead pellets.
4. Gently shake the mixture for 2 h at room temperature.
5. Centrifuge the mixture at 1000 × *g* for 30–60 s and collect the supernatant comprising desorbed proteins (eluate “a”).
6. Wash the bead pellet once with 300 μL distilled water and collect by centrifugation the supernatant (eluate “b”).
7. Submit the resulting bead pellet to the third elution by adding 500 μL of a solution composed of 6% SDS and 2% 2-Mercaptoethanol (SDS+ 2-Me).
8. Gently shake the mixture for 2 h at room temperature.
9. Centrifuge the mixture at 1000 × *g* for 30–60 s and collect the supernatant (eluate “c”).
10. In order to ensure complete protein stripping from the beads, perform a fourth elution step by adding the UCA solution to the bead pellet.
11. Gently shake the mixture for 2 h at room temperature.
12. Centrifuge the mixture at 1000 × *g* for 30–60 s and collect the supernatant (eluate “d”).
13. Pool the four eluates together (PUN eluate).
14. Adjust the pH to 7.0 and dialyze (cutoff 3500 Da) PUN eluate against 3 M urea overnight (*see* **Note 8**).

15. Perform the Bradford–Lowry standard spectrophotometric method and store eluates at −20 °C for further proteomic analysis (*see* **Note 9**).

4 Notes

1. It is recommended to use freshly prepared buffer when conducting all experiments.
2. Because of its particular structural features and physicochemical composition, cypress pollen is one of the most difficult pollens to analyze in terms of protein content and therefore allergens. In aqueous media with pH 7.5 [15] the external wall (exine) cracks in a few minutes under the effect of swelling of the intine (inner wall) particularly rich in polysaccharide. Few proteins are then extracted in aqueous conditions. The dry milling may be a good alternative for the extraction of cypress pollen proteins and to generate fragments of smaller sizes for experiments and ultrastructural analysis of immunoreactivity. Good results were achieved by using Minilys homogenizer and the Precellys kit: 03961-1-003 (1.4 mm ceramic beads) from Bertin Instruments (Montigny-le-Bretonneux, France) [15].
3. In this step the ionic strength of PBS extracts could be reduced by dialysis against a solution containing a lower concentration of sodium chloride, for instance 50 mM instead of 150 mM. This condition improves the ability of the beads to absorb proteins and is especially advised when the concentration of the protein is very low.
4. Beads should be washed with the same solution used for specific protein extraction. After vortexing, the suspension is centrifuged to remove the excess of supernatant and the bead-preserving agents. Repeat the washing procedure three times to be sure that all undesired preserving agents are removed. For powdered dry beads, it is recommended to slurry 100 mg of dry beads in 2 mL methanol for 30 min under shaking and then add 2 mL of phosphate buffer pH 7.5. The rehydration is performed overnight at room temperature. The rehydrated beads are then washed extensively, with the same buffer used for protein extraction as described above.
5. The optimal protein concentration of treated extracts should be between 1 and 10 mg/mL. For a treatment with 100 μL of hexapeptide, the total amount of protein should be larger than 50 mg.
6. Since the pH influences largely the protein capture by CPLLs, it is possible to capture proteins from a biological sample at different pHs in order to increase the efficacy of the CPLL

enrichment [16]. This approach is particularly recommended for biological samples with low protein content.

7. After protein absorption by CPLL, beads can be stored at 4 °C for a few days before desorbing and protein elution. In this case, the beads should be equilibrated prior to elution at the same temperature at which the protein capture has been carried out.
8. In parallel 100 μL of ProteoMiner beads was washed with 200 μL PBS solution. PBS extract (12 mL) was treated using the same protocol described above for the PUN extract. Here also the four eluates (a, b, c, and d) were pooled together (PBS eluate), neutralized, desalted, lyophilized, and stored at −20 °C after the Bradford protein assay.
9. For 2-DE separation, the desired volume of non-treated sample and the mixture of all eluates were solubilized in the "2-DE sample buffer" (TUC, 40 mM Tris-acetate) to a final concentration of 2 mg/mL protein.

Acknowledgments

Collaboration with Drs. Egisto Boschetti and Pier Giorgio Righetti, the pioneer developers of the CPLL technology, is gratefully acknowledged and was at the basis of this protocol and its adaptation to allergomics studies.

References

1. Boschetti E, Righetti PG (2013) Low-abundance proteome discovery: state of the art and protocols. Newnes
2. Shahali Y, Sutra JP, Peltre G, Charpin D, Sénéchal H, Poncet P (2010) IgE reactivity to common cypress (*C. sempervirens*) pollen extracts: evidence for novel allergens. World Allergy Organ J 3:229–234
3. Shahali Y, Sutra JP, Fasoli E, D'Amato A, Righetti PG, Futamura N et al (2012) Allergomic study of cypress pollen via combinatorial peptide ligand libraries. J Proteome 21:101–110
4. Righetti PG, Fasoli E, D'Amato A, Boschetti E (2014) The "dark side" of food stuff proteomics: the CPLL-marshals investigate. Foods 3:217–237
5. D'Amato A, Bachi A, Fasoli E, Boschetti E, Peltre G, Sénéchal H et al (2009) In-depth exploration of cow's whey proteome via combinatorial peptide ligand libraries. J Proteome Res 8:3925–3936
6. D'Amato A, Bachi A, Fasoli E, Boschetti E, Peltre G, Sénéchal H et al (2010) In-depth exploration of *Hevea brasiliensis* latex proteome and "hidden allergens" via combinatorial peptide ligand libraries. J Proteome 73:1368–1380
7. Fasoli E, Pastorello EA, Farioli L, Scibilia J, Aldini G, Carini M et al (2009) Searching for allergens in maize kernels via proteomic tools. J Proteome 72:501–510
8. Pedreschi R, Nørgaard J, Maquet A (2012) Current challenges in detecting food allergens by shotgun and targeted proteomic approaches: a case study on traces of peanut allergens in baked cookies. Nutrients 4:132–150
9. D'Amato A, Kravchuk AV, Bachi A, Righetti PG (2010) Noah's nectar: the proteome content of a glass of red wine. J Proteome 73:2370–2377
10. Fekkar A, Pionneau C, Brossas JY, Marinach-Patrice C, Snounou G, Brock M, Mazier D

(2012) DIGE enables the detection of a putative serum biomarker of fungal origin in a mouse model of invasive aspergillosis. J Proteome 75:2536–2549

11. Martos G, López-Fandiño R, Molina E (2013) Immunoreactivity of hen egg allergens: influence on in vitro gastrointestinal digestion of the presence of other egg white proteins and of egg yolk. Food Chem 136:775–781
12. Hartwig S, Lehr S (2012) Combination of highly efficient hexapeptide ligand library-based sample preparation with 2D DIGE for the analysis of the hidden human serum/plasma proteome. Methods Mol Biol 854:169–180
13. Shahali Y, Nicaise P, Brazdova A, Charpin D, Scala E, Mari A et al (2014) Complementarity between microarray and immunoblot for the comparative evaluation of IgE repertoire of French and Italian cypress pollen allergic patients. Folia Biologica (Prague) 60:192
14. Danti R, Della Rocca G, Calamassi R, Mori B, Mariotti Lippi M (2011) Insights into a hydration regulating system in Cupressus pollen grains. Ann Bot 108:299–306
15. Shahali Y (2011) Etude analytique de l'allergie au pollen de cyprès: aspects moléculaires et particulaires, Thesis Université Paris VI, Pierre et Marie Curie, Paris, France, p 220
16. Fasoli E, Farinazzo A, Sun CJ, Kravchuk AV, Guerrier L, Fortis F et al (2010) Interaction among proteins and peptide libraries in proteome analysis: pH involvement for a larger capture of species. J Proteome 73:733–742

Chapter 22

Efficient Extraction and Digestion of Gluten Proteins

Haili Li, Keren Byrne, Crispin A. Howitt, and Michelle L. Colgrave

Abstract

Coeliac disease (CD) is a T-cell mediated autoimmune disorder triggered by ingestion of cereal gluten found in wheat (gliadins and glutenins), barley (hordeins), and rye (secalins). As the only treatment for CD is a lifelong gluten-free diet, the measurement of gluten in raw ingredients and processed food products is critical to protecting people with CD or gluten intolerance. The most commonly employed method is the enzyme-linked immunosorbent assay (ELISA), but more recently mass spectrometry has been employed wherein the extracted gluten proteins are digested to peptides that are then directly measured. To achieve the goal of accurate gluten quantitation, gluten must be efficiently extracted from the ingredient or food matrix and then digested to yield the peptides that are monitored by LC-MS. In this chapter, a rapid, simple, and reproducible protocol for extraction and digestion of gluten proteins is described.

Key words Flour, Gluten, Digestion, Trypsin, Chymotrypsin, Mass spectrometry

1 Introduction

Coeliac disease (CD) is a disease of the small intestine that occurs in genetically susceptible subjects triggered by the ingestion of cereal gluten proteins. The only treatment is strict adherence to a lifelong gluten-free diet [1, 2]. Gluten is the collective name for a class of proteins found in wheat, rye, and barley that are the elicitors of CD, an autoimmune disorder with a prevalence of ~1% worldwide [3–5]. Establishing accurate methods for gluten measurement is of critical importance to the health of those affected by CD or non-coeliac gluten sensitivity (NCGS) [6]. Enzyme-linked immunosorbent assays (ELISAs) are the currently accepted method for the detection of gluten in food [7]. One drawback of ELISAs is that they cannot adequately quantify gluten that has been highly hydrolyzed [8]. A number of approaches have been developed with mass spectrometry (MS) showing great promise in gluten measurement owing to its specificity, sensitivity, ability to multiplex and identify hydrolyzed gluten [2, 9–13]. The successful application of bottom-

Xing Wang and Matthew Kuruc (eds.), *Functional Proteomics: Methods and Protocols*, Methods in Molecular Biology, vol. 1871, https://doi.org/10.1007/978-1-4939-8814-3_22,

up proteomics to the analysis of gluten critically depends upon the efficiency and reproducibility of proteolytic digestion of gluten from the grain or highly processed food products.

In this chapter, a rapid, simple, and reproducible method for extraction and digestion of gluten proteins is described. Using this method, raw ingredients and processed food products may be analyzed enabling detection and relative quantitation using liquid chromatography-mass spectrometry (LC-MS/MS). This method has been found to be applicable to all food matrices tested to date.

2 Materials

Prepare all solutions using MilliQ water (prepared by purifying deionized water to attain a resistance of 18 MΩ-cm at 25 °C) and analytical grade reagents. Prepare and store all reagents at room temperature (unless indicated otherwise). Follow all waste disposal regulations when disposing of waste materials and check Material Safety Data Sheets (MSDS) for reagents prior to use. All reagents are prepared immediately prior to use. The isopropanol (IPA) is HPLC grade. All other reagents are the highest possible commercial grade available (*see* **Note 1**). The volumes described below are suitable for the analysis of 10 samples with four replicates per sample.

2.1 Chemicals

1. Extraction buffer: 55% isopropanol (IPA)/2% dithiothreitol (DTT) (*see* **Note 2**). Prepare 200 μL per sample (replicate). Prepare 10 mL of 55% IPA/2% DTT by combining 5.5 mL IPA with 4.5 mL water and 200 mg DTT.
2. Urea (UA) buffer: 8 M urea in 0.1 M Tris–HCl (*see* **Note 3**). Prepare 1 mL per 1 sample (replicate). Prepare 50 mL of UA buffer by weighing 24 g of urea and dissolving in 45 mL of water. To this solution, add 5 mL of 1 M Tris–HCl, pH 8.5 (*see* **Note 4**).
3. Iodoacetamide (IAM) solution: 0.05 M IAM in UA (*see* **Note 5**). Prepare 0.1 mL per 1 sample (replicate). Prepare 5 mL by weighing 46.2 mg and dissolving in UA buffer.
4. Ammonium bicarbonate: 50 mM NH_4HCO_3 in water, pH 8.0 (*see* **Note 6**). Prepare 0.5 mL per 1 sample (replicate).
5. Trypsin: 0.25 mg/mL in 50 mM ammonium bicarbonate, 1 mM $CaCl_2$ (*see* **Note 7**). Prepare 0.2 mL per sample (replicate).
6. Chymotrypsin: 0.25 mg/mL in 50 mM ammonium bicarbonate, 1 mM $CaCl_2$ (*see* **Note 8**). Prepare 0.2 mL per sample (replicate).

2.2 Equipment

1. 10 kDa MWCO filters (e.g., Millipore: catalog number UFC5010BK).
2. Benchtop centrifuge (e.g., Eppendorf, model: 5415R), temperature 25 °C.

3. Wet chamber with a rack for eppendorf tubes.
4. Thermal mixer set to 50 °C.

2.3 HPLC Components

1. HPLC buffer A: 0.1% formic acid, 99.9% water. Mix by inversion (*see* **Note 9**).
2. HPLC buffer B: 0.1% formic acid, 90% acetonitrile, 9.9% water. Mix by inversion.

3 Methods

3.1 Gluten Extraction

1. Weigh 20 mg of flour (or milled food product) into a 1.5 mL eppendorf tube.
2. Add 200 μL of 55% IPA/2% DTT and vortex the tube until flour is properly mixed with solution (*see* **Note 10**).
3. Place the tube in an ultrasonic bath for 5 min at room temperature.
4. Put the tube in a dry block heater at 50 °C for 30 min (*see* **Note 11**).
5. Centrifuge the suspension for 10 min at 20,800 × *g*.
6. Transfer the supernatant, containing the gluten, into a fresh tube.

3.2 Protein Digestion

1. Transfer 100 μL of gluten extract to a 10 kDa MWCO filter, add 100 μL of UA buffer, and centrifuge at 20,800 × *g* for 15 min.
2. Wash the protein by addition of 200 μL of UA buffer to the filter unit. Centrifuge at 20,800 × *g* for 15 min.
3. Add 100 μL of IAM solution and incubate at room temperature for 20 min in the dark (*see* **Notes 5** and **12**).
4. Centrifuge the filter units at 20,800 × *g* for 15 min.
5. Add 200 μL of UA to the filter unit and centrifuge at 20,800 × *g* for 15 min to remove excess IAM. Discard the flow-through from the collection tube.
6. Exchange the buffer (*see* **Note 13**) by adding 200 μL of 50 mM ammonium bicarbonate to the filter unit and centrifuge at 20,800 × *g* for 15 min. Repeat.
7. Transfer the filter units to new collection tubes. Digest the protein by adding 200 μL of 0.25 mg/mL trypsin or chymotrypsin (in 50 mM ammonium bicarbonate, 1 mM $CaCl_2$) and mix briefly at low speed (400 rpm). Incubate the units in a wet chamber at 37 °C overnight (~18 h).
8. Centrifuge the filter units at 20,800 × *g* for 15 min to collect the digested peptides. Wash the filter by adding 200 μL of 50 mM ammonium bicarbonate and centrifuge the filter units at 20,800 × *g* for 15 min.
9. Lyophilize the filtrate in a vacuum centrifuge and store at −20 °C until analysis.

3.3 Assessment of Digestion Efficiency

1. Reconstitute samples in 100 μL of 1% formic acid immediately prior to analysis.
2. The peptide fractions (5.0 μL) can be chromatographically separated on a reverse-phase (RP) high-performance liquid chromatography (HPLC) system. In this example, we describe the use of a Shimadzu Nexera UHPLC system. The HPLC eluate can be directly coupled to a mass spectrometer. In this example, we describe the use of a QTRAP 6500 MS/MS (SCIEX, Redwood City, CA, USA).
3. The peptides were separated on a Phenomenex Kinetex C18 (1.7 μm, 100 Å, 150 mm × 2.1 mm) column at a flow rate of 400 μL/min. A linear gradient from 5% to 45% solvent B over 10 min was employed followed by 45–80% B over 1 min, a 1 min hold at 80% B, return to 5% B over 0.1 min, and a 3 min re-equilibration.
4. Relative quantitation was achieved using scheduled multiple reaction monitoring (MRM) scanning experiments (Table 1) using a 40 s detection window for each MRM transition and a 0.3 s cycle time (*see* **Note 14**). The ion spray voltage was set to 5500 V, the curtain gas was set to 35 psi, ion source gas 1 and 2 (GS1 and GS2) to 40 and 50 psi, and the heated interface was set to 500 °C (*see* **Note 15**). Spectra were acquired using the manufacturer's rolling collision energy (CE) based on the size and charge of the precursor ion for optimum peptide fragmentation (*see* **Note 16**).

Table 1
Gluten peptide markers used in the assessment of digestion efficiency. All transitions for a given peptide are summed to give the peak area

#	Peptide sequence	Uniprot accession	RT (min)	Q1 *m/z* (z)	Q3 *m/z* (fragment, z)	CE (V)
G1	ELQESSLEAC(cam)R	Barley: I6TRS8 Wheat: P10387 Rye: Q94IL1	3.15	661.296 (2+)	735.325 (y6, 1+) 822.357 (y7, 1+) 951.400 (y7, 1+)	33.1
G2	AQQLAAQLPAMC (cam)R	Barley: I6TRS8 Wheat: P08489 Rye: D3XQB7	4.90	729.361 (2+)	747.343 (y6, 1+) 946.439 (y8, 1+) 1017.476 (y9, 1+)	36.5
G2*	AQQLAAQLPAM(ox)C (cam)R	Barley: I6TRS8 Wheat: P08489 Rye: D3XQB7	4.00	737.361 (2+)	763.343 (y6, 1+) 962.439 (y8, 1+) 1033.476 (y9, 1+)	36.9

RT retention time (min), *Q1* precursor ion *m/z* with charge, z, *Q3* product ion *m/z* with fragment ion assignment and charge, z, *CE* collision energy (V)

5. Peaks were integrated using MultiQuant v3.0 (SCIEX) (*see* **Note 17**) wherein all three transitions were required to co-elute at the same retention time (RT, min) with a signal-to-noise (S/N) > 3 for detection and a S/N > 5 and intensity >1000 counts per second (cps) for quantitation. Figure 1 shows an example LC-MS chromatogram for the detection of the peptides in barley (a), wheat (b), and rye (c).

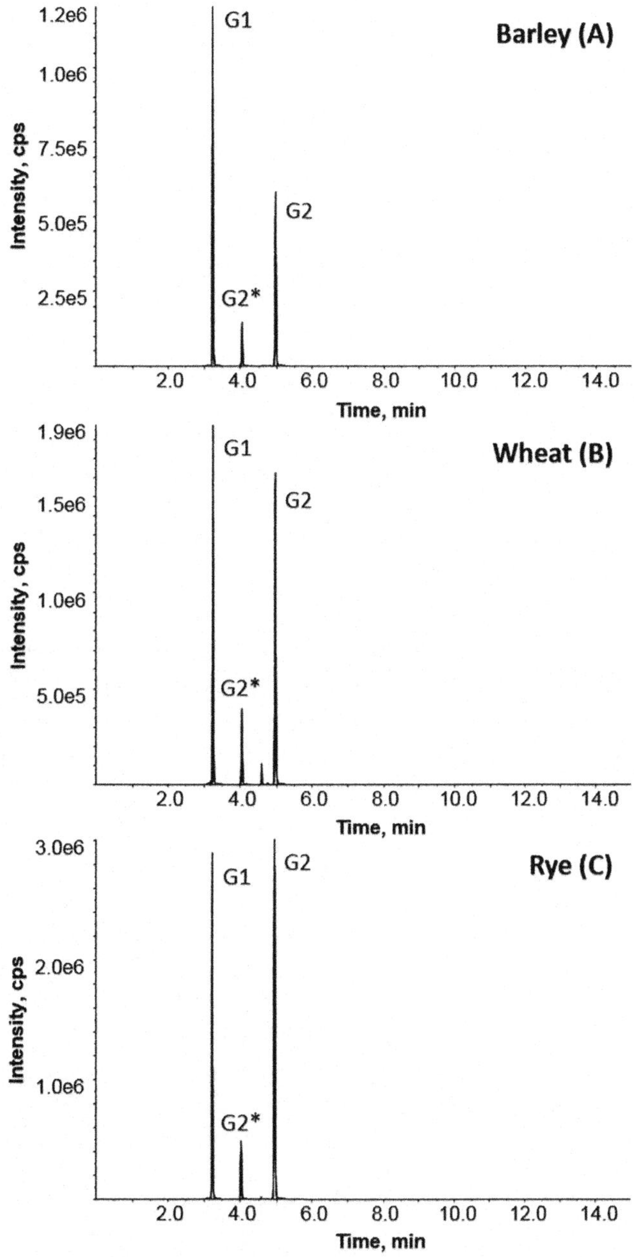

Fig. 1 LC-MRM-MS analysis of gluten-derived peptides commonly derived from barley (**a**), wheat (**b**), and rye (**c**). The peptides are ELQESSLEACR (G1), AQQLAAQLPAMCR (G2), and its oxidized form AQQLAAQLPAM(ox)CR (G2*)

6. The peak areas of the three MRM transitions monitored were summed and the data for the four replicates should be assessed by examining the mean, standard deviation (SD), and coefficient of variation (CV), wherein the CV should be $<10\%$.

4 Notes

1. Isopropanol (propan-2-ol, IPA) is highly flammable (keep away from naked flames) and is toxic by inhalation, ingestion, or in contact with skin. DTT and ammonium bicarbonate are harmful by ingestion—do not breathe dust. All chemicals used here should be used in well-ventilated areas and avoid contact by using personal protective equipment.
2. Always prepare fresh extraction buffer.
3. Urea is a powerful protein denaturant that disrupts noncovalent bonds in the proteins. Urea is used to increase the solubility of proteins. Urea is a hygroscopic material (absorbs water) and as such we recommend purchasing and storing it in small quantities to avoid large discrepancies in the concentration of prepared solutions.
4. A stock solution of 1 M Tris–HCl may be prepared and stored for 1 month. Prepare by weighing 78.8 g of Tris–HCl and dissolving in 450 mL of water. Adjust the pH of the solution to 8.5 using HCl and then make up to a final volume of 500 mL.
5. IAM is light sensitive and should be prepared immediately prior to use and stored wrapped in foil until use.
6. Ammonium bicarbonate may be prepared and stored for 1 month. Prepare by weighing 1.98 g and dissolving in 450 mL of water. Adjust the pH of the solution to 8.0 using HCl and then make up to a final volume of 500 mL. Both trypsin and chymotrypsin are maximally active over the pH range 7–9.
7. The recommended range for protein digestion by trypsin is 100:1 to 20:1 with incubation at 30–37 °C. A protein-to-enzyme ratio of 20:1 is employed for gluten digestion owing to its recalcitrant nature.
8. Chymotrypsin is both activated and stabilized by Ca^{2+}. The recommended range for protein digestion by trypsin is 200:1 to 20:1 with incubation at 30–37 °C. A protein-to-enzyme ratio of 20:1 is employed for gluten digestion owing to its recalcitrant nature.
9. Formic acid is a strong acid, is corrosive, and may cause burns by inhalation, ingestion, or contact. Use in a fume hood with

appropriate personal protective equipment. The HPLC buffers should be degassed if the HPLC to be used is not fitted with a degassing module. This can be achieved by sonication of the solutions in an ultrasonic bath for 10 min.

10. The mixing step is critical to gluten extraction. It is important to vortex mix the sample immediately after solvent addition to wet the flour and avoid clumping of flour.
11. The extraction step also serves to reduce the disulfide bonds between the cysteines (intra- and intermolecular linkages). It is preferable to use a thermal mixer for the extraction step at a low speed, e.g., 400 rpm.
12. The addition of IAM serves to irreversibly alkylate the cysteines preventing reoxidation.
13. It is important to wash away excess reagents. IAM can modify additional sites, for example, tyrosine. Urea interferes with protein digestion above concentrations of 2 M.
14. MRM methods comprise lists of precursor ion *m/z* values that represent the peptides (also known as the Q1 ions) and product ion *m/z* values that represent the peptide fragment ions (also known as the Q3 ions). Mass selection at two levels renders greater specificity to MRM methods.
15. The source conditions (gas and voltage settings) used are dependent on the instrument, the flow rate, and the solvents employed. The parameters defined here may act as a starting point but should be optimized using standards to achieve the best sensitivity.
16. The collision energy (CE) is increased in a linear manner as the *m/z* increases for each charge state. Doubly charged ions require a higher CE than triply charged ions.
17. A number of software packages that allow peak integration are available. MultiQuant™ is available from SCIEX as a licensed product; however, there are software packages available free of charge, such as Skyline as available on https://skyline.ms.

Acknowledgments

This work was supported by a fellowship to HL "International Training for High-level Talent in 2016" (YUWAIZHUAN [2016] No.8) from Foreign Experts Bureau of Henan Province.

References

1. Guandalini S, Assiri A (2014) Celiac disease: a review. JAMA Pediatr 168(3):272–278
2. Colgrave ML, Byrne K, Blundell M et al (2016) Comparing multiple reaction monitoring and sequential window acquisition of all

theoretical mass spectra for the relative quantification of barley gluten in selectively bred barley lines. Anal Chem 88(18):9127–9135

3. Stamnaes J, Sollid LM (2015) Celiac disease: autoimmunity in response to food antigen. Semin Immunol 27(5):343–352
4. Fasano A, Berti I, Gerarduzzi T et al (2003) Prevalence of celiac disease in at-risk and not-at-risk groups in the United States: a large multicenter study. Arch Intern Med 163 (3):286–292
5. Lionetti E, Gatti S, Pulvirenti A et al (2015) Celiac disease from a global perspective. Best Pract Res Clin Gastroenterol 29(3):365–379
6. Catassi C, Bai JC, Bonaz B et al (2013) Non-celiac gluten sensitivity: the new frontier of gluten related disorders. Nutrients 5 (10):3839–3853
7. Koerner TB, Abbott M, Godefroy SB et al (2013) Validation procedures for quantitative gluten ELISA methods: AOAC allergen community guidance and best practices. J AOAC Int 96(5):1033–1040
8. Thompson T, Mendez E (2008) Commercial assays to assess gluten content of gluten-free foods: why they are not created equal. J Am Diet Assoc 108(10):1682–1687
9. Colgrave ML, Byrne K, Blundell M et al (2016) Identification of barley-specific peptide markers that persist in processed foods and are capable of detecting barley contamination by LC-MS/MS. J Proteome 147:169–176
10. Colgrave ML, Goswami H, Byrne K et al (2015) Proteomic profiling of 16 cereal grains and the application of targeted proteomics to detect wheat contamination. J Proteome Res 14(6):2659–2668
11. Fiedler KL, McGrath SC, Callahan JH et al (2014) Characterization of grain-specific peptide markers for the detection of gluten by mass spectrometry. J Agric Food Chem 62 (25):5835–5844
12. Gomaa A, Boye J (2015) Simultaneous detection of multi-allergens in an incurred food matrix using ELISA, multiplex flow cytometry and liquid chromatography mass spectrometry (LC-MS). Food Chem 175:585–592
13. Sealey-Voyksner JA, Khosla C, Voyksner RD et al (2010) Novel aspects of quantitation of immunogenic wheat gluten peptides by liquid chromatography-mass spectrometry/mass spectrometry. J Chromatogr A 1217 (25):4167–4183

Chapter 23

Glycosylation Profiling of Tumor Marker in Plasma Using Bead-Based Immunoassay

Hongye Wang, Zheng Cao, Hu Duan, and Xiaobo Yu

Abstract

As one of the most important posttranslational modifications, glycosylation plays critical roles in protein folding, trafficking, cell differentiation, immune recognition, etc. The alteration of glycosylation is closely associated with the pathological processes during and after caner development, and thus holds great value in cancer detection. In this chapter, we describe a protocol on the glycosylation profiling of tumor marker in plasma using bead-based immunoassay with CA125 as a model, including bead coupling, coupling control, glycosylation assay, as well as the plasma screening for breast cancer patients. This protocol can be used to profile the glycosylation of protein markers in clinical plasma or serum samples for different human cancers.

Key words Posttranslational modification, Glycosylation, Tumor marker, Plasma, Lectin, Bead-based immunoassay

1 Introduction

As a major public health issue in the world, cancer has remained as the leading cause of death in China [1]. The early detection is probably the optimal approach to increase the opportunity for the treatment and control of this disease [2–4]. However, the sensitivity and specificity of most cancer biomarkers are still limited and it is urgent to identify new biomarkers that can improve the cancer detection [5, 6]. Glycosylation is one of the most important posttranslational modifications that plays critical roles in protein folding, trafficking, cell differentiation, immune recognition, etc. [7]. The alteration of glycosylation is closely associated with the pathological processes during and after the cancer development. Many tumor protein markers have been found that contain altered glycosylation expression, such as CA19-9, CA15-3, and AFP-L3 [8, 9]. Thus, the method that can detect and quantify glycosylation

Hongye Wang and Zheng Cao contributed equally to this work.

Xing Wang and Matthew Kuruc (eds.), *Functional Proteomics: Methods and Protocols*, Methods in Molecular Biology, vol. 1871, https://doi.org/10.1007/978-1-4939-8814-3_23, © Springer Science+Business Media, LLC, part of Springer Nature 2019

on tumor markers would hold great value in improving cancer diagnosis and prognosis [6, 10, 11].

Lectins preferably bind to a specific terminal mono- or polysaccharide structure and have been widely used as affinity reagent for the capture and detection of glycosylated proteins [12]. Chen et al. developed a multiplexed assay that enables the detection of protein expression and glycosylation in high-throughput on antibody microarrays using protein detection antibody and lectins, respectively. The results indicated that the changes of protein glycosylation on MUC1 and CEA are associated to pancreatic cancer [13]. Using similar strategy, Li et al. coupled the antibody to magnetic beads and detected the expression and glycosylation of serological TIMP-1 (tissue inhibitor of metallopeptidase 1) and DPP-4 (membrane metallo-endopeptidase and dipeptidyl peptidase-IV) using biotinylated antibody and lectins, respectively. The system provides a useful tool for the validation of glycoproteins as biomarkers for cancers [14].

Our research is focused on the glycosylation of tumor markers in the improvement of cancer detection, prognosis, as well as therapeutic treatment [6, 15]. Using CA125 as a model, here we describe a protocol on the glycosylation profiling of tumor marker in plasma using bead-based immunoassay, including beads coupling, coupling control, glycosylation assay, as well as the plasma screening for breast cancer patients (Fig. 1). This assay described in this chapter can be used to profile the glycosylation of protein markers in clinical plasma or serum samples for different human cancers.

2 Materials

2.1 Reagents

1. Beads: Luminex MegPlex beads (Luminex Inc., Austin, TX, USA).
2. Activation buffer: 0.1 M NaH_2PO_4 (Sigma-Aldrich, St. Louis, MO, USA), pH 6.2.
3. EDC solution: 50 mg/mL EDC (1-ethyl-3-[3-dimethylaminopropyl] carbodiimide hydrochloride, Thermo Fisher Scientific, IL, USA).
4. Sulfo-NHS solution: 50 mg/mL sulfo-NHS (Thermo Fisher Scientific, IL, USA).
5. Coupling buffer: 50 mM MES, pH 5.0 (Sigma-Aldrich, St. Louis, MO, USA).
6. Assay/Washing buffer: 0.05% (v/v) Tween 20 in 0.01 M PBS.
7. Blocking buffer: PBS-TBN, 1% BSA in washing buffer.
8. Lectin: Biotinylated Aleuria Aurantia Lectin (AAL), Biotinylated Phaseolus vulgaris Erythroagglutinin (PHA-E),

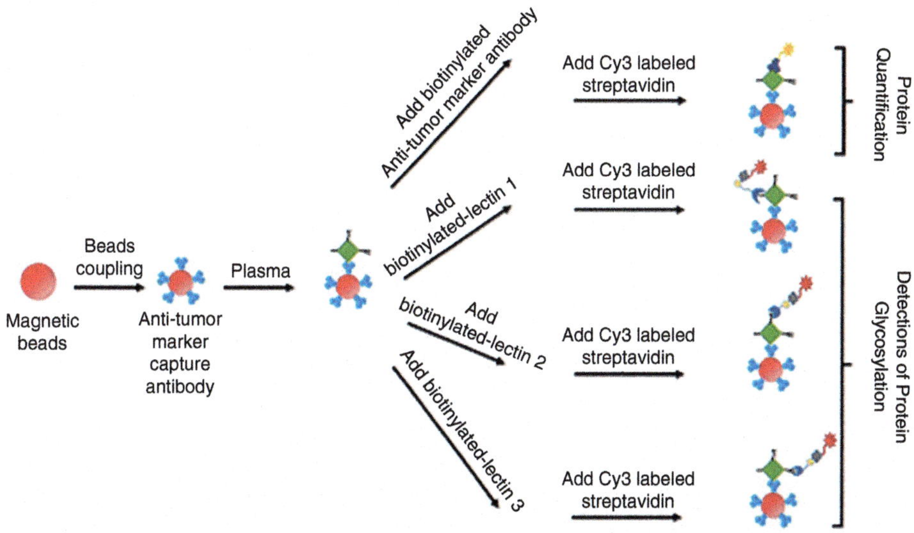

Fig. 1 Schematic illustration of bead-based immunoassay on glycosylation profiling of tumor marker in plasma. The antitumor marker antibody is coupled to the magnetic beads modified with carboxyl groups through EDC/NHS chemistry, which is then incubated with the plasma or serum from cancer patients. The tumor proteins captured to the beads can be detected by either biotinylated anti-protein antibody for protein quantification or biotinylated lectins for glycosylation profiling, respectively. At last, the fluorescent signal can be detected after incubation with Cy3-labeled streptavidin through Luminex instrument or flow cytometry

Biotinylated Vicia villosa Lectin (VVL, VVA) (Vector Laboratories Inc., Burlingame, CA, USA).

9. Cy3-Streptavidin (Jackson ImmunoResearch Inc., West Grove, PA, USA).
10. Recombinant Human CA125/MUC16 Protein (R&D systems Inc., Minneapolis, MN, USA).
11. Anti-CA125/MUC16 mAb (Abcam Inc., Cambridge, MA, USA).
12. Goat anti-Mouse IgG-Alexa555 (Thermo Fisher Scientific, IL, USA).

3 Methods

3.1 Beads Coupling

1. Resuspend the stock beads by vortex for 30 s and followed with sonication for 20 s (*see* **Note 1**).
2. Take 100 μL of 1.25×10^7 beads into a microcentrifuge tube. Put the tube into a magnetic separator and let it stay for 1 min (*see* **Note 2**).
3. Remove the supernatant and resuspend the beads in 100 μL dH_2O by vortex for 30 s.

4. Keep the tube in a magnetic separator for 1 min and remove the supernatant without disturbing the beads.
5. Resuspend the beads in 80 μL activation buffer by vortex for 30 s.
6. Prepare 50 mg/mL Sulfo-NHS and EDC diluted in dH_2O.
7. Add 10 μL 50 mg/mL Sulfo-NHS to the beads and mix gently by vortex.
8. Add 10 μL 50 mg/mL EDC to the beads and mix gently by vortex.
9. Incubate for 20 min at room temperature with gentle mixing.
10. Put the tube into a magnetic separator for 1 min and remove the supernatant.
11. Wash the beads with 250 μL coupling buffer twice.
12. Resuspend the beads in 100 μL washing buffer and add 5 μg anti-CA125 antibody. Incubate for 2 h at room temperature with gentle mixing (*see* **Note 3**).
13. Wash the coupled beads with 500 μL blocking buffer, then incubate for 30 min at room temperature.
14. Wash the blocked beads with 500 μL washing buffer and resuspend the beads in 1000 μL washing buffer.
15. Count the number of beads using BD FACSVerse (*see* **Note 4**).

3.2 Quality Control of Anti-CA125 Antibody-Coupled Magnetic Beads

1. Dilute the mouse anti-CA125 antibody-coupled beads to a final concentration of 50 beads/μL with assay buffer and add 50 μL to a 96-well plate (pre-blocked with PBS-TBN buffer) in duplicate.
2. Incubate beads with Alexa555 conjugated goat anti-Mouse IgG with 0.125, 0.25, 0.5, 1, 2, 4 μg/mL for 30 min at room temperature in a 96-well plate (pre-blocked with PBS-TBN buffer), with assay buffer as a blank control (*see* **Note 5**).
3. Put the 96-well plate on a magnetic separator for 3 min and remove the supernatant.
4. Wash the beads with 150 μL washing buffer twice and resuspend the beads in 100 μL washing buffer.
5. Submit the resulting beads to a Luminex200 instrument for the fluorescent measurement. The representative result of coupling control is shown in Fig. 2.

3.3 Detection of CA125's Expression in Plasma Using Sandwich Immunoassay

1. Dilute the coupled beads to a final concentration of 50 beads/μ L with assay buffer and add 50 μL to a 96-well plate (pre-blocked with PBS-TBN buffer) with duplicate.
2. Incubate beads with a series of concentrations of CA125 protein in matrix for 2 h at room temperature in 96-well plate (pre-blocked with PBS-TBN buffer) in duplicate.

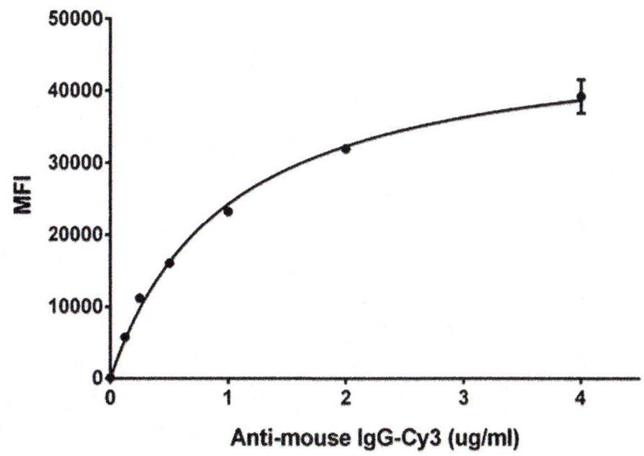

Fig. 2 Quality control of antibody-coupled magnetic beads. MFI is the median fluorescence intensity of 100 counted magnetic beads

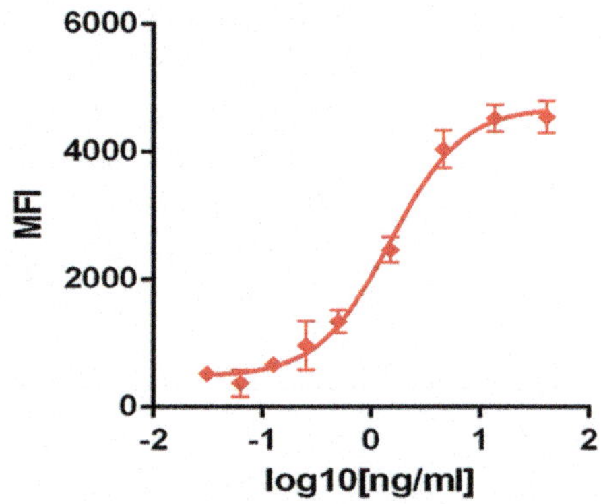

Fig. 3 Dose-response curve for CA125 detection using bead-based sandwich immunoassay. MFI is the median fluorescence intensity of 100 counted magnetic beads

3. Put the 96-well plate on a magnetic separator for 3 min and remove the supernatant.
4. Wash the beads with 150 μL washing buffer three times.
5. Add 100 μL CA125 detection antibody (4 μg/mL) and incubate for 1 h at room temperature.
6. Wash the beads with 150 μL washing buffer three times.
7. Add 100 μL goat anti-mouse IgG-Alexa555 (4 μg/mL) and incubate for 1 h at room temperature.
8. Wash the beads with 150 μL washing buffer twice and resuspend the beads in 100 μL washing buffer.
9. Submit the resulting beads to a Luminex200 instrument for the fluorescent measurement. The detection of different concentrations of CA125 using sandwich immunoassay is shown in Fig. 3.

3.4 Glycosylation Profiling of CA125 in Plasma for Breast Cancer Patients

1. The plasma samples are thawed at 4 °C and centrifuged for 10 min at 14,000 × *g* (*see* **Note 6**).
2. Dilute the coupled beads to a final concentration of 50 beads/μL with assay buffer and add 50 μL to a 96-well plate (pre-blocked with PBS-TBN buffer) in duplicate.
3. Add 50 μL plasma sample and mix with Eppendorf MixMate at 390 × *g*. Incubate for 2 h at room temperature.
4. Wash the beads with 150 μL washing buffer three times.
5. Add 100 μL biotinylated-lectin(2 μg/ml) and incubate for 1 h at room temperature.
6. Wash the beads with 150 μL washing buffer three times.
7. Add 100 μL Cy3-Streptavidin (4 μg/mL) to each well and incubate at room temperature. For 1 h on an Eppendorf MixMate at 390 × *g*.
8. Wash the beads with 150 μL washing buffer twice and resuspend the beads in 150 μL washing buffer.
9. Submit the resulting beads to a Luminex200 instrument for the fluorescent measurement (*see* **Note 7**). The representative results of plasma screening are shown in Fig. 4, in which differential profiling of CA125 glycosylation can be observed with the lectins of AAL, PHA-E, and VVA.

4 Notes

1. This step is essential to prevent beads aggregation.
2. Minimize exposure time, as prolonged exposure to light might result in fluorescent quenching. During incubation, the 96-well plate should be covered with aluminum foil.
3. Do not make use of solution containing Tris or other amine-based buffers, which could decrease coupling efficiency.
4. The number of coupled beads can also be counted by cell counter or hemocytometer.
5. During incubation, 96-well plate should be put on the shaker to reduce the precipitation and aggregation of the beads.
6. We highly recommend that the plasma or serum samples are thawed at 4 °C, and followed with centrifugation for 10 min at 14,000 × *g* to remove the precipitation in samples.
7. This method is suitable for the detection of different glycosylation on tumor protein markers. The color-coded beads enable researchers to screen up to 100 parameters in a single experiment.

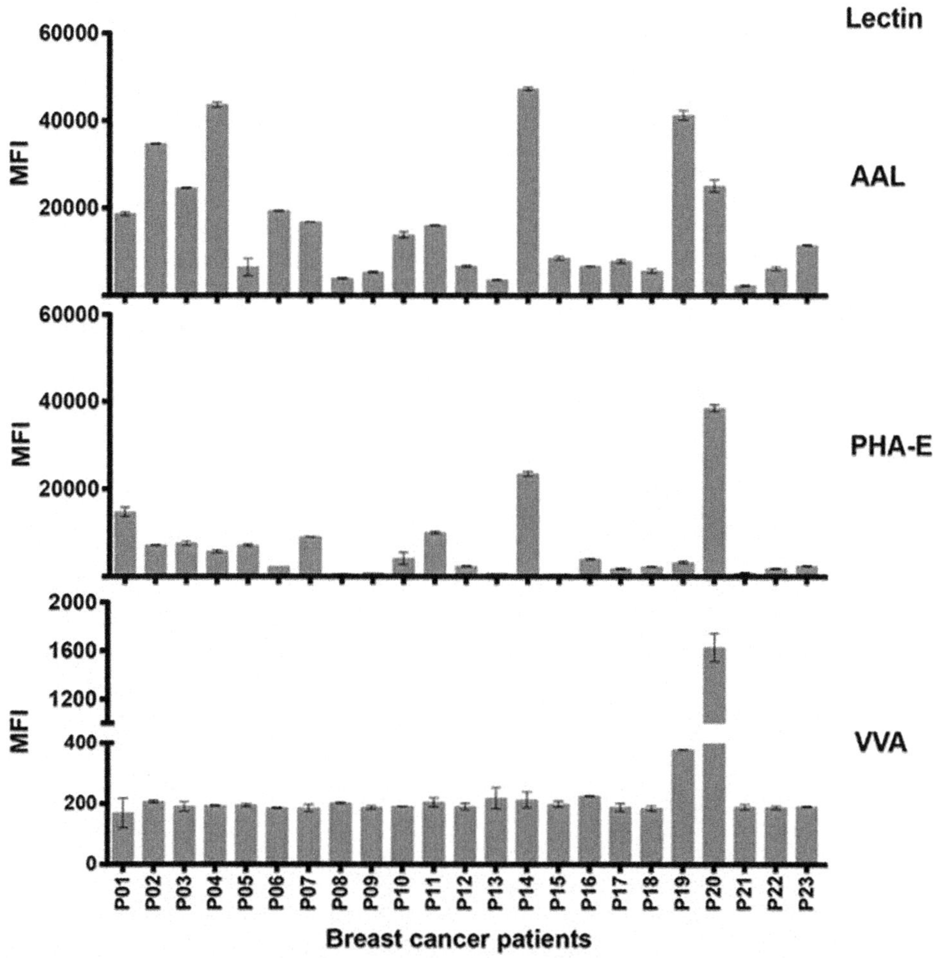

Fig. 4 Glycosylated profiling of CA125 in the plasma of breast cancer patients using biotinylated lectins. All plasma samples were collected with the written informed consent under the approval of intuitional review board (IRB) from Beijing Proteome Research Center

Acknowledgments

This project is financially supported from the National Natural Science Foundation of China (81673040), the State Key Laboratory of Proteomics (SKLP-O201504 and SKLP-K201505) to X.Y.

References

1. Chen W et al (2016) Cancer statistics in China, 2015. CA Cancer J Clin 66:115–132
2. Levin B et al (2008) Screening and surveillance for the early detection of colorectal cancer and adenomatous polyps, 2008: a joint guideline from the American Cancer Society, the US multi-society task force on colorectal cancer, and the American College of Radiology. CA Cancer J Clin 58:130–160
3. Cuzick J et al (2014) Prevention and early detection of prostate cancer. Lancet Oncol 15:e484–e492

4. Wang D, Yang L, Zhang P, et al (2017) AAgAtlas 1.0: a human autoantigen database. Nucleic acids research 45 (D1): D769–D776
5. Hanash SM, Baik CS, Kallioniemi O (2011) Emerging molecular biomarkers—blood-based strategies to detect and monitor cancer. Nat Rev Clin Oncol 8:142–150
6. Yu X, Schneiderhan-Marra N, Joos TO (2010) Protein microarrays for personalized medicine. Clin Chem 56:376–387
7. Wang JR et al (2017) A method to identify trace sulfated IgG N-glycans as biomarkers for rheumatoid arthritis. Nat Commun 8:631
8. Li D, Mallory T, Satomura S (2001) AFP-L3: a new generation of tumor marker for hepatocellular carcinoma. Clin Chim Acta 313:15–19
9. Kirwan A, Utratna M, O'Dwyer ME, Joshi L, Kilcoyne M (2015) Glycosylation-based serum biomarkers for cancer diagnostics and prognostics. Biomed Res Int 2015:490531
10. Hanash SM, Pitteri SJ, Faca VM (2008) Mining the plasma proteome for cancer biomarkers. Nature 452:571–579
11. Wu L, Qu X (2015) Cancer biomarker detection: recent achievements and challenges. Chem Soc Rev 44:2963–2997
12. Syed P et al (2016) Role of lectin microarrays in cancer diagnosis. Proteomics 16:1257–1265
13. Chen S et al (2007) Multiplexed analysis of glycan variation on native proteins captured by antibody microarrays. Nat Methods 4:437–444
14. Li D, Chiu H, Chen J, Zhang H, Chan DW (2013) Integrated analyses of proteins and their glycans in a magnetic bead-based multiplex assay format. Clin Chem 59:315–324
15. Yu X, Petritis B, Duan H, Xu D, LaBaer J (2018) Advances in cell-free protein array methods. Expert Rev Proteomics 15:1–11

Chapter 24

Protein-Specific Analysis of Invertebrate Glycoproteins

Alba Hykollari, Daniel Malzl, Iain B. H. Wilson, and Katharina Paschinger

Abstract

N-Glycans are posttranslational modifications of proteins attached to the amide side chains of asparagine residues, with possible heterogeneity due to different structures being possible at the same glycosylation site. In contrast to the mammalian systems, invertebrate N-glycosylation presents a challenge in analysis as there exist unfamiliar epitopes and a high degree of structural and isomeric variation between different species. A simple analytical approach to analyze N-glycans on specific glycoproteins is presented, which involves a combination of tryptic peptide mass spectrometry and "off-line" RP-HPLC MALDI-TOF MS/MS complemented by blotting to recognize specific epitopes. An additional N-glycan enrichment and labeling step can facilitate the analysis of single structures and even provide isomeric separation of N-glycans from specific proteins.

Key words Glycosylation, Glycoproteomics, Mass spectrometry, "off-line" MALDI-TOF MS(MS)

Abbreviations

DTT	Dithiothreitol
HRP	Horseradish peroxidase
MALDI-TOF	MS matrix-assisted laser-desorption/ionization time-of-flight mass spectrometry
NPGC	Nonporous graphitized carbon
PC	Phosphorylcholine
PE	Phosphoethanolamine
RP-HPLC	Reversed phase high pressure liquid chromatography
SDS-PAGE	Sodium dodecyl sulfate polyacrylamide gel electrophoresis

Full data on the N-glycans of honeybee royal jelly will appear in a forthcoming paper: Hykollari et al. 2018. Isomeric separation and recognition of anionic and zwitterionic N-glycans from royal jelly glycoproteins.

Xing Wang and Matthew Kuruc (eds.), *Functional Proteomics: Methods and Protocols*, Methods in Molecular Biology, vol. 1871, https://doi.org/10.1007/978-1-4939-8814-3_24,

1 Introduction

Among the posttranslational modifications, the various forms of glycosylation represent a big challenge in terms of analysis. Most commonly, glycans are N- or O-linked to amino- or hydroxyl-side chains of proteins, although C- and S-linkages are also known [1]. Probably N-glycans are the most studied forms and are known from bacteria, archaea, and almost all eukaryotes; in the latter case, asparagine residues are modified with an oligosaccharide via a core *N*-acetylglucosamine residue [2]. Certainly, N-glycans from mammals are quite well studied, whereas for invertebrate organisms the N-glycan structures and their functions remain rather unknown. In contrast to their simple body shape or size, recent studies on N-glycosylation have proven that invertebrate organisms synthesize complicated N-glycomes, which compete in terms of complexity with those of vertebrates [3–6].

Due to the high variety of N-glycan core and antennal modifications in invertebrates, their analysis is challenging and more time consuming, as most of the available bioinformatics tools are based on mammalian structures and so have limited utility when considering invertebrate glycopeptide and N-glycan data. On the other hand, invertebrate glycoproteins are biomedically relevant either due to their immunogenicity (e.g., in venoms), immunomodulatory activity and relevance as vaccine targets (e.g., parasite glycoproteins) or in the use of invertebrate cell lines (e.g., for baculovirus-based systems) for production of recombinant biopharmaceuticals [7–9]. Thus, there is a need to adequately determine invertebrate N-glycan structures at the glycomic and glycoprotein levels. Clearly, glycan annotations on the basis of *m/z* alone are insufficient and are often misleading. Orthogonal proofs are therefore necessary, including the use of specific detection reagents, MS/MS fragmentation, chemical or exoglycosidase treatments or reference to in-depth glycomic analyses from the same organism [10, 11].

The analysis of N-glycans is often required from either purified single proteins or low amounts of biological material. Here we describe procedures for protein-specific N-glycan purification, enrichment, and analysis, successfully used with invertebrate glycoproteins. The protein is first screened for N-glycan epitopes with Western blot, whereby the affinity of specific antibodies, lectins and pentraxins, can give first impressions as to the modifications of the oligosaccharides attached to proteins (e.g., core fucose, terminal galactose, phosphorylcholine, phosphoethanolamine). The tryptic digest of the protein and subsequent peptide mass spectrometry fingerprinting help to identify the protein in silico and provide (glyco)peptides for further analysis. Peptide:N-glycosidases can then be used to cleave the N-glycans from the peptides prior to mass spectrometric analyses; as amounts allow, these can be fluorescently labeled and subject to HPLC and MS/MS analyses. Thus, we can go beyond the typical glycoproteomic procedures in order to more firmly define N-glycan structures on specific glycoproteins.

2 Materials

2.1 Equipment

1. Probe sonifier, e.g., Branson Sonifier 250.
2. Porcelain mortar with pestle; tight fitting glass homogenizer (Wheaton; customize as required).
3. Vacuum centrifuge (e.g., Speedvac, Thermo).
4. Microcentrifuge (e.g., Heraeus, Thermo).
5. Lyophilizer (Labconco).
6. Mini Protean® Tetra cell and Power Pac power supply (Bio-Rad).
7. Trans blot SD semi dry electrophoretic transfer cell (Bio-Rad).
8. Glass columns of 1 cm diameter and 50 cm length (Bio-Rad).
9. Multifunctional microtiter plate reader (e.g., Infinite M200, Tecan); black 96-well microtiter plates, e.g., Microfluor™ 1 or LumiNunc (Thermo).
10. HPLC liquid chromatograph with fluorescence detector (e.g., Shimadzu Nexera); reverse phase chromatography column, e.g., Ascentis® Express RP-Amide (150 mm × 46 mm, 2.7 μm, Supelco).
11. MALDI-TOF-TOF-MS: Autoflex Speed or UltrafleXtreme MALDI-TOF-TOF; appropriate MALDI polished or ground steel target plate (Bruker Daltonics, Billerica, MA). Alternatives are available commercially from Shimadzu or Applied Biosystems.

2.2 Reagents, Buffers, and Columns (See Notes 1 and 2)

2.2.1 Disruption of Biological Material and SDS-PAGE Sample Preparation

1. 2× SDS-PAGE reducing sample buffer containing 200 mg SDS, 154 mg DTT, 5 mL stacking gel buffer, 3.6 mL 87% glycerol (make up to 10 mL with water, then add a few crystals of bromophenol blue).

2.2.2 SDS-PAGE and Western Blotting

1. 12% SDS-PAGE gel (using 40% acrylamide stock from Bio-Rad, diluted with either stacking gel buffer with 0.5 M Tris/HCl pH 6.8 or separation gel buffer with 1.5 M Tris/HCl pH 8.8).
2. SDS-PAGE running buffer (25 mM Tris, 192 mM glycine, 0.1% SDS; components from either VWR or Roth).

3. Western blotting transfer buffer (25 mM Tris, 192 mM glycine, 10% methanol; VWR or Roth).
4. SDS-PAGE protein standard ladder (e.g., Thermo PageRuler™).
5. Nitrocellulose membrane (BioTrace™ NT from Pall Life Science).
6. Extra thick blotting paper (Bio-Rad).
7. 0.5% (w/v) Ponceau S (Sigma) in 1% (v/v) acetic acid solution.
8. Membrane washing buffer: Tris buffered saline (TBS, i.e., 0.1 M Tris/HCl, pH 7.4, 0.1 M NaCl; typically made as a tenfold concentrated stock) with 0.05% Tween (Sigma).
9. Membrane blocking and antibody/lectin dilution buffer: Tris buffered saline with 0.05% Tween and 0.5% BSA (Roth).
10. Primary, secondary antibodies, lectins and pentraxins (Sigma or Vector Laboratories) (*see* Table 1).
11. SigmaFAST BCIP/NBT or SigmaFAST 3,3′-diaminobenzidine tetrahydrochloride tablets (Sigma), dissolved in 10 mL and 5 mL, respectively.

2.2.3 Tryptic Peptide Mapping

1. Colloidal Coomassie Blue staining solution: 0.02% (w/v) Coomassie Brilliant Blue G-250 (Bio-Rad), 5% aluminum sulfate-(14-18)-hydrate [e.g., $Al_2(SO_4)_3.16H_2O$ (Roth)], ethanol 96% (VWR), phosphoric acid 85% (Roth). Weigh in 100 g of aluminum sulfate and dissolve it in 1500 mL of water; add 200 mL ethanol and mix well; add 0.4 g of Coomassie Brilliant Blue G-250 and mix well for at least 30 min; add slowly 47 mL of phosphoric acid and mix well; make up to 2000 mL with water (*see* **Note 3**).
2. Acetonitrile LC-MS grade (VWR), ammonium bicarbonate (Roth), water HPLC super gradient grade (VWR), iodoacetamide (Sigma), dithiothreitol (Roth), trifluoroacetic acid (Fluka).
3. Sequencing grade modified trypsin dissolved to 0.1 mg/mL in 50 mM acetic acid (Promega); typically cleaves after Arg and Lys residues.

2.2.4 N-Glycome Release and Analysis

1. Peptide:N-glycosidase F (PNGase F, recombinant from *Flavobacterium meningosepticum*; Sigma).
2. Peptide:N-glycosidase A (either purified native from almond meal, PNGase A from Sigma, or recombinant Endo H-treated from *Oryza sativa* and expressed in *Pichia pastoris*, PNGase Ar from NEB).
3. For PNGase F: 100 mM McIlvaine phosphate/citrate buffer (pH 7.5) or 50 mM ammonium hydrogen carbonate (pH 8;

Table 1
List of selected antibodies, lectins, and pentraxins for N-glycan epitope screening (*see* Note 4)

	Dilution	Epitope [13, 14]	Supplier
Antibody (1st)			
Anti-HRP from rabbit, 10 mg/mL	1:10000	Core α1,3-Fuc/ core β1,2-Xyl	Sigma
Anti-PC (TEPC-15 mouse IgA), 10 mg/mL	1:200	PC-Hex(NAc)	Sigma
Antibody (2nd)			
Anti-rabbit IgG from goat conjugated with alkaline phosphatase	1:2000		Vector Labs
Anti-mouse IgA from goat conjugated with alkaline phosphatase	1:1000		Sigma
Pentraxin			
C-reactive Protein (CRP) from human plasma ($CaCl_2$ 2.5 mM added)	1:200	PC-Hex(NAc)	MP Biochemicals
Amyloid P component from human serum (SAP)	1:200	PE-Hex(NAc)	Sigma
Pentraxin recognition			
Anti-human C-reactive protein from rabbit	1:1000		Dako
Anti-Amyloid P human serum component IgG from rabbit (anti SAP)	1:1000		Calbiochem
Lectin			
Biotinylated *Aleuria aurantia* lectin	1:1000	Core α1,6-Fuc/ Le^x	Vector Labs
Biotinylated wheat germ agglutinin	1:1000	β1,4HexNAc/ α2,3Sia	Vector Labs
Biotinylated peanut agglutinin	1:1000	Galβ1,3GalNAc	Vector Labs
Lectin recognition			
Anti-biotin from goat conjugated with alkaline phosphatase	1:10000		Sigma

mixture of ammonium carbonate and ammonium hydrogen carbonate).

4. For PNGase A: 50 mM ammonium acetate (pH 5).
5. 1–3 mL solid-phase extraction column and frits (Supelco)
6. Acetonitrile, isopropanol, acetic acid, water (as above).
7. Dowex AG® 50 W-X8 200–400 mesh H^+ form (Bio-Rad, biotechnology grade; washed serially with 0.1 M NaOH, water, 0.1 M HCl, water, 1 M ammonium acetate and water)

and pre-equilibrated with 2% acetic acid prior to usage; C18 material (Lichroprep, Merck); nonporous graphitized carbon material (NPGC; e.g., Supelco ENVICarb™).

8. MALDI matrices: α-cyanocinnamic acid (ACH, Sigma; 10 mg/mL in 0.1% trifluoroacetic acid/50% acetonitrile) or 6-aza-thiothymine (ATT, Sigma; 3 mg/mL ATT dissolved in 50% ethanol).
9. Glycan labeling: 2-aminopyridine (PA, >99% purity, Sigma), sodium cyanoborohydride (95% purity, Sigma), hydrochloric acid (37% HCl, Roth).
10. Gel filtration: Sephadex G-15 and G-25 coarse (GE Healthcare).

2.2.5 Proteome and N-Glycan Data Analysis

Database Search Programs

1. Matrix Science web server (www.matrixscience.com/cgi).
2. ProteinProspector (prospector.ucsf.edu/prospector/cgi-bin).

Peptide/Protein Utility Program

1. Theoretical peptide mass calculator (www.expasy.org).

N-Glycan Analysis

1. Glycoworkbench (www.glycoworkbench.org).
2. FlexAnalysis Bruker software.

3 Methods (*See* Flow Chart and Example Data in Figs. 1 and 2)

3.1 Sample Preparation and Glycoepitope Recognition

The purification procedure of the (glyco)protein of interest depends on the biological material which can be whole organisms, cells, tissues, cyst fluid, semi-purified proteins or secreted (glyco) proteins in culture media or buffer.

3.1.1 Sample Preparation for Glycoprotein Analysis (See Notes 1 and 2)

1. Heat inactivate the biological material in boiling water for 10 min. After cooling the sample (cells, worms, royal jelly, etc.), homogenize the material using a sonifier or mortar and pestle or tight fitting glass homogenizer; for tissues or fungal mycelium, lyophilize after heat inactivation and ground to a fine powder in liquid nitrogen. Large volumes of proteinaceous liquid samples are concentrated with an additional precipitation step (e.g., methanol). Alternatively, after homogenization, purify a subset of glycoproteins by an additional enrichment step using affinity chromatography (if available), e.g., agarose immobilized monoclonal antibodies [12].
2. Prior to SDS-PAGE, precipitate an aliquot of the samples with a fivefold excess volume of methanol, incubate at −80 °C for 1 h and centrifuge for 10 min at 4 °C, 21,000 × *g*. Dry the protein pellet at 65 °C for several minutes to evaporate excessive

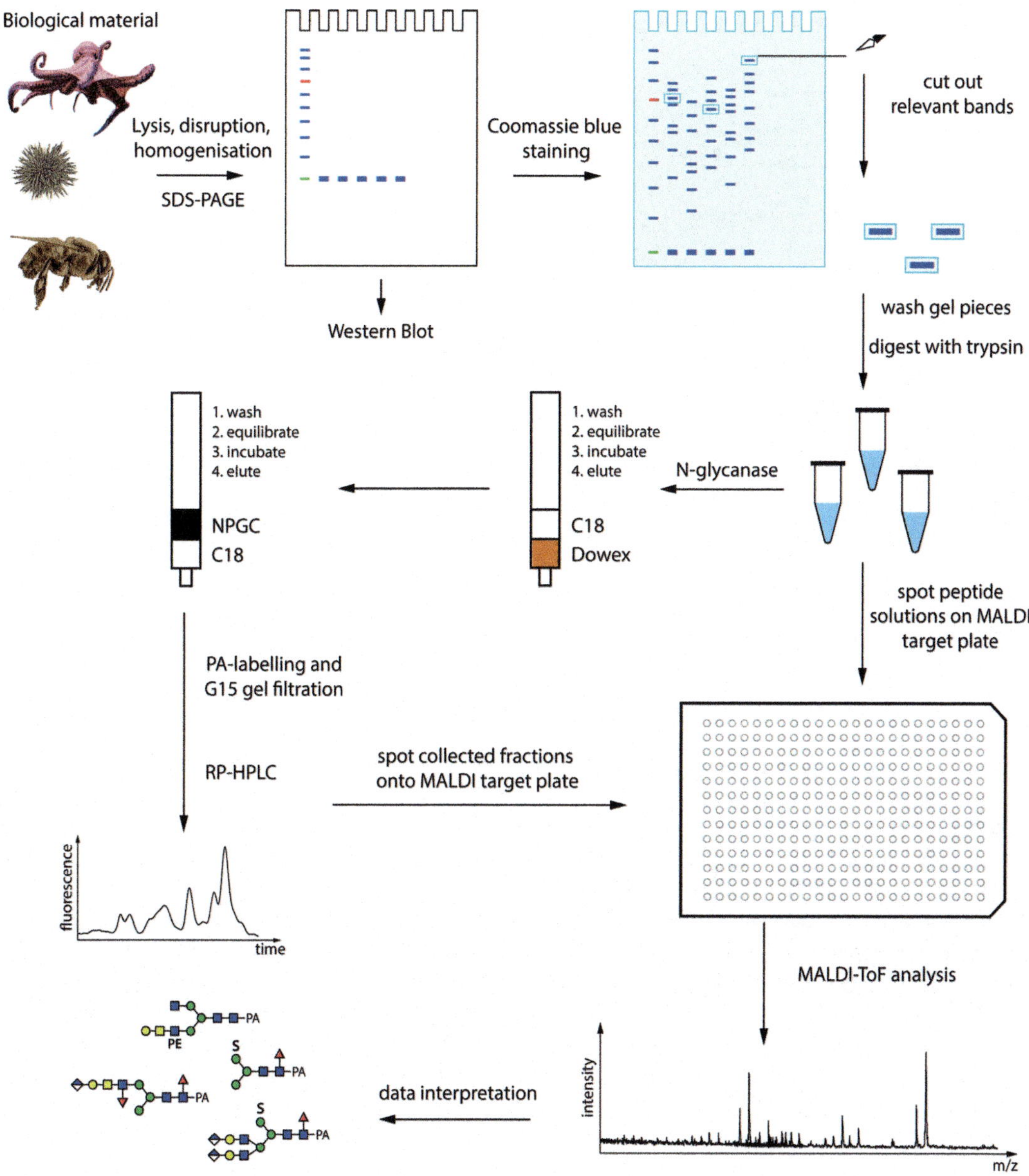

Fig. 1 A potential glycome and glycoproteomic workflow. Starting from biological material, proteins can be separated by SDS-PAGE prior to Western blotting or peptide map fingerprinting. The peptides and glycopeptides are analyzed directly by mass spectrometry; the glycans are released by an N-glycanase such as PNGase Ar and purified by two rounds of solid-phase extraction prior to mass spectrometry and/or HPLC. Glycans (examples from honeybee royal jelly) are depicted according to the Symbol Nomenclature for Glycans, whereby circles, squares, triangles and diamonds respectively represent hexose (here Man or Gal), *N*-acetylhexosamine (GalNAc or GlcNAc), deoxyhexose (Fuc) or hexuronic acids (GlcA); S, sulfate; PE, phosphoethanolamine

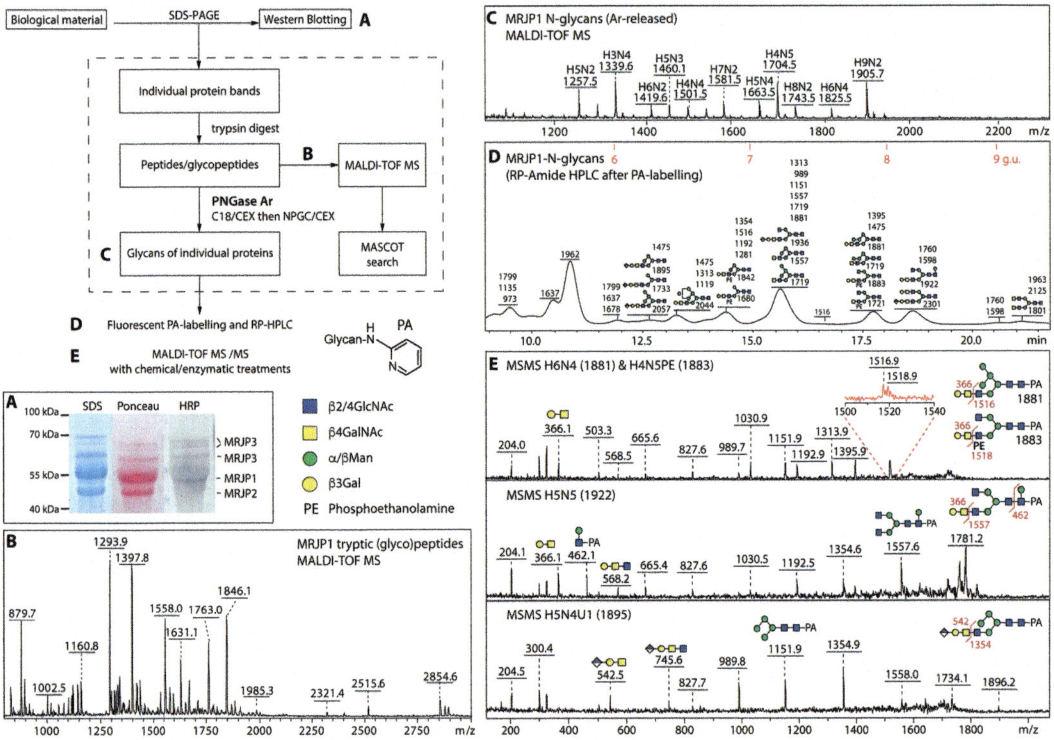

Fig. 2 Example of a glycoproteomic study of honeybee glycoprotein MRJP1 found in royal jelly. The flow chart scheme shows the glycoproteomic workflow for the biological sample; the letters refer to steps exemplified by the following data. (**a**) N-Glycan epitope detection by Western blotting of the royal jelly glycoproteins after incubation with anti-HRP antibodies. (**b**) Tryptic peptide mapping of one of the major royal jelly glycoproteins as measured with MALDI TOF MS in positive ion mode. (**c**) Free N-glycans from MRJP1 measured with MALDI-TOF MS after deglycosylation with PNGase Ar with the $[M + Na]^+$ ions annotated with abbreviations of the form HxNy, where H is hexose and N is *N*-acetylhexosamine. (**d**) RP-HPLC chromatogram of reductively aminated N-glycans from royal jelly glycoprotein MRJP1 with fractions annotated with the detected glycan *m/z* values and calibrated in terms of glucose units. (**e**) MALDI-TOF MS/MS data of fractionated MRJP1 glycans, whereby $[M + H]^+$ ions were fragmented and key B and Y ions are annotated. Note that the precursor ions for the H6N4 and H4N5PE structures cannot be separated ($\Delta m/z = 2$), but the zoom shows that Y ions derived from both are present; the structure of the PA (pyridylamino) reducing-terminal label as well as the symbol nomenclature are shown beneath the flowchart

methanol and redissolve the pellet in 20 μL SDS-PAGE sample buffer. In addition, heat treat the mixture for 10 min at 95 °C, and after cooling, centrifuge again for 5 min at room temperature, 21,000 × *g*.

3.1.2 SDS-PAGE and Western Blotting

For initial screening of the N-glycan epitopes, approx. 2 μg of proteins are subject to SDS-PAGE under reducing conditions, followed by protein transfer to a nitrocellulose membrane (Western blotting).

1. Check the quality of the successful transfer by incubating the membrane with Ponceau S staining solution for 1 min. After

de-staining with water (protein bands will stain red), block the membrane with Tris buffered saline containing 0.05% Tween and 0.5% BSA for 1 h at room temperature under smooth shaking.

2. Wash the membrane three times using Tris buffered saline with 0.05% Tween (washing buffer).
3. Incubate with biotinylated lectins, pentraxins or primary antibodies in blocking/dilution buffer for 60 min (*see* Table 1 and **Note 4**).
4. Wash the membrane again thrice as above and incubate with the relevant peroxidase or alkaline phosphatase conjugated secondary antibodies in blocking/dilution buffer for 60 min.
5. Again wash the membrane three times as above.
6. Develop the Western blots respectively for peroxidase or phosphatase conjugates with either SigmaFAST 3,3′-diaminobenzidine tetrahydrochloride or SigmaFAST BCIP/NBT (dissolve tablets first in water). Chemiluminescence or other detection methods can also be used.

3.2 Tryptic Peptide Mapping (See Note 5)

1. For the peptide mass fingerprinting identification of proteins with MALDI-TOF MS, apply 10 μg of protein to the SDS-PAGE and stain with Coomassie Blue.
2. After de-staining the gel with water, excise the protein bands in small pieces on glass plates using a clean scalpel.
3. Wash/destain the gel pieces twice with 50% acetonitrile in water and successively once with 1:1 0.1 M ammonium bicarbonate/acetonitrile and 100% acetonitrile only, prior to drying in a Speedvac.
4. In addition, reduce the gel pieces with 10 mM DTT for 1 h at 56 °C and alkylate for 45 min at room temperature with iodoacetamide (55 mM in 0.1 M ammonium bicarbonate) in the dark. Subject the gel pieces to a second round of serial washing (twice 50% acetonitrile, 1:1 0.1 M ammonium bicarbonate/acetonitrile and 100% acetonitrile only) and drying in a Speedvac.
5. For proteolytic digestion, cover the gel pieces with a 1:2 mixture of 0.1 M ammonium bicarbonate /trypsin (100 ng/μL) and incubate overnight at 37 °C.
6. Extract the peptides at room temperature three times using acetonitrile/water/trifluoroacetic acid in a ratio of 660:330:1 (v/v/v). Dry the enriched glycopeptides using a vacuum centrifuge and redissolve them in 5 μL water, prior to spotting on a target plate for MALDI-TOF MS analysis.
7. Spot 0.5 μL of the peptides before applying the matrix (either α-cyanocinnamic acid (ACH) or 6-aza thiothymine (ATT).

The peptides are typically measured in the positive ion mode and 2000 shots are summed for MS and 4000 for MSMS. The spectra are processed with the manufacturer's software; for the Bruker Flexanalysis software, this includes the SNAP algorithms with corresponding signal to noise thresholds.

3.3 (Glyco)Peptide Analysis (See Note 6)

1. Predict/identify the corresponding proteins with, e.g., the MASCOT program (Matrix Science web server) or MS-Fit (ProteinProspector server) using the peptide masses obtained from tryptic digest and MALDI-TOF MS results. Use one of the sequence databases available online such as Swissprot or Uniprot. In parallel, the list of theoretical peptide masses can be generated by online software (e.g., MS-digest at prospector.ucsf.edu or web.expasy.org/peptide-mass).
2. Verify the selected "peptide-hits" with the sequences of the single masses when subject to MALDI-TOF MS/MS. In order to obtain optimal sequence coverage, allow a mass tolerance of 0.5 Da, one missed cleavage site and consider all fixed modifications (e.g., carbamidomethylation of Cys residues if alkylated) and potentially known contaminants for mass spectrometric fingerprint analysis (e.g., human keratin). Include in the results the protein accession number, number of successfully assigned peptides and the percentage of sequence coverage, the software version, number of database entries, and number of species selected for the software search. Glycosylated peptides will not be identified unless subject to PNGase digestion, whereby Asn residues will be converted to Asp ($\Delta m/z = +1$ Da); controls or digestion in $^{18}O\text{-}H_2O$ may be necessary to assess for non-PNGase-mediated deamination.

3.4 N-Glycome Release and Analysis

3.4.1 N-Glycome Release from Intact Glycoproteins (See Note7)

1. The peptide:N-glycosidase F (PNGase F) can release N-glycans from undigested proteins. Denature approximately 8 μg of protein first in 10 μL 0.5% SDS in water for 5 min at 95 °C. Alternatively, recombinant PNGase Ar from rice may be partly effective and release also any core α1,3-fucosylated structures, which commonly occur in invertebrates and plants.
2. After cooling, add 3 μL of 100 mM McIlvaine phosphate/citrate buffer pH 7.5 and 2 μL of PNGase F to the sample prior to incubation for 2 days at 37 °C.
3. Mix approximately 2 μg of either glycosylated or deglycosylated protein with 2× SDS-PAGE buffer and after heat denaturation and a short centrifugation step, apply both samples to SDS-PAGE and Western blotting to estimate the degree of deglycosylation and N-glycan epitope removal.

3.4.2 N-Glycome Release from Glycopeptides (See Notes 7, 8 and 9)

1. After protein identification, heat treat the glycopeptides to inactivate the protease and incubate 90% of the sample with either PNGase F (*see* Subheading 3.4.1, **Step 2**) or PNGase Ar (*see* **Note 7**).
2. Optimal conditions for the PNGase Ar activity (use approximately 5 U/reaction) are 20 mM ammonium acetate buffer, pH 5 for 2 days at 37 °C.
3. Purify the released N-glycans using two different columns packed with Lichroprep C18/Dowex AG 50 and nonporous graphitized carbon/Lichroprep C18 (*see* **Note 8**). Wash first the Lichroprep C18/Dowex AG 50 column with 2% acetic acid and 60% isopropanol and equilibrate with 2% acetic acid. Apply the glycopeptide sample after acidifying with 10% acetic acid and collect immediately the unbound released N-glycans in the flow-through and wash fractions (three column volumes of 2% acetic acid).
4. Apply the flow-through/wash from the Lichroprep/Dowex column directly to a nonporous graphitized carbon/Lichroprep C18 column (prewashed and pre-equilibrated with first 100% acetonitrile then water). After sample application, wash the column with water and elute the N-glycans with 40% acetonitrile containing 0.1% trifluoroacetic acid. Due to the presence of TFA, this sample contains a mixed pool of the neutral and anionic N-glycans.
5. Lyophilize the purified N-glycans overnight and after dissolving them in water, spot an aliquot for MALDI-TOF MS/MS analysis with 6-azathiothymine (ATT); regarding acquisition and interpretation of mass spectra, refer to **Notes 6** and **9**. In comparison to peptides, higher laser power and detector gain settings are necessary to detect glycans. For a more detailed analysis, label the N-glycans by reductive amination using 2-aminopyridine and in addition subject them to HPLC and MALDI-TOF MS analysis as described below.
6. Fluorescent labeling is performed as follows: dissolve 100 mg 2-aminopyridine in 76 μL concentrated HCl and 152 μL water; add 80 μL of this solution to the dried glycan sample, prior to incubation in boiling water for 15 min. Then prepare a solution of 4.4 mg of sodium cyanoborohydride in a mixture of 9 μL of the aforementioned 2-aminopyridine solution and 13 μL water; add 4 μL of this cyanoborohydride-aminopyridine solution to the sample and continue the incubation overnight at 90 °C.
7. Removal of excess labeling reagent is performed immediately the following day by gel filtration. Dilute the sample in 1.5 mL of 0.5% acetic acid (i.e., no more than 5% of the gel filtration column volume), apply to a 30 mL Sephadex G-15 column

(1 × 40 cm) equilibrated in 0.5% acetic acid, and collect 1.5 mL fractions. Transfer aliquots of fractions (80 μL) to a 96 F black plate and detect fluorescence in a microtiter plate reader (excitation/emission: 320/400 nm). Pool fluorescent glycans eluting before the excess labeling reagent and lyophilize.

8. Dissolve dried sample by washing the flask four-times with 20 μL of water and transfer to a microcentrifuge tube; re-lyophilize as required and analyze an aliquot by MALDI-TOF MS.
9. Inject the major portion of the sample onto an Ascentis® Express RP-Amide column pre-equilibrated with 100 mM ammonium acetate (pH 4; buffer A); elute at 0.8 mL/min using a linear gradient of 30% (v/v) MeOH (buffer B) from 0% B up to 35% B over 35 min (higher percentages of B generate higher pressure). The glycans are detected by fluorescence using excitation/emission wavelengths of 320/400 nm and the column is calibrated in terms of glucose units with a fluorescently labeled oligoglucose standard (partial dextran hydrolysate). Collect fractions based on fluorescence intensity and lyophilize prior to another round of MALDI-TOF MS and MS/MS to identify the glycans in the fractions (for example, data, refer to Fig. 2). Normal phase or non-fused core reversed phase columns can also be used [15].

4 Notes.

1. The quality of water and other reagents (acetonitrile, methanol, isopropanol) used for analytical purposes should be high and free of ionic and microbial contaminants.
2. In general, contaminants should be avoided; to prevent analysis of "foreign" components from the food/nutrition source or media (e.g., fetal calf serum), the material (whole organisms or cells) should be washed several times before the heat treatment and homogenization. After collection, the biological material should be stored at −80 °C, if not immediately homogenized. To prevent hydrolysis of the anionic or zwitterionic residues (e.g., phosphate, sialic acid, PC, or PE), the samples should be heat treated only in water and not in acidic buffers; however, heat inactivation is necessary to prevent degradation of the glycans by endogenous glycosidases. For small amounts of biological samples, also a lysis buffer supplemented with protease inhibitor cocktail (Sigma) can be used prior to SDS-PAGE. The methanol precipitation step after cell lysis helps to desalt the sample and so avoids smearing upon electrophoresis.

3. Colloidal Coomassie aggregates and tiny blue dots are visible. Make sure that the staining solution is mixed well (e.g., with a magnetic mixer) before each use.
4. Results obtained from antibody or lectin binding are no structural proof of the N-glycans on the glycoprotein as their specificities are sometimes wide or not fully determined. Positive and negative controls and pull-downs to "pre-clear" endogenous biotinylated proteins, as well as Western blots with and without lectins/antibodies (i.e., just secondary reagents) or after glycosidase digestions should be considered for data interpretation. The "mini-description" of the epitopes in Table 1 is based on determination of binding of the antibodies, lectins or pentraxins to standard ligands; these determinations are by no means exhaustive as invertebrate standards are rarely tested [13, 16]. Nevertheless, anti-horseradish peroxidase is valuable for screening of core β1,2-xylose and core α1,3-fucose [17], but the anti-xylose and anti-fucose components of the antisera are difficult to properly separate. Phosphorylcholine (PC) epitopes can be detected with either the TEPC-15 antibody or by human C-reactive protein [12].
5. The peptides measured "off-line" by MALDI-TOF MS can be sometimes suppressed by contaminant ions generating from the protease itself (e.g., trypsin), which is in part autohydrolyzed. It is recommended to generate online the theoretical peptide masses of the protease as well as the target protein using the MS-digest software.
6. The method described here is a simple and initial procedure for glycoprotein identification by off-line peptide mass fingerprinting and (glyco)peptide analysis of the selected proteins before and after PNGase F or A digestion followed by MALDI-TOF MS. For qualitative/quantitative peptide studies several "online" methods such as LC-ESI MS/MS can also be employed. Invertebrate N-glycomes can dramatically differ from those of mammalian systems, so the N-glycan assignment on the defined glycopeptides should be based at least on MS/MS data analysis as compositions based on mass alone can be misleading: for instance, a difference of 324 Da can either correspond to two hexoses or one methylaminoethylphosphonate-modified HexNAc as seen, e.g., in molluscs. Also, a difference of 176 Da may be either a methylated hexose or a glucuronic acid [11]. Nevertheless, mass differences of 146, 162, or 203 can suggest the presence of fucose, hexose, and *N*-acetylhexosamine residues. There are various bioinformatics tools for automated glycopeptide and glycan identification and the following software can be applied for glycopeptide MS: "GlycoMod," "GlycoX," "GlycopepDB," "Massy tools," and "GlycoSpectrumScan" and

MSMS "GlycoMiner," "Protein Prospector," "GlycopepID," "GlycoMasterDB," and many more [18]. As these are generally applied to mammalian glycomes and glycoproteomes, caution is required when using search engines to annotate invertebrate glycans. For publication, consider the MIRAGE guidelines for presentation of glycomic data [19] as well as use of the diagrammatic Symbol Nomenclature for Glycans [20].

7. PNGase F can release N-glycans from both glycoproteins and glycopeptides, whereas recombinant PNGase Ar still works best on peptides. PNGase F does not release N-glycans with core α1,3-fucose modification (but does release core α1,6-fucosylated or β1,3-mannosylated structures), while recombinant PNGase Ar can release substituted core α1,3-fucosylated glycans [6]. The degree of protein deglycosylation can be monitored with SDS-PAGE (reduced size of the protein after deglycosylation) and Western blotting (reduced or abolished N-glycan epitope binding).
8. After PNGase F or A digest of the glycopeptides one aliquot of the sample should be analyzed with MALDI-TOF MS to verify deglycosylated peptides and potential "occupation" of an N-glycosylation site of the protein. The released glycopeptides should be acidified with 10% acetic acid before Dowex cation exchange chromatography. For N-glycan recovery protocol after PNGase F or A release, refer to our recent protocol on "Analysis of invertebrate and protist N-glycans" [15]. For O-glycosylation, there is no single universal de-O-glycosylation enzyme available; O-glycanase has a restricted substrate specificity and will not remove most extended GalNAc-Ser/Thr (mucin-type) or other O-glycan structures.
9. The released N-glycans should be measured in positive and negative ion mode for the identification of potential anionic residues as sulfate (+80 Da), phosphate (+80 Da), glucuronic acid (+176 Da), phosphoethanolamine (+123 Da), and aminoethylphosphonate (+107 Da; +121 Da if methylated) [11]. Sialic acids are rare in invertebrates and have been only convincingly proven in *Drosophila* or in echinoderms [21, 22], but are absent, e.g., from nematodes.

Acknowledgments

This work was supported by the Austrian Fonds zur Förderung der wissenschaftlichen Forschung (FWF; grants P26662, P25058, and P23922 to A.H., K.P, and I.B.H.W.).

References

1. Spiro RG (2002) Protein glycosylation: nature, distribution, enzymatic formation, and disease implications of glycopeptide bonds. Glycobiology 12:43R–56R
2. Aebi M (2013) N-linked protein glycosylation in the ER. Biochim Biophys Acta 1833:2430–2437
3. Schiller B, Hykollari A, Yan S, Paschinger K, Wilson IBH (2012) Complicated N-linked glycans in simple organisms. Biol Chem Hoppe Seyler 393:661–673
4. Eckmair B, Jin C, Abed-Navandi D, Paschinger K (2016) Multi-step fractionation and mass spectrometry reveals zwitterionic and anionic modifications of the N- and O-glycans of a marine snail. Mol Cell Proteomics 15:573–597
5. Stanton R, Hykollari A, Eckmair B, Malzl D, Dragosits M, Palmberger D, Wang P, Wilson IBH, Paschinger K (2017) The underestimated N-glycomes of lepidopteran species. Biochim Biophys Acta 1861:699–714
6. Yan S, Vanbeselaere J, Jin C, Blaukopf M, Wols F, Wilson IBH, Paschinger K (2018) Core richness of N-glycans of Caenorhabditis elegans: a case study on chemical and enzymatic release. Anal Chem 90:928–935
7. Tretter V, Altmann F, Kubelka V, März L, Becker WM (1993) Fucose α1,3-linked to the core region of glycoprotein N-glycans creates an important epitope for IgE from honeybee venom allergic individuals. Int Arch Allergy Immunol 102:259–266
8. Prasanphanich NS, Mickum ML, Heimburg-Molinaro J, Cummings RD (2013) Glycoconjugates in host-helminth interactions. Front Immunol 4:240
9. Geisler C, Mabashi-Asazuma H, Jarvis DL (2015) An overview and history of glyco-engineering in insect expression systems. Methods Mol Biol 1321:131–152
10. Hykollari A, Malzl D, Yan S, Wilson IBH, Paschinger K (2017) Hydrophilic interaction anion exchange for separation of multiply modified neutral and anionic Dictyostelium N-glycans. Electrophoresis 38:2175–2183
11. Paschinger K, Wilson IBH (2016) Analysis of zwitterionic and anionic N-linked glycans from invertebrates and protists by mass spectrometry. Glycoconj J 33:273–283
12. Paschinger K, Gonzalez-Sapienza GG, Wilson IBH (2012) Mass spectrometric analysis of the immunodominant glycan epitope of Echinococcus granulosus antigen Ag5. Int J Parasitol 42:279–285
13. Iskratsch T, Braun A, Paschinger K, Wilson IBH (2009) Specificity analysis of lectins and antibodies using remodeled glycoproteins. Anal Biochem 386:133–146
14. Mikolajek H, Kolstoe SE, Pye VE, Mangione P, Pepys MB, Wood SP (2011) Structural basis of ligand specificity in the human pentraxins, C-reactive protein and serum amyloid P component. J Mol Recognit 24:371–377
15. Hykollari A, Paschinger K, Eckmair B, Wilson IBH (2017) Analysis of invertebrate and protist N-glycans. Methods Mol Biol 1503:167–184
16. Purohit S, Li T, Guan W, Song X, Song J, Tian Y, Li L, Sharma A, Dun B, Mysona D, Ghamande S, Rungruang B, Cummings RD, Wang PG, She JX (2018) Multiplex glycan bead array for high throughput and high content analyses of glycan binding proteins. Nat Commun 9:258
17. Paschinger K, Rendić D, Wilson IBH (2009) Revealing the anti-HRP epitope in Drosophila and Caenorhabditis. Glycoconj J 26: 385–395
18. Tsai PL, Chen SF (2017) A brief review of bioinformatics tools for glycosylation analysis by mass spectrometry. Mass Spectrom (Tokyo) 6:S0064
19. York WS, Agravat S, Aoki-Kinoshita KF, McBride R, Campbell MP, Costello CE, Dell A, Feizi T, Haslam SM, Karlsson N, Khoo KH, Kolarich D, Liu Y, Novotny M, Packer NH, Paulson JC, Rapp E, Ranzinger R, Rudd PM, Smith DF, Struwe WB, Tiemeyer M, Wells L, Zaia J, Kettner C (2014) MIRAGE: the minimum information required for a glycomics experiment. Glycobiology 24:402–406
20. Varki A, Cummings RD, Aebi M, Packer NH, Seeberger PH, Esko JD, Stanley P, Hart G, Darvill A, Kinoshita T, Prestegard JJ, Schnaar RL, Freeze HH, Marth JD, Bertozzi CR, Etzler ME, Frank M, Vliegenthart JF, Lutteke T, Perez S, Bolton E, Rudd P, Paulson J, Kanehisa M, Toukach P, Aoki-Kinoshita KF, Dell A, Narimatsu H, York W, Taniguchi N, Kornfeld S (2015) Symbol nomenclature for graphical representations of glycans. Glycobiology 25:1323–1324
21. Aoki K, Perlman M, Lim JM, Cantu R, Wells L, Tiemeyer M (2007) Dynamic developmental elaboration of N-linked glycan complexity in the Drosophila melanogaster embryo. J Biol Chem 282:9127–9142
22. Miyata S, Sato C, Kumita H, Toriyama M, Vacquier VD, Kitajima K (2006) Flagellasialin: a novel sulfated α2,9-linked polysialic acid glycoprotein of sea urchin sperm flagella. Glycobiology 16:1229–1241

Chapter 25

The Use of Proteomics Studies in Identifying Moonlighting Proteins

Constance Jeffery

Abstract

Proteomics studies that characterize hundreds or thousands of proteins in parallel can play an important part in the identification of moonlighting proteins, proteins that perform two or more distinct and physiologically relevant biochemical or biophysical functions. Functional assays, including ligand-binding assays, can find a surprising second function for a protein that was previously identified as performing a different function, for example, a DNA-binding ability for an enzyme in amino acid metabolism. The results of large-scale assays of protein–protein interactions, gene knockouts, or subcellular protein localizations, or bioinformatics analysis of amino acid sequences and three-dimensional structures, can also be used to predict that a protein has additional functions, but in these cases it is important to use biochemical and biophysical methods to confirm the protein can perform each function.

Key words Moonlighting proteins, Multifunctional proteins, Protein function prediction, Proteomics

1 Introduction

The goal of many proteomics studies is to identify protein functions. Complicating this task is the ability of a single protein to have different functions in different cellular processes, with different ligands or different protein partners, in different cell types, and/or in different subcellular locations. Hundreds of proteins have been identified as these moonlighting proteins, which comprise a subset of multifunctional proteins that perform two or more distinct and physiologically relevant biochemical or biophysical functions that are not due to gene fusions, multiple RNA splice variants, or pleiotropic effects [1]. Some of the first moonlighting proteins to be identified were taxon-specific crystallins [2, 3], proteins that are found in high concentration in the lens of the eye but function as enzymes in other cell types. For example, zeta-crystallin from the guinea pig lens is identical to the enzyme quinone oxidoreductase [4]. Over 300 moonlighting proteins are described in the

Xing Wang and Matthew Kuruc (eds.), *Functional Proteomics: Methods and Protocols*, Methods in Molecular Biology, vol. 1871, https://doi.org/10.1007/978-1-4939-8814-3_25,

online MoonProt Database [5]. As a group, the known moonlighting proteins perform a large variety of functions and combinations of functions and don't share sequence or structural motifs or other physical characteristics that enable easy identification. Although interpreting the results of proteomics projects might be complicated by the presence of moonlighting proteins, the diversity of moonlighting proteins means that these large-scale projects to characterize proteins, without a prior hypothesis about each protein's function, can be the best way for finding more moonlighting proteins and their multiple functions.

2 Methods

Proteomics experiments can be used to help identify proteins with multiple functions in both direct and indirect ways. Projects based on a functional assay can identify a second function for a protein that already has a known function. Projects that test other protein characteristics can also be used to suggest that some proteins have a second function, although they might or might not provide information about what the function is. Bioinformatics analyses used alone or in combination with proteomics projects can be used to suggest which other proteins might also have multiple functions.

2.1 Experimental Methods

The proteomics methods that have been the most useful in finding proteins with multiple functions include those that test binding to a specific molecule, protein–protein interactions, the results of gene knockout experiments, and cellular localization.

2.1.1 Binding Studies

Proteomics studies that involve screening hundreds or thousands of proteins to find those that bind to DNA, extracellular matrix, or other macromolecules have identified several dozen proteins that were already known to have a different function. This is not too surprising because dozens of the known moonlighting proteins have at least one function that involves binding to another molecule—as a cell surface receptor for a soluble ligand or the extracellular matrix, as a secreted ligand binding to a receptor on another cell, or as a DNA- or RNA-binding protein.

The use of microarrays of proteins, DNA oligonucleotides, or RNA oligonucleotides enables screening of vast numbers of proteins to find those that bind to a chosen macromolecule. Hall and coworkers screened a microarray of yeast proteins for binding to DNA oligonucleotides and identified mitochondrial Arg5,6 (N-acetylglutamate kinase/N-acetylglutamyl-phosphate reductase), an enzyme in the arginine biosynthetic pathway, as having DNA-binding activity. Complementary chromatin immunoprecipitation experiments and gene deletion experiments confirmed that

Arg5,6 is a transcription regulator for several specific nuclear and mitochondrial genes [6].

Assays to identify proteins that bind to a specific protein also identified several dozen yeast and bacterial proteins that perform one function when expressed inside the cell and a second function when displayed on the cell surface. A proteomic study of cell wall proteins from the pathogenic fungus *Candida albicans* was used to identify eight cytosolic proteins (phosphoglycerate mutase, alcohol dehydrogenase, thioredoxin peroxidase, catalase, transcription elongation factor, glyceraldehyde-3-phosphate dehydrogenase, phosphoglycerate kinase, and fructose bisphosphate aldolase) as cell surface receptors for host plasminogen [7]. A similar study of the intestinal "pro-biotic" bacterium *Bifidobacterium lactis* identified the cytosolic enzymes bile salt hydrolase, glutamine synthetase, and phosphoglycerate mutase and the chaperone DnaK also to bind host plasminogen when expressed on the cell surface [8].

2.1.2 Protein–Protein Interactions

Proteomics-scale studies of protein–protein interactions, such as yeast 2-hybrid assays, often yield results that are more complex than expected, with a single protein being found to interact with proteins acting in multiple biochemical pathways, molecular machines or multiprotein complexes. These results are sometimes interpreted as being due to false positives, but interacting with multiple groups of proteins from different cellular processes is a common characteristic of moonlighting proteins, and these protein–protein interaction results could be due to physiologically relevant interactions [9]. In a study of human proteins, Chapple and coworkers combined protein–protein interaction information with analysis of the protein functional annotations to identify 430 proteins that they described as being extreme multifunctional proteins [10]. Further analysis of protein–protein interaction networks from humans and other species could lead to predictions of additional proteins that might interact with different groups of protein partners to perform different functions, along with suggestions of the types of additional functions based on the identities of the interacting proteins. Follow-up testing through biochemical or biophysical assays would be needed to confirm the observed interactions are due to a second function and not due to false positives, or to proteins that interact with multiple proteins as part of a single function, such as in a signaling pathway, or proteins that perform the same function in different cellular locations.

2.1.3 Gene Knockouts

When a single protein participates in several cellular processes, deletion of the gene encoding the protein can result in phenotypes that are more complex than can be explained by the loss of a single function. Several labs made use of yeast genetics to look for enzymes for which replacement of the wild-type enzyme with a catalytic deficient mutant does not recapitulate the results of the complete gene knockout. Because the mutant protein exhibits only part of the deletion phenotype, the wild-type protein must have a second function. The *S. cerevisiae* Bat2 transaminase in sugar and amino acid metabolism and the isoleucine/valine biosynthetic enzymes Ilv1 and Ilv2 were found to have second functions in addition to their respective catalytic functions [11]. The Alt1 alanine transaminases from two other yeast species, *Lacchancea kluyveri* (LkAlt1) and *Kluyveromyces lactis* (KlAlt1), were also found to be moonlighting proteins [12].

2.1.4 Expression Patterns/Cellular Localization

Proteomics projects that determine a protein's cellular location (s) can be used to suggest that it might have multiple functions. Many of the known moonlighting proteins perform their different functions in different subcellular locations or cell types. For example, several dozen cytosolic proteins, like the plasminogen-binding proteins mentioned above, have a second function as a receptor or adhesin on the cell surface in bacteria, humans, and many other species. Glyceraldehyde 3-phosphate dehydrogenase (GAPDH) was the first cytoplasmic protein found to be attached to the surface of pathogenic streptococci [13], and several dozen cytosolic enzymes in multiple species have also been found to be displayed on the cell surface, where they play roles in signaling, adhesion, or acquiring nutrients. Studies to identify all the proteins on the cell surface of dozens of bacterial species, through cell fractionation and isolation of proteins followed by identification through mass spectrometry, found that many other cytoplasmic proteins are also attached to the cell surface, and some may be additional moonlighting proteins [14, 15]. Other methods of studying protein localization can also be scaled up and used in proteomics studies. In a recent study using antibody-based immunofluorescence of 12,003 human proteins in 30 subcellular structures and 13 organelles, about half of the proteins were found in more than one compartment and might also include candidates for moonlighting proteins [16].

The taxon-specific crystallins mentioned above clearly have two functions because these enzymes are found in a high concentration in the lens of the eye where their catalytic substrates are not found, and the known intracellular/cell surface moonlighting proteins have been tested through binding studies to confirm the presence of a second function. In other cases, finding a protein in a subcellular location where it is not expected to perform its known function can suggest that the protein has a second function, but experimental evidence that the protein is performing a different function in each location is needed to confirm that the protein is multifunctional. For example, proteins that move between cellular compartments as part of a signaling pathway would not be considered moonlighting proteins.

2.2 Bioinformatics Analysis

The lack of common sequence or structural characteristics among the moonlighting proteins has made it difficult to develop a universal computational method to predict that a protein has more than one function, but several labs are developing bioinformatics methods by using collections of known moonlighting proteins, such as the MoonProt Database, as a positive control set. Unfortunately, a true negative control set of proteins that don't have multiple functions is not available because it's currently not possible to know if a protein has only one function or if it has additional functions that have not yet been identified.

Large-scale searches of the literature and database annotation, including searches for proteins with diverse GO terms in UniProt [17, 18], have shown some success in identifying proteins with multiple functions in diverse processes [19, 20]. Searches for amino acid sequence or structural motifs known to be associated with specific protein functions can help identify proteins that have motifs corresponding to multiple functions. For example, the use of an X-ray crystal structure revealed the *Streptomyces coelicolor* albaflavenone synthase also has a terpene synthase active site [21]. However, the use of motifs as a tool for prediction of function is limited by the lack of known motifs for many classes of functions, for example, protein–protein interactions. As more information becomes available about sequence motifs, protein–protein interaction surfaces, the constellations of amino acids that make up catalytic sites, and the three-dimensional structures of moonlighting proteins, the ability of these methods to find the multiple functions of moonlighting proteins might be improved [22, 23].

Some recent methods combine analysis of protein sequences and structures with information about the results from proteomics projects (protein–protein interactions, cellular locations, etc.) [24, 25]. Because of the challenges in identifying moonlighting proteins, these combined methods might have the most success in the future.

3 Conclusions

The presence of moonlighting proteins adds to the complexity of the results of proteomics studies, but these large-scale methods are valuable for identifying more examples of these proteins, which are otherwise usually found through serendipity. One caution, however, is that only the proteomics methods that directly test for a specific function, such as the plasminogen-binding assays mentioned above, provide evidence that the protein performs that function. The results of the other proteomics and bioinformatics methods described herein can usually only be used to predict that a protein has a second function. It is necessary to use biochemical and biophysical methods to confirm that a protein performs both functions.

References

1. Jeffery CJ (1999) Moonlighting proteins. Trends Biochem Sci 24(1):8–11. PMID: 10087914
2. Piatigorsky J, Wistow GJ (1989) Enzyme/crystallins: gene sharing as an evolutionary strategy. Cell 57:197–199
3. Wistow GJ, Kim H (1991) Lens protein expression in mammals: taxon specificity and the recruitment of crystallins. J Mol Evol 32:262–269
4. Huang QL, Russell P, Stone SH, Zigler JS Jr (1987) Zeta-crystallin, a novel lens protein from the Guinea pig. Curr Eye Res 6:725–732. PMID: 3595182
5. Mani M, Chen C, Amblee V, Liu H, Mathur T, Zwicke G, Zabad S, Patel B, Thakkar J, Jeffery CJ (2015) MoonProt: a database for proteins that are known to moonlight. Nucleic Acids Res 43:D277–D282
6. Hall DA, Zhu H, Zhu X, Royce T, Gerstein M, Snyder M (2004) Regulation of gene expression by a metabolic enzyme. Science 306:482–484
7. Crowe JD, Sievwright IK, Auld GC, Moore NR, Gow NA, Booth NA (2003) *Candida albicans* binds human plasminogen: identification of eight plasminogen-binding proteins. Mol Microbiol 47:1637–1651. PMID: 12622818
8. Candela M, Bergmann S, Vici M, Vitali B, Turroni S, Eikmanns BJ, Hammerschmidt S, Brigidi P (2007) Binding of human plasminogen to Bifidobacterium. J Bacteriol 189:5929–5936. https://doi.org/10.1128/JB.00159-07
9. Gómez A, Hernández S, Amela I, Piñol J, Cedano J, Querol E (2011) Do protein-protein interaction databases identify moonlighting proteins? Mol BioSyst 7:2379–2382. https://doi.org/10.1039/c1mb05180f
10. Chapple CE, Robisson B, Spinelli L, Guien C, Becker E, Brun C (2015) Extreme multifunctional proteins identified from a human protein interaction network. Nat Commun 6:7412. https://doi.org/10.1038/ncomms8412
11. Espinosa-Cantú A, Ascencio D, Herrera-Basurto S, Xu J, Roguev A, Krogan NJ, DeLuna A (2018) Protein moonlighting revealed by noncatalytic phenotypes of yeast enzymes. Genetics 208:419–431. https://doi.org/10.1534/genetics.117.300377
12. Escalera-Fanjul X, Campero-Basaldua C, Colón M, González J, Márquez D, González A (2017) Evolutionary diversification of alanine transaminases in yeast: catabolic specialization and biosynthetic redundancy. Front Microbiol 8:1150. https://doi.org/10.3389/fmicb.2017.01150
13. Pancholi V, Fischetti VA (1992) A major surface protein on group a streptococci is a glyceraldehyde-3-phosphate-dehydrogenase with multiple binding activity. J Exp Med 176:415–426
14. Olaya-Abril A, Jiménez-Munguía I, Gómez-Gascón L, Rodríguez-Ortega MJ (2014) Surfomics: shaving live organisms for a fast proteomic identification of surface proteins. J Proteome 97:164–176. https://doi.org/10.1016/j.jprot.2013.03.035
15. Wang W, Jeffery CJ (2016) An analysis of surface proteomics results reveals novel candidates for intracellular/surface moonlighting proteins in bacteria. Mol BioSyst 12:1420–1431
16. Thul PJ, Åkesson L, Wiking M, Mahdessian D, Geladaki A, Ait Blal H, Alm T, Asplund A, Björk L, Breckels LM, Bäckström A, Danielsson F, Fagerberg L, Fall J, Gatto L, Gnann C, Hober S, Hjelmare M, Johansson F, Lee S, Lindskog C, Mulder J, Mulvey CM, Nilsson P, Oksvold P, Rockberg J, Schutten R, Schwenk JM, Sivertsson Å, Sjöstedt E, Skogs M, Stadler C, Sullivan DP, Tegel H, Winsnes C, Zhang C, Zwahlen M, Mardinoglu A, Pontén F, von Feilitzen K, Lilley KS, Uhlén M, Lundberg E (2017) A subcellular map of the human proteome. Science 356:eaal3321. https://doi.org/10.1126/science.aal3321
17. Consortium GO (2015) Gene ontology consortium: going forward. Nucleic Acids Res 43: D1049–D1056
18. UniProt Consortium (2015) UniProt: a hub for protein information. Nucleic Acids Res 43: D204–D212
19. Khan IK, Bhuiyan M, Kihara D (2017) DextMP: deep dive into text for predicting moonlighting proteins. Bioinformatics 33: i83–i91. https://doi.org/10.1093/bioinformatics/btx231
20. Pritykin Y, Ghersi D, Singh M (2015) Genome-wide detection and analysis of multifunctional genes. PLoS Comput Biol 11: e1004467. https://doi.org/10.1371/journal.pcbi.1004467
21. Zhao B, Lei L, Vassylyev DG, Lin X, Cane DE, Kelly SL, Yuan H, Lamb DC, Waterman MR (2009) Crystal structure of albaflavenone monooxygenase containing a moonlighting terpene synthase active site. J Biol Chem 284:36711–36719. https://doi.org/10.1074/jbc.M109.064683
22. Khan I, Chitale M, Rayon C, Kihara D (2012) Evaluation of function predictions by PFP,

ESG, and PSI-BLAST for moonlighting proteins. BMC Proc 6(Suppl 7):S5. https://doi.org/10.1186/1753-6561-6-S7-S5

23. Hernández S, Franco L, Calvo A, Ferragut G, Hermoso A, Amela I, Gómez A, Querol E, Cedano J (2015) Bioinformatics and moonlighting proteins. Front Bioeng Biotechnol 3:90. https://doi.org/10.3389/fbioe.2015.00090

24. Khan IK, Kihara D (2016) Genome-scale prediction of moonlighting proteins using diverse protein association information. Bioinformatics 32:2281–2288. https://doi.org/10.1093/bioinformatics/btw166

25. Khan I, McGraw J, Kihara D (2017) MPFit: computational tool for predicting moonlighting proteins. Methods Mol Biol 1611:45–57. https://doi.org/10.1007/978-1-4939-7015-5_5

Chapter 26

Two-Dimensional Biochemical Purification for Global Proteomic Analysis of Macromolecular Protein Complexes

Reza Pourhaghighi and Andrew Emili

Abstract

A high-resolution two-dimensional (2-D) proteomic fractionation technique for the systematic purification and subsequent mass spectrometry-based identification of endogenous protein macromolecular complexes is described. The method hyphenates preparative isoelectric focusing (IEF) with mixed-bed ion exchange chromatography (IEX) to efficiently separate cell- or tissue- derived soluble protein mixtures, allowing for more effective and less biased physiochemical characterization of stable multiprotein assemblies. After comprehensive 2D fractionation of cell-free lysates, each fraction is subjected to quantitative tandem mass spectrometry (MS/MS) and subsequent computational analysis to map high-confidence protein--protein interactions (PPIs). Herein, the experimental component (workflow protocols) for this global "interactome" network mapping platform is described.

Key words Protein–protein interaction, Protein complexes, Isoelectric focusing, High-performance liquid chromatography (HPLC), Biochemical separation, Ion exchange chromatography, Fractionation, Nanoflow liquid chromatography tandem mass spectrometry (nLC-MS/MS)

1 Introduction

Since stable macromolecular assemblies are responsible for many, if not most, of the key biochemical processes operating inside living cells, the comprehensive experimental analysis of multiprotein complexes represents a significant goal of the field of systems (network) biology. To date, several methodologies for systematic large-scale analysis of the composition (physically associated components) of cellular multiprotein complexes and protein–protein interaction networks have been reported [1–4]. In this context, we have developed a flexible platform for the global study of endogenous protein complexes from diverse cells and tissue samples based on the extensive biochemical pre-fractionation of native protein assemblies prior to in-depth quantitative nLC/MS/MS-based detection [5]. Recently, in order to further improve the analytical dynamic range, we have devised complementary

Xing Wang and Matthew Kuruc (eds.), *Functional Proteomics: Methods and Protocols*, Methods in Molecular Biology, vol. 1871, https://doi.org/10.1007/978-1-4939-8814-3_26,

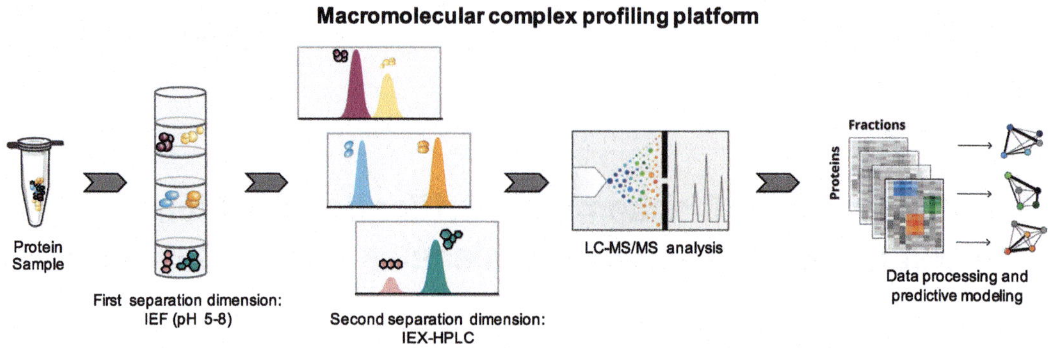

Fig. 1 Two-dimensional macromolecular complex profiling platform

biochemical separation methods that enable rapid and efficient fractionation of soluble cell-free mixtures to isolate stably associated protein complexes with the highest possible resolution without disturbing macromolecular integrity [5]. These include a novel hyphenated 2-D separation workflow based on hyphenation of non-denaturing preparative IEF with orthogonal IEX-based HPLC separations. In this approach, native soluble protein complexes are first gently extracted from a biological specimen and then selectively enriched by IEF over pH range of 5–8 into five fractions, which are each then subsequently subjected to a more extensive salt-gradient mixed-bed IEX-based fractionation. The collected protein factions are precipitated and digested into peptides with trypsin, in order to identify stably associated interacting proteins which reproducibly co-elute through 2-D IEF-IEX protein separations platform, resulted fractions are quantitative analyzed by tandem mass spectrometry.

In this chapter, we provide a detailed experimental protocol and an illustrative example showing the application of this 2-D proteomic fractionation technique to resolve protein assemblies in the microbe *E. coli* for a comprehensive assessment of the microbial interactome analysis. Figure 1 illustrates the described protein complex profiling platform. Key steps and troubleshooting issues are explained, while complementary computational data analysis strategies are described in [6].

2 Materials

It is essential to consult the appropriate Material Safety Data Sheets and the institutional Environmental Health and Safety Office for proper handling of the potentially hazardous material used in this protocol. Use of analytical (HPLC) grade water and solvents to prepare buffers and reagents is required.

2.1 Protein Extraction from *E. coli* Cells

1. Lysis buffer: Modified B-PER protein extraction buffer (Thermo Scientific) supplemented with 10% v/v glycerol, 0.5 mM Dithiothreitol (DTT), 0.2 mg/mL Lysozyme, 2 μL/mL DNase I, and Ethylenediaminetetraacetic acid (EDTA)-free protease inhibitor (Thermo Scientific) (*see* **Note 1**).
2. Bradford reagent: Store in dark at 4 °C.

2.2 Non-denaturing Isoelectric Focusing

1. MicroRotofor cell (Bio-Rad).
2. High voltage power supply (*see* **Note 2**).
3. Vacuum source and tubing.
4. MicroRotofor focusing chamber (Bio-Rad).
5. Sealing tape.
6. 3-mL syringe.
7. Anodic and cathodic ion exchange membranes (each one piece).
8. Anodic electrolyte solution: 0.1 M H_3PO_4.
9. Cathodic electrolyte solution: 0.1 M NaOH.
10. Carrier ampholytes (Bio-Lyte), pH 5-8 (Bio-Rad).
11. Glycerol.

2.3 HPLC Fractionation

1. HPLC-IEX Column: Pre-packed mixed-bed PolyCATWAX column (PolyLC Inc.) 200 × 2.1 mm, 5 μm, 1000-Å (*see* **Note 3**).
2. HPLC MES mobile phase-A: 10 mM 2-(N-morpholino)ethanesulfonic acid (MES) buffer pH 6, 5% Glycerol and 0.01% NaN_3 to inhibit bacterial growth.
3. HPLC MES mobile phase-B: 10 mM MES buffer pH 6, 1.5 M NaCl, 5% Glycerol, 0.01% NaN_3.
4. HPLC Tris mobile phase-A: 10 mM tris(hydroxymethyl)aminomethane (Tris) buffer pH 8, 5% Glycerol and 0.01% NaN_3.
5. HPLC Tris mobile phase-B: 10 mM Tris buffer pH 8, 1.5 M NaCl, 5% Glycerol, 0.01% NaN_3.

2.4 Sample Preparation and Protein Digestion

1. Microcentrifuge with temperature control.
2. Trichloroacetic acid (TCA). Store at 4 °C.
3. Acetone. Store at −20 °C.
4. Incubating shaker with temperature control.
5. DTT stock solution: Dissolve 7.7 mg DTT in water to obtain a final concentration of 0.5 M (*see* **Note 4**).

6. Iodoacetamide (IAA) stock solution: Dissolve 9.2 mg of IAA in 500 μL of 50 mM NH_4CO_3 pH 8 solution to obtain a IAA stock solution of 0.1 M (*see* **Note 5**).
7. Sequencing grade trypsin (Promega).

2.5 nLC-MS/MS Analysis

1. nLC system coupled online with a high-resolution tandem MS system.
2. 1% Formic Acid (FA).
3. nLC mobile phase-A: 0.1% FA.
4. nLC mobile phase-B: 80% ACN, 0.1% FA.

3 Methods

Timing is critical throughout the protocol. Work quickly and consistently. Try to minimize the time especially before the protein sample is fractionated. Keep sample on ice unless otherwise stated.

3.1 Soluble Protein Extraction

1. Pellet *E. coli* cells by centrifugation at 5000 × *g* for 10 min (*see* **Note 6**).
2. Add 4 mL of modified B-PER protein extraction buffer per one gram of cell pellet and mix gently until the pellet is fully dissolved.
3. Lightly mix the lysate for 30 min at 4 °C using a rotatory shaker.
4. Centrifuge lysate at 15,000 × *g* for 10 min at 4 °C to separate soluble proteins from insolubles and the cell debris.
5. Remove and filter the supernatant with 0.45 μm centrifugal filters according to manufacturer's recommended time and speed.
6. Use Bradford protein assay to measure the protein concentration in lysate obtained.

3.2 Isoelectric Focusing (IEF)

1. Place the MicroRotofor unit in cold room (4 °C) at least 15 min before beginning the focusing run.
2. Equilibrate the ion exchange anode (red) and cathode (black) membranes overnight in 0.1 M H_3PO_4 and 0.1 M NaOH solutions, respectively.
3. The sample volume for the separation chamber in a MicroRotofor unit is ~2.5 mL. Prepare IEF sample at a protein concentration of ~2 mg/mL. Add carrier ampholyte and glycerol to protein sample solution to the final concentrations of 2 and 10% v/v, respectively (*see* **Note 7**).

4. Rinse the equilibrated anodic and cathodic membranes with deionized water and securely mount them at two ends of the focusing chamber.
5. Anodic and cathodic electrode assemblies are color coded in red and black. Assemble them at corresponding ends of the focusing chamber. Align the vents on electrode assemblies with one row of ports on the focusing chamber and tight the threaded sleeve around the assemblies.
6. The other row of ports of the focusing chamber which are not aligned with the vents on the electrode chambers are used for harvesting the fractionated material. Seal with a strip of sealing tape before loading the sample.
7. Using a 3-mL syringe, gradually load the sample solution through the centermost loading port of the focusing chamber.
8. Make sure that all the channels of focusing chamber are filled with sample solution and there is no bubble inside them (*see* **Note 8**).
9. If outside surface of the focusing chamber is wet, gently dry it and seal the row of focusing chamber used for sample loading with a strip of sealing tape. Make sure all the ports are fully covered and the tape does not overlap with the harvesting port sealing tape.
10. Place the focusing assembly on MicroRotofor unit with the vents on the electrode chambers facing up.
11. Using two syringes, add 6 mL 0.1 M H_3PO_4 through the vent hole of the anodic electrolyte chamber (left/red) and 6 mL 0.1 M NaOH through the vent hole of the cathodic electrolyte chamber (right/black).
12. Close the cooling block cover. Make sure the focusing chamber and the sealing tapes can rotate without any obstruction.
13. Place the green lid on top of the chassis and turn on the power switch.
14. Attach the electrodes from the MicroRotofor lid to the power supply and perform focusing at 1 W constant power. Typical run is usually completed in 2–3 h (*see* **Note 9**).
15. After the IEF run is complete, collect the fractions as quickly as possible and avoid sample diffusion.
16. Turn the power supply and the MicroRotofor unit off.
17. Apply a vacuum to the MicroRotofor chassis.
18. Open the cooling block cover and gently remove the sealing tape from the sample loading ports (top row).
19. Place the focusing assembly in the harvesting station with the row of sample loading ports facing up. Press down it firmly so

that the needles could pass sealing tape and penetrate the harvesting ports.

20. After the fractions are aspirated into the harvesting tray, turn off the vacuum source and remove the harvesting tray.
21. Transfer the fractions to 1.5 mL tubes and store them at 4 °C.

3.3 Ion Exchange Chromatography (IEX-HPLC)

NOTE: No sample clean-up and/or buffer exchange step is required prior to IEX-HPLC separation of IEF fractions collected. Depending of the pH of the IEF fractions, Tris–HCl pH 8 (for acidic IEF fractions with the pH up to 7) or MES pH 6 buffer (for basic fractions with pHs above 7) are used for IEX separations of IEF focused protein complexes. During IEX separation, proteins can be eluted from the IEX-HPLC column with a salt (NaCl) gradient and recovered in biologically active forms.

1. Equilibrate the IEX-HPLC column by running two blank gradients using the buffer systems preferred before running the protein sample. Likewise, re-equilibrate the column after each gradient with at least 30 min running 100% buffer-A.
2. Set the injection volume of the HPLC method to inject the entire sample collected from IEF chamber into the IEX column (normally between 200 and 250 μL).
3. The recommended flowrate for a 2.1-mm i.d. column as described here is 0.2 mL/min.
4. A typical gradient as below could be employed for IEX-HPLC separation:
   (a) 100% A from 0 to 3 min, followed by a shallow gradient to 10% B from 3 to 45 min, a linear gradient to 35% B from 45 to 65 min and then a gradient to 100% B from 65 to 80 min followed by an isocratic hold at 100% B until 90 min.
5. Monitor the protein elution by UV absorption signal at 280 nm.
6. Set the method to collect the fractions with 2-min intervals during IEX separation run. Depending on the chromatogram, the collected fractions could be further merged to reduce the total number of fractions for subsequent analysis.

The HPLC-IEX parameters described are summarized in Table 1. Figure 2 shows a representative IEX-HPLC chromatogram recorded after running combined IEF fractions 1 and 2 (pH 5–5.6) using a Tris pH 8 buffer system and using the parameters described in Table 1.

Table 1
IEX-HPLC parameters

Flow rate	0.2 mL/min
Time (min)	**LC gradient (%B)**
0–3	0
3–45	0–10
45–65	10–35
65–80	35–100
80–90	100
Detection	280 nm
Fraction collection intervals	2 min

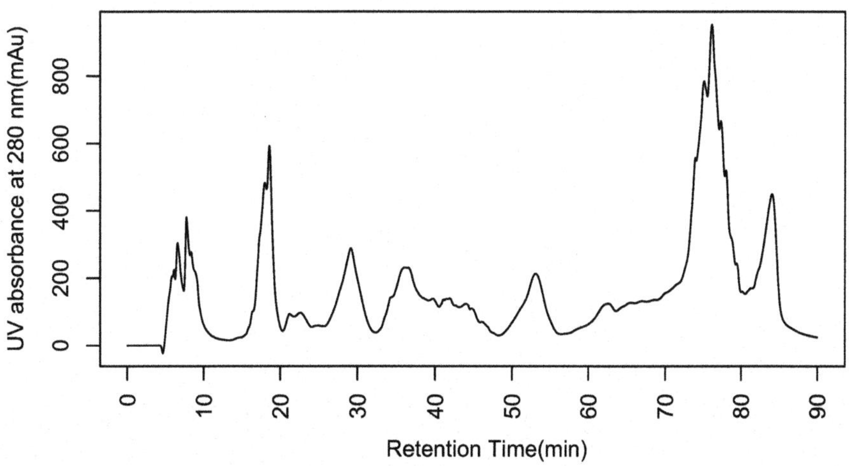

Fig. 2 Representative chromatogram of IEX-HPLC fractionated *E. coli* protein extract, generated after combining IEF pre-fractions 1 and 2 (pH 5–5.6). Detailed experimental parameters are listed in Table 1

3.4 Trichloroacetic Acid (TCA) Precipitation

1. If fractions are collected in a 96-well plate, carefully transfer them into individual 1.5-mL tubes.
2. Precipitate the proteins by adding 10% v/v cold TCA to each tube, briefly vortex, and incubate them at 4 °C overnight.
3. Centrifuge the protein samples at 15,000 × *g* at 4 °C for 30 min to pellet the proteins.
4. Remove supernatant gently leaving protein pellet intact. Consider that the protein pellet might not be visible in dilute fractions.
5. Add 200 μL ice-cold acetone to wash the white protein pellet. Briefly vortex and incubate the sample for an hour at −20 °C (*see* **Note 10**).

6. Spin the samples with 15,000 × *g* at 4 °C for 30 min.
7. Repeat the acetone wash steps for one more time. Add acetone, incubate, and centrifuge the sample as before.
8. Remove supernatant and leave the protein pellet to air-dry for about 30 min.

3.5 Trypsin Digestion

1. Dissolved each dried pellet in 90 μL of 5 mM DTT, 50 mM NH_4CO_3 pH 8 solution and incubate the sample for 15 min at 50 °C with gentle agitation (*see* **Note 11**).
2. Bring the protein solution to room temperature. Add 10 μL of 100 mM IAA solution to reach a final concentration of 10 mM and incubate the samples for 15 min in the dark with gentle agitation.
3. In order to quench possible excess of IAA, add 1 μL DTT from 0. 5 M DTT stock solution.
4. Add sequencing grade trypsin at a 1:50 enzyme:protein ratio and incubate the samples overnight at 37 °C with gentle agitation.
5. Quench the digestion by acidifying the samples by adding formic acid to a final concentration of 1% (v/v).
6. Use a vacuum centrifuge to lyophilize the peptides to dryness. Then dissolve the dried samples in 1% formic acid for subsequent LC-MS/MS analysis (*see* **Note 12**).

3.6 LC-MS/MS

1. Set the method to inject about 1–2 μg for digested protein samples into the nanoLC column for MS analysis.
2. A 60-min LC gradient as outlined below is generally appropriate for nLC-MS/MS analysis of the fractions obtained: A linear gradient from 5% to 30% B from 0 to 46 min and then a gradient to 100% B from 46 to 50 min followed by an isocratic hold at 100% B until 60 min.
3. The parameters for MS analysis of peptide fractions depend on the type and performance specifications of MS instrument used. As an illustrative example, the recommended MS parameters for a 60-min method on an Orbitrap Q-Exactive HF instrument are listed in Table 2.

3.7 Computational Proteomics Analysis

Search all MS/MS spectra against an appropriate fasta file using a search engine (e.g., MaxQuant). The calculated related intensity of proteins in each IEX fraction can then be used to calculate similarity profiles, to predict protein associations and co-complex memberships, using computational algorithms (e.g., machine learning classifiers) and statistical filtering to identify high-confidence physical interactions among the co-eluting proteins as described in [6].

Table 2
nLC-MS/MS parameters

Time (min)	LC gradient (%B)
0–46	5–30
46–50	30–100
50–60	100–100
Full-MS	
Microscans	1
Resolution	60,000
Automatic gain control target	3e6
Maximum ion time	70 ms
Number of scans	1
Scan range	300–1650 *m/z*
dd-MS^2	
Microscans	1
Resolution	15,000
Automatic gain control target	1e5
Maximum ion time	25 ms
Loop count	15
Isolation window	1.4 *m/z*
Normalized collision energy	27
dd setting	
Charge exclusion	Unassigned, 1
Exclude isotopes	On
Dynamic exclusion	6 s

4 Notes

1. The indicated values are final concentrations of each reagent in B-PER buffer. Prepare the lysis buffer freshly before use and keep it on ice.
2. The power supply must be capable of power control at 1 W constant power. If constant power mode is not available, it must be programmable to run multiple step constant voltage methods and capable of supplying up to 1000 V as well as operating at low currents.

3. Alternatively, an IEX column with 4.6 i.d. could be used. However, experimental parameters like sample loading capacity and gradient flow rate should be adjusted accordingly.
4. DTT is susceptible to oxidation and should be prepared freshly before use. Keep the DTT solution on ice and store the remaining at −20 °C for later use.
5. IAA is sensitive to light. Prepare it freshly and keep in the dark.
6. The bacterial cell pellet could be stored frozen at −80 °C. The method descried in this chapter is also applicable to extracts generated from frozen cell pellets.
7. The final concentration of ampholyte in sample solution depends on the protein concentration and may be increased up to 3% to maintain protein solubility. The pH and concentration of ampholyte is critical and is recommended to be optimized according to sample and for each fractionation experiments.
8. Air bubbles will disturb the electric field which result to a poor IEF separation. Make sure to remove all air bubbles by aspirating the sample from a channel and reloading it.
9. During an IEF run under constant power, the voltage increases over time. The run is typically complete when the voltage stabilizes. Allow the run to continue for 30 min after this point before harvesting the fractions.
10. Incubate the acetone in −20 °C for at least 1 h before adding to sample.
11. Make sure that the pH of the solution is above 7.5 to avoid alkylation of lysine and histidine.
12. Using small C18 tips (Ziptips) for further cleaning digested samples is recommended.

References

1. Krogan NJ, Cagney G, Yu H et al (2006) Global landscape of protein complexes in the yeast Saccharomyces cerevisiae. Nature 440:637–643. https://doi.org/10.1038/nature04670
2. Tarassov K, Messier V, Landry CR et al (2008) An in vivo map of the yeast protein interactome. Science 320:1465–1470. https://doi.org/10.1126/science.1153878
3. Uetz P, Giot L, Cagney G et al (2000) A comprehensive analysis of protein-protein interactions in Saccharomyces cerevisiae. Nature 403:623–627. https://doi.org/10.1038/35001009
4. Kristensen AR, Gsponer J, Foster LJ (2012) A high-throughput approach for measuring temporal changes in the interactome. Nat Methods 9:907–909. https://doi.org/10.1038/nmeth.2131
5. Havugimana PC, Hart GT, Nepusz T et al (2012) A census of human soluble protein complexes. Cell 150:1068–1081. https://doi.org/10.1016/j.cell.2012.08.011
6. Zhong Ming Hu L, Goebels F, Wan C, et al EPIC: elution profile-based inference of protein complex membership. Nat Methods

Chapter 27

A Data Analysis Protocol for Quantitative Data-Independent Acquisition Proteomics

Sami Pietilä, Tomi Suomi, Juhani Aakko, and Laura L. Elo

Abstract

Data-independent acquisition (DIA) mode of mass spectrometry, such as the SWATH-MS technology, enables accurate and consistent measurement of proteins, which is crucial for comparative proteomics studies. However, there is lack of free and easy to implement data analysis protocols that can handle the different data processing steps from raw spectrum files to peptide intensity matrix and its downstream analysis. Here, we provide a data analysis protocol, named *diatools*, covering all these steps from spectral library building to differential expression analysis of DIA proteomics data. The data analysis tools used in this protocol are open source and the protocol is distributed at Docker Hub as a complete software environment that supports Linux, Windows, and macOS operating systems.

Key words Proteomics, Mass spectrometry, DDA, DIA, SWATH-MS, Spectral library, Data analysis

1 Introduction

The current method of choice for large-scale identification and quantification of proteins is liquid chromatography tandem mass spectrometry (LC-MS/MS) [1]. In addition to data-dependent acquisition (DDA) mode of mass spectrometry, there is an increased interest in data-independent acquisition (DIA) mode, such as Sequential Windowed Acquisition of All Theoretical Fragment Ion Mass Spectra (SWATH-MS) [2].

DIA has been suggested to combine the advantage of the high-throughput of DDA proteomics with the benefit of the high reproducibility of targeted analysis, such as selective reaction monitoring (SRM) [2, 3]. In SWATH-MS, all precursors generated from a sample are systematically fragmented within a predetermined mass-to-charge ratio (m/z) and retention-time range. Since the spectra are generated without explicit association between peptide precursors and corresponding fragments, a spectral library is used for the identification of peptides from the data. For building the

Xing Wang and Matthew Kuruc (eds.), *Functional Proteomics: Methods and Protocols*, Methods in Molecular Biology, vol. 1871, https://doi.org/10.1007/978-1-4939-8814-3_27,

spectral library, data from samples produced by mass spectrometry in DDA mode can be used.

Here, we provide a comprehensive data analysis protocol, named *diatools*, together with its open source implementation for analyzing DIA data. The protocol covers all steps from raw spectrum files to a final result of differentially expressed proteins, with focus on SWATH-MS data. After installation of the required software and preparation of folder structure for the data (Subheading 3.1), the raw mass spectrum files are converted to required open formats (Subheading 3.2) and a database FASTA is constructed, which should contain sequences of all possible proteins that can be potentially found from the whole data set (Subheading 3.3). Subheading 3.4 then discusses the optional customization of the parameters of the protocol and Subheading 3.5 illustrates how to run the protocol to build the spectral library, to produce an intensity matrix of the identified peptides for each sample, and to perform differential expression analysis between sample groups. The spectral library is built as described by Schubert et al. [4]. The SWATH-MS data is processed using the OpenSWATH software [5] including a TRIC alignment step [6].

The *diatools* protocol is distributed as a Docker image at Docker Hub (compbiomed/diatools). Docker [7] is a software technology that provides light-weight virtualized software environment enabling easy implementation of the data analysis protocol. The protocol supports Linux, Windows, and macOS operating systems, with the exception that Windows is needed to convert the raw spectrum files to open formats.

2 Materials

The raw mass spectrum files produced by mass spectrometers are typically in proprietary vendor-specific formats, which need to be converted to open formats before data analysis. The format conversion of the raw mass spectrum files can be done with the ProteoWizard software [8] on a Windows platform. Otherwise, our *diatools* data analysis protocol and all the required software are distributed as a Docker image and, therefore, can be run on any platform that supports Docker.

The *diatools* Docker image is available at Docker Hub (compbiomed/diatools). Additionally, the source code of our implementation is released under open source General Public Licence (GPL) 3.0 and can be downloaded from GitHub https://github.com/computationalbiomedicine/diatools.git. The Docker image is based on Ubuntu 17.04 operating system and it contains multiple proteomics tools, including OpenMS version 2.3 [9], Trans-Proteomics Pipeline (TPP) version 5.0 [10], msproteomicstools version 0.6.0, and ProteoWizard version 3.0.11252 [8]. The downstream statistical analyses are performed using R. The R 3.3.2 environment with the appropriate packages are also included in the Docker image.

For running the *diatools* data analysis protocol, we recommend having at least 128 GB of RAM depending on the number of samples and the size of the sequence database (FASTA). The Docker image requires at least 30 GB of free disk space. Additionally, mass spectrometry raw data take typically a lot of space and, therefore, depending on the data, multiple terabytes of disk space might be required to store the input files.

The *diatools* protocol assumes that Biognosys iRT kit peptides are used in the laboratory protocol. The default settings are for the Q Exactive HF mass spectrometer (Thermo Fisher Scientific, Waltham, Massachusetts, USA) but data from other instruments can be used as well by adjusting the parameters accordingly.

3 Methods

This chapter describes the steps of our *diatools* data analysis protocol, including data conversion, peptide identification, quantification, and differential expression analysis for the acquired data. A schematic illustration of the protocol is shown in Fig. 1.

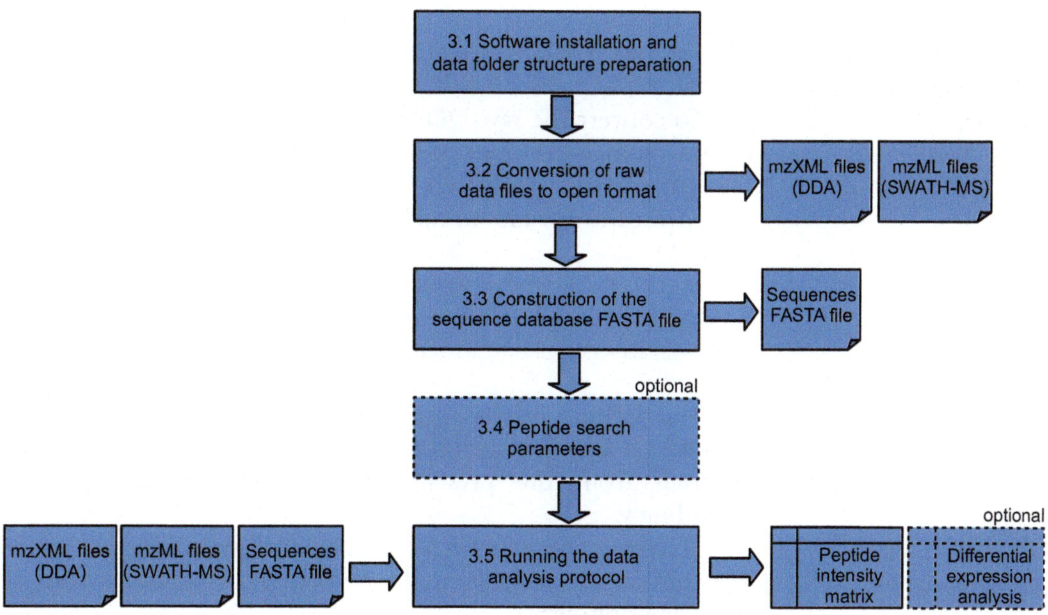

Fig. 1 The *diatools* data analysis protocol for quantitative data-independent acquisition proteomics data. Protocol steps are shown with their corresponding section numbers and their input/output files. Optional steps are marked with dashed line

3.1 Software Installation and Data Folder Structure Preparation

Install the ProteoWizard software to the Windows machine that is used to convert the raw data files to an open format. The ProteoWizard software can be downloaded from http://proteowizard.sourceforge.net/.

Install Docker on the machine on which the data analysis protocol will be run. Docker installation package for Linux, Windows, and macOS can be downloaded from www.docker.org. If the Linux distribution is Ubuntu, RedHat, or CentOS Linux, the Docker can be installed from their respective software repositories (*see* **Note 1**). For Windows, the latest version 10 is recommended. On Windows, allow Docker to access the needed drive (for example, C:) from the Docker settings.

Once Docker is installed, download the data analysis environment with the command:

```
docker pull compbiomed/diatools
```

On the machine where the data analysis is done, create a folder called *dataset* with the following subfolders under it:

- config
- DDA
- DIA
- ref
- out

3.2 Conversion of Raw Data Files to Open Format

Use the ProteoWizard tool to convert the raw data files to an open format.

To convert the raw DDA files to mzXML format, open Windows Command Prompt and go to the folder containing the DDA raw data. Run *qtofpeakpicker* from ProteoWizard to pick peaks and to convert the raw files to mzXML format:

```
FOR %i IN (*.raw) DO "\Program Files\ProteoWizard\ProteoWizard
3.0.11252\qtofpeakpicker.exe" --resolution=2000 --area=1
--threshold=1 --smoothwidth=1.1 --in %i --out %~ni.mzXML
```

The default settings in the protocol are according to those by Schubert et al. [4]. If the ProteoWizard install location or version is different from the present protocol, modify the command accordingly.

To convert the raw SWATH-MS files to mzML format, use the MSConvert program from the ProteoWizard software with the following options:

- Output format: mzML
- Extension: empty

- Binary encoding precision: 64bit
- Write index: checked
- TPP compatibility: checked
- Use zlib compression: unchecked
- Package in gzip: unchecked
- Use numpress linear compression: unchecked
- Use numpress short logged float compression: unchecked
- Use numpress short positive integer compression: unchecked
- Only titleMaker filter

Copy the DDA mzXML files to the *dataset/DDA* folder and the SWATH-MS mzML files to the *dataset/DIA* folder.

3.3 Construction of the Sequence Database FASTA File

Create a sequence database in FASTA format that consists of all proteins that may exist in the sample set under analysis. The FASTA file is used to construct the spectral library by searching the DDA files against it. Create a FASTA file that contains the following protein or peptide sequences:

- Proteins of interest (for example Swiss-Prot Human)
- IRT peptides[1] (Biognosys|iRT-Kit_WR_fusion)
- Peptides related to lysis (Uniprot ID: Q7M135)
- Digestion enzyme (typically Trypsin (Uniprot ID: P00761))
- Possible contaminants

Do not generate decoy sequences to the FASTA file manually. They are automatically generated by the protocol by reversing the peptide/protein sequences. Copy the FASTA file to the *dataset/ref* folder and name it as sequences.fasta.

3.4 Peptide Search Parameters

The default parameters of the protocol are for the nanoflow HPLC system (Easy-nLC1200, Thermo Fisher Scientific) coupled to the Q Exactive HF mass spectrometer (Thermo Fisher Scientific) equipped with a nano-electrospray ionization source. The device and lab protocol specific default settings are listed below:

- Precursor mass tolerance: 10 ppm
- Fragment ion tolerance: 0.02 Da
- Cleavage site: Trypsin_P
- Fixed modification: Carbamidomethyl (C)
- Variable modification: Oxidation (M)

If another type of instrument is used, these settings need to be customized (*see* **Note 2**).

[1] https://biognosys.com/media.ashx/irtfusion.fasta

3.5 Running the Data Analysis Protocol

Open terminal prompt and set the working directory to *dataset/out* folder, where the LOCALPATH refers to the path to the previously created folder structure:

```
cd /LOCALPATH/dataset/out
```

Run the data analysis protocol with the following command:

```
docker run --rm \
-v /LOCALPATH/dataset/:/dataset \
--workdir /dataset/out \
-u $(id -u):$(id -g) \
compbiomed/diatools \
/opt/diatools/dia-pipeline.py \
--in-DDA-mzXML ../DDA/*.mzXML \
--in-DIA-mzML ../DIA/*.mzML \
--db ../ref/sequences.fasta \
--use-comet \
--use-xtandem
```

On a Windows platform, the path to the dataset is given in the following form: "`-v//c/LOCALPATH/dataset:/dataset`" where c is the drive letter. On Linux platform, Docker might be available only to superusers. In that case, add sudo command before the docker command.

To perform the optional differential expression analysis between sample groups, the groups must be provided using an additional parameter in the command:

```
--design-file <designFilename>
```

The design file must be defined as a tab-separated file (*see* Table 1 for an example), where the column *Filename* refers to the SWATH-MS filename of a sample, the column *Condition* is the group to which the sample belongs, the column *BioReplicate* refers to the biological replicate, and the column *Run* to the MS run.

By default, the false discovery rate (FDR) used by the *diatools* protocol for the peptide identifications is 0.01. However, the FDR threshold can be adjusted by the user (*see* **Note 3**). The number of parallel processing threads used is four by default, but the user can choose different numbers of threads according to hardware resources (*see* **Note 4**). For the downstream differential expression analysis, the data are median normalized and differential expression analysis is performed for all possible pairs of sample groups using the PECA R-package [11] available from Bioconductor [12].

Once the data analysis run has completed successfully, the output folder *dataset/out* contains two tab-separated data files: DIA-peptide-matrix.tsv and DIA-protein-matrix.tsv. These files contain the peptides and proteins with their respective intensity

Table 1
Example design file

Filename	Condition	BioReplicate	Run
Sample1.mzML	Treatment	1	1
Sample2.mzML	Treatment	2	2
Sample3.mzML	Treatment	3	3
Sample4.mzML	Control	1	4
Sample5.mzML	Control	2	5
Sample6.mzML	Control	3	6

values for each sample. The files can be opened with MS Excel or with LibreOffice Calc. The output folder contains also files of intermediate results written by various external tools run by the protocol as well as a log.txt file which includes details on the run. The log can be used for troubleshooting if the run fails.

If the optional differential expression analysis is performed, those results are stored as tab-separated files in the *dataset/out* folder with the compared groups as filenames. For each identified protein, the result files contain the protein name, the value of the test statistic (t), the number of peptides per protein (n), the significance p-value (p), and the estimated false discovery rate ($p.fdr$). In addition to performing the differential expression analysis by running the *diatools* protocol, it is possible to perform the differential expression analysis separately using the peptide intensity file produced by the protocol (*see* **Note 5**).

4 Notes

1. **Docker installation under Ubuntu, RedHat, or CentOS.** If the analysis is done on Ubuntu, RedHat, or CentOS Linux distributions, the Docker can be installed from the software repository.

   In Ubuntu, use the following shell command:

   ```
   apt-get install docker.io
   ```

   For Ubuntu, it is also convenient to add user to a system group called *docker*, which makes it possible to run the Docker without a *sudo* command.
   In RedHat/CentOS, use the following commands:

   ```
   yum install docker
   systemctl enable docker
   systemctl start docker
   ```

2. **Customization of peptide search parameters.** The default parameters of the *diatools* protocol are for the nanoflow HPLC system (Easy-nLC1200, Thermo Fisher Scientific) coupled to the Q Exactive HF mass spectrometer (Thermo Fisher Scientific) equipped with a nano-electrospray ionization source. If another type of instrument is used, the settings need to be customized by editing the Comet and X!Tandem search engine parameters. This can be done by modifying the Comet and X!Tandem configuration files comet.params.template and xtandem_settings.xml, respectively, which are distributed with the protocol. Copy the modified files to the *dataset/config* folder and give the following extra parameters when running the protocol:

```
--comet-cfg-template config/comet.params.template
--xtandem-cfg-template config/xtandem_settings.xml
```

3. **Adjusting the false discovery rate (FDR) for the peptide identifications.** By default, the protocol uses 0.01 as a FDR threshold for the spectral library building. For the TRIC alignment step, 0.01 is used as target and 0.05 as max threshold. These values can be adjusted by adding the following extra parameters when running the *diatools* protocol:

```
--library-FDR
--feature-alignment-FDR
```

For instance, the following parameter instructs to use 0.05 as the FDR threshold for spectral library building:

```
--library-FDR 0.05
```

Similarly, the following parameter instructs to use 0.01 as target and 0.02 as max threshold for the TRIC alignment:

```
--feature-alignment-FDR 0.01 0.02
```

4. **Adjusting the number of parallel processing threads.** Currently, the protocol uses a maximum of four threads by default to process the data. If the protocol is run on a high-end desktop or on a server, the number of threads can be increased to correspond to the CPU core count. It speeds up the analysis, but also increases the amount of consumed RAM. For example, the following parameter increases the thread count to 20:

```
--threads 20
```

5. **Separate differential expression analysis using the peptide intensity file.** In addition to performing the differential expression analysis by running the *diatools* protocol, it is possible to perform the analysis separately using the peptide intensity file produced by the protocol (peptide-intensity-matrix.tsv).

First, the peptide intensity data are transformed into a suitable format using the R/Bioconductor package SWATH2stats. To install SWATH2stats, open R and enter:

```
source("https://bioconductor.org/biocLite.R")
biocLite("SWATH2stats")
```

To read in the peptide intensity data, the following commands can be used:

```
library(data.table)
library(SWATH2stats)
   data <- data.frame(fread(file=" peptide-intensity-
matrix.tsv",
sep='\t', header=TRUE))
```

Next, get rid of unneeded columns (line 2), remove the iRT peptides that are used for retention time normalization (line 3), filter out rows corresponding to multiple proteins (line 4), and shorten the protein names (line 5):

```
data$run_id <- basename(data$filename)
data <- reduce_OpenSWATH_output(data)
data<-data[grep('iRT', data$ProteinName, invert=TRUE),]
data <- data[grep('^1/', data$ProteinName),]
data$ProteinName<-sapply(strsplit(data$ProteinName, "\cr|"),
function(x) unlist(x)[2])
```

To define the sample groups, read in the design matrix, after which the data can be converted to a suitable format for the statistical analysis with PECA:

```
design <- read.table('design.txt', sep="\t", header=TRUE,
stringsAsFactors=FALSE)
data <- sample_annotation(data, design)
data <- convert4PECA(data)
```

PECA determines differential protein expression using directly the peptide-level measurements, instead of the common practice of using pre-calculated protein-level values. Differential expression statistic is first calculated for each measured peptide, after which the protein-level significance is estimated by taking the number of identified peptides per protein into account. To install PECA, open R and enter:

```
source("https://bioconductor.org/biocLite.R")
biocLite("PECA")
```

For DIA data, the reproducibility-optimized test statistic (ROTS) [13] is the suggested statistic to be used within PECA [14]. This is done by setting the test parameter as `rots` when calling the function. The following commands perform the differential expression analysis for all possible pairs of sample groups:

```
library(PECA)
comb <- combn(unique(design$Condition),2)
for(i in 1:ncol(comb)) {
   group1 <- paste(design$Condition,
      design$BioReplicate,sep="_")
      [design$Condition==comb[1,i]]
   group2 <- paste(design$Condition,
      design$BioReplicate,sep="_")
      [design$Condition==comb[2,i]]
   peca.out <- PECA_df(data, group1, group2,
      id="ProteinName", normalize="median",
      test="rots", progress=TRUE)
   write.table(peca.out,
     file=paste("PECA_",comb[1,i],
     "-",comb[2,i],".txt",sep=""),
   sep="\t", quote=FALSE, row.names=TRUE, col.names=NA)
}
```

References

1. Aebersold R, Mann M (2003) Mass spectrometry-based proteomics. Nature 422:198–207
2. Gillet LC, Navarro P, Tate S et al (2012) Targeted data extraction of the MS/MS spectra generated by data-independent acquisition: a new concept for consistent and accurate proteome analysis. Mol Cell Proteomics 11: O111.016717
3. Huang Q, Yang L, Luo J et al (2015) SWATH enables precise label-free quantification on proteome scale. Proteomics 15:1215–1223
4. Schubert OT, Gillet LC, Collins BC et al (2015) Building high-quality assay libraries for targeted analysis of SWATH MS data. Nat Protoc 10:426–441
5. Röst HL, Rosenberger G, Navarro P et al (2014) OpenSWATH enables automated, targeted analysis of data-independent acquisition MS data. Nat Biotechnol 32:219–223
6. Röst HL, Liu Y, D'Agostino G et al (2016) TRIC: an automated alignment strategy for reproducible protein quantification in targeted proteomics. Nat Methods 13:777–783
7. Merkel D (2014) Docker: lightweight Linux containers for consistent development and deployment. Linux J
8. Chambers MC, Maclean B, Burke R et al (2012) A cross-platform toolkit for mass spectrometry and proteomics. Nat Biotechnol 30:918–920
9. Sturm M, Bertsch A, Gröpl C et al (2008) OpenMS—an open-source software framework for mass spectrometry. BMC Bioinformatics 9:163

10. Deutsch EW, Mendoza L, Shteynberg D et al (2010) A guided tour of the trans-proteomic pipeline. Proteomics 10:1150–1159
11. Suomi T, Corthals GL, Nevalainen OS et al (2015) Using peptide-level proteomics data for detecting differentially expressed proteins. J Proteome Res 14:4564–4570
12. Huber W, Carey VJ, Gentleman R et al (2015) Orchestrating high-throughput genomic analysis with bioconductor. Nat Methods 12:115–121
13. Elo LL, Filén S, Lahesmaa R et al (2008) Reproducibility-optimized test statistic for ranking genes in microarray studies. IEEE/ACM Trans Comput Biol Bioinform 5:423–431
14. Suomi T, Elo LL (2017) Enhanced differential expression statistics for data-independent acquisition proteomics. Sci Rep 7:5869

INDEX

Xing Wang and Matthew Kuruc (eds.), *Functional Proteomics: Methods and Protocols*, Methods in Molecular Biology, vol. 1871, https://doi.org/10.1007/978-1-4939-8814-3, © Springer Science+Business Media, LLC, part of Springer Nature 2019

C

X

Y

Z

Printed by Printforce, the Netherlands